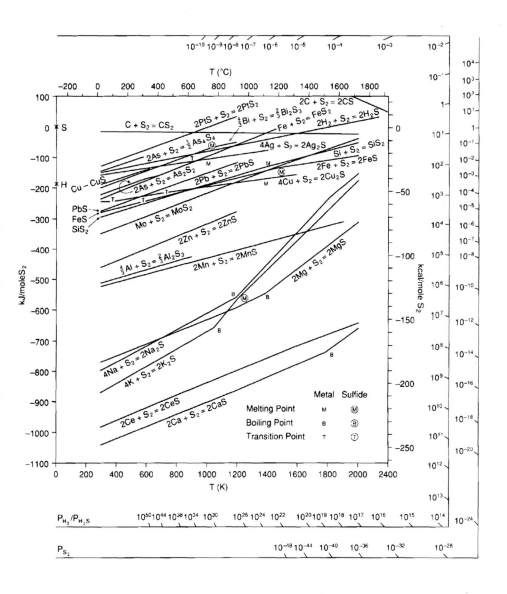

The Production and Processing of Inorganic Materials

The Production and Processing of Inorganic Materials

James W. Evans
Lutgard C. De Jonghe

Professors
Dept. of Materials Science and Engineering
University of California, Berkeley

With a CD of thermodynamic data and
spreadsheet examples prepared by
Arthur Morris
Professor Emeritus of Metallurgical Engineering
University of Missouri, Rolla

CD files can now be found by entering in ISBN 978-0873395410 on
booksupport.wiley.com

A Publication of
TMS (The Minerals, Metals & Materials Society)
184 Thorn Hill Road
Warrendale, Pennsylvania 15086-7528
(724) 776-9000

Visit the TMS web site at
http://www.tms.org

Library of Congress Catalog Number 2002109114
ISBN 978-0-87339-541-0

If you are interested in purchasing a copy of this book, or if you would like to receive
the latest TMS publications catalog, please telephone
1-800-759-4867 (U.S. only) or 724-776-9000, EXT. 270.

To our families:

Sylvia Johnson
James
Hugh
Claire

Lynn
Erika
Jessica

Preface

The past few decades have seen the emergence of materials science (and engineering) as a discipline. The roots of that growth lay initially in the long established fields of metallurgy and ceramics and the discipline is nowadays seen as embracing these traditional fields plus electronic, optical, composite and polymeric materials, as well as biomaterials. Materials science and engineering examines materials over a large range of scales, from the nanometers of quantum dots to the meters of the steel girders used for building construction The discipline is currently growing in popularity and likely to remain so. New materials for aerospace applications, optical fibers and new applications of materials in bioengineering stir the imaginations of students and stimulate a flow of research funds .

The character of materials science is evolving as the growth continues and as scientists and engineers from other disciplines (notably physics, chemistry, chemical engineering and electrical engineering) enter the field. Nevertheless most would agree that materials science is a field at the interface between more traditional sciences and engineering fields (those listed already plus civil and mechanical engineering). It is both a scientific discipline concerned with fundamentals (in the same way as the adjacent topic of solid state physics) and an engineering subject concerned with the application of knowledge to the production and use of materials.

Another view of the field is that it has the structure of a tetrahedron with vertices labeled "Structure", "Properties", "Performance" and "Processing" and the edges representing the important interactions between these four topics. While alternative geometries have been suggested, an opinion held by many is that, until recently, the processing vertex has been neglected, at least in the USA, in favor of the other vertices. Certainly, if processing is taken to include everything from synthesis in laboratory "test tube" experiments to industrial manufacturing, then this opinion is supported by even the most casual inspection of trade figures or the business pages.

This neglect of processing has been reflected in the materials curricula at many of our universities. Courses on processing have frequently been regarded as optional or secondary and have failed to embrace the unity of materials that is recognized, for example, in other courses on fracture or characterization of microstructure. Our hope is that the production of all types of (inorganic) materials can be taught as one. Naturally there are differences between the production of a silicon nitride engine component and that of a copper wirebar but they appear to us no larger than that between producing the wirebar and producing a titanium ingot. In any event we claim that these differences are insignificant compared to the commonality of thermodynamics, kinetics, transport phenomena, phase equilibria and transformation, process engineering and surface chemistry on which all three rely. Naturally, those who find such integration implausible can use the book selectively, by skipping chapters on "other" materials.

This book, then, places its emphasis on fundamentals and on how those fundamentals can be applied to understand how the major inorganic materials are produced and the initial stages of their processing. These should equip the student for engineering future processes for producing materials or for studying the processing of the many less common materials not examined in this text.

This emphasis on fundamentals has posed a question. Can other texts and previous courses be relied upon to provide the necessary fundamentals for this text and a course based on it? The answer will of course depend on the students and the instructor will always have the option of skipping parts or the whole of the first few chapters. However, we argue that, because so many of the practitioners of materials science enter from different disciplines, it is wise to establish a common ground. For example, those entering from a mechanical engineering background would probably have a knowledge of chemical thermodynamics not as complete as that expected of chemists, but a considerable knowledge of process engineering. Even those from a materials science background frequently have not come across some fundamental topics important in materials processing. For example, undergraduate materials courses on diffusion seldom treat the convective mass transport that is important in the many materials processing operations where gases or liquids play a part.

Previous courses and texts encountered by the student may well have probed deeper and ranged wider into the topics of the first seven chapters but those chapters serve to emphasize what is important from those courses in terms of materials production. Selective reading of these chapters is encouraged but few will be the students who have an understanding of all the subject matter contained therein. The student who reads and grasps these chapters will be well equipped for the remaining chapters and work in the field of materials processing.

A second question we have faced is the extent to which references should be included in the text. Usually we have not included references although each chapter concludes with a list of suggested additional reading. Much of the material in this, and many other undergraduate texts, is "well known" in the sense of being scattered throughout other texts, or widely known to scientists or engineers in the field. Examples would be Fick's laws of diffusion or the statement that the electrolytic cells used to produce aluminum typically operate at 950–960°C. Little is gained by including original references to substantiate these facts. The exception is in some of the later chapters which on occasion touch on developments that are too recent to be regarded as well known. At those points we have inserted a limited number of references, sufficient to serve as the origin for a literature survey.

We have used SI units in most cases. However, if present progress is any indication, it will be well into this century before the older units disappear, particularly in the United States. Consequently a few of the problems at the ends of the chapters have used older units.

Our intention has been to write a book suitable for a one semester (or two quarter) course at the Junior or Senior level. Furthermore it is hoped that graduate students, or practicing scientists or engineers, entering the field will find the book a useful introductory and reference work. This selection of the level of presentation has not been allowed to constrain our writing. Some subjects are best described using mathematics that may tax some less accomplished juniors. Indeed some of the problems have been written so as to challenge the better seniors. Conversely, other topics, for example some of the process descriptions, are so easily grasped that we have striven to avoid dwelling on them.

We emphasize the scope of the text as implied by the title. The production and primary processing of inorganic materials (ceramics, metals, silicon and some composite materials) is the subject. Although much in the first few chapters applies equally to the production of polymeric materials, their production entails topics (organic chemistry and rheology, for example) that are peculiar to them. We have therefore thought it wise to stay within the bounds of inorganic materials.

At a late stage in the production of this second edition, Arthur Morris, Professor Emeritus of Metallurgical Engineering, University of Missouri, Rolla, suggested the inclusion of a CD, to include thermodynamic data and a few Excel® based examples of how to use them. This excellent suggestion led to the CD included with this text. That CD contains files that are the thermodynamic data for approximately one hundred elements and compounds (all those encountered in the problems and more). This is a subset of the large data set FREED that is mostly the thermodynamic database of the US Bureau of Mines and is available from http://home.att.net/~thermart/ at nominal cost to students. Prof. Morris has also included on the CD some sample solutions to problems in the book, plus other problems. Some of these solutions were obtained using the software THERBAL (also available from http://home.att.net/~thermart/). Solutions obtained by such computational techniques are much more powerful instructional tools than the usual printed solutions. The student can easily change the parameters of the problem and the consequences are immediately and graphically apparent. We are greatly indebted to Art Morris

We also wish to thank Eve Edelson, who was very helpful in the preparation of the book. The late David Johnstone, editor at MacMillan, patiently steered us toward the finished manuscript that became the first edition published by that company. Stephen J. Kendall of TMS has been our aid in producing this second edition. Claire M. Evans is thanked for help in preparing the index. We would also express our appreciation to our students, who have contributed to our research and thereby to some of the body of knowledge that we have incorporated in the book. Thanks are also due to our reviewers, particularly Professors Dave Gaskell, Dennis Readey, George St. Pierre, and Nickolas Themelis for constructive criticism. We express special gratitude to our families for enduring our preoccupation with this book for a lengthy period.

JWE
LCDJ
Berkeley, March, 2002

Contents

CHAPTER 1

Significance, History, and Sources of Materials

1.1 What Materials Are and Their Importance

To most engineers and scientists a material is a solid with useful structural, electrical (including electronic or magnetic), optical or corrosion-resistance properties. If we accept this definition then it is clear that the products of most of the major manufacturing industries are either materials or are fabricated from materials. The exceptions are the food, beverage, chemical and fuels industries; even within these industries materials play a key role. Examples are the steel, aluminum and glass used for packaging food and drink, corrosion-resistant materials used to build chemical reactors, and the bits used to drill oil wells. Materials, then, have a great importance in the economies of industrialized nations. In this text we shall be concerned with materials of mineral origin, so that steel, glass, silicon, and ceramic materials are included, but materials of animal or vegetable origin (wood, rubber, leather, etc.) are excluded. As extracted from the earth, the minerals ("raw materials" in Fig. 1.1) from which these materials are produced have a value of 0.4 percent of the U.S. gross domestic product. However, processing of these minerals (coupled with recycling of old materials) to produce metals, ceramics, etc. brings their value to over 4 percent of the GDP.[1]

This direct contribution to the economy of every industrialized nation is just one reason why materials are politically significant. Another reason is the indirect dependence of the economy on materials. For example, the aircraft industry would not be what it is today without aluminum, nor the computer industry without high-purity silicon. If we think of the dependence of our society on jet aircraft and the rapid processing of large amounts of information it is clear that materials play a key role.

A third socio-political aspect of materials arises from the fact that the US, Japan, Europe, in fact nearly all of the Western industrialized nations are dependent on imports for a number of key materials (or the minerals to make them). For example in 2000 the US imported 100 percent of the bauxite and alumina it used for producing aluminum (the importance of which has already been stressed), 74 percent of its cobalt (an important element in the production of materials for jet engines) and 83 percent of its platinum group metals (used largely for catalysts in the chemical industry and automobile exhaust systems). Furthermore, the reliability of some of the sources of these imports is at

---

[1] Strictly speaking Fig. 1.1 includes data for minerals processed into fertilizers and chemicals but elimination of those data does not alter the conclusion that materials are economically important.

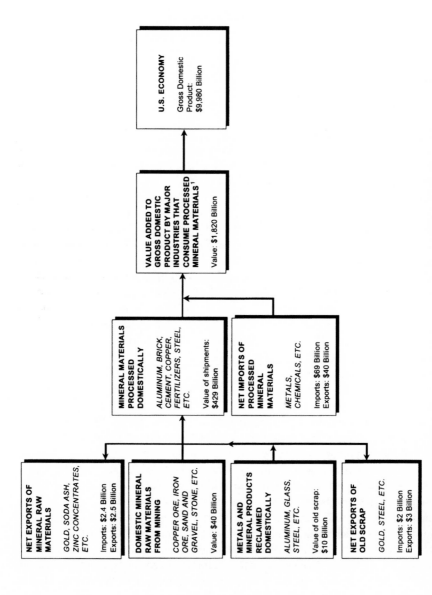

FIGURE 1.1 Role of nonfuel minerals in the U.S. economy (estimated values for 2000). (U.S. Geological Survey)

present open to question; for example most of the platinum imports originate from the former USSR or South Africa.

Yet another social impact of materials concerns the energy employed in their production. For example, the production of aluminum consumes several percent of the electrical energy generated in the US. At a time when the nation is concerned about energy consumption, this represents a significant burden and has caused some "export" of energy-intensive materials production to countries with an energy surplus (construction of aluminum plants abroad, for example).

Finally materials have a social and political significance because of the impact of their production on the local environment and employment. Eastern cities such as Pittsburgh, that have ceased to be major centers of steel production, are now cleaner than in the past but the loss of jobs in the steel industry required considerable adjustment. Fig. 1.2 illustrates that in the late sixties and early seventies more than one million people were employed in the production of metals, more than in many other important industries. Although this number has declined significantly in recent years it still represents a major source of employment, ranging from 600-700 thousand in the past decade, about the same number of people as are employed in the production of electronic components.

We see then that there are good reasons to be concerned with materials and with their production. There are additional reasons for the materials scientist or engineer to learn of materials production in addition to learning of materials properties. First, materials are seldom pure and the properties are affected (in some cases markedly) by impurities left in, or introduced, during processing. An example is provided by non-metallic inclusions which are introduced during the processing of steel in a step which is intended to remove dissolved oxygen and sulfur impurities from the steel. These inclusions can,

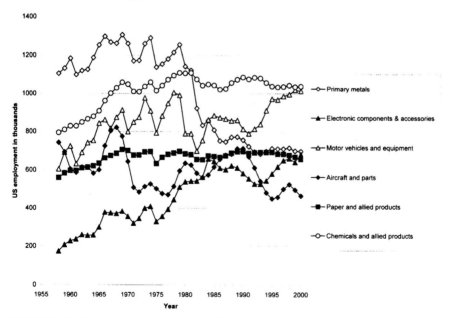

FIGURE 1.2 Employment in the United States in various industrial activities.

to a large extent, be removed by additional processing but at additional cost. The materials scientist working on the design of new steels should comprehend both the adverse impact of non-metallic inclusions on the steel and the means available for their elimination. Another example is provided by fiber-optic cables. These cables will handle much more than the communication traffic possible with copper cable, and provide much greater security than can be achieved with satellite communication. They would have been impossible without both significant advances in the science of optical communication materials and the development of means to produce optical fibers reliably and economically. The development of materials with new desirable properties is a futile exercise unless the means to produce the material at reasonable cost, from raw materials available in sufficient supply and with acceptable impact on the environment, can be found.

1.2 Historical Summary

1.2.1 The Neolithic and Chalcolithic Period

In a sense, our first use of materials occurred when a human being first picked up a stick or stone to be used as weapon, tool or source of fire. If, however, we exclude such rudimentary naturally-occurring materials from our scope then the first materials that were produced by processing were probably mud bricks and lime plaster. The archaeological record of the former is, no doubt, partially obliterated by the relative weakness of this material and its similarity to the nearby Earth but it is known that they were in use in the Near East by 8000 BC. The use of lime plaster, made by roasting limestone to lime which was then mixed with water and other additives (such as clay or, in some cases, reinforcing fibers) and formed into the desired shape can be dated to around 12,000 BC.

The production of mud bricks would entail finding clays or soils that past experience had taught would yield a brick, mixing with water to the required consistency, shaping into bricks and drying in the sun. The application of this primitive technology to producing other items (vessels for storing food for example) evolved thereafter but the utility of these items would have been severely limited by the weakness and permeability to water of the sun-dried material. As we shall see later in this text, bringing a ceramic material to a high temperature ("firing") is an important step in its processing and it appears that this way of improving the properties of a ceramic was one of man's earliest discoveries. Fired molded figures from an archaeological site in what was until recently Czechoslovakia have been dated to around 22,000 BC (Kingery, 1988). However, this knowledge does not appear to have been widespread and it was not until 8000 BC to 7000 BC that the practice of making clay vessels harder and impervious to water by firing became common. Thus pottery, the forerunner of modern ceramics, was established as one of man's most important technologies. Although fires similar to cooking fires would have been used in the early days of pottery production (and are still used in parts of East Africa and Central America), the technology evolved until, by 6000 BC, specialized pottery kilns capable of achieving higher temperatures and consequently superior products had been developed.

The first human use of metals was also of naturally occurring materials. Gold, silver, copper, and iron occur as elements in nature, the last in the form of meteorites that have reached the earth from space. These "native metals", although uncommon, are easy to recognize in nature and were readily worked by primitive man, perhaps using stone hammers. For example, at Catal Huyuk, in what is now Turkey, ornamental hammered copper pieces dated to 6000 BC have been found and objects made of native copper have been found at sites in Iran and Turkey dating to the 9th to 7th millenia BC (Tylecote, 1976).

Native copper is more abundant than native gold or silver but native copper is far less common than the compounds of copper, such as its oxides or sulfides. The first widespread use of metals except as trinkets therefore had to await the discovery of how to obtain metals from the compounds which constitute their major occurrence in the Earth, an operation known as "smelting". This may have been achieved first with copper although there is some evidence that lead was the first metal smelted (Krysko, 1980). Malachite (copper carbonate), which is easily recognizable by its green color, had long been used for jewelry before it was discovered how, at high temperatures, it could be converted into metallic copper. One suggestion of how this discovery took place is that malachite was accidentally or serendipitously placed in a cooking fire and copper subsequently observed in the ashes. This appears unlikely; to produce copper metal in a form recognizable to primitive man two things are necessary. The first is a reducing atmosphere (in this case rich in carbon monoxide from the partial combustion of the wood) which would convert the carbonate to the metal. The second is a sufficiently high temperature (above the 1084°C melting point of copper) to cause the grains of copper, formed by reduction, to coalesce into globules or an actual pool of molten metal. The atmosphere of a campfire is only occasionally reducing as air currents eddy amongst the burning wood and temperatures seldom approach a thousand degrees.

An alternative explanation is that the necessary reducing atmosphere and high temperatures were first achieved in primitive potters' kilns that had been developed by 6000 BC. It is even possible that the introduction of malachite into such kilns was deliberate, the malachite being used for decoration or glazing of the pots. As we shall see in Chapter VII, achieving a high temperature by burning a fuel requires that air be supplied at a sufficient, but not excessive, rate and that heat losses to the surroundings be minimized. In primitive kilns the former was achieved by a chimney (or "flue") up which hot gases from the kiln would rise, drawing air into the furnace; heat losses were minimized by thick insulating walls.

No smelting furnaces from this period have been discovered, although a copper mace head dated from approximately 5000 BC and cast from molten metal can be seen in a museum in Ankara. Furthermore, a malachite mine at Rudna Glava in what was until recently Yugoslavia appears to have been worked since before 4000 BC. It therefore remains uncertain exactly how smelting was discovered. In fact the discovery may have been made many times at different dates and places. The oldest smelting furnace discovered by archaeologist at Timna in the Sinai dates from approximately 3500 BC and appears to have been intermediate between a pottery kiln and a cooking fire. Probably air was forced into the furnace using bellows. The fuel was charcoal, rather than wood, presumably because higher temperatures could be achieved with charcoal combustion. The furnace was in the form of a bowl half a meter or so in diameter with walls of stone surrounding the fire to a height of close to one meter. Smelting of copper had reached as far as Britain by the late third millenium BC and southern Russia by the early second millenium BC, but by that time copper was being overtaken by a far superior material, namely bronze, an alloy of copper and tin.

1.2.2 The Bronze Age

Bronze first appeared in Mesopotamia around 3000 BC. The significance of this new material and its subsequent widespread use were such that we now know the period as the Bronze Age. When mixed with tin, arsenic or antimony, copper forms this alloy that is more easily cast than copper (having a melting point more than 100°C less) and is harder than copper when solidified. Early bronzes contained arsenic or antimony but tin was the principal alloying agent later in the Bronze Age. The source of the tin for bronze production is uncertain. Tin deposits in Italy, Sardinia and Cornwall are possibilities but deposits further afield, in Afghanistan, for example, have been suggested.

The development of bronze appears to have occurred elsewhere besides the Near East. Bronze production flourished in northern Thailand, where both copper and tin are found, by 2000 BC, and there is strong evidence that this technology was locally developed rather than imported from the Near East. By the end of the second millenium BC, bronze technology had advanced in the Far East to the point that a cauldron weighing almost a ton, now on display in the National Historical Museum in Beijing, was cast from this metal (Fig. 1.3).

Nor was the Far East devoid of developments in ceramics, for example porcelain invented in China. This material is produced by firing clay with two other minerals (feldspar and quartz) at temperatures above 1300°C. We see therefore that the parallel developments in metals production and ceramic production were both dependent on the ability to achieve high temperatures. That dependency continues to the present day as we shall see throughout this book.

1.2.3 The Iron Age

The Bronze Age may have been ended by interruption of the tin supplies necessary to make this alloy, although shortages of copper or fuel have also been suggested. The cause was the blocking of trade routes through the Near East by invasions from the North, starting in 1200 BC. Metalworkers of the time had no alternative but to turn to a material with more abundant ores but, at that time, inferior properties, namely iron. In the furnaces available in the Mediterranean area at that time, iron ores could be reduced with charcoal but temperatures were not high enough to melt the resulting metal, which was pulled from the furnace as a *bloom*. The bloom was a lump of a few kilograms consisting of metallic iron with unreduced material (*slag*). The iron could be consolidated, the slag removed and the piece shaped to a final product by hammering on an anvil while hot, a tedious process entailing frequent reheating of the bloom. The resultant *wrought iron* was softer than bronze (unless subjected to further processing known as *cold working*) and had an undesirable tendency to rust.

Fortunately, methods soon were established for improving on the mechanical properties of wrought iron. One method of improving wrought iron is by what is now

FIGURE 1.3 Bronze cauldron, weighing almost a ton, cast in China in the second millennium BC. (Photograph courtesy of Robert Raymond.)

known as *carburizing*. Wrought iron is almost pure iron; by contacting it with carbonaceous matter (e.g., charcoal) in a furnace it is possible to dissolve carbon into the solid iron. At a level of between roughly one half and one percent carbon, the resulting metal is both harder and stronger than bronze; it is close to steel as we know it today. By 1200 BC selective carburizing was being carried out, the cutting edge of a sword for example having its carbon content raised considerably. In this way a hard edge which would retain its sharpness was produced on a blade of lower carbon content. The hardness of steel can be further increased by *quenching*, and there is evidence that this technique was in use early in the Iron Age. The red-hot steel is plunged into water and rapidly cooled. Unfortunately quenching also tends to make steel brittle but this difficulty can be avoided, with little loss of hardness, by *tempering*. This involves heating the steel to a dull red heat for a period of the order of one hour, after which it is allowed to cool naturally in air.

This technology of carburizing, quenching and tempering which was to be central to the blacksmith's skills for the next three thousand years and become the root of the field that we now call physical metallurgy, diffused throughout Europe and the Near East over the next several hundred years. In the Far East, however, a very different technology gained ascendancy. The high temperatures achieved in Chinese pottery furnaces have already been mentioned, as have early skills in casting bronze. With this background the Chinese were able to produce *liquid iron* from its ores in early versions of what we now call an iron blast furnace. This was achieved in the sixth or fifth century BC and relied upon the fact that iron produced in these furnaces, as we shall see in Chapter VIII, contained a high percentage (4%-4.5%) of carbon, lowering the melting point to approximately 1150°C. The iron could therefore be run from the furnace into molds, the labor-consuming operation of hammering blooms was therefore avoided, and making iron objects became virtually mass production in these Chinese operations.

Unfortunately the cast iron produced in this way is virtually the antithesis of wrought iron; the high carbon content made this material hard but extremely brittle. However, by the first century AD, Chinese metallurgists had learned to overcome this difficulty by *puddling* the molten iron. This operation, an early form of the steelmaking that we shall treat in Chapter 8, consisted of stirring the molten iron in an oxidizing environment, causing the carbon to oxidize and reducing the carbon content to less than 1%. In this way steel was produced by decarburizing molten metal rather than by the less productive method of carburizing solid metal practiced in the West. It was not until the industrial revolution 1700 years later that this superior technique for steelmaking of the Chinese was rediscovered in Europe.

The early history of two important metals, copper and iron (steel) has now been summarized. The typical reader will already be aware of the roles played later in history by these metals and by gold. We refer to the colonization, first by Spain and Portugal of North America, then by Britain and other European countries of many other parts of the world; that colonization was driven in part by the quest for gold and metal ores. The reader will also know something of the significance of metal production, particularly steelmaking, in the Industrial Revolutions of Europe and the US, together with the stimulus that gold and silver provided for the development of the American West and parts of Australia.

The development of ceramics paralleled the advances in metallurgy; their common reliance on furnaces achieving higher temperatures has already been stressed. China exported porcelain to the middle East during the T'ang dynasty (618-907 AD) and large quantities of this porcelain were reaching as far West as Turkey by the middle of the first millenium AD. Japan also started to export porcelain during this period. Preliminary attempts at making porcelain in Europe occurred in Florence in the late sixteenth century but the first successful production appears to have been at St. Cloud near Paris at the end of the seventeenth century (Kingery, 1988). This was followed not long after by the establishment of production in Saxony by von Tschirnhaus and Bottger. This last effort is noteworthy in that experiments leading to successful production were frequently carried out in an early version of a solar furnace that von Tschirnhaus had developed.

The industrial revolution saw the introduction of machines and larger, more efficient kilns for the mass production of pottery, particularly in the English Midlands. A related improvement was the advancement of brickmaking from the mere shaping and baking of clay bricks to the production of bricks from other more "refractory" materials that could withstand exposure as the lining material for the new high-temperature furnaces. Examples of these materials are lime (CaO), calcined dolomite (CaO•MgO) or magnesia (MgO) which became available for use in steelmaking furnaces in the 1870s.

Much of the incentive for the development of ceramics capable of withstanding high temperatures arose from the introduction of incandescent lighting. By the middle of the nineteenth century, stages were lit by lamps using cylinders of high-purity calcium oxide heated to incandescence by oxygen-hydrogen flames (hence the term *limelight*). Gas lighting was common and, because natural or coal-gas flames are not very luminous, relied upon mantles that were brought to a bright white heat by these flames. These mantles, of rare earth oxides, are still in use today in Coleman lanterns. Silicon carbide, used primarily as an abrasive material, became readily available with the development of the Acheson process (treated in Chapter 11) at the end of the nineteenth century.

It is appropriate to conclude with a synopsis of the history of aluminum production. This metal, second only to iron and steel in terms of tonnage or value of production in the US, is a comparative newcomer to the field of metallurgy, having been in widespread use for roughly a century compared to over two thousand years for steel and even longer for copper.

Although Oersted produced impure aluminum, and Wohler a few globules of nearly pure metal in the first half of the nineteenth century, it was not until the middle of the century that the French chemist Sainte-Claire Deville produced sufficient quantities to be put to use. Deville reacted sodium with aluminum chloride, the latter being first formed by reacting aluminum oxide with carbon in the presence of chlorine, while the sodium was produced by electrolysis of fused sodium carbonate. Electricity of the time came from voltaic cells, resulting in a high cost for the sodium and therefore of the aluminum product.

Fortunately for the history of this metal, Charles Hall in the US and Paul Héroult in France independently discovered a superior method of making aluminum later in the century, both filing patents in 1886. The method involved dissolving aluminum oxide, obtained from a mineral called bauxite, which we shall encounter again in this book, into a molten fluoride of sodium and aluminum. Passage of a direct electric current through the melt resulted in the generation of liquid aluminum. The modern version of this electrolytic cell, the Hall-Héroult cell, is treated in detail in Chapter 9. The invention relied upon the recently developed electric generator which provided electricity at much lower cost than voltaic cells. Later on the "favor" was to be returned as the use of aluminum in electric transmission lines enhanced the application of electricity in modern societies. This symbiosis of industrial development and metallurgy occurs elsewhere in history. For example, the development of the steam engine would have been much impeded without the availability of iron and steel for pistons, crankshafts, valves, etc., while the growth of steelmaking in the 19th century relied upon steam engines for pumping water from coal mines, supplying the fuel for steelmaking, transporting raw materials and driving machinery.

1.3 Materials at Present

We are now in a period when materials are evolving more rapidly than at any time in their history. This is evident in the large number of materials that have become commercially available over the past two or three decades. One example, already

FIGURE 1.4 Ceramic turbocharger for automobile engine, made by the Japanese company NGK. The low specific density of the ceramic improves the response of the engine during acceleration compared to the heavier metal turbocharger. (Copyright 1988 NGK Insulators, Ltd.)

mentioned, is the high-purity silicon which is the principal raw material of most of the semiconductor industry. Mention should also be made of newer semiconductor materials such as gallium arsenide and cadmium telluride that may soon challenge silicon. The field of ceramics has also had recent significant advances. Two in particular spring to mind: high T_c superconductors, which to date have all been ceramic rather than metallic, and silicon nitride which has found application in some automobile engines (Fig. 1.4). Silicon nitride, and also silicon carbide, are attractive materials for use in improved combustion engines because they possess remarkable resistance to high-temperature deformation. Together with their good thermal shock resistance and high toughness these materials seem particularly suited to such applications.

Modern technical ceramics constitute an ever expanding list. Some examples are the titanates, from which a variety of devices can be made due to the remarkable electro-optical, electromechanical and dielectric properties of this class of ceramics. Devices are now commercial, based on these materials, and include boundary layer capacitors, positive temperature coefficient thermistors, optical storage devices, and transducers that can be micromechanical controllers, ultrasonic actuators, or sonar detectors. Some tough ceramics are based on dispersions of a second phase, such as zirconia, in a ceramic matrix such as mullite, to give materials with exceptionally high mechanical strength and toughness (Fig 1.5). A mixed oxide of aluminum and sodium has found an application since its remarkably high sodium ion conductivity makes it useful as a ceramic solid electrolyte (Kummer and Weber, 1968). Another important development is the advent

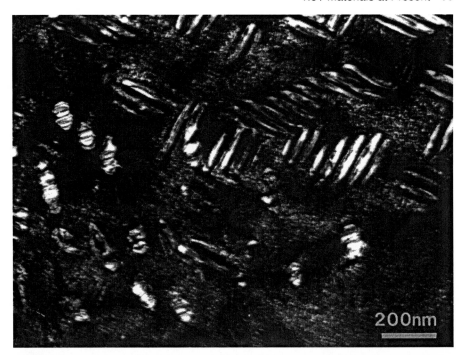

FIGURE 1.5 Transmission electron micrography of the internal structure of a high-toughness yttrium oxide-zirconium alloy ceramic. The light-colored particles are tetragonal zirconia precipitates and play an important role in the increased toughness of this ceramic alloy. [From A. Heuer and M. Ruhle, Phase Transformation in ZrO_2-Containing Ceramics: I, Adv. Ceram. 12(1), 6, (1984). Reprinted by permission of the American Ceramic Society, Columbus, OH]

of high-speed cutting tools tipped with carbides or nitrides, allowing much higher rates of machining than possible with metallic tools.

The pace at which new metallic materials have been introduced into the market has been less frenetic, perhaps because the major metals (steel, copper and aluminum) are well entrenched and their markets established. However, evolutionary improvements such as "cleaner" steels (less contaminated with impurity particles), or more widespread use of less common metals, such as titanium, have been numerous in the recent past and one completely new group of alloys has become important for aerospace applications. That group of alloys is the aluminum-lithium alloys which provide an opportunity for a 10% saving in weight compared to a component of the same stiffness and strength in a conventional aluminum alloy. The interest here is in increasing aircraft payload or reducing fuel consumption.

New classes of materials are emerging which are blurring the old distinctions between metals and ceramics or between organic and inorganic materials. One of these classes is the composite materials, perhaps most familiar in the form of fiberglass-reinforced plastics, but now including composites where a ceramic or metal matrix is reinforced by fibers of carbon or a (second) ceramic. A second class is the so-called *engineered*

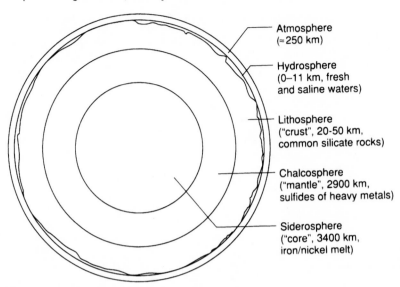

Atmosphere
(≈ 250 km)

Hydrosphere
(0–11 km, fresh
and saline waters)

Lithosphere
("crust", 20-50 km,
common silicate rocks)

Chalcosphere
("mantle", 2900 km,
sulfides of heavy metals)

Siderosphere
("core", 3400 km,
iron/nickel melt)

FIGURE 1.6 Simplified drawing of the earth, indicating major geological zones.

materials where materials (for example a film) are fashioned from component parts designed to bring a distinctive property (e.g., tensile strength or corrosion resistance or gas impermeability or opacity) to the material.

The evolution of materials brings with it an increased interest in the production/ processing of materials, which is the subject of this book.

1.4 The Earth as the Source of Materials

In this book we are concerned with the production of materials that are derived from minerals (rather than vegetation or animals, where material production is more an activity of nature than of man). We shall examine briefly the occurrence of minerals in the earth before an equally brief treatment of how they are physically removed from the earth by mining.

It is important to grasp at the outset that the minerals from which we obtain our materials are nearly always chemical compounds, rather than elements. Thus a typical "copper" mine does *not* contain metallic copper, but rather the sulfides and oxides of copper. However, the major use of copper is as the metal and we shall therefore be concerned in this text with the chemical steps involved in the conversion of sulfides and oxides to metals. Another example is silicon which occurs in nature as oxides or silicates but is used in the semiconductor industry as highly purified elemental silicon. In fact, with very few exceptions, materials are used in a different chemical state than they occur in nature and much of the text will be concerned with the chemistry of this conversion from one state to another.

Fig. 1.6 is a simplified drawing of the Earth indicating the major geological zones.

The atmosphere contains no minerals (although it is the source of nitrogen for ammonia-based fertilizers and oxygen and argon for many industrial purposes, including steelmaking) while the chalcosphere and siderosphere are inaccessible (the deepest mine at present is approximately 5200 meters). Consequently we must rely on the hydrosphere and lithosphere, and on materials recycling, for the minerals from which to produce materials. The major part of the magnesium production is from the hydrosphere, namely the ocean and the Great Salt Lake in the US. Apart from this, the hydrosphere presently has little importance as a source of materials and we must therefore look more closely at the lithosphere.

1.4.1 The Composition of the Lithosphere

Table 1.1 gives the estimated abundances of the elements in the crust of the Earth; the reader will be aware that the majority of these elements will be present in the form of chemical compounds such as sulfixes or oxides. Table I presents the *average* composition of the crust. It is clear that some important elements (e.g., silicon and aluminum) are relatively abundant but that a large number of important materials (chromium, manganese, cobalt, nickel, copper, zinc, tin, lead, uranium) are present in this average composition only at the level of a few tens or hundreds of parts per million; other important elements (molybdenum, silver, tungsten and gold) are present in even smaller amounts. The lower the concentration of a particular element, the more difficult (and therefore more expensive) it is to obtain that element from the crust. In fact, for each element there is a minimum concentration below which the cost of extraction makes the processing uneconomic; these *minimum grades* fluctuate slightly with the price of the metal (or other material) produced and with other factors. As an example the minimum grade for copper extraction is rarely below 0.4 percent (40,000 ppm) while that for aluminum is rarely below 25 percent. Clearly these minimum grades are well above the abundances in Table I. In fact it is true that at present (and for the foreseeable future) no metals can be economically produced from any part of the earth's crust that has the average composition! We are saved, of course, by the fact that the earth's crust is extremely heterogeneous, i.e., there are enormous variations in composition from location to location. Therefore we can find locations where the concentration of one (or more) element is well above the minimum grade and we call these locations "ore bodies". The geologist refers to the phenomena that produce these enormous local variations in composition as *mineralization*, and we list next some of these phenomena.

1.4.2 Mineralization Phenomena

Selective crystallization from the lithosphere during its solidification

Consider the rocks that now form a region of the Earth's crust as they solidified from molten rock during the Earth's early stages. Higher melting-point constituents would tend to solidify first and appear in those locations which were the first to cool, while lower melting-point constituents would appear in those regions that were the last to cool.

Selective crystallization from aqueous solutions

During the cooling or evaporation of aqueous solutions the least soluble compounds in

TABLE 1.1 Estimated Abundance of Some Elements in the Earth's Crust*

Element	Crust	Element	Crust
Li	20	Ag	0.07
Be	2.8	Cd	0.2
B	10	In	0.1
F	625	Sn	2
Na	2.4%	Sb	0.2
Mg	1.95%	Te	
Al	8.2%	I	0.5
Si	28.2%	Cs	3
P	1,050	Ba	425
S	260	La	30
Cl	130	Ce	60
K	2.1%	Pr	8.2
Ca	4.2%	Nd	28
Sc	22	Sm	6
Ti	0.57%	Eu	1.2
V	135	Gd	5.4
Cr	100	Tb	0.9
Mn	950	Dy	3
Fe	5.6%	Ho	1.2
Co	25	Er	2.8
Ni	75	Tm	0.5
Cu	55	Yb	3
Zn	70	Lu	0.5
Ga	15	Hf	3
Ge	1.5	Ta	2
As	1.8	W	1.5
Se	0.05	Re	
Br	2.5	Os	
Rb	90	Ir	
Sr	375	Pt	
Y	33	Au	0.004
Zr	165	Hg	0.08
Nb	20	Tl	0.45
Mo	1.5	Pb	12.5
Ru		Bi	0.17
Rh		Th	9.6
Pd		U	2.7

*Unless otherwise indicated, the values are in parts per million (1ppm=0.0001%). Arranged in order of atomic number. *Source:* Ahrens 1965.

solution are the first to crystallize. Consider for example a hot spring where the water issuing from the spring cools, as it flows away from the spring, by exposure to lower surface temperatures. Least soluble species in solution would be deposited at the spring while more soluble species might be carried downstream. This mechanism of mineralization can describe phenomena in the distant past (during cooling of the Earth's lakes and oceans) or the present (mineralization from evaporation of surface brines, volcanic action, etc.). Under this heading we may include crystallization of minerals from aqueous solutions that have flowed along fissures in the rock; in this way veins of mineralizations are produced

Precipitation from aqueous solutions by reaction

The reader is probably aware that if an aqueous solution of some metal compounds (e.g., iron or magnesium) is made basic, then hydroxides tend to precipitate. Changes of pH or other chemical parameters can bring about similar precipitation in nature.

Selective natural leaching

The major example of this kind of mineralization is the formation of bauxite (largely oxides of aluminum and some silicon, and the major source of aluminum). Bauxite deposits are typically found in tropical countries with abundant rainfall and vegetation, such as Jamaica and Surinam. Over geological time periods the original aluminosilicate host mineral (which might have been similar to the average composition of Table 1.1) has been "leached" by downward-trickling humic acid formed by the action of rainwater on decaying vegetation. This weak acid dissolved out iron (and some other metal) compounds, leaving behind a region enriched in aluminum. Additional enrichment of aluminum may occur in bauxites by deposition of aluminous dust derived from weathered rocks upwind.

Of course, this does not explain why bauxites are low in silicon (typically less than 10 percent), the oxide of which is *not* soluble under acid conditions. The low silicon content can be explained by seasonal variation in the pH of the solution in contact with the rock, from slightly acid to slightly alkaline, with the oxide of silicon dissolving during the alkaline periods. These seasonal variations might be caused by rainfall variations with high rainfall injecting considerable amounts of humic acid into the nascent ore body and low rainfall allowing the solution to become alkaline by contacting alkaline carbonate rock. In this way there is an alternate leaching of acid-soluble and alkaline-soluble constituents, leaving behind a region enriched in hydrated aluminum oxide.

Downward migration of surface water

The example here is the formation of copper sulfide ores illustrated in Fig. 1.7. Close to the Earth's surface where the minerals are accessible to atmospheric oxygen, the mixed copper-iron sulfide $CuFeS_2$ can be oxidized in the presence of water to yield a solution of copper sulfate and insoluble iron oxide which is left behind as the solution runs downward. In lower regions (less accessible to oxygen and therefore less oxidizing) the solution can react with $CuFeS_2$ to precipitate the copper from solution (as CuS). The end result is an enriched zone where copper is present at levels much higher than in the original rock.

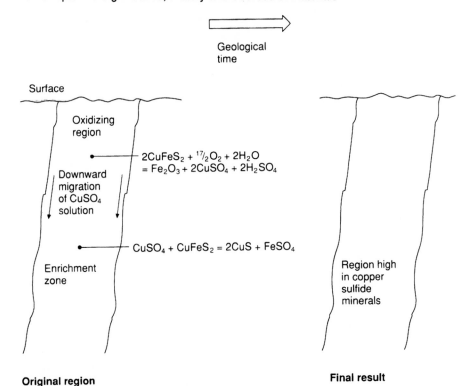

FIGURE 1.7 Formation of copper sulfide ores by downward migration of surface water.

Weathering, transportation and settling

Imagine a mountain range which is being "weathered" by the action of wind, rain, frost and snow (see Fig. 1.8). Streams and rivers leaving this range will carry particles of rock from the mountain range. Because rocks are heterogeneous on a small scale (i.e., show variations in composition in distances of tenths of millimeters or less), these particles may differ from each other in composition. If the river enters a lake, or other calm water, the particles will tend to settle out, starting with the denser and larger particles (such as any containing gold); *a placer deposit* of heavy mineral particles is formed at this point. Such a deposit gave rise to the name of the town "Placerville" in the foothills of California's Sierra Nevada, where a placer deposit was mined.

The fortuitous concentration of one or more valuable minerals into an ore body by one or more of these mineralization phenomena enables us to produce materials economically from these regions of the lithosphere. Of course these ore bodies have first to be found using methods of geology, geochemistry and geophysics that are beyond the scope of this text (a book on exploration is included in the suggested reading list at the end of this chapter). The next step is to remove the minerals from the ore body by mining.

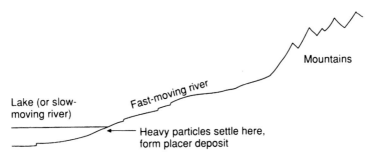

FIGURE 1.8 Schematic diagram showing how placer deposits are formed.

1.5 Mining

The usual object of mining is to obtain the ore from the ground at minimum cost with due regard for the health and safety of the miners and any adverse impacts on the environment. Mining techniques are divided into *surface mining* and *underground mining* with the depth of the ore body below the surface being the primary factor in selecting between the two. Deep deposits are typically mined by underground methods. Today the most widely practiced technique of surface mining is *open pit (open cut)* mining. Examples of such mines can be found throughout the US, particularly in the West and mountain states. The heart of the mine is a large excavation comprised of several *benches* as shown in Fig. 1.9. The benches are advanced downwards as follows. First holes are drilled down through a section of the bench, the holes are filled with explosive and the explosive detonated. Power shovels are then used to load the broken rock onto huge trucks which carry the rock up *haul roads* out of the pit. If the top of the ore body lies beneath the ground (the usual case since most outcropping deposits have long been discovered and already mined) then it is necessary to remove an *overburden* of worthless rock before the ore can be reached. The greater the depth of the top of the ore body, the greater is the cost of overburden removal and open cut mining is therefore not economic for deep deposits. In some instances open pit mining is carried out selectively. By analyzing the ore in the benches ahead of time, the richest ore can be mined first, with the lean ore set aside for processing later, or rock containing less than the minimum grade is identified and dumped immediately.

A limited amount of surface mining is still done by *dredging*. In this technique a dredge is used to scoop up the bottom of a natural or manmade lake. As practiced on a large scale the technique uses a large floating dredge with a chain of buckets similar to the dredge used to free ship channels of silt. Smaller hydraulic dredges are used in small mining operations; these employ compressed air to bring about a flow of water capable of sucking particles off the bottom of a lake or river.

The techniques of underground mining are too numerous and varied to be detailed here. Most of them can be loosely described as follows. Vertical shafts are sunk to one or more *levels* within or adjacent to the ore body. Some of these shafts, served by elevators, transport miners and machinery between the surface and the levels, others serve for transporting ore. From the vertical shafts, horizontal shafts (adits) are bored at several levels and give access to the ore body. Excavation of the ore is again by blasting and shoveling with the ore transported by gravity through *ore chutes* to conveyor

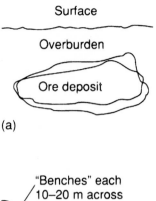

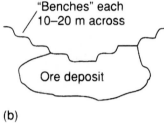

FIGURE 1.9 Open-pit mining: (a) the ore deposit before mining; (b) during mining.

belts or small rail cars. Two great needs in underground mining are to ensure the stability of the openings in the rock (hence a great interest in rock mechanics) and to provide ventilation. The latter involves removal of the considerable amount of heat generated by machines underground, the supply of fresh air, the control of humidity and the avoidance of accumulation of toxic or explosive gases.

One mining technique that is radically different from the others is *solution mining* or *in-situ leaching*, as it is sometimes called. In this technique the ore is left in the ground and solutions capable of dissolving the valuable minerals are pumped through it by means of wells (a few inches in diameter) drilled from the surface. The technique is only suitable if the ore body is permeable (or can be rendered so by underground explosion, for example). It has been used at several locations in Texas for uranium extraction using aqueous ammonium carbonate/bicarbonate solutions as the leaching agent.

From our earlier discussion it should be clear that ores from most mines will contain only a small amount of the compound that will ultimately become a material. A cobalt ore might contain only 5 percent of cobalt compounds, a copper ore perhaps 1 percent of copper compounds and a gold ore a few tenths (or even hundredths) of an ounce of gold per ton. The rest of the ore is worthless *gangue*. The reader might guess, correctly, that immediately subjecting the ore to chemical processing, such as adding reagents and raising the temperature, might result in considerable wastage of reagents and energy on the gangue. This waste would carry a high economic penalty, because of the cost of most chemicals and fuels; hence ores are subjected first (with some notable exceptions) to *physical* processes to remove inexpensively as much as possible of the gangue. This is done by "mineral concentrating" (known sometimes by the older term *mineral beneficiation*) which is part of the field of *mineral processing*.

1.6 Mineral Processing

1.6.1 Liberation

The ore from a mine would consist of lumps perhaps as much as a meter or two across, down to quite fine particles. Fig. 1.10 is a sketch of a lump somewhere in the middle of this size range. The dark areas are intended to represent the valuable compound (valuable mineral) that is to become a material eventually, and the rest is worthless gangue. The valuable mineral particles within this lump are approximately 2 mm across. Now consider what happens if this lump is broken up into a number of small particles. Our original lump is a mixed particle, i.e., it contains both valuable mineral and gangue. After breaking it up into smaller particles it is virtually certain that some of those particles are entirely gangue, some entirely valuable mineral, while some are mixed. The reader is invited to simulate this by making a photocopy of Fig. 1.10 and cutting it up "randomly" into twenty or thirty pieces, preferably with the photocopy turned over so as to ensure an unbiased experiment! If we were to determine the fraction of mixed particles as a function of mean particle size on breaking our lump into finer and finer particles we should get the result sketched in Fig. 1.11. For mean particle sizes not much less than the original size the majority of particles are still mixed; the first random cut through Fig. 1.10, for example, is likely to yield two mixed particles. As the mean particle size starts to approach the valuable mineral particle size, or the separation between valuable mineral particles (whichever is the greater), particles that are not mixed appear and when we are well below the valuable mineral particle size almost none of the particles are mixed. This physical *liberation* of the valuable mineral particles is also aided by the tendency of the ore to fracture along the interfaces between mineral and gangue.

Typically liberation is carried out by first *crushing* and then *grinding* the ore. *Crushing* is the breaking of the (predominantly) large lumps of ore from the mine into particles approximately 1 cm across. One device for doing this is the *jaw crusher*

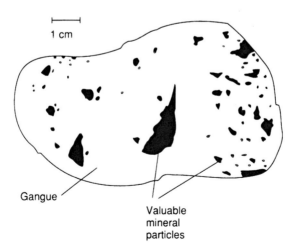

FIGURE 1.10 Sketch of a lump of ore consisting of gangue in which valuable mineral particles are embedded.

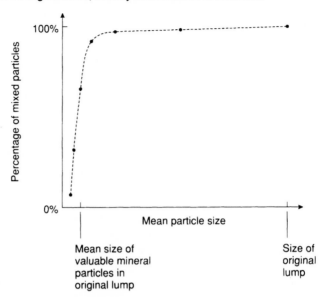

FIGURE 1.11 How the percentage of particles containing both gangue and valuable mineral particles declines during comminution.

depicted schematically in Fig. 1.12. Ore is dropped from the truck, which delivered it from the mine, between massive steel plates. One plate is fixed while the other is pivoted at the top and moves back and forward at the bottom, driven by a linkage connected via a flywheel to a powerful electric motor. As the jaws open the lump descends until it jams and it is then crushed as the jaws close. A similar principle applies in other crushing devices (gyratory crushers and cone crushers) which consist of an inner cone inside a fixed hollow with the inner cone driven by an eccentric connection to a flywheel.

Following crushing, the ore is reduced further in size by grinding, typically in a *ball mill* or *rod mill*. These similar devices consist of drums (2-5 meters in diameter) that are rotated about a horizontal axis at a few revolutions per minute. Fig. 1.13 is a sketch of a ball mill. Crushed ore, usually with water and a recycle stream consisting of ore particles in water, is fed continuously into one end of the drum which contains numerous steel balls as a *grinding medium*. The balls are tumbled over each other and particles trapped between impacting balls are smashed into smaller particles. The suspension (*slurry*) of ground particles in water flows continuously out of the other end of the mill. Rod mills are similar except that steel rods, rather than balls, are used as the grinding medium. Typically the product of a mill has a wide range of particle size, reflecting the size distribution in the feed and the stochastic nature of the grinding phenomena. Usually it is desirable to have the particles of a similar size and this is achieved by connecting the mill to a particle size classification device as shown in Fig. 1.14. One commonly used classification device is the hydrocyclone shown in Fig. 1.15. Slurry from the mill is pumped into the hydrocyclone tangentially and swirls around the vertical axis before exiting, partly through the top and partly through the bottom. Larger (heavier) particles are flung to the outside and exit from the bottom while smaller particles remain suspended

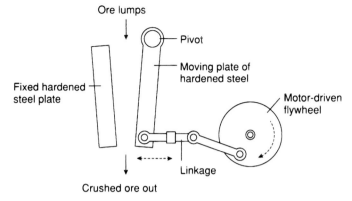

FIGURE 1.12 Jaw crusher.

and exit from the top. In Chapter XI we will return to the grinding process when we discuss the production of fine powders.

Finally we mention *autogenous* grinding where large lumps of ore, rather than a steel grinding medium, are used to grind the ore. *Semi-autogeneous* grinding is more common and the meaning is obvious from the name.

Liberation of the mineral particles from the gangue is just the first step. At this point we have a mixture of particles perhaps 50-500 μm in size, few of which are mixed, and we must now separate the mineral particles from the gangue particles to the extent that

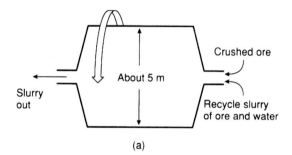

(a)

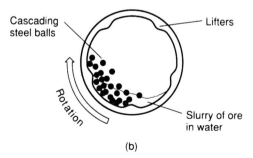

(b)

FIGURE 1.13 (a) Ball mill; (b) section showing the tumbling steel balls.

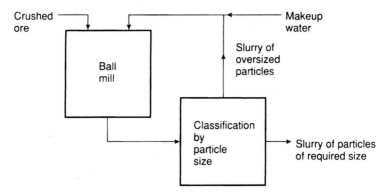

FIGURE 1.14 How a size classification device is linked to a ball mill.

it is possible. To do so we exploit some difference in physical properties between the valuable mineral particles and the gangue particles. The properties most commonly exploited are density, magnetic susceptibility and wettability.

1.6.2 Gravity Separation

There are several devices exploiting differences in density to separate particles and they are usually known as gravity separation devices. One common device is illustrated schematically in Fig. 1.16. This is a *heavy medium separator or sink-float separator* and the principle of operation is simple. The particles are continuously fed to a gently agitated tank containing a liquid with a density intermediate between those of the two types of particles. The denser particles sink while the less dense float.

1.6.3 Magnetic Separation

Magnetic separation is a convenient separation method when one type of particle is magnetic, and is commonly used in separating iron ore (magnetite) from gangue. A rotating horizontal steel cylinder containing an electromagnet is partly immersed in a

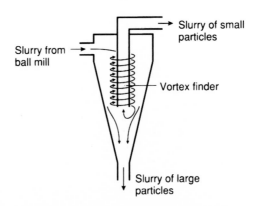

FIGURE 1.15 Hydrocylone for classification of particles by size.

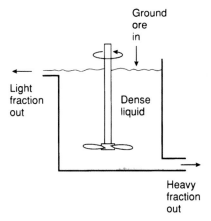

FIGURE 1.16 Separation of particles into dense and less dense fractions by the sink-float technique.

slurry of the particles (see Fig. 1.17). The magnetic particles adhere to the cylinder and are scraped off and removed; the non-magnetic particles remain in the slurry and flow out of the bottom of the tank.

1.6.4 Flotation

Flotation is perhaps the most important separation technique in that it is widely used on a variety of ores, particularly sulfide ores. The method relies on differences in surface chemistry that make one of the types of particles more readily wetted by an aqueous solution than another type of particle. For example, by adding inexpensive reagents to a slurry containing copper sulfide particles and silica (SiO_2) gangue particles, we can make the sulfide particles *hydrophobic* (preferring not to be in contact with water),

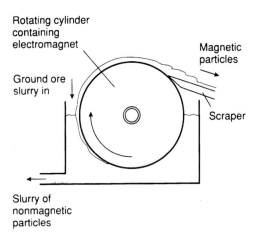

FIGURE 1.17 Magnetic separation.

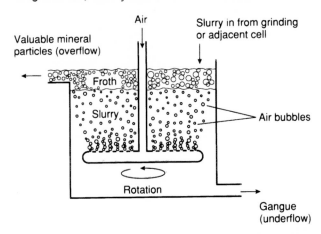

FIGURE 1.18 Separation of particles into hydrophilic and hydrophobic fractions by flotation.

while the gangue particles are *hydrophilic* (preferring to remain in contact with water). If we now contact the slurry with an air bubble, sulfide particles will attach to the bubble (thereby freeing part of their surface from contact with water) while the silica particles will not. If the bubble is allowed to float upwards it will reach the surface forming a froth rich in sulfide particles which can then be allowed to overflow from the apparatus. A flotation *cell* in which this is carried out is sketched in Fig. 1.18; such a device is operated continuously. The technique depends on adjustments to the chemistry of the slurry to *selectively* render one class of particles hydrophobic. This is done by addition of "collectors" (which selectively adsorb on one type of particle, making it hydrophobic) and "conditioners" (to adjust pH for optimum surface chemistry). "Frothers" are added to yield a stable froth at the top of the flotation cell. It is even possible to separate more than two types of particles. For example, some copper sulfide ores also contain molybdenum sulfide, as well as gangue. By adjusting the solution chemistry it is possible to first render the molybdenum sulfide particles hydrophobic and float them. Then further flotation reagents are added to make the copper sulfide particles hydrophobic and float them. The reader should be careful to distinguish between flotation and the sink-float technique described in Section 1.6.2.

1.6.5 Separation Systems

The three separation methods described above (and other separation techniques of mineral processing) are highly imperfect in that just one application of one of the methods would result in little separation. For example the underflow from a flotation unit may contain many hydrophobic particles that simply have not chanced to encounter an air bubble, or the overflow may contain hydrophilic particles that were simply caught in the wake of an air bubble or in a cluster of hydrophobic particles attached to a bubble. It is therefore common to run many cells in complicated series-parallel arrangements, perhaps with adjustment of chemistry at some point. A simple arrangement of this nature is shown in Fig. 1.19. A series of *rougher cells* (with the underflow passing from

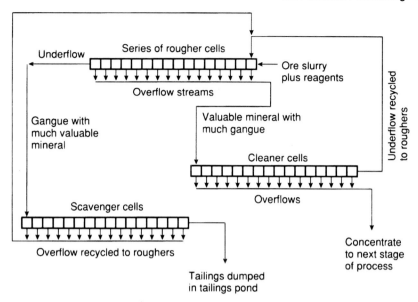

FIGURE 1.19 One way in which a number of flotation cells can be arranged into a flotation system.

one cell to the next) makes a rough separation of the slurry from grinding into two streams. The mineral-rich stream then goes to a second series of *cleaner cells* whose purpose is to remove as much residual gangue as possible, producing a slurry of *concentrate* particles or *heads* that contains much of the valuable mineral in the ore and a second stream that (in Fig. 1.19) is recycled to the rougher cells. The other stream from the rougher cells, containing less valuable mineral (but still too much to throw away), passes to *scavenger cells* where as much as possible of the valuable material is recovered from the *tailings* stream going to the waste pond. This scavenged valuable mineral is recycled to the rougher cells in the scheme of Fig. 1.19. Similar schemes can be designed for other separation techniques.

At this point our concentrate is in the form of mineral particles suspended in water. Much of the gangue will have been removed but much will still remain in the concentrate; the concentration of valuable mineral in the concentrate might be one order of magnitude greater than in the ore. In some instances we can now start to chemically process but usually this concentrate must first be separated from the water and frequently agglomerated into a form different from the small particles (50-500 μm diameter) from separation. One obvious need for drying is if we wish to ship the concentrate to a distant plant for chemical processing; there is little virtue in shipping water. The need for agglomeration can arise from the nature of the chemical process itself. For example the iron blast furnace that we shall encounter in Chapter 8 must be fed iron oxide in the form of particles a centimeter or so in diameter.

1.6.6 Filtering, Drying and Agglomeration

Filtering of the slurry from separation is usually done continuously, on disc or drum filters. The principle of operation is obvious from a casual inspection and we devote no more space to these. Drying is usually carried out continuously in rotary kilns heated by natural gas or oil; these are similar to kilns treated later in the text and are given no further attention here.

The *pelletization* of iron ore concentrates is a representative agglomeration technique and we provide some discussion of this. *Green* pellets are formed in either *balling drums* (to be distinguished carefully from ball mills) or on *pellet wheels*. The former are large drums, perhaps 5 meters in diameter and 10 meters long, that are rotated about their axes, which are slightly inclined to the horizontal; the rotational speed is kept below the speed at which the contents would centrifuge but is sufficient to cause the contents to constantly tumble over each other. Dry concentrate and *binders* are fed continuously into the high end of the drum. The binders (typically a clay called bentonite, plus some water) cause the particles to stick together as they tumble over each other. Hence, as the contents move along the drum they are agglomerated into nearly spherical pellets perhaps 1 centimeter in diameter.

The pellet wheel might be 2-5 meters in diameter and saucer-shaped. It is rotated about an axis at approximately 45° to the horizontal. Concentrate and binder are fed onto the wheel and pellets are flung off the edge into a circumferential chute when they reach a certain size.

Green pellets formed in this way are too weak for all but the gentlest handling and in order to ship them, or feed them into chemical processes, they must be strengthened. This is done by *heat induration*; the pellets are brought to a high temperature and the sintering phenomenon described in later chapters takes place, considerably increasing the strength of the pellets. This heat induration is typically carried out on a *travelling grate furnace* where flames from overhead burners are sucked through a bed of the pellets a few centimeters deep as it travels continuously through the furnace on a moving grate. These *fired* pellets typically have enough strength to drop through several meters without much breakage and to withstand a load of several meters of their fellows. They are therefore suitable for shipping and feeding into units such as the iron blast furnace.

An alternative agglomeration technique that is widely used for iron ore concentrates is *sintering*. This *process* should be distinguished from the *phenomenon* sintering that has already been mentioned as occurring during heat induration and which, because of its importance in ceramic processing, is treated in detail later in this book. To add to the possible confusion, the sinter process is carried out on a traveling grate furnace (known as a *sinter strand*). Concentrate particles, fine coke particles and *flux* (to promote partial melting) are fed continuously onto one end of the travelling grate. The coke is ignited by overhead burners and air is sucked down through the mass to continue the coke combustion and raise the temperature to a point where some of the concentrate constituents (in conjunction with the flux) melt. Towards the end of the machine, where all the coke has burned, the mass starts to cool down and solidify to a porous hard material known as *sinter*. This breaks up into lumps as it falls from the end of the machine and is a suitable feed for the blast furnace. This sinter process has two distinct advantages; it uses little fuel, apart from the coke fines (which are essentially a waste product of coke production), and some iron-bearing wastes (e.g., dust from the BOFs used in steelmaking) can also be fed to the sinter strand. Sinter strands have been used for agglomeration of non-ferrous concentrates and in the case of sulfide concentrates

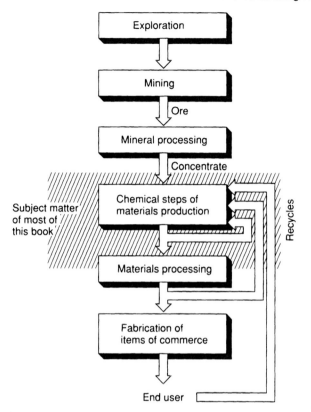

FIGURE 1.20 Overview of the subject matter of this book.

(e.g., lead sulfide) sufficient heat may be generated by oxidation of the sulfides to oxides, to make coke unnecessary.

1.7 Concluding Remarks

Fig. 1.20 presents an overview of the field of minerals and materials production. We have already covered the upper three boxes in this figure to the extent necessary in this text and much of what follows will be concerned with the chemical steps involved in converting a concentrate into a material as we define materials in the opening sentence of this chapter. These chemical steps will usually result in the material produced being in the required chemical form but not usually the desired physical form. For example the chemical processes entailed in producing steel result in molten steel and further *materials processing* is required to cast it and roll it to produce steel strip, say, for automobile manufacture (one of the manufacturing activities belonging to the last box of Fig. 1.20). We shall therefore treat some of these materials processing operations in this book, although many will be left for other, more specialized texts. In fact, the boundaries between the various boxes in Fig. 1.20 are not sharp. For example, do we put solution mining in the second or fourth box? Do we categorize sinter strands as falling in box three or four? Is chemical vapor deposition (to be discussed in a later chapter) a topic of the fourth, fifth or sixth box? This difficulty in classifying operations into materials production (sometimes called *materials synthesis*) or materials processing is particularly evident in the case of ceramic materials. In many cases (e.g., the fabrication of reaction-bonded silicon nitride parts) a chemical step occurs in the final stage of ceramic processing. Consequently many ceramists may prefer to label large parts of this text as ceramic processing rather than ceramic production. We prefer to regard the labels as no more than that, a convenience but not one that should restrict our thinking

REFERENCES AND FURTHER READINGS

L. H. Ahrens, *Distribution of the Elements in Our Planet*, McGraw-Hill, New York, 1965.

D. Altenpohl, *Materials in World Perspective*, Springer-Verlag, New York, 1980.

U.S. Geological Survey, *Mineral Commodity Summaries and Mineral Facts and Problems* (www.usgs.gov).

Cameron, E.N., ed., *The Mineral Position of the United States, 1975-2000*, University of Wisconsin Press, Madison, WI, 1973.

P. T. Craddock, P. T., and M. J. Hughes, eds., *Furnaces and Smelting Technology in Antiquity*, British Museum Publications, London, 1985.

J. Diamond, *Guns, Germs and Steel: The Fates of Human Societies*, W. W. Norton and Co., 1997.

Dorr, A., *Minerals – Foundations of Society*, League of Women Voters of Montgomery County, Maryland, Inc., Rockville, MD, 1984.

Evans, J. W., and J. Szekely, *Newer Versus Traditional Industries: A Materials Perspective*, J. Met., 37 (1985).

Hartman, H. L., *Introductory Mining Engineering*, Wiley- Interscience, New York, 1987.

Kearey, P. and M. Brooks, *An Introduction to Geophysical Exploration*, Blackwell, London, 1984.

Kingery, W. D., "Some Aspects of the History of Ceramic Processing", W. D. Kingery in *Proc. 3d Intl. Conf. Ultrastructure Processing of Advanced Ceramics*, J. D. Mackenzie and D. R. Ulrich (eds.), Wiley-Interscience, New York, 1988.

Krysko, W. W., J. Met., 32, 1980.

Kummer, J., and N. Weber, *Trans. Soc. Auto. Eng.* 76 , 1003 (1968).

"Materials for Economic Growth", October 1986 edition of *Scientific American*.

Raymond, R., *Out of the Fiery Furnace*, The Pennsylvania State University Press, University Park, PA, 1986 (and the videotape series that this volume augments).

Spottiswood, D. J. and E. G. Kelley, *Introduction to Mineral Processing*, Wiley, New York, 1982.

R. F. Tylecote, *A History of Metallurgy*, The Metals Society, *London*, 1976.

Wills, B.A., *Mineral Processing Technology*, 4th ed., Pergamon Press, *Oxford, 1987.*

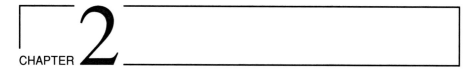

Chemical Thermodynamics

2.1 Introduction

It is anticipated that the readers of this book will have had at least a beginning course in thermodynamics. This chapter will therefore serve as a brief review of those aspects of thermodynamics that are particularly relevant to the production of materials. In addition, this chapter will treat some topics (e.g., Ellingham diagrams) that are frequently given limited coverage in undergraduate courses but that are of great value to the practicing engineer or scientist (e.g., in rapidly determining whether a chemical reaction, proposed as a way of producing a material, will actually take place).

As a starting point we shall assume that the reader is familiar with the concepts of an ideal gas, internal and external energy, enthalpy, heat capacity, equilibrium and the laws of thermodynamics. We begin our review with an examination of the reversibility or irreversibility of changes.

2.2 Reversibility (or Otherwise) of Changes

A *reversible* change is one where the system under investigation is at equilibrium with its surroundings throughout the change. If at any point during the change there is a departure from this equilibrium, the change is *irreversible* .

A common illustration of this concept appears in Fig. 2.1. A reservoir of water is held in place by a wall which is somehow contrived to be both leakproof and movable in the horizontal direction without friction. Our system consists of the reservoir and movable wall together with the other walls and bottom of the reservoir. Initially the wall is stationary with the force, F, due to hydrostatic pressure balanced by a force of equal magnitude but opposite direction exerted by the surroundings. First consider a reversible change where the latter force is reduced by an infinitesmal amount, dF. There is then a net (infinitesmal) force on the movable wall to the right and the wall will accelerate to the right along with the water behind it. Because the net force is infinitesmal the acceleration will also be infinitesmal and the wall will take an infinite amount of time to move through a finite distance Δx. The reader will note that if the wall moves through a finite distance then the force exerted by the water will drop by a finite amount. However, we can compensate for this by adjusting the external force so that it is always

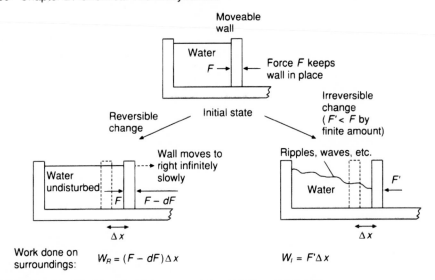

FIGURE 2.1 Difference between reversible and irreversible changes.

a differential amount less than the force exerted by the water. Intuitively it is obvious that the water will be relatively undisturbed during this change and, because departure from equilibrium between the forces on the wall is not by any finite amount, we can regard the change as a reversible one. It is clear that at any point we can restore the system to its initial state by increasing the external force F to $F + dF$ and waiting another infinity of time. The work done by the system on the surroundings as the wall moves to the right we shall call W_R, the subscript indicating a reversible change.

$$\mathrm{W_R} = (\mathrm{F} - \mathrm{dF})\Delta\mathrm{x} \qquad (2.1)$$

which assumes that Δx is sufficiently small that we can neglect the variation in F. Note that we are treating work done by the system on the surroundings as positive.

Now let us consider what happens when the force initially holding the wall in place is reduced by a finite amount to F'. The net force moving the wall to the right is no longer differential, the system is no longer in equilibrium with its surroundings, and the wall will accelerate to the right at an appreciable rate causing waves and other disturbances of the water in the reservoir. Intuitively it is clear that if we attempt to restore the system to its initial state we can do so by increasing the external force on the wall by a finite amount. When the wall is back in place the water behind it will remain in motion for some time until eventually its viscosity dissipates the mechanical energy of the water's motion into heat, raising the temperature of the water. Restoring the wall to its original position therefore does *not* reverse the change since the system does not return to its initial state but rather to one at higher temperature. This second change is therefore irreversible. The work done by the system on the surroundings as the wall moves irreversibly a distance Δx to the right is

$$\mathrm{W_I} = \mathrm{F}'\Delta\mathrm{x} \qquad (2.2)$$

Since $F - dF > F'$,

$$W_R > W_I$$

That is, more work is done by the system if the change is reversible (other things being equal). (The difference W_R-W_X is "degraded" to heat in an irreversible change and we shall take up this topic again in Section 2.4). This is an example of a general principle: *When a system undergoes a change so as to do work on its surroundings, the maximum work is done if the change occurs reversibly.* Note, however, that the *power* that can be obtained from the system is not a maximum when the change is reversible, because reversible changes are infinitely slow, and power (the rate of doing work) is therefore infinitesmal for a reversible change.

This characteristic of reversible changes, that they occur only at infinitesmal rates, means that they can never be achieved in the laboratory or anywhere else in the real world. Nevertheless, this idealization is one of the most useful ones in thermodynamics and we can approach this ideal limit in the laboratory by making changes over longer and longer periods of time.

Spontaneous changes are a special category of irreversible change. We attach this label to irreversible changes taking place in an *isolated system*. An isolated system is one that does not exchange either matter or energy with its surroundings and should be distinguished from a *closed system* which is a system that does not exchange matter but may exchange energy.

2.3 Entropy

It has been the authors' experience that entropy is the first concept in thermodynamics that presents difficulty to the student, perhaps because entropy is not an intuitive concept and perhaps because of its apparent failure to obey "common sense" conservation laws. At an early age (usually painfully) we grasp the concept of temperature, and enthalpy is merely an extensive embodiment of this intensive property. However, our first encounter with entropy is usually in the form of a mathematical equation (such as equation (2.6) below) which does little to provide an understanding of the physical/chemical significance of entropy. In this section we have borrowed from an approach to entropy based on information theory which is fully presented in an excellent book by Tribus (1961).

One view of *entropy* is that it is a measure of the irreversibility of a change within an isolated system. Thus if an isolated system undergoes a reversible change, its entropy does not change (although there may be a transfer of entropy from one part of the system to another). If the isolated system undergoes an irreversible change (i.e., a spontaneous change) then its entropy increases. The reader is asked to consider the implication of this last statement with respect to the entropy of the universe.

Suppose the system is not isolated (but closed) and a differential amount of heat dq is added reversibly to the system at some point where the temperature is T. The entropy of the system is, by definition, increased by a differential amount

$$dS = \left(\frac{dq}{T}\right)_R \qquad (2.3)$$

where the subscript R serves to remind us that the heat is added reversibly. From (2.3) we can generalize to the case where finite amounts of heat are added by exploiting the fact that a sum of differential amounts is an integral. The increase in the entropy of the system, ΔS, is given by

$$\Delta S \equiv S_{final} - S_{initial} = \int_{initial}^{final} \left(\frac{dq}{T}\right)_R \qquad (2.4)$$

(Throughout this book we shall use $\Delta x \equiv$ increase in $x =$ final x - initial $x = -$ decrease in x, with the implication that the increase is finite rather than differential.)

Consider a change to a closed system containing unit mass of a particular chemical species which involves no chemical reaction, no change of phase and no change in pressure (i.e., the only change is a change in temperature). Then on adding an amount of heat dq the temperature of the system increases by an amount $C_P dT$ where C_P is the heat capacity at constant pressure. Substituting in equation (2.4) yields

$$\Delta \overline{S} = \int_{initial}^{final} \frac{C_P dT}{T} \qquad (2.5)$$

where now we can drop the subscript indicating reversibility because C_P is a unique function of temperature for the species in question and therefore the right-hand side is independent of how the heat is added (provided pressure is held constant). The overbar on S indicates that we are examining increases in entropy per unit mass of the species in question.

If we let S_0 be the entropy of a species at absolute zero then we can exploit (2.5) to calculate the entropy at temperature T^*. In so doing we must revert to equation (2.4) and recognize that between zero and T^* there might be phase changes (melting, for example) and at each phase change we will have an additional increase in entropy given by $\Delta H_{PC}/T_{PC}$ where H_{PC} is the latent heat of the phase change at temperature T_{PC}. The result is

$$\overline{S} = \int_0^{T^*} \frac{C_P dT}{T} + \overline{S}_0 + \sum_{\substack{phase \\ changes}} \frac{\Delta H_{PC}}{T_{PC}} \qquad (2.6)$$

$\overline{S}_0$ will be zero (by the third law) if the species forms a perfect crystal at 0 K. Examining the quantities on the right of equation (2.6) we see that they are all state variables (i.e., variables that depend only on temperature, pressure and composition and not on the path followed by the system to arrive at that state) and consequently entropy is a state variable.

Equation (2.6) provides us with a means of determining entropy for a species if we can obtain its specific heat (and latent heat data) between zero and T^*. The integral on the right can be calculated analytically (if a suitable equation for C_P as a function of temperature can be found), numerically or graphically. The last is achieved by plotting C_P/T versus T. The integral is then the area under the curve from $T=0$ to $T=T^*$. A difficulty here is that $1/T$ in the integrand tends to infinity as $T \rightarrow 0$. Fortunately C_P also tends to zero so that the integrand C_P/T remains finite and the integration can be carried out.

Let us finally try to illustrate the formal equations regarding entropy to arrive at an understanding that can serve us in the rest of this book. We can view the entropy as being a measure of disorder in the system. This disorder may be regarded as the possibility of forming the system in different ways that are thermodynamically indistinguishable. For example, in a liquid there are myriads of positions that the molecules might take, while still being in exactly the same thermodynamic state. Practically, we have no way of determining the coordinates of the various atoms or molecules and their corresponding velocities from thermodynamic measurements. This leads to the view that entropy, being a measure of *disorder*, can be regarded as the *uncertainty* about the positions and velocities of individual atoms or molecules making up our system. To illustrate this viewpoint let us consider (Fig. 2.2) a box containing a perfect crystal at absolute zero. At this temperature the atoms in the crystal have no vibrations and, given instruments of sufficient precision (atomic resolution microscopes, etc.) and sufficient resources we could, in theory (and perhaps in practice) determine the positions and velocities (zero) of the atoms to an exactness that is limited only by the imperfections of our instruments[1]. We regard the uncertainty as zero, consistent with equation (2.6). If now we raise the temperature (but stay well below the melting point of the crystal) then equation (2.6) tells us that there is a moderate increase in entropy. Correspondingly there is an increase in uncertainty in the positions of the atoms due to the fact that they are now vibrating about their previous positions in the lattice of the crystal, or possibly diffusing within the crystal, and the crystal has acquired defects, such as vacant lattice sites. All of these processes are random ones in the sense that we can no longer be sure of the whereabouts and velocities of the atoms at any instant, even assuming we could determine them a few seconds previously.

Now let us raise the temperature in the box above the melting point. An increase in entropy occurs both because of the temperature increase and because of the melting. A significant increase in the uncertainty, or disorder, also occurs; the atoms that had previously mostly been "centered" about lattice points are now free to take many other positions throughout the liquid pool.

Finally let us bring the temperature to above the boiling point. Equation (2.6) predicts a large increase in entropy because latent heats of boiling are typically large (in Joules per mole they are approximately 88 times the boiling point in Kelvin, which is one order of magnitude greater than the latent heats of fusion). Concomitantly, there is a large increase in uncertainty or disorder because our atoms, previously largely confined to the liquid pool in the bottom of the box, now wander throughout the whole volume of the box. We can say little more than that the atoms are somewhere in the box.

From this discussion it is immediately clear that for any particular chemical species

$$\bar{S}_{gas} \gg \bar{S}_{liquid} > \bar{S}_{solid}$$

Of course we have only shown that, qualitatively, entropy and uncertainty go hand-in-hand. Fortunately, it is possible to establish rigorously a quantitative relationship between entropy and uncertainty; this is beyond the scope of this review and the reader is referred to the book by Tribus (1961).

---

[1] No violations of the Heisenberg uncertainty principle are implied here because the atoms are stationary.

As an exercise the reader is asked to predict whether the following (carried out at constant temperatures) are associated with an entropy increase or decrease, using the uncertainty concept of entropy, and where possible, to check the prediction by examining equation (2.6).

1. Expansion of a gas
2. $C + CO_2 = 2CO$
   (with carbon a solid and the other species gases)
3. The *devitrification* (crystallization) of glass
4. $C + O_2 = 2CO$ [as in (2)]
5. $C + O_2 = CO_2$ [as in (2)]
6. $(C_2H_5)_3Al = 3/2\ H_2 + 3C_2H_4 + Al$
   (with aluminum a solid and the other species gases)
7. Polymerization of ethylene

2.4 Free Energy

Much of thermodynamics, particularly as taught to mechanical engineers, is concerned with engines and with obtaining "useful work" from systems as they undergo changes of state (such as expansion). Let us consider a change where there is a decrease in internal energy, $-\Delta E$, with no change in external energy. What fraction of this decrease can be converted to useful work? First we must recognize that, if the change is carried out irreversibly (as it will be in all practical devices) then there will be a *degradation* of some of the internal energy reduction into heat. This degradation was described in connection with the reservoir of Fig. 2.1 but is present in all systems undergoing irreversible changes and is greater the greater the departure from reversibility. Quantitatively an amount $T\Delta S$ of internal energy is so degraded (or $\int T dS$ if the change is not isothermal). The system therefore does an amount of work, $-\Delta E - T\Delta S$, on its surroundings during a change but this is still not the answer to our question because part of that work may be work done as the system expands against its surroundings; we do not usually regard that work as part of the useful work. Let us take a circuitous route to the answer by examining the state function

$$G \equiv E + PV - TS \qquad (2.7)$$

with P the pressure of the system and V its volume. G is known as the *Gibbs free energy* of the system. Consider a system undergoing a reversible change at constant temperature and pressure where it does work W_R on its surroundings and a quantity of heat q_R is added to the system. From the first law

$$\Delta E = q_R - W_R \qquad (2.8)$$

Because the temperature is constant, $q_R = T\Delta S$, hence

$$\Delta E = T\Delta S - W_R \qquad (2.9)$$

During this change the increase in G is given by (using the defining equation (2.7))

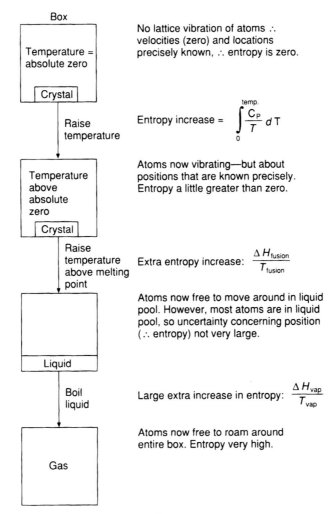

Box

Temperature = absolute zero

Crystal

No lattice vibration of atoms ∴ velocities (zero) and locations precisely known, ∴ entropy is zero.

Raise temperature

Entropy increase = $\int_{0}^{\text{temp.}} \frac{C_P}{T} \, dT$

Temperature above absolute zero

Crystal

Atoms now vibrating—but about positions that are known precisely. Entropy a little greater than zero.

Raise temperature above melting point

Extra entropy increase: $\dfrac{\Delta H_{\text{fusion}}}{T_{\text{fusion}}}$

Atoms now free to move around in liquid pool. However, most atoms are in liquid pool, so uncertainty concerning position (∴ entropy) not very large.

Liquid

Boil liquid

Large extra increase in entropy: $\dfrac{\Delta H_{\text{vap}}}{T_{\text{vap}}}$

Atoms now free to roam around entire box. Entropy very high.

Gas

FIGURE 2.2 Increase in entropy as a material is brought from absolute zero to a temptemperature above its boiling point.

$$\Delta G = \Delta E + \Delta(PV) - \Delta(TS) \tag{2.10}$$

which, because P and T are constant, becomes

$$\Delta G = \Delta E + P\Delta V - T\Delta S \tag{2.11}$$

Substituting from (2.9) yields

$$\Delta G = P\Delta V - W_R \tag{2.12}$$

Let us split the work into useful work (W_{RU}) and expansion work (W_{RE}) and use the fact that at constant pressure the expansion work is simply the pressure times the increase in volume, that is,

$$W_R = W_{RU} + W_{RE} = W_{RU} + P\Delta V \tag{2.13}$$

Substituting (2.13) in (2.12) we get the final result

$$-\Delta G = W_{RU} \tag{2.14}$$

But remember that the work obtained from a reversible change is greater than that obtained from any irreversible change (between the same initial and final states) and we can therefore conclude *that the maximum useful work obtainable from any change undergone by a system at constant temperature and pressure is equal to the decrease in the Gibbs free energy of the system.*

In equation (2.7) we defined Gibbs free energy in terms of internal energy E, but we can exploit the relationship between internal energy and enthalpy H,

$$H = E + PV \tag{2.15}$$

to obtain

$$G = H - TS \tag{2.16}$$

Because H and S are state functions, so is G. Just as H is tabulated for all common substances as a function of temperature, for example in the JANAF Tables (Chase, 1998), so is G. Note that, as with H, only *changes* in Gibbs free energy can be measured and so G is defined only with respect to its value at some arbitrary reference state. For example, the enthalpy and Gibbs free energy of water are given in Table 2.1 with respect to their values at 298.15 K. Note that the values given are for the most stable phase at the temperature in question (liquid for temperatures between the melting point and boiling point, etc.). The enthalpy shows discontinuities (equal to the latent heat) at the phase changes but the Gibbs free energy is continuous at these points.

PROBLEM

Calculate the entropy of 1 mol of water at 1000 K and 1 atm pressure if the entropy of water at 300 K is 70.35 J/mol°K.

Solution

From equation (2.16),

$$\Delta G = \Delta H - \Delta(TS)$$

where Δ indicates the increase in going from 300 K to 1000 K maintaining the pressure at 1 atm. Hence

$$(1000 \times \tilde{S}_{1000}) - (300 \times \tilde{S}_{300}) = (\tilde{H}_{1000} - \tilde{H}_{300}) - (\tilde{G}_{1000} - \tilde{G}_{300})$$
$$= [(\tilde{H}_{1000} - \tilde{H}_{298}) - (\tilde{H}_{300} - \tilde{H}_{298})] -$$
$$[(\tilde{G}_{1000} - \tilde{G}_{298}) - (\tilde{G}_{300} - \tilde{G}_{298})]$$

or, substituting the given value for $\tilde{S}_{300}$ and values for H and G from Table 2.1, we obtain

$$\tilde{S}_{1000} = \frac{1}{1000}\left(69.99 - 0.14 + 141.74 - 0.13 + \frac{300 \times 70.35}{1000}\right)\frac{kJ}{mol \cdot K}$$

$$= 0.232 \ \ kJ / mol \cdot K$$

TABLE 2.1 Enthalpy and Gibbs Free Energy of Water with respect to Values for Liquid Water at 298.15K

Temperature (K)	$\tilde{H} - \tilde{H}_{298}$ (kJ/Mol)	$\tilde{G} - \tilde{G}_{298}$ (kJ/mol)
300	0.14	- 0.13
373.15	5.66	- 5.90
373.15	46.50	- 5.90
400	47.46	- 11.15
500	50.93	- 31.42
600	54.51	- 52.39
700	58.20	- 73.98
800	62.00	- 96.10
900	65.94	-118.70
1000	69.99	- 141.74
1200	78.49	-189.06
1400	87.46	-237.82
1600	96.86	- 287.91
1800	106.62	- 339.20
2000	116.70	- 391.59
2500	142.98	- 526.88
3000	170.37	- 667.60

Source: (Adapted from L. B. Pankratz, "Thermodynamic Properties of Elements and Oxides", U.S. Bureau of Mines *Bulletin* 672, 1982.)

2.5 Equilibrium

The result in italics in the previous section leads to a very powerful conclusion. Consider some system which can exist in several "states" (some of which might be unstable) at a fixed temperature and pressure. For example, if care is taken to avoid disturbances and dirt particles, many liquids (including water) can be cooled to well be-

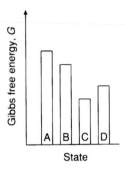

FIGURE 2.3 Gibbs free energy of a system related to the formation of reactants and products.

low their normal freezing points without solidification. Liquid water at below 0 °C would then be an example of an unstable state. Let us label these states A, B, etc., and suppose we can determine the Gibbs free energy of each state. A bar chart of these free energies might appear as in Fig. 2.3. Imagine the system initially in state A. If the system were to change to state B it could do so without the surroundings doing work on the system, i.e., without intervention from the surroundings. Indeed, because the change is associated with a decrease in Gibbs free energy, we can abstract useful work from the system during the change and lift weights, pump water, drill teeth, launch projectiles or whatever we wish. In a nutshell, the system will proceed from state A to state B "under its own steam", perhaps doing useful work for us en route. A similar statement can be made about a change from B to C. However, a change from C to D (or back to B) is different. It is associated with an *increase* in G (i.e., negative useful work) and can only be achieved if the surroundings do work on the system. The minimum in Gibbs free energy represented by state C is therefore a special state; the system will remain there until the surroundings "interfere" by doing work on the system (or the temperature or pressure change). This leads to the conclusion that *at equilibrium a system is at a minimum Gibbs free energy compared to other states of the system at the same temperature and pressure* .

This principle is a powerful one. A major goal of chemical thermodynamics is to determine quantitatively the composition that exists when a particular chemical reaction has reached equilibrium. We can now see how this might be achieved; if we can calculate G as a function of composition (at specified constant temperature and pressure) we can then determine the composition that gives the minimum G and this will be the equilibrium composition. Much of the rest of this chapter is concerned with how this is done. Let us start by examining the dependence of G on temperature, pressure and composition.

2.5.1 Change of Gibbs Free Energy with T, P and Composition

First we will examine the simpler case of fixed composition and varying temperature and pressure. Any change in both temperature and pressure can be carried out by first changing the temperature at constant pressure and then changing the pressure at constant temperature. Let us first consider the case where these changes are small (differential) ones, dT and dP. The first change increases the Gibbs free energy by $(\partial G/\partial T)_P dT$, where the subscript indicates constant pressure, while the second increases it by $(\partial G/\partial T)_T dP$. Because G is a state variable (in this instance determined solely by T and P) the total change in G is given by

$$dG = \left(\frac{\partial G}{\partial T}\right)_P dT + \left(\frac{\partial G}{\partial P}\right)_T dP \tag{2.17}$$

From the defining equation (2.7) we can write

$$\begin{aligned} dG &= dE + d(PV) - d(TS) \\ &= dE + PdV + VdP - TdS - SdT \end{aligned} \tag{2.18}$$

If the only work done is expansion work and the change is reversible,

$$dE = q_R - W_{RE} = TdS - PdV \tag{2.19}$$

where we have used equation (2.3) to get q_R and expressed the expansion work as pressure times volume increase. Substituting (2.19) in (2.18), we have

$$dG = -SdT + VdP \tag{2.20}$$

We are examining an arbitrary change (i.e., dT and dP can have any ratio to each other) and we can therefore compare coefficients in equations (2.17) and (2.20) to arrive at

$$\left(\frac{\partial G}{\partial T}\right)_P = -S \tag{2.21}$$

and

$$\left(\frac{\partial G}{\partial P}\right)_T = V \tag{2.22}$$

We assumed a reversible change in deriving equations (2.21) and (2.22). However, because all the quantities involved are state variables, that assumption is not necessary, and (2.21) and (2.22) are applicable to all variations of pressure and temperature (at constant composition).

Now we examine a system of variable composition containing n_1 moles of a species that we label 1, n_2 moles of 2, ... and n_m moles of species m. As before, any change can be broken up into $m+2$ partial changes where one variable at a time is changed and

$$\begin{aligned} dG &= \left(\frac{\partial G}{\partial T}\right)_{P,n_1,n_2,\dots,n_m} dT + \left(\frac{\partial G}{\partial P}\right)_{T,n_1,n_2,\dots,n_m} dP \\ &+ \left(\frac{\partial G}{\partial n_1}\right)_{P,T,n_2,\dots,n_m} dn_1 + \dots + \left(\frac{\partial G}{\partial n_m}\right)_{P,Tn_1,n_2,\dots,n_{m-1}} dn_m \end{aligned} \tag{2.23}$$

Note that the first two terms on the right of this equation are quantities at constant composition and we can therefore immediately exploit equations (2.21) and (2.22). We therefore can rewrite (2.23) as

$$dG = -SdT + VdP + \sum_{i=1}^{m} \mu_i dn_i \qquad (2.24)$$

where

$$\mu_i = \left(\frac{\partial G}{\partial n_i} \right)_{T,P,n_1,\dots,n_{i-1},n_{i+1},\dots,n_m} \qquad (2.25)$$

and is called the *partial molar Gibbs free energy*. The complicated subscripts indicate that temperature, pressure and the quantities of all species *except i* are being held constant.

2.5.2 Chemical potential

We now use equation (24) to develop some understanding of the meaning of μ_i. We seek to comprehend how Gibbs free energy varies with T, P and composition. If we were creatures who could imagine $(m+3)$-dimensional space our task would be an easy one; we should merely imagine a surface representing G for a particular system as a function of the m+2 independent variables T, P, n_1, n_2... n_m. Because we are human beings and are presently confined to two-dimensional paper we can only catch glimpses of this hyperspace by holding all but one of the independent variables constant and varying that one. For example, Fig. 2.4 shows a plot of G versus temperature at fixed pressure and composition. We know the slope is negative because it is given by equation (2.21) and entropy is always positive. Similarly Fig. 2.5 shows how the Gibbs free energy increases with increasing pressure.

In Fig. 2.6 we have shown schematically the variation of G with n_1 at fixed T, P, n_2, etc. The slope of the line is μ_1 and this quantity may be positive or negative. Now consider a closed system where species 1 is capable of reacting to produce some other species which is present in the system. Suppose μ_1 is positive and much larger than the partial molar free energies of any other species in the system. Under these circumstances great reductions in Gibbs free energy will be achieved if n_1, the number of moles of species 1, is reduced by chemical reaction. The price that has to be paid for this great reduction in G is an offset of this reduction due to the increase in the contribution to G from the product(s) of the chemical reaction if μ_i is positive, but we have

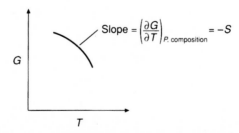

FIGURE 2.4 Variation of Gibbs free energy with temperature.

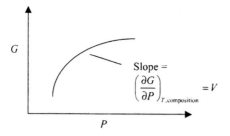

FIGURE 2.5 Variation of Gibbs free energy with pressure.

just stipulated that this contribution is small and therefore the net effect is still a great decrease in Gibbs free energy. Recalling that equilibrium is a state of minimum Gibbs free energy (with respect to other states at the same temperature and pressure) we see that μ_1 is *a driving force* tending to compel the system towards a lower G by reducing the amount of species 1 through chemical reaction. This leads to the alternative, shorter and much more descriptive name of μ_1; it is known as the *chemical potential* of species 1. If the chemical potential of species 1 is high there will be a considerable tendency for that species to react. Similarly, for the other chemical species, μ_i is the chemical potential of species i.

There are other ways besides chemical reaction in which the system might lower its Gibbs free energy by eliminating molecules that have a high chemical potential (we will continue to label these the molecules of species 1). For example, suppose one of the walls is permeable so that molecules of species 1 can move from our system (which we call system A) to an adjacent system (call this B) which initially contains no molecules of species 1. Clearly system A can lower its Gibbs free energy by transferring molecules to system B. The chemical potential of species 1 is therefore a driving force for the transfer of species 1 to system B. As that transfer proceeds, the chemical potential of 1 in system B will increase while that in system A will decrease. Eventually the chemical potentials of species 1 in the two systems will be equal and no further net transfer will occur; we then have equilibrium. If equilibrium exists with respect to the transfer of a species from one system to another, then the chemical potential of that species is the same in both systems. For example, if water in contact with air has reached equilibrium with respect to the dissolution of nitrogen, then the chemical potential of nitrogen in the water will equal that in the gas phase above the water.

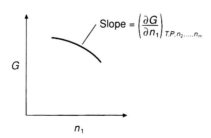

FIGURE 2.6 Variation of Gibbs free energy with the number of moles of species 1 present.

In some cases in this book we may find it more convenient to use the chemical potential per molecule, rather than per mole; the former is obtained from the latter merely by dividing by Avogadro's number.

Sometimes other factors besides temperature, pressure and composition can affect the chemical potential. For example, if the species we are examining is an ion in a solution, then its potential is influenced by the electric field that we might apply across the solution on electrolysis. This subject is examined in Chapter 9.

The atoms at an interface (e.g., a free surface) have their chemical potential altered by the curvature of the interface and this is significant in the densification of ceramic materials, a topic that we shall encounter again in Chapter 6.

2.5.3 Quantitative Description of Chemical Equilibrium

Let us examine a closed system at constant temperature and pressure where chemical species A, B and C are present and reacting according to

$$aA + bB = cC \tag{2.26}$$

Let n_A be the number of moles of A, etc. If chemical equilibrium has been reached then the Gibbs free energy of the system will be at a minimum. Mathematically this means that the differential increase in Gibbs free energy, dG, when a differential amount of A, dn_A, reacts to form products, is zero. That differential amount of reaction will change the number of moles of B by dn_B and of C by dn_C. From the stoichiometry of the reaction

$$dn_B = \frac{b}{a} dn_A \tag{2.27}$$

and

$$dn_C = -\frac{c}{a} dn_A \tag{2.28}$$

Writing equation (2.24) for this state of equilibrium (and remembering that T and P are constant) we get

$$0 = dG = \mu_A dn_A + \mu_B dn_B + \mu_C dn_C \tag{2.29}$$

Substituting (2.27) and (2.28) and then dividing by dn_A gives

$$a\mu_A + b\mu_B - c\mu_C = 0 \tag{2.30}$$

This is a quantitative statement concerning the relationship between the chemical potentials of reactants and products at equilibrium. It is easy to generalize it to other reactions, for example, if the reaction is

$$aA + bB + cC + dD + \cdots = mM + nN + \cdots \tag{2.31}$$

we would obtain an equation corresponding to (2.30), that is,

$$a\mu_A + b\mu_B + c\mu_C + d\mu_D \cdots -m\mu_M - n\mu_M - \cdots = 0 \qquad (2.32)$$

We have now made another major step toward our goal of a quantitative description of the composition of a reacting mixture when equilibrium is reached. The final step is to establish a connection between chemical potential and variables that are more familiar to us and more readily measured, such as concentration.

2.5.4 Relating Chemical Potential to More Familiar Variables; Equilibrium Constants

To establish a connection between chemical potential and variables that are familiar to us, let us start with the simplest possible case, a system containing only one pure element or compound. Let us increase the number of moles in the system from n to $n+dn$, keeping the pressure and temperature constant. The increase in the Gibbs free energy of the system,

$$dG = \tilde{G}dn \qquad (2.33)$$

where $\tilde{G}$ is the Gibbs free energy per mole of the chemical species at the temperature and pressure of interest. From equation (2.24)

$$dG = \mu dn \qquad (2.34)$$

and comparing equations (2.33) and (2.34)

$$\mu = \tilde{G} \qquad (2.35)$$

Thus we see that the chemical potential of a pure element of compound equals its Gibbs free energy per mole at the temperature and pressure in question.

Now consider one mole of a pure ideal gas and increase the pressure of that gas from P to $P+dP$ at constant temperature. From equation (2.24) the increase in G is

$$d\tilde{G} = VdP = \frac{RT}{P}dP \qquad (2.36)$$

where the tilde over the G is to remind us that we are considering one mole. If we make a finite, rather than differential, change in pressure from P_0 to P_1 the finite increase in $\tilde{G}$ is

$$\tilde{G}_1 - \tilde{G}_0 = \int_{P_0}^{P_1} RT \frac{dP}{P} = RT \ln P_1 - RT \ln P_0 = \mu_1 - \mu_0 \qquad (2.37)$$

or, dropping the "1" subscript

$$\mu = RT \ln P - RT \ln P_0 + \mu_0 \qquad (2.38)$$

This equation gives us the chemical potential of a pure ideal gas at pressure P, temperature T, relative to its value (μ_0) at some other (reference) pressure (P_0) and the same temperature. In section 2.3 we mentioned that the Gibbs free energy of a system is not absolute but is expressed with respect to its value at some arbitrarily chosen reference state; from equation (2.35) we see that this means the same is true of chemical potential. This is now an appropriate point to make a choice of a reference state or *standard state* as it is more commonly known. *The usual choice of standard state is atmospheric pressure and the specified temperature of interest* . We see that this is really a number of choices rather than one choice, because there is a different standard state for each temperature. Letting atmospheric pressure P_{atm} be the pressure P_0 in equation (2.38), we get

$$\mu = RT\ln\frac{P}{P_{atm}} + \mu_{atm} = RT\ln\frac{P}{P_{atm}} + \tilde{G}_{atm} \qquad (2.39)$$

When 1 atmosphere was a commonly accepted unit of pressure, equation (2.39) could be simplified to

$$\mu = RT\ln P + \tilde{G}_{atm} \qquad (2.40)$$

with P in equation (2.40) being in atmospheres. However, atmospheres are not SI units and if we were to stick rigorously to SI units then equation (2.39) rather than (2.40) should be used. [The alternative would be to use a pressure of one Pascal for the standard state but this is a low pressure and adoption of it as a standard state pressure does not seem likely.] At the time of writing this text many scientists and engineers still prefer to use atmospheres; for this reason and to simplify the algebra we will use atmospheres in the rest of this text. The reader who prefers SI units may replace P by (P/P_{atm}) in all terms $RT\ln P$ to follow or simply use the conversion

$$1 \text{ atmosphere} \equiv 1.0133 \times 10^5 \text{ pascals}$$

We have now chosen our standard state and we shall henceforth indicate the standard state by a subscript 0. We have one other choice to make. What is to be the value of the chemical potential ($\mu_0 = \tilde{G}_0$) at the standard state? Again we follow convention and put

$$\tilde{G}_0 = \begin{cases} 0 \text{ for all elements} \\ \Delta\tilde{G}_{F0} \text{ for all compounds} \end{cases} \qquad (2.41)$$

$\Delta\tilde{G}_{F0}$ is the *standard free energy of formation of the compound in question; it is the free energy that must be added to a system when one mole of that compound is formed in the system from its elements in order that the system is in the standard state before and after reaction.* Care should be taken to distinguish $\Delta\tilde{G}_{F0}$ from $\Delta\tilde{H}_{F0}$, the standard heat (enthalpy) of reaction. This last is the *enthalpy* that must be added to a system when one mole of the compound is formed in order that the system is at standard state before and after reaction. $\Delta\tilde{G}_{F0}$ is available in thermodynamic tables for a large number of compounds and Table 2.2 provides an example of such a tabulation.

TABLE 2.2 Standard enthalpy of formation and standard free energy of formation of water

Temperature (K)	$\Delta\tilde{H}_{F0}$ (kJ/Mol)	$\Delta\tilde{G}_{F0}$ (kJ/mol)
298.15	- 285.8	- 237.2
300	- 285.8	- 236.9
373.15	-283.5	-225.2
373.15	-242.6	-225.2
400	-242.8	- 223.9
500	-243.8	- 219.1
600	-244.8	- 214.0
700	-245.6	- 208.8
800	-246.5	- 203.5
900	-247.2	-198.1
1000	-247.9	- 192.6
1200	-249.0	-181.5
1400	-249.9	-170.1
1600	-250.7	- 158.7
1800	-251.2	- 147.1
2000	-251.7	- 135.6
2500	-252.5	- 106.4
3000	-253.2	- 77.1

Source: (Adapted from L. B. Pankratz , "Thermodynamic Properties of Elements and Oxides", U.S. Bureau of Mines *Bulletin* 672, 1982.)

The reader will observe that $\Delta\tilde{G}_{F0}$ is negative (as it is for all stable compounds) and (unlike $\Delta\tilde{H}_{F0}$) is continuous as we pass in temperature through a phase change. By definition, $\Delta\tilde{G}_{F0}$ = zero for all elements and we can therefore rewrite equation (2.41) as

$$\tilde{G}_0 = \Delta\tilde{G}_{F0} \qquad (2.42)$$

and, substituting in equation (2.40),

$$\mu = \Delta\tilde{G}_{F0} + RT\ln P \qquad (2.43)$$

Note that our choice of $\Delta\tilde{G}_0$ will frequently result in the chemical potential being negative (because $\Delta\tilde{G}_0$ is negative for stable compounds) at low or moderate pressures). We have so far been restricting our attention to a pure ideal gas but let us now generalize equation (2.43) to the case of a mixture of ideal gases. In particular, we will examine ideal gas A in a mixture of ideal gases. One characteristic of an ideal gas is that the gas molecules act independently of each other. In other words, the molecules of gas A are oblivious to the presence of the other gas (or gases) and equation (2.43) should still apply, provided we use, for the pressure, the pressure that gas A would exert in the absence of the other gases. That pressure is the partial pressure of A for which we use the symbol P_A. This partial pressure is related to the total pressure by

$$P_A = x_A P \tag{2.44}$$

where

$$x_A \equiv \frac{\text{moles of A in gas mixture}}{\text{total number of moles in gas mixture}} \tag{2.45}$$

x_A is the mole fraction of A. Hence for a component A in an ideal gas mixture

$$\mu_A = \Delta \tilde{G}_{FOA} + RT \ln P_A \tag{2.46}$$

with P_A in atmospheres.

Now let us suppose that our ideal gas mixture can react according to chemical equation (2.31). Substituting (2.46) in (2.32) along with similar equations for the other gas components, we obtain

$$a\left(\Delta \tilde{G}_{FOA} + RT \ln P_A\right) + b\left(\Delta \tilde{G}_{FOB} + RT \ln P_B\right)$$
$$+ \cdots - m\left(\Delta \tilde{G}_{FOM} + RT \ln P_M\right) - n\left(\Delta \tilde{G}_{FON} + RT \ln P_N\right) - \cdots = 0 \tag{2.47}$$

Rearranging gives

$$m\tilde{G}_{FOM} + n\tilde{G}_{FON} + \cdots - a\tilde{G}_{FOA} - b\tilde{G}_{FOB} - \cdots = -RT \ln \frac{P_M^m P_N^n \cdots}{P_A^a P_B^b \cdots} \tag{2.48}$$

If the temperature is constant (but not necessarily the total pressure) then the left hand side of (2.48) is constant, as is the factor RT, hence

$$\frac{P_M^m P_N^n \cdots}{P_A^a P_B^b \cdots} = \text{a constant, which we call } K_P \tag{2.49}$$

K_P is an "equilibrium constant". As an example let us consider

$$2CO + O_2 = 2CO_2 \tag{2.50}$$

and regard all three gases as ideal. When this reaction reaches equilibrium the partial pressures must be related by

$$\frac{P_{CO_2}^2}{P_{CO}^2 P_{O_2}} = K_P \tag{2.51}$$

We can return to equation (2.49) and exploit equation (2.44) to get

$$\frac{x_M^m x_N^n \cdots}{x_A^a x_B^b \cdots} = \text{another constant, which we call } K_X \tag{2.52}$$

provided temperature and total pressure are kept constant. [As an exercise the reader is asked to determine when this restriction on total pressure can be lifted.] Finally, because we are treating ideal gases we can write

$$C_A = \frac{P_A}{RT} \tag{2.53}$$

where C_A = concentration of A in gas mixture in moles per unit volume. Substituting (2.53) in (2.49) and rearranging

$$\frac{C_M^m C_N^n \cdots}{C_A^a C_B^b \cdots} = \text{another constant, which we call } K_C \tag{2.54}$$

provided the temperature is constant.

For ideal gas mixtures we have now almost achieved our goal; we have quantitative descriptions of equilibrium in terms of familiar variables that are easily measured (partial pressure, mole fraction or concentration), provided we can obtain the equilibrium constant. Obtaining the equilibrium constant is straightforward because (usually) all the quantities on the left hand side of equation (2.48) are available from thermodynamics tables. Let us look more closely at this, starting with the free energy diagram of Fig. 2.7. In this figure the products of reaction have been formed by two possible routes (1) directly from the elements of M, N, etc. and (2) first by forming A, B, etc. from those same elements and then reacting A, B, etc., according to equation (2.31) to form M, N,

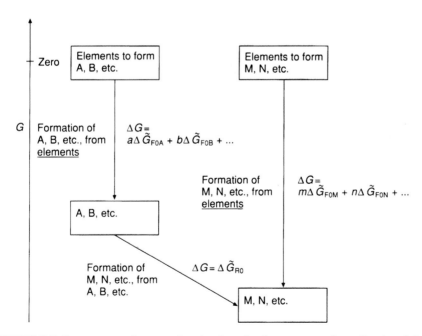

FIGURE 2.7 Free-energy diagram showing how the free energy of reaction is related to the formation of reactants and products.

etc. The free energies of the starting elements are set equal to their standard free energies of formation (i.e., zero) which will make the free energies of the compounds negative. We really seek $\Delta \tilde{G}_{R0}$ which is the free energy increase when reaction (2.31) takes place in our system starting entirely with A, B,... (but no M, N, etc.) in the standard state, and finishing with M, N,... (but no A, B,..., etc.), also in the standard state. By subtraction

$$\Delta \tilde{G}_{R0} = m\Delta \tilde{G}_{0M} + n\Delta \tilde{G}_{0N} + \cdots - (a\Delta \tilde{G}_{0A} + b\Delta \tilde{G}_{0B} + \cdots) \qquad (2.55)$$

that is, $\Delta \tilde{G}_{R0}$ is the left-hand side of equation (2.48).

A neater equation for $\Delta \tilde{G}_{R0}$ is

$$\Delta \tilde{G}_{R0} = \sum_{products} \left(\begin{array}{c} \text{stoichiometric} \\ \text{coefficient} \end{array} \right) \times \Delta \tilde{G}_{F0} - \sum_{reactants} \left(\begin{array}{c} \text{stoichiometric} \\ \text{coefficient} \end{array} \right) \times \Delta \tilde{G}_{F0} \quad (2.56)$$

which parallels the equation that we can use to calculate the enthalpy of reaction under standard conditions:

$$\Delta \tilde{H}_{R0} = \sum_{products} \left(\begin{array}{c} \text{stoichiometric} \\ \text{coefficient} \end{array} \right) \times \Delta \tilde{H}_{F0} - \sum_{reactants} \left(\begin{array}{c} \text{stoichiometric} \\ \text{coefficient} \end{array} \right) \times \Delta \tilde{H}_{F0} \quad (2.57)$$

Note again that the implication of the tilde over these free energies and heats of reaction is that these quantities have dimensions of energy *per mole*. Where possible we write chemical equations so that at least one of the chemical species in the equation has a stoichiometric coefficient of unity; $\Delta \tilde{G}_{R0}$ and $\Delta \tilde{H}_{R0}$ are then in energy per mole of that species.

Equations (2.56) and (2.57) are powerful ones. $\Delta \tilde{G}_{F0}$ and $\Delta \tilde{H}_{F0}$ are tabulated for a large number of compounds and we can therefore use these equations to calculate $\Delta \tilde{G}_{R0}$ and $\Delta \tilde{H}_{R0}$ for a vast number of possible reactions.

Substituting (2.55) and (2.49) in (2.48) we obtain

$$RT \ln K_p = -\Delta \tilde{G}_{R0} \qquad (2.58)$$

or

$$K_p = \exp\left(-\frac{\Delta \tilde{G}_{R0}}{RT} \right) \qquad (2.59)$$

and we now have, for most reactions we will encounter, the ability to calculate a value of K_p which, using equation (2.49), completes our quantitative description of chemical equilibrium for reacting ideal gases. In the next section we will examine how we handle the case of reacting mixtures that are not ideal gases.

2.5.5 Activity

Real gases are not ideal but for these gases we can replace the partial pressure P_A in equation (2.46), etc., with a variable known as the "fugacity", f_A. The fugacity can then be thought of as an "effective" pressure introduced to preserve the simplicity of the chemical potential concept. f_A would be different from the pressure calculated using ideal gas formula, for example using equation (2.44), to an extent that increases the less ideal the gas is. For real gas mixtures at close to atmospheric pressure, departure from ideal gas behaviour is not large and we can usually approximate the fugacity by the partial pressure. In fact this approximation is a good one up to quite high pressures and throughout the rest of this book we shall use equation (2.46) for real gas mixtures as well as ideal ones. Equation (2.58) or (2.59), can then be used for real gases. Text books on thermodynamics should be referred to when the approximation of fugacity by partial pressure is no longer valid; such texts provide methods for calculating the fugacity as a function of pressure, temperature and critical point data.

We are left, then, with the task of assigning meaning to the chemical potential of species in liquid or solid solutions. We approach this task indirectly by first introducing a quantity known as the activity of species i, a_i, which is defined by

$$\mu_i = \Delta \tilde{G}_{F0i} + RT \ln a_i \qquad (2.60)$$

First note the similarity between equations (2.60) and (2.46). Equation (2.60) will obviously embrace equation (2.46) if we put

$$a_i = \left\{ \begin{array}{l} f_i \text{ for a real gas} \\ P_i \text{ for an ideal gas (and approximately for a real gas)} \end{array} \right\}$$

[We note in passing that equation (2.60) could be applied to the chemical potential of an ion in an electrolyte, but an extra term is required on the right-hand side if, as is frequently the case, an electric field is present. We shall return to the thermodynamics of ions in Section 5.3 and, in more detail, in Section 9.4.]

Turning now to liquid and solid solutions, we must first select a standard state. Regrettably there are a number of standard states in widespread use. Even worse, it is not unusual for different standard states to be used for different components of a solution. It is therefore essential to define precisely the standard state when presenting activity data or to determine what standard state is referred to when using others' data. A common choice of standard state for solvents is the pure solvent at one atmosphere pressure (and the stated temperature). A common choice for solutes is the infinitely dilute solution (1 atmosphere and stated T). Of course in many instances (e.g., a solution of ethyl alcohol and water) it is not possible to distinguish between solute and solvent. Whatever the choice of standard state, the activity of a species will equal unity at its standard state. To fix our ideas, in the rest of this text we will use the pure species as the standard state for liquid and solid solutions, which will mean that the activities of pure liquids and solids are unity. For other concentrations we can write

$$a_i = \gamma_i C_i \qquad (2.61)$$

where C_i is concentration and γ_i is the *activity coefficient* and is defined by this equation. What units should we use for concentration? Note that if we choose our units so that the concentration is one at the standard state, this will force the activity coefficient also to be one at the standard state. For our choice of standard state this means that the concentration should be expressed in mole fraction

$$a_i = \gamma_i x_i \qquad (2.62)$$

Close to the standard state we expect $\gamma_i = 1$ and therefore, approximately

$$a_i = x_i \qquad (2.63)$$

For cases where this is exact we say the solution is *ideal* or *Raoultian*.

Let us examine the equilibrium between a solution and the vapor that is in equilibrium with it in order to establish the link between vapor pressures and activity as well as to see how the term *Raoultian solution* arose. We shall start with the pure solvent, i (activity of the liquid phase equals unity). The vapor pressure of the solvent in a gas phase with which it is in equilibrium we shall call P_{oi}. At equilibrium the chemical potential of the solvent molecules in the liquid phase equal their chemical potential in the gas phase. If this were not so, the system would be able to reduce its free energy by vaporization or condensation and therefore we would not be at equilibrium. Using equation (2.60) for the liquid phase (with $a_i = 1$) and (2.43) for the gas phase (with $P = P_{oi}$)

$$\Delta \tilde{G}_{F0iL} = \Delta \tilde{G}_{F0iG} + RT \ln P_{oi} \qquad (2.64)$$

where new subscripts (L for the liquid and G for the gas) have been introduced because the free energy of formation of the liquid equals that of the gas only at the normal boiling point. Now turning to the gas phase in equilibrium with the solution, where the vapor pressure of the solvent is P_{vapi}, we can use the equality of the chemical potentials again and the same two equations as before to get

$$\Delta \tilde{G}_{F0iL} + RT \ln a_i = \Delta \tilde{G}_{F0iG} + RT \ln P_{vapi} \qquad (2.65)$$

Subtracting equation (2.64) from equation (2.65) and rearranging

$$a_i = \frac{P_{vapi}}{P_{0i}} \qquad (2.66)$$

and we see then that the activity of a species, i, in solution is given by the ratio of the vapor pressure of the species in a gas in equilibrium with the solution to the vapor pressure of the pure solvent, i, at the same temperature, provided the standard state is defined as the pure solvent. For many solutions, particularly when the solution is close to being pure solvent, the vapor pressure $P_{vapi} = x_i P_{oi}$. This is Raoult's law and for solutions for which it holds we can substitute in equation (2.66) to get equation (2.63).

Further from the standard state most solutions become very non-ideal and we must resort to experimental measurements (or theory that is beyond the scope of this text) to determine the activity (or activity coefficient). For example, Fig. 2.8 presents data on the activity of carbon (pure carbon as the standard state) at 1500°C in liquid iron. Note that the activity reaches one well before the mole fraction does. If we try to exceed this activity the result will be that we start to form particles of solid carbon in the liquid. As an exercise the reader is asked to replot the data in terms of an activity coefficient rather than an activity.

For dilute solutions it is frequently found that the activity of the solute is propor-

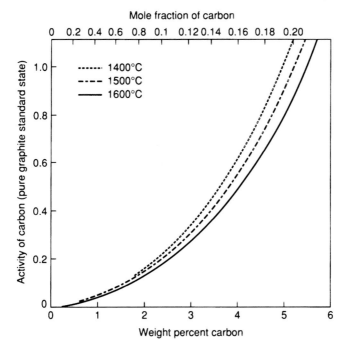

Mole fraction of carbon

FIGURE 2.8 Activity of carbon in liquid iron at 1400, 1500, and 1600°C, with pure graphite as the standard state. Calculated from equations of Benz and Elliot [*Trans. Metall. Soc. AIME 221,* 323 (1961)] that correlate experimental data.

tional to concentration, i.e., the activity coefficient is constant; such solutions are described as obeying Henry's law. Of course if the solution contains more than two species then there exists the possibility that the activity of species i may be influenced by the concentration of another species (j). For example, the activity data of Fig. 2.8 would be altered if silicon were also present in solution. In some instances (e.g., the industrially important solutions of C, Si and other elements in molten iron) an allowance for these effects can be made using *published interaction parameters*. These interaction parameters, ε_i^j are defined by

$$\varepsilon_i^j \equiv \lim_{x_{\text{solvent}} \to 1} \frac{\partial \ln \gamma_i}{\partial x_j} \tag{2.67}$$

that is, the parameter represents the influence of the mole fraction of species j on the (logarithm of) the activity coefficient of species i at infinite dilution. We can then approximate the activity coefficient in a solution containing m solutes by

$$\ln \gamma_i = \ln \gamma_i^* + x_1\varepsilon_i^1 + \cdots + x_i\varepsilon_i^i + \cdots + x_m\varepsilon_i^m \tag{2.68}$$

where γ_i^* is the activity coefficient of species i in an infinitely dilute solution containing no solute other than i. Note that the term $x_i\varepsilon_i^i$ on the right hand side of (2.68) allows for the departure from Henry's law as the concentration of solute is increased. It can be shown that $\varepsilon_i^j = \varepsilon_j^i$ which simplifies the task of measuring (or calculating from theory) these interaction parameters. Nevertheless that task remains a challenging one when the large number of possible solvents, solutes and temperatures is considered. Also it should be recognized that (2.68) is an approximation that will become increasingly inexact as concentrations are increased.

Using equation (2.60) to express the chemical potential for our equilibrium condition given by equation (2.32), and carrying out the same manipulations as we did in section 2.4.4 to arrive at equation (2.49), we get

$$\frac{a_M^m a_N^n \cdots}{a_A^a a_B^b \cdots} = K_a \tag{2.69}$$

and now the equilibrium constant is expressed in terms of activities and is given by [compare equation (2.58)]

$$RT \ln K_a = -\Delta \tilde{G}_{R0} \tag{2.70}$$

or

$$K_a = \exp\left(-\frac{\Delta \tilde{G}_{R0}}{RT}\right) \tag{2.71}$$

Equation (2.70), or equation (2.71) its equivalent, is perhaps the most important one in this text; it is the goal towards which we have been striving in most of this chapter; namely, the tool which enables us to predict the composition of a reacting system when it reaches equilibrium. Let us use a worked example to illustrate the power of this equation.

PROBLEM

Fe_2O_3 is reduced to Fe_3O_4 by hydrogen at 1000 K to produce water vapor. What is the composition of the gas mixture when chemical equilibrium is reached if the pressure in the reactor is atmospheric and no other gases are present? If the reactor initially contains two moles of hydrogen, how much heat must be added to (removed from) the reactor to maintain 1000 K? Assume no other reactions take place and there is an excess of Fe_2O_3.

Solution

The reaction is[2]

$$3Fe_2O_3 + H_2 = 2Fe_3O_4 + H_2O \tag{2.72}$$

---

[2] It is the authors' experience that students frequently attempt to incorporate information on the moles of reactant placed in the reactor into the chemical equation, i.e., to double the stoichiometric coefficients in the present example. The folly of this can be seen if we were to examine an industrial-scale reactor containing thousands or perhaps millions of moles. Worse yet, the inexperienced student immediately leaps to the wrong conclusion that 2 mol of H_2 must react. Chemical equations are best left in their simplest form (i.e., as identifying reactants and products and ratios of moles in reactions); material balance equations can then be written separately.

and we shall assume that the Fe_2O_3 and Fe_3O_4 are present as two separate pure solids, which will make their activities unity. We will also make the usual assumption of this text that gas activities are equal to their partial pressures (in atm.). From Table 2.2 we have $\Delta \tilde{G}_{F0}$ at 1000 K = -192.6 kJ/mole for H_2O; we shall also need the standard free energy of formation of Fe_2O_3 (= -561.0 kJ/mole) and Fe_3O_4 (= -792.4 kJ/mole) at this temperature and 1 atm. pressure from thermodynamic tables. $\Delta \tilde{G}_{F0}$ of hydrogen is zero since it is an element. Hence, $\Delta \tilde{G}_{R0}$, from equation (2.56), is

$$\Delta \tilde{G}_{R0} = 2 \times (-792.4) + 1 \times (-192.6) - 3 \times (-561.0)$$
$$= -94.4 \text{ kJ/mole of } H_2 \text{ (or per mole of } H_2O)$$

Using equation (2.71) and $R = 8.314$ J/(mole K), from tables of physical constants,

$$K_a = \exp\left(\frac{+94,400 \text{J} / \text{mol}}{8.314 \text{J} / \text{mol} \cdot \text{K} \times 1000 \text{K}}\right) = 8.54 \times 10^4$$

Writing the equilibrium constant in terms of activities and then partial pressures, we have

$$K_a = 8.54 \times 10^4 = \frac{a_{Fe_3O_4}^2 a_{H_2O}}{a_{Fe_2O_3}^3 a_{H_2}} = \frac{P_{H_2O}}{P_{H_2}}$$

But we know that $P_{H2} + P_{H2O} = 1$ atm (because no other gas is present), hence

$$P_{H_2O} = (1 - P_{H_2O}) \times 8.54 \times 10^4$$

or

$$P_{H_2O} = \frac{8.54 \times 10^4}{8.54 \times 10^4 + 1} = 0.9999882 \text{ atm}$$

and

$$P_{H_2} = 1 - P_{H_2O} = 1.17 \times 10^{-5} \text{atm}$$

From equation (2.44) we see that these are also the mole fractions of H_2O and H_2 at equilibrium, i.e., there are $2 \times 1.17 \times 10^{-5}$ moles of H_2 left unreacted at equilibrium, or $2 - 2 \times 1.17 \times 10^{-5}$ moles of the initial hydrogen have reacted. Of course we have been assuming, in this example, that there is sufficient Fe_2O_3 for equilibrium to be reached (i.e., Fe_2O_3 is in excess). If there is less Fe_2O_3 than $3 \times (2 - 2 \times 1.17 \times 10^{-5})$ moles of Fe_2O_3 initially, then reaction (2.72) will stop when all the Fe_2O_3 is used up, i.e., before equilibrium is reached. It should be noted that, in reality, other reactions (the reduction to FeO and Fe) might also take place.

To determine the heat involved in reaction we need standard heats of formation. For water vapor, Table 2.2 gives $\Delta \tilde{H}_{F0}$ at 1000 K = -247.9 kJ/mole and the values for the other species are obtained at this temperature from thermodynamic tables: For Fe_2O_3 = -806.5 kJ/mole, for Fe_3O_4 = -1088.0 kJ/mole. Using equation (2.57) gives

$$\Delta \tilde{H}_{R0} = 2 \times (-1088.0) + 1 \times (-247.9) - 3 \times (-806.5) = -4.4 \text{kJ} / \text{mol } H_2$$

Because $(2 - 2 \times 1.17 \times 10^{-5})$ moles of H_2 react, the heat involved in reaction = -8.8 kJ. The implication of the minus sign is that the reaction is exothermic (i.e., that heat must be removed from the reactor to keep the temperature constant).

2.5.6 The Extent To Which Reactions Take Place

Consider the reaction (2.72) in the worked example above. The standard free energy of reaction at 1000 K is -94.4 kJ/mole hydrogen, and substituting this value into equation (2.71) yields a very large vale for the equilibrium constant and, consequently, a very low value for the amount of reactant (hydrogen) left when chemical equilibrium is reached. We say "reaction (2.72) proceeds almost completely to the right at 1000 K" or use similar words[3] to describe the fact that at equilibrium very little of the reactants (other than any reactant present in excess) have not been converted into products for a reaction with this large *negative* free energy of reaction. We can see in Table 2.3 how the equilibrium constant for a reaction varies with the free energy of reaction at a particular temperature (1000 K in the case of the table).

We see that the equilibrium constant is strongly dependent on the standard free energy of reaction; a strong dependence exists at other practical temperatures. Let us make the following generalizations: For reactions with $\Delta \tilde{G}_{R0}$ more negative than roughly -20 kJ/mole, the equilibrium constant becomes large and we can think of these reactions as reaching equilibrium when they have proceeded almost completely to the right.

Table 2.3 Dependence of equilibrium constant on free energy of reaction at 1000K

$\Delta \tilde{G}_{R0}$ (kJ/mol)	K_a
-100	1.76×10^4
-50	419
-10	3.35
0	1.0
10	0.299
50	0.0024
100	5.76×10^{-6}

For reactions with $\Delta \tilde{G}_{R0}$ between approximately -20 kJ/mole and +20 kJ/mole, the equilibrium constant will be within an order of magnitude of one. Considering equation (2.69), this means that the activities of the reactants and products will be in the same range when equilibrium is reached, i.e., there will be considerable amounts of unreacted reactants left at equilibrium. We say that the reaction proceeds partially to the right, or that it is "reversible". For reactions with $\Delta \tilde{G}_{R0}$ greater than roughly +20 kJ/mole the equilibrium constant is small and therefore the activities of the products will

---

[3] "The reaction is irreversible" is also common terminology (although the connection with "irreversible" as used in Section 2.2 is tenuous) and will be used in this book for reactions with large equilibrium constants.

be small (compared to the reactants) at equilibrium and we can say reaction hardly proceeds at all.

A much cruder generalization is that reactions take place to an appreciable extent when $\Delta \tilde{G}_{R0}$ becomes negative. This latter generalization is satisfactory for many purposes of this book, but its oversimplification should be recognized. For example, $\Delta \tilde{G}_{R0}$ for

$$C + CO_2 = 2CO$$

is positive at lower temperatures and becomes negative at approximately 700°C; nevertheless this reaction is capable of generating the small concentrations of carbon monoxide necessary to kill you at temperatures well below 700°C! The reader will now recognize why $\Delta \tilde{G}_{F0}$ for common chemical compounds is negative.

2.5.7 Variation of Equilibrium Constants With Temperature

We can rewrite $\Delta \tilde{G}_{R0}$ in terms of $\Delta \tilde{H}_{R0}$ and $\Delta \tilde{S}_{R0}$, where the last is the standard entropy of reaction (entropy increase associated with reaction at the standard state)

$$\Delta \tilde{G}_{R0} = \Delta \tilde{H}_{R0} - T\Delta \tilde{S}_{R0} \tag{2.73}$$

and substituting in (2.70), we have

$$\ln K_a = -\frac{\Delta \tilde{H}_{R0}}{R}\left(\frac{1}{T}\right) + \frac{\Delta \tilde{S}_{R0}}{R} \tag{2.74}$$

If there are no phase changes (melting, etc.) of either reactants or products over the temperature range of interest, then $\Delta \tilde{H}_{R0}$ and $\Delta \tilde{S}_{R0}$ are approximately independent of temperature. Consequently a plot of $\ln K_a$ versus reciprocal temperature (Fig. 2.9) yields a straight line from which we can determine the heat of reaction. More significant, as far as this text is concerned, is the subject discussed next.

2.5.8 The Variation of Standard Free Energy of   Reaction with Temperature

Examining equation (2.73), we expect $\Delta \tilde{G}_{R0}$ to depend linearly on temperature, over temperature ranges where phase changes do not occur, as in Fig. 2.10. The slope of the

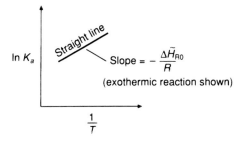

FIGURE 2.9 Plot of the logarithm of the equilibrium constant versus reciprocal absolute temperature.

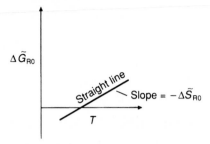

FIGURE 2.10 Plot of free energy of reaction against temperature in range where there are no phase changes.

line may be positive or negative, depending on whether the standard entropy of reaction is negative or positive. For example, for the oxidation of a solid metal by oxygen (gas) to form a solid oxide

$$metal + oxygen = oxide$$

we expect (from the uncertainty concept of section 2.3) a *decrease* in entropy to accompany reaction, i.e., $\Delta \tilde{S}_{R0}$ is negative and the slope of the plot is positive. Now let us consider a slightly higher temperature which is above the melting point of the metal (but below the melting point of the oxide, we shall suppose). $\Delta \tilde{S}_{R0}$ should now be slightly more negative, i.e., the slope should increase slightly as shown in Fig. 2.11. A much greater increase in slope will occur if further increase in temperature takes the metal through its boiling point (but not the oxide, we suppose). Fig. 2.12 shows these changes in slope for a number of metal/oxides (e.g., Mn or Zn) although the slope changes at the melting points are barely distinguishable. Fig. 2.12 is an *Ellingham diagram* for oxides. It is a plot of the standard free energy of formation of numerous oxides versus temperature *except* that, to facilitate comparisons between oxides (see below), the quantities plotted are in free energy per mole of *oxygen*, rather than free energy per mole of oxide. We shall see shortly that this (and similar diagrams for sulfides, chlorides, etc.) is one of the most useful diagrams in this text. The reader will observe that in some cases (e.g., copper oxidized to Cu_2O) there is a decrease in slope, or the slope actually becomes negative (e.g., lead oxidized to PbO). These correspond to the points at which

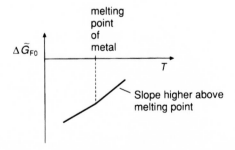

FIGURE 2.11 Plot of free energy of reaction against temperature over range where one of the reactants melts.

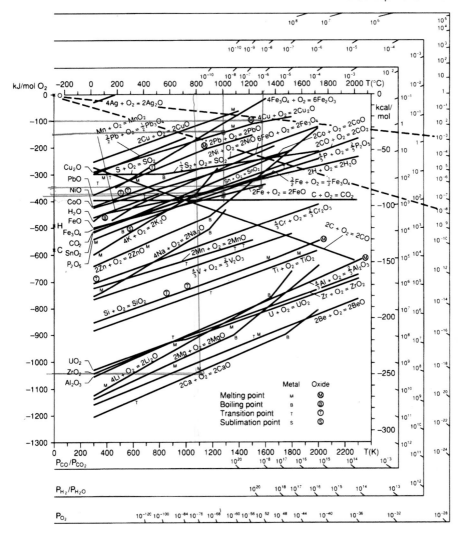

FIGURE 2.12 Ellingham diagram for oxides.

the *oxide* melts or boils and the change of slope can be deduced by the reader from the uncertainty concept of section 2.2. The reader will also note that lines corresponding to oxidation of non-metals (e.g., sulfur or carbon) appear in Fig. 2.12 and even lines for some important reactions that are not the oxidation of an element (e.g., the oxidation of CO to CO_2). One line (the oxidation of carbon to CO) has a negative slope throughout the diagram and it is left as an exercise for the reader to deduce why.

For some classes of chemical species (usually sulfates and carbonates) the Ellingham diagram is occasionally plotted using the free energy of formation from the oxide, rather than the element, e.g., for

$$ZnO + SO_2 + 1/2\ O_2 = ZnSO_4 \qquad (2.75)$$

and care must be exercised when using data from such plots (e.g., to calculate an equilibrium constant).

2.6 Significance of Ellingham Diagrams

Ellingham diagrams, introduced at the end of the last section, are extremely useful in exploring the feasibility of alternative chemical routes to materials. They are also of value in studying the suitability of materials for use in hostile environments, although this is outside the scope of this text.

2.6.1 Thermal Stability (Equilibrium Oxygen Partial Pressure)

Consider the thermal decomposition of an oxide

$$2Ag_2O = 4Ag + O_2 \tag{2.76}$$

$\Delta \tilde{G}_{R0}$ for this reaction is the negative of the value for oxidation of silver to Ag_2O, which appears towards the top of Fig. 2.12. For reaction (2.76), starting with pure Ag_2O and recognizing that pure Ag (solid or liquid) will be formed

$$K_a = P_{O_2} = \exp\left(-\frac{\Delta \tilde{G}_{R0}}{RT}\right) \tag{2.77}$$

That is, P_{O_2} will reach one atmosphere at $\Delta \tilde{G}_{R0} = $ zero, i.e., at the temperature on the Ellingham diagram where the line for Ag/Ag_2O crosses the horizontal axis (approximately 200°C). The implication is that if we heat Ag_2O in a reactor which is open to an atmosphere of pure oxygen at 1 atmosphere, then oxygen will start to flow out of the reactor (thermal decomposition takes place) at approximately 200°C. If the atmosphere outside the reactor is air at 1 atm rather than pure oxygen, the decomposition will start at a slightly lower temperature, namely the temperature at which P_{O_2} reaches 0.21 atm.; however, at this lower temperature, thermal decomposition will be rather slow (because egress of the oxygen from the reactor will be by a slow diffusion process). Consequently even if the atmosphere outside the reactor is air, rather than oxygen, we still need to reach 200°C before decomposition becomes vigorous (i.e. egress of oxygen is by hydrodynamic flow driven by the gas pressure in the reactor exceeding one atmosphere). We see that the vast majority of metal oxides require very high temperatures to thermally decompose them in this way; i.e., the oxides of everyday metals are thermodynamically stable at practical temperatures.

The inquisitive reader may ask what temperature is necessary to start thermal decomposition if the oxide is in contact with an atmosphere containing oxygen at partial pressure P_{O2}. Of course, this number can be obtained by trial and error but there is a much more convenient way of using Fig. 2.12 to answer the question. Remember that equilibrium is achieved when the reactants and products of reaction have the same Gibbs free energy. Considering reaction (2.76), for example, the free energy of the Ag and Ag_2O are negligibly affected by pressure and are therefore equal to their standard free energies of formation, multiplied by four and two, respectively. The former number is zero, silver being an element, while the latter number is simply that plotted in

Fig. 2.12. The free energy of oxygen *is* affected by pressure and increases by an amount $RT\ln P_{O2}$ as the oxygen partial pressure is raised. As an example, broken lines in Fig. 2.12 show $RT\ln P_{O2}$ for $P_{O2} = 10^{-3}$ and 10^{-8} atm. Consequently, equilibrium for reaction (2.76) when the oxygen partial pressure in the surrounding atmosphere is 10^{-8} atm. is achieved at the point where the lower broken line crosses the line for reaction (2.76) on the Ellingham diagram (approx. 0°C). This therefore is the temperature at which silver oxide will start to thermally decompose in that environment. The nomographic scale for oxygen appearing on the edge of the drawing can be used in a similar way to determine the thermal decomposition temperature of any of the oxides in any environment, or, conversely, the temperature below which the oxide is stable. Does reducing the pressure and thermally decomposing the oxide seem like a practical way of producing a metal from its oxide? The other nomographic scales in Fig. 2.12 are discussed in the next section.

2.6.2 Reduction of Metal Oxides

Consider first the reaction

$$2H_2 + 2NiO = 2Ni + 2H_2O \tag{2.78}$$

at 1000°C. Recalling that we are dealing with state variables we can calculate $\Delta\tilde{G}_{R0}$ for reaction (2.78) by equating it to the sum of $\Delta\tilde{G}_{R0}$ for two other reactions:

(I) $2H_2 + O_2 = 2H_2O$
(II) $2NiO = O_2 + 2Ni$

The ΔG_{R0-I} (i.e., for reaction I) can be read immediately off the Ellingham diagram as -360 kJ/mole. ΔG_{R0-II} is merely minus the value read from the diagram for the oxidation of Ni because that oxidation is the opposite of reaction II, i.e., $\Delta G_{R0-II} = +250$ kJ/mole. Hence ΔG_{R0} for reaction (2.78) is $-360 + 250$ kJ/mole $= -110$ kJ/mole which means that reaction (2.78) should proceed almost all the way to the right before equilibrium is reached. Hydrogen is seen therefore as a very effective reducing agent for nickel oxide at this temperature. Now consider

$$2H_2 + 2MnO = 2H_2O + 2Mn \tag{2.79}$$

We can calculate ΔG_{R0} for reaction (2.79) proceeding as before. Reaction I remains unchanged and II is replaced by

$$2MnO = O_2 + 2Mn$$

for which ΔG_{R0-II} is $+585$ kJ/mole, resulting in ΔG_{R0} for (2.79) being $-360 + 585$ kJ/mole $= +225$ kJ/mole. In other words, reaction (2.79) hardly proceeds at all before equilibrium is achieved and hydrogen is not effective in reducing MnO. As soon as a minute amount of H_2O gas is formed, the reaction stops altogether.

The difference between NiO and MnO, in this instance, is a consequence of their position on the Ellingham diagram. In Fig. 2.12 the line for Ni/NiO lies *above* the line for H_2/H_2O, but that for Mn/MnO is *below* . We can therefore formulate a rule (to which there are exceptions to be discussed later): *An element low in the diagram can reduce*

---

[4] A useful mnemonic here is that "1" occurs in "element" and "low," while "i" occurs in "oxide" and "high."

an oxide high in the diagram but not vice versa .[4] The Ellingham diagram now becomes an extremely useful "road map" for possible (or impossible) chemical reactions. The reader can see this by answering the following in a few seconds each:

1. Which material would be more suitable for containing molten aluminum - silica (SiO_2) or magnesia (MgO)?
2. Which material would you select for a steam pipe if attack of the pipe were the sole criterion - aluminum or nickel?
3. Will carbon reduce chromium oxide (producing CO) at 1600°C? Will hydrogen?
4. At what temperature will the reaction $CO_2 + C = 2CO$ proceed?
5. On oxidizing an alloy of aluminum and magnesium, which should be the first metal to be oxidized?

There are two lines for the oxidation of carbon in Fig. 2.12 and deliberation on question (4) will show the reader that the appropriate one at any temperature is the lower one.

Although the rule formulated above is a very useful one it is important to realize that it can sometimes be misleading, for either of the two following reasons: The rule (like the rest of chemical thermodynamics) describes the nature of chemical equilibrium, i.e., whether significant reaction will take place before equilibrium is reached. It is mute on the topic of *how long* is needed to reach equilibrium. Thus the rule may predict that reaction will take place, under given circumstances, but, in reality, the reaction may be too slow for appreciable amounts of product to appear in a reasonable time. For example, the diagram predicts that reaction (2.78) will occur at room temperature; in fact, if nickel oxide and hydrogen are contacted at room temperature, no reaction is discernible. The chemical kinetics of this reaction are simply too slow at room temperature. These kinetic limitations are usually less significant at higher temperatures, as will be seen in Chapter 3.

Secondly, the rule, because it pertains to equilibrium, can be circumvented if we interfere with the reaction so as to prevent equilibrium being reached (or maintained). For example, if we contact Cr_2O_3 and H_2 in a closed reactor at 1600°C then the rule (correctly) predicts that only miniscule amounts of H_2O and the metal chromium will be formed before equilibrium is reached [Cr/Cr_2O_3 lying below H_2/H_2O on the Ellingham diagram.] However, we could then replace the gas in the reactor with fresh hydrogen and produce an additional miniscule amount of chromium and, by doing this an enormous number of times, eventually reduce all the Cr_2O_3 to metal. The same effect is achieved if we pass hydrogen through the reactor for a sufficiently long time at high flow rates. This is not a practical way of producing chromium but the concept has been exploited to produce magnesium by the reaction

$$Si + 2MgO = 2Mg + SiO_2 \qquad (2.80)$$

The reaction should not go according to the rule, as a glance at Fig. 2.12 shows. However, above 1100°C, the magnesium formed is a vapor and leaves the reactor (the other three species remaining in the reactor) and this prevents equilibrium from being reached. Reduction then proceeds until all the Si or MgO is used up. The magnesium vapor is condensed as it leaves the reactor.

We see, therefore, that our rule is a very useful approximation to the truth. The approximation is of the same order as saying New Orleans is south of Chicago. If you flew exactly due south from Chicago you would not pass over New Orleans but nevertheless most people would accept this geographic statement as a practical approximation. If we wish to be more precise, then we can take a quantitative approach to the Ellingham diagram. Consider for example the question of whether a given gas *mixture* will reduce a certain oxide at a particular temperature. For example we might be interested in whether a mixture of hydrogen and water vapor, in the molar ratio $10^3:1$, will reduce zinc oxide at 800°C. The rule suggests that the reactions will *not* take place because the line for the oxidation of hydrogen is above that for oxidation of zinc. However, the lines are sufficiently close that a second look is called for. Here we use the nomographic scale for H_2/H_2O around the edge of Fig. 2.12. This scale is used in a similar way to the oxygen nomograph treated above, i.e., a straight line is drawn from the point on the nomograph representing the composition to the point marked H on the left side of the figure. This line passes *below* the line for zinc oxidation at 800°C and so this gas mixture is capable of carrying out reduction. Practically speaking, however, only a little reaction will take place (sufficient to lower the H_2/H_2O ratio to about 300 to 1) before equilibrium is reached and reaction stops (unless we maintain the high H_2/H_2O ratio by, for example, flowing the gas through the reactor). Note that Fig. 2.12 also has a nomographic scale for CO/CO_2 mixtures. This scale, which is used in conjunction with the point marked C on the left axis, tells us, for example, that a CO/CO_2 gas mixture is theoretically capable of reducing zinc oxide at 800°C provided the CO/CO_2 ratio is greater than approximately 200:1.

The Ellingham diagram immediately leads to conclusions about some common reducing agents. Carbon (coke is its common form) will reduce most oxides if the temperature is made high enough. Carbon monoxide and hydrogen will reduce only a few oxides but appear to be more effective reducing agents than carbon at lower temperatures. Sodium and silicon, falling lower on the Ellingham diagram than hydrogen or CO, are effective in reducing a greater number of oxides. Aluminum and magnesium, at the bottom of the diagram, are particularly powerful reducing agents.

When any reducing agent participates in a reaction then there must be physical contact between the reducing agent and the oxide being reduced. If the reducing agent is a gas then this is readily achieved. If the reducing agent and the oxide are both solid then there is some difficulty in achieving the necessary contact, i.e., even in well-mixed solid powders there will be only limited solid-solid touching. In the case of reduction by coke, silicon, magnesium or aluminum then reduction would typically occur much more readily above the melting point of the reducing agent or the oxide. One reacting phase can then flow into contact with the other. Magnesium and aluminum, having relatively low melting points, are particularly convenient in this respect.

2.7 Predominance (Pourbaix) Diagrams

The Ellingham diagrams are drawn using standard state data (i.e., they apply to atmospheric pressure), and the question arises as to what happens if we change to some other total pressure. To focus our thinking let us examine the reaction

$$2C + O_2 = 2CO \tag{2.81}$$

and calculate the free energy of reaction $\Delta \tilde{G}_{R1 + dP}$ at a small pressure increment dP

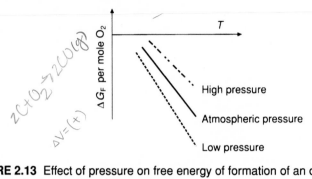

FIGURE 2.13 Effect of pressure on free energy of formation of an oxide formed with a increase in the number of gas molecules.

above 1 atm

$$\Delta \tilde{G}_{R1+dP} = \Delta \tilde{G}_{R0} + \frac{\partial}{\partial P}\left(\Delta \tilde{G}_R\right)_0 dP \tag{2.82}$$

where the subscript zero on the bracket indicates that this derivative is to be evaluated at standard state. If we use equation (2.22) we can rewrite this as

$$\Delta \tilde{G}_{R1+dP} = \Delta \tilde{G}_{R0} + \Delta \tilde{V}_{R0} dP \tag{2.83}$$

where $\Delta \tilde{V}_{R0}$ is the volume increase accompanying reaction at standard state = $(2\tilde{V}_{CO} - 2\tilde{V}_C - \tilde{V}_{O_2})_0$, where $\Delta \tilde{V}_{CO}$ is the volume of one mole of carbon monoxide, and so on. The volume of the reaction products (two moles CO) would be much greater than that of the reactants (two moles solid carbon plus one mole of oxygen) and so

$$\Delta \tilde{G}_{R1+dP} > \Delta \tilde{G}_{R0} \tag{2.84}$$

Or, if we were to draw a line on the Ellingham diagram for higher pressures for reaction (2.80), it would be above that for 1 atm (see Fig. 2.13).

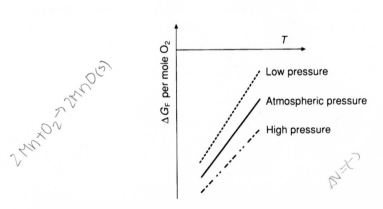

FIGURE 2.14 Effect of pressure on free energy of formation of an oxide formed with a decrease in the number of gas molecules.

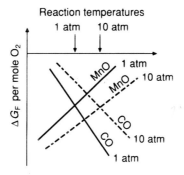

FIGURE 2.15 Superposition of Figures 2.13 and 2.14, showing the increase in temperature at which reaction becomes significant with increasing pressure.

Generalizing the result of equation (2.83) yields $G = RT \ln(P)$

$$\frac{\partial \Delta \tilde{G}_R}{\partial P} = \Delta \tilde{V}_{R0} \tag{2.85}$$

where $\Delta \tilde{V}_{R0}$ is the volume increase per mole accompanying reaction at standard state. Applying this to a metal oxidation reaction such as

$$2Mn + O_2 = 2MnO_{(s)} \tag{2.86}$$

we see that $\Delta \tilde{G}_R$ would be lowered (made more negative) by an increase in O_2 pressure, as illustrated in Fig. 2.14. Now superimposing these two plots as in Fig. 2.15 in order to examine the reduction reaction $2MnO + 2C = 2Mn + 2CO$, we observe that at the higher pressure the *reaction temperature* (the temperature at which $\Delta \tilde{G}_R$ becomes zero) is higher. We can regard the two intersection points on Fig. 2.15 as two points on a plot of reaction temperature versus pressure. Such a plot appears in Fig. 2.16 and an

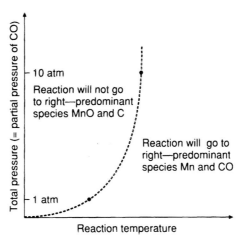

FIGURE 2.16 Predominance diagram for manganese-oxygen-carbon.

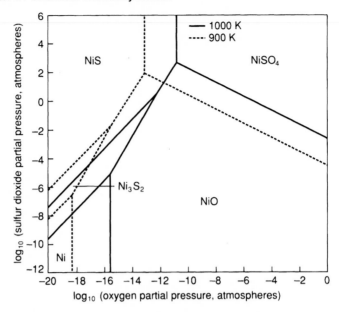

FIGURE 2.17 Predominance diagram for nickel-sulfur-oxygen.

alternative label for such a plot is a "predominance diagram" (sometimes called a *Pourbaix-Ellingham diagram*). This name arises because it tells us what chemical species will be dominant, when equilibrium is reached, if we fix the temperature and pressure in our reactor. If the point representing the temperature and pressure is above the curve then MnO and C will be dominant at equilibrium; below the curve Mn and CO will dominate.

Of course, there is no need to restrict ourselves to temperature and pressure as manipulated variables in constructing predominance diagrams; we could equally fix the temperature and pressure and manipulate the partial pressures (composition) of a gas in contact with a solid. Fig. 2.17 provides an example of this for the nickel-sulfur-oxygen system. The diagram tells us, for example, that the solid in equilibrium at 1000 K with a gas containing high partial pressures of both oxygen and sulfur dioxide is nickel sulfate. What should be expected on oxidizing NiS at 1000 K if the partial pressure of oxygen is kept high but that of sulfur dioxide is kept low? How difficult would it be to oxidize NiS to Ni_3S_2? Is it likely that nickel metal will form on oxidizing nickel sulfide in any practical reactor?

Another important predominance diagram is that shown in Fig. 2.18, for the oxides of iron in contact with CO/CO_2 mixtures. The diagram can be used to answer the following questions:

1. What should form on contacting Fe_2O_3 with a gas mixture such that the composition and temperature are held at point A?
2. What is the minimum temperature at which we expect to form wustite (FeO) on reducing Fe_3O_4 with CO/CO_2 mixtures?
3. If iron is oxidized with CO_2 at 1200 K what reaction product is expected first?

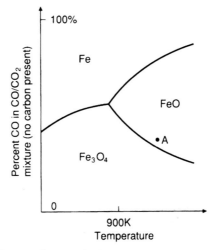

FIGURE 2.18 Predominance diagram for iron and its oxides in contact with carbon monoxide/dioxide mixtures.

Fig. 2.18 is important because of the industrially important reduction of iron ores by carbon, a step in the product of steel, and this figure is frequently studied with a line superimposed to indicate the equilibrium in reaction

$$C + CO_2 = 2CO \tag{2.87}$$

at 1 atm. as is done in Fig. 2.19 by a broken line. This broken line gives the gas composition in equilibrium with solid carbon at the temperature in question. Therefore any point off the broken line will represent a composition that is *not* in equilibrium with

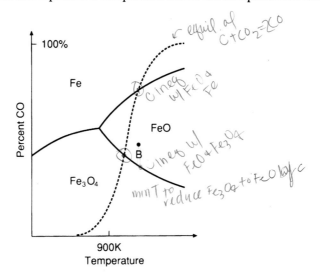

FIGURE 2.19 Predominance diagram of Figure 2.18 with the equilibrium line for carbon in contact with its oxides superimposed.

carbon, and reaction (2.87) would take place, in the forward or reverse direction. In a closed reactor and with no other reactions taking place, the composition of the gas would shift down (or up) towards the broken line. However, if iron and its oxides are present then there are a number of other reactions that can accompany (2.87). For example suppose a gas of composition and temperature indicated by point B in Fig. 2.19 is contacted with carbon and Fe_3O_4 in a reactor. Clearly the stable oxide at this point is FeO and the reaction

$$Fe_3O_4 + CO = 3FeO + CO_2 \qquad (2.88)$$

takes place, consuming CO and tending to move the gas composition down from point B. Simultaneously reaction (2.87) occurs, generating CO and tending to move the gas composition *up* from point B. In reality some shift of the composition will take place (dependent on the rates of reactions) and (2.88) plus (2.87) take place until either all the Fe_3O_4 or all the carbon has been used up. Considerations such as this lead to the conclusion that only at the two points marked by intersection of solid and broken lines is carbon in equilibrium with more than one solid phase. Below the lower temperature intersection, carbon is in equilibrium with Fe_3O_4; above the higher temperature one, it is in equilibrium with Fe (ignoring for the moment the solubility of carbon in iron); between the two intersections it is in equilibrium with FeO. The lower intersection therefore gives the minimum temperature for reduction of Fe_3O_4 to FeO by C (at 1 atm.) and the upper gives the minimum temperature for reduction of FeO to iron.

Later in the text we shall come across the application of predominance diagrams to aqueous solutions when we examine their importance in the hydrometallurgical production of metals.

2.8 Physical Equilibria

Throughout most of the chapter we have been concerned with chemical equilibria. However, many of the concepts that have been discussed can be applied to changes which are more physical than chemical. Examples are the vaporization of a liquid (or solid), the dissolution of a partly soluble solid into a solution and the dissolution of a common diatomic gas (such as H_2) into a liquid metal.

Let us pick the last and write it (for nitrogen) as

$$N_2 \text{ (gas)} = 2N \text{ (dissolved in metal)} \qquad (2.89)$$

"Reaction" (2.89) has been written to show the dissociation into atoms that takes place when gases dissolve in metals.

We can write an equilibrium constant

$$K_a = \frac{a_N^2}{P_{N_2}} \qquad (2.90)$$

and if we assume that the solution obeys Henry's law we can write

$$a_N = \text{constant } x_N$$

leading to

$$x_N = \left(P_{N_2}\right)^{1/2} \times \text{ constant} \tag{2.91}$$

This is Sievert's law which tells us that the amount of nitrogen dissolved in a metal should be proportional to the square root of the partial pressure of that gas above the metal. The reader is asked to show (using the approach of section 2.5.7) that, provided the activity coefficient does not vary with temperature, the heat of solution of the gas can be found by measuring the solubility at various temperatures.

Turning to the other two physical phenomena, it is instructive to show how the saturation vapor pressure of a liquid varies with temperature, how the solubility product of a dissociating species partially soluble in a liquid arises, and how solubility product data might be used to determine heats of solution. These are left as exercises for the reader.

2.9 Concluding Remarks

Much of thermodynamics has been left out of this short review; the justification is that the material left out is either already familiar to the reader or is less relevant to the topic of materials production. We hope that the chapter has served to stress the utility of thermodynamics in practical engineering; we have dealt with some topics (interaction coefficients, Ellingham diagrams, predominance diagrams) that are given little or no treatment in many undergraduate thermodynamics courses. A considerable part of the rest of this book is founded on this chapter which provides the basis for understanding why certain reaction sequences are employed in materials production (whereas others are not).

LIST OF SYMBOLS

a_i	activity of species i
C_i	concentration of species i
C_P	specific heat at constant pressure
E	internal energy
F, F', dF	forces and differential change in force (see Section 2.2)
f_i	fugacity of species i
G	Gibbs free energy
H	enthalpy
K_a, K_c, K_p, K_x	equilibrium constants written in terms of activity, concentration, pressure, mole fraction
n_i	number of moles of species i
P, P_i	pressure, partial pressure of i
dq	differential heat added to system
R	gas constant
S	entropy
T, T_{pc}	temperature, temperature of phase change

V	volume
W	work done by system
x_i	mole fraction of species i
Δ	increase in following quantity (finite)
ε_i^j	interaction coefficient (see Section 2.5.5)
γ_i	activity coefficient of species i
μ_i	chemical potential of species i

Subscripts

$A,B,C...$	for chemical species A,B,C, etc.
atm	at atmospheric pressure
F	of formation
P	at constant pressure
R	of reaction
T	at constant temperature
0	at standard conditions

Overbars ($^-$)	indicate per unit mass
Tildes ($^\sim$)	indicate per mole

PROBLEMS

2.1 Using the concept of entropy discussed in section 2.2, would you expect the following to be accompanied by a decrease or increase in entropy (if carried out at constant temperature)?
(a) $CaCO_3 + 1/2\ O_2\ (g) + SO_2\ (g) = CaSO_4\ (s) + CO_2\ (g)$
(b) The dissolution of solid carbon in molten iron
(c) Oxidation of a solid metal to a solid oxide by air
(d) Condensation of a gas to form a liquid
(e) The sublimation of ammonium chloride
(f) The sublimation of iodine
(g) $Ni(CO)_6\ (g) = Ni\ (s) + 6CO\ (g)$
(h) The rusting of iron
(i) The combustion of ethane in air to produce carbon dioxide and water vapor

2.2 A solid substance has a specific heat given by the equation

$$C_P = aT + bT^2 + cT^3$$

where C_P is the specific heat (J/kg K), T the temperature (K), and a, b and c are constants. The substance has no phase changes between 0 K and its melting point, T_{mp}. Its latent heat of fusion is ΔH_L (J/kg). Write down an equation for the entropy of one mol of the liquid at T_{mp} in terms of *a, b, c,* etc.

2.3 How would you write the equilibrium constants for the following reactions (in terms of partial pressures, concentrations, etc. rather than activities)?

(a) $2H_2S$ (g) + SO_2 (g) = $2H_2O$ (g) + 3S (s)

(b) $2H_2S$ (g) + SO_2 (g) = $2H_2O$ (l) + 3S (s)

(c) $H^+ + OH^- = H_2O$ in dilute solution in water

(d) C (dissolved in iron) + CO_2 (g) = 2CO (g)

(e) C (s) + Al_2O_3 + $2SiO_2$ = 2Al + 2Si + 7CO (g) with Al_2O_3 and SiO_2 as two mutually soluble liquids which form an oxide phase, and Al and Si as two mutually soluble liquids which form a separate metal phase

2.4 Examining the curves in Figs. 2.4 and 2.5 you will see that their slopes are not constant (i.e., the curves are not straight lines). Explain why the slopes change as shown.

2.5 Suppose we were to rank the ability of an element or compound to act as a reducing agent by the number of oxides that it is capable of reducing. Using this ranking, place the following in order of decreasing ability as reducing agents for the circumstances (a) and (b) described below: Si, Mg, Na, CO, H_2, C, Al.

(a) In a reactor operated close to equilibrium and at a temperature and pressure of 500°C and 1 atm, respectively.

(b) In a reactor at 1200°C and 1 atm from which all gaseous species can readily escape but which retains all liquid and solid species. All reactants are charged to the reactor before reaction begins.

2.6 It should be clear that water vapor can be thermally dissociated into hydrogen and oxygen if the temperature is high enough:

$$H_2O = H_2 + 1/2\ O_2$$

An analytical device is capable of detecting hydrogen in excess of one microgram in any volume of a gas mixture. At 1500 K and 1 atmosphere pressure, what is the volume (in cubic meters) of an equilibrium mixture of the above gases in which the device can just detect hydrogen. [Standard free energy of formation of water vapor at 1500 K = -164 kJ/mol.]

2.7 Consider the temperature at which the line for a particular metal/metal oxide crosses the horizontal axis (ΔG_F of oxide = zero) in the Ellingham diagram. Call this temperature T_1. At T_1 the partial pressure of oxygen = 1 atmosphere in a gas which is in equilibrium with the metal and oxide. For an environment in which the oxidation partial pressure is held at 1 atmosphere the oxide is unstable (and the metal stable) at temperatures above T_1, while for temperatures below T_1 the oxide is stable (and the metal unstable). You may recall that for most metals in everyday use, T_1 is a high temperature (several hundred °C). Does this mean that if we were to establish a colony on a planet with the same atmospheric pressure as that on Earth, but with an atmosphere consisting almost entirely of oxygen, our present metals would be useless to that colony? Explain your answer.

2.8 An iron concentrate consisting of iron oxides, together with some silicon dioxide and manganese dioxide left over from the gangue, is to be reduced using coke (carbon). Based on the Ellingham diagram for oxides, what would you consider to be a suitable temperature range for carrying out the reaction if it is desired to reduce the iron oxide to metal, but *not* to produce silicon or manganese that would contaminate the iron? [Note: the first step in the reduction of manganese dioxide would be reduction of MnO_2 to MnO.]

2.9 Consider reactions of the type

$$A(s) + B(g) = AB(s)$$

For reactions of this type that occur to significant extents (i.e., have negative free energies of reaction) the reaction is always found to be exothermic. Explain this.

2.10 "Thermodynamics is perhaps the most important topic in materials production. From thermodynamics we can calculate whether or not any reaction will actually take place. This enables us to choose suitable metals for use in oxidizing environments, or to eliminate many suggested processes for making metals, on the basis of the fact that the reactions entailed are not thermodynamically feasible." Comment on the above in not more than ten sentences, with examples.

2.11 An unfortunate aspect of our present patent laws is that patent applications receive little scrutiny from persons with an appropriate scientific or engineering training before a patent is issued. A case in point is a patent issued a few years ago to a major U.S. corporation that is a household name. That patent is for a process for treating municipal garbage (largely organic matter, glass, steel and aluminum). It is a high-temperature operation in which the organic matter is burned and the glass, steel and aluminum end up as liquids at the bottom of the apparatus. The three liquids are in contact with each other, the glass floating on the aluminum which, in turn, floats on the molten steel; they are then drained separately from the apparatus. Show that this patent is nonsensical. [Glass may be regarded as a mixture of the oxides of silicon, sodium and calcium.]

2.12 Calculate the activity coefficient of silicon in liquid iron at 1600 °C containing 0.1 weight % carbon, 0.05 weight % silicon and negligible quantities of other elements. The activity coefficient of silicon in iron at infinite dilution is $\gamma^* = 0.0072$. The interaction coefficients are $\varepsilon_{Si}^{C} = 12.7$ and $\varepsilon_{Si}^{Si} = 37.6$.

2.13 Silicon carbide, an important ceramic material, is manufactured from coke and sand by the reaction

$$SiO_2 + 3C = SiC + 2CO$$

The manufacturing process is a crude one in which a mixture of coke (carbon) and sand is heated electrically in a batch reactor that is open to the atmosphere. At a high enough temperature the reaction starts and carbon monoxide flows out of the reacting mass. Calculate the temperature at which this happens, based on thermodynamic data (e.g., the JANAF tables). Assume that the sand, coke and silicon carbide are separate pure solid phases.

2.14 Using the table of free energy of formation given below, calculate and plot the partial pressure of oxygen in equilibrium with both nickel and nickel oxide in the temperature range 1500 - 2500°C. Another way of viewing the plot is as a Pourbaix-Ellingham diagram, in other words, as a plot showing the most stable phase (nickel oxide or nickel) as a function of the prevailing conditions of temperature and oxygen partial pressure. Mark the region of stability of nickel oxide and the region of stability of nickel on your diagram.

Temperatures (K)	Free energy of formation of NiO (kJ/mol)
1726	-89.3
2300	-34.7
2800	+13.4

2.15 In a coal combustion process a nickel part at 600°C is constantly exposed to a gas containing oxygen at a partial pressure of 0.1 atmosphere and sulfur dioxide at 0.05 atm. The nickel part is likely to undergo attack by the gas. Based on the thermodynamic data below, would you expect the product of reaction to be nickel sulfate or nickel oxide? Neither of these compounds are gases at the temperature in question. Assume that the other constituents of the gas will not react with nickel or its compounds and that the possibility of SO_3 formation can be neglected. (In truth this last assumption is not a good one but make it for the purposes of the present calculation.) Standard free energy of formation of NiO at 600°C = −163 kJ/mol. Standard free energy of reaction at 600°C for the formation of nickel sulfate, from nickel oxide, oxygen and *sulfur dioxide* = −90 kJ/mol $NiSO_4$.

2.16 Hematite (Fe_2O_3) is to be reduced to wustite (which we can take as FeO for the purposes of this problem), without the formation of metallic iron. The reducing agent is to be a mixture of hydrogen and water vapor. What are the upper and lower limits on the % water vapor in the reducing gas if reduction is to be carried out at 1000 K. [*Hint*: the reduction will proceed first by the formation of magnetite (Fe_3O_4).] Standard free energies of formation at 1000 K (kJ/mol):

$$H_2O = -192$$
$$FeO = -199$$
$$Fe_2O_3 = -560$$
$$Fe_3O_4 = -790$$

2.17 It is proposed to reduce stannic oxide (SnO_2) to stannous oxide (SnO) by the following procedure. The stannic oxide will be placed in a reactor and brought up to 1000 K. A mixture of hydrogen and water vapor will then be passed through the reactor while maintaining the 1000 K temperature. The composition of the gas is to be adjusted so that the reduction of stannic oxide to stannous oxide can take place but so that the reduction of stannous oxide to metal is prevented. Using the thermodynamic data below, decide whether the proposed operation will work. Free energies of formation at 1000 K (kJ/mol):

$$H_2O = -192$$
$$SnO_2 = -373$$

$$SnO = -182$$

2.18 The reduction of FeO by carbon monoxide is a reversible reaction. This means that in attempting to reduce FeO by a flowing stream of carbon monoxide, a great deal of the carbon monoxide is wasted since the exit gases must necessarily contain a great deal of carbon monoxide. A way around this difficulty might be to reduce iron oxide in the presence of lime, which would then absorb the carbon dioxide produced by the reduction reaction:

$$FeO + CaO + CO = Fe + CaCO_3$$

Calculate the minimum total pressure in an apparatus in which this reaction can take place at 1000 K, recognizing that there is a tendency for the calcium carbonate to decompose by the reaction

$$CaCO_3 = CaO + CO_2$$

Obtain the data necessary to work this problem from thermodynamic tables.

2.19 Mercuric oxide (HgO) is placed in a vessel which is then evacuated, filled with nitrogen and heated to 327 °C. At this temperature the total pressure in the vessel is observed to be 2 atmospheres and part of the mercuric oxide has decomposed, according to the reaction

$$2HgO = 2Hg + O_2$$

The mercuric oxide is a pure solid, the mercury produced is present as a *gas*, as is the oxygen.

(a) Calculate the equilibrium constant for the reaction if the standard free energy of formation of HgO at 327 °C is -26.4 kJ/mol HgO.

(b) Calculate the composition of the atmosphere in the vessel, assuming it is in equilibrium with the HgO. (*Hint*: the stoichiometry of the reaction tells you a relationship between the partial pressure of oxygen and the partial pressure of mercury.)

2.20 Tungsten carbonyl is a gas with the formula $W(CO)_6$. Like other metal carbonyls, it is an intermediate compound used in the preparation of very pure metal powders, or in gas plating processes. Make a thermochemical analysis of the preparation of tungsten metal powder at 1200 K from $W(CO)_6$. The process will operate by blending a $CO_2/W(CO)_6$ gas mixture together at room temperature or a bit above, and heating the gas mixture rapidly to 1200 K in an electric furnace. Show how the formation of WC or WO_2 can be prevented. Pressure = 1 atm throughout.

2.21 The standard heat and entropy of reaction for

$$C(\text{dissolved in steel}) + O(\text{dissolved in steel}) = CO(g)$$

are 22.4 kJ/mol CO and 39.6 J/(mol CO K), respectively, for a standard state consisting of one weight percent of carbon and one weight percent of oxygen in steel. This reaction is important in the final stages of steelmaking when it is desired to reduce the content of dissolved carbon and oxygen by removing them as carbon monoxide. If a steel contains 0.3 weight percent carbon, calculate the amount of oxygen in the steel if the steel is in equilibrium with a partial pressure of 1 atmosphere CO at 1600 °C. Assume that the activity coefficients of carbon and oxygen in steel are both unity (solute concentrations expressed as weight percent). Sketch a plot of ln(carbon content) versus ln(oxygen content) for higher and lower partial pressures of CO and write on your plot the slope of the lines produced.

2.22 From tables of thermodynamic data construct a predominance diagram similar to Fig. 2.18 but for the oxides of cobalt.

REFERENCES AND FURTHER READINGS

CHASE, M.W., Jr, Ed., NIST-JANAF Thermochemical Tables (Journal of Physical and Chemical Reference Data), Fourth edition, 1998, available from the American Institute of Physics.

GASKELL, D. R., *Introduction to Metallurgical Thermodynamics*, 2nd ed McGraw Hill, New York, 1981.

KUBASCHEWSKI , O., E. L. EVANS, and C. B. ALCOCK, *Metallurgical Thermochemistry*, 5th ed., Pergamon Press, Oxford, 1979.

MODELL, M., and R. C. REID, *Thermostatics and Its Applications* , 2nd ed., Prentice Hall, Englewood Cliffs, N.J., 1983.

TRIBUS, M., *Thermostatics and Thermodynamics*, D. Van Nostrand Co. Inc., Princeton, N.J., 1961.

CHAPTER 3

Reaction Kinetics

3.1 Introduction And Some Definitions

In Chapter 2 we reviewed the principles of chemical thermodynamics that enable us to predict whether or not a particular reaction will take place, under specified conditions, and the quantities of reactants and products present when chemical equilibrium is reached and reaction stops. Thermodynamics therefore provides us with a first test of any scheme of reactions for producing a material: are the reactions feasible at practical temperatures and do they produce the desired material in sufficient yield? An equally important consideration, from the viewpoint of economics, is how rapidly the reaction takes place. A reaction may be thermodynamically feasible but take place so slowly that the productivity of a process involving the reaction is too low. "Too low" here means that the revenue derived from the material produced is too low to justify the investment of capital to build the plant or perhaps even too low to meet the operating costs (labor, energy, raw materials, etc.) of the process. This chapter is concerned with the rate at which reactions take place.

3.1.1 Homogeneous and Heterogeneous Reactions

We shall find it useful to distinguish between two types of reactions. *Homogeneous reactions* involve only one phase (i.e., a liquid phase, or a solid phase, or a gas phase). One example is the neutralization of an acid by a base in an aqueous solution:

$$H^+ + OH^- = H_2O \qquad (3.1)$$

where only a liquid phase is involved.

Another example is the reaction between iodine vapor and hydrogen to form hydrogen iodide:

$$I_2 + H_2 = 2HI \qquad (3.2)$$

which involves only a gas phase.

Homogeneous reaction kinetics has been an important subject in chemistry for many decades and most scientists and engineers have been exposed to this topic in undergraduate chemistry courses. Homogeneous reactions are seldom encountered in reactions used to produce materials.

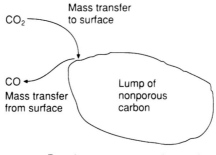

FIGURE 3.1 Lump of nonporous carbon being oxidized by carbon dioxide to yield carbon monoxide.

More common are *heterogeneous reactions,* which involve more than one phase. An example is the reaction between the iron oxide FeO and hydrogen. At temperatures where this reaction is practical, the water produced by reaction is a gas, while the FeO and the iron produced may be solid or liquid:

$$FeO + H_2 = Fe + H_2O \tag{3.3}$$

This reaction therefore involves three phases (one gas phase, the solid FeO and the solid or liquid Fe).

A second example (used in one process for producing nickel) is the thermal decomposition of nickel carbonyl vapor:

$$Ni(CO)_4 = Ni(s) + 4CO \tag{3.4}$$

Here two phases are present, a gas and a solid.

A third example is instructive. Consider the reaction between the gases nitrogen and hydrogen to produce ammonia (also a gas at temperatures and pressures where this reaction is practical)

$$3H_2 + N_2 = 2NH_3 \tag{3.5}$$

This is a step in the economically important synthesis of ammonia-based fertilizers. Because the reaction involves only a gas phase we might be inclined to categorize it as a homogeneous reaction. However, the reaction proceeds at useful rates only on the surface of a solid catalyst introduced into the chemical reactor; the reaction is therefore better categorized as a heterogeneous one because it involves two phases (the gas phase and a solid catalyst).

There is a very significant difference between the way homogeneous and heterogeneous reactions occur. Homogeneous reactions occur throughout the volume where the reactants are mixed. Heterogeneous reactions, however, involve more than one phase; they therefore take place at the interface between phases. As an example consider the oxidation of solid carbon by carbon dioxide gas illustrated in Fig. 3.1. In the bulk of the gas there is no carbon present so that the reaction

$$C + CO_2 = 2CO \tag{3.6}$$

does not take place. Similarly within the carbon particle (provided it is non-porous) there is no carbon dioxide so no reaction takes place there either. The reaction is confined to the interface between gas and solid phases, i.e., the surface of the carbon particle.

We shall see in section 3.1.2 how this difference in the way reaction takes place leads to a distinction between the definitions of reaction rates for the two categories of reaction.

3.1.2 Definition of Reaction Rates

Although we may have an intuitive feeling for what constitutes a fast or slow reaction we must now decide on quantitative definitions which will serve us in the rest of this chapter and throughout the book. Starting with homogeneous reactions, imagine a sealed vessel containing a mixture of hydrogen and iodine vapor reacting according to equation (3.2). Let N_{H2} be the number of moles of hydrogen, etc. One measure of the rate of reaction is the rate of decrease of the moles of hydrogen (mathematically, $- dN_{H2}/dt$) or of iodine vapor ($ - dN_{I2}/dt$, which is numerically equal to $- dN_{H2}/dt$ by the stoichiometry of the reaction). Alternatively we could use the rate of increase of moles of hydrogen iodide (dN_{HI}/dt, which numerically is twice $- dN_{H2}/dt$). The difficulty with these choices for reaction rate is that they are extensive rather than intensive quantities. That is, they depend both on the chemical behaviour of the molecules participating in reaction and *on the scale of the system (being larger for larger numbers of molecules)* . However, for homogeneous reactions it is a simple matter to get an intensive measure of reaction rate by dividing by the volume of the vessel (V). We shall use the symbol $\Re$ for an intensive reaction rate defined in this way

$$\Re_{H_2} \equiv - \frac{1}{V} \frac{dN_{H_2}}{dt} \tag{3.7}$$

Similar definitions of reaction rate can be set up using iodine vapor ($\Re_{I2}$) or hydrogen iodide

$$\Re_{HI} \equiv \frac{1}{V} \frac{dN_{HI}}{dt} \tag{3.8}$$

Note that this last number is twice $\Re_{H2}$ (from stoichiometry) and that the subscript is important in specifying which chemical species we are referring to when we define the reaction rate. Note that if the volume of the reactor is kept constant then

$$\Re_{H_2} \equiv - \frac{dC_{H_2}}{dt} \tag{3.9}$$

where C_{H2} is the concentration of hydrogen in moles per unit volume. Equation (3.9) should be used with caution; it is really a definition of rate for a homogeneous reaction in a closed vessel of constant volume (a restrictive set of circumstances).

The question arises as to how we would define reaction rate in a reactor that is not closed (e.g., a reactor where reactants are continuously fed). $- dN_{H2}/dt$ is no longer simply connected to the reaction rate because it depends on how rapidly we feed hydrogen to the reactor as well as on how rapidly hydrogen is consumed by reaction. Equations (3.7), (3.8) and (3.9) are no longer valid definitions (although, regrettably, they have been used as such in earlier texts). However, we can still, theoretically at least, count the number of molecules of hydrogen that react in a short time interval. If we divide that number by the time interval, by Avogadro's number (to convert molecules into moles) and by the reactor volume, we obtain an intensive measure of reaction rate which we can call $\Re_{H_2}$. [In practical terms that rate could be determined at any instant by closing the inlets and outlets to the reactor, then measuring - dN_{H2}/dt and using equation (3.7).] This definition of reaction rate is universally accepted for homogeneous reactions.

Now, turning our attention to heterogeneous reactions, we can imagine a closed reactor in which reaction (3.6) is taking place. Again, a first measure of reaction rate might be - dN_{CO2}/dt or - dN_C/dt or even dN_{CO}/dt, the last quantity being twice either of the others. Again this measure of reaction rate is unsatisfactory because it is an extensive quantity. The temptation now is to divide by the reactor volume as we did for homogeneous reactions. However, because heterogeneous reactions take place at the interface between phases (surface of the carbon particle in our present example), yielding to this temptation leads to an unsatisfactory definition, albeit one that is sometimes used. This definition gives us a rate which depends on the surface-to-volume ratio in the reactor, merely for mathematical (rather than physical) reasons. A far better definition is to divide - dN_{CO2}/dt by the surface area, A, at which the reaction is taking place:

$$\Re_{CO_2} \equiv -\frac{1}{A}\frac{dN_{CO_2}}{dt} \tag{3.10}$$

Two facts should be clear immediately. The dimensions of the rate appearing in equation (3.10) are different (moles per unit time per unit area) from those for the homogeneous reaction appearing in equation (3.7) (moles per unit time per unit volume). Furthermore, we cannot immediately re-express the rate in terms of concentrations as we did in equation (3.9).

Note that again we need to modify our definition of rate if the reactor is not closed. Again quantities such as - dN_{CO2}/dt cannot be used in the definition and again we can arrive at a correct definition by (theoretically) counting the number of molecules reacting in a brief interval. That number is divided by the interval, Avogadro's number and the area for reaction to give the rate.

Frequently it will be found that the area, A, at which reaction is taking place is changing with time. In our example, A is decreasing as the carbon particle is consumed by reaction.

3.2 Rate Equations

3.2.1 Factors Affecting Reaction Rate

The reader may already be aware that the following determine reaction rates:
1.   The chemical nature of the reactants. For example

$$3H_2 + N_2 = 2NH_3 \tag{3.11}$$

is very slow at practical temperatures and pressures (if no catalyst is present),

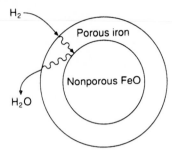

FIGURE 3.2 Lump of iron oxide being reduced by hydrogen to produce water vapor and porous iron.

whereas

$$2H_2 + O_2 = 2H_2O \qquad (3.12)$$

proceeds with explosive rapidity at temperatures over a few hundred degrees Celsius.

2. *Temperature.* For example, although thermodynamics predicts that nickel oxide can be reduced by hydrogen at all practical temperatures (see the Ellingham diagram of chapter 2), the heterogeneous reaction

$$NiO + H_2 = Ni + H_2O \qquad (3.13)$$

is immeasurably slow at room temperature. Above approximately 250°C, lumps of NiO a few centimeters in size can be completely reduced in a few tens of minutes or less.

3. Concentrations *of reactants.* For example the dissolution of cupric oxide in acids

$$CuO + 2H^+ = Cu^{2+} + H_2O \qquad (3.14)$$

is a heterogeneous reaction whose rate is very dependent on pH.

4. Presence *of catalysts.* For example, the gasification of carbon by water vapor

$$H_2O + C = CO + H_2 \text{ (+ other gaseous products)} \qquad (3.15)$$

is greatly accelerated in the presence of potassium hydroxide.

5. *In the case of heterogeneous reactions, the transport of reactants and products.* For example if reaction (3.3) is carried out at temperatures where the iron produced is solid, a sectioned, partly reacted lump of FeO might appear as in Fig. 3.2. As shown schematically, for the hydrogen to reach the FeO with which it reacts, it first has to diffuse through a layer of iron (usually porous). Furthermore, the water vapor produced by reaction has to be transported away from the Fe-FeO interface by diffusion or its accumulation at the interface would cause chemical equilibrium to be reached and the reaction to stop.

6. *The transport of heat.* An example is the endothermic thermal decomposition of limestone to produce lime. This heterogeneous reaction is

$$CaCO_3 = CO_2 + CaO \qquad (3.16)$$

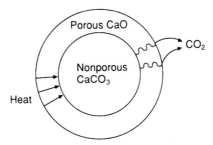

FIGURE 3.3 Sketch of a thermally decomposing calcium carbonate particle showing that transport of heat is necessary for reaction.

A partially reacted and sectioned piece of limestone would appear as Fig. 3.3. For reaction to be sustained it is necessary that heat flow from outside the particle to the reaction interface which is the surface of the "shrinking core" of $CaCO_3$. This heat transport (by conduction in the pores and solid of the CaO) can have an influence on reaction rate.

3.2.2 Rate Equations

A *rate equation* is a correlation of the rate of a particular reaction with the concentrations of the species involved in reaction. For example triethyl amine and ethyl bromide are both soluble in benzene and if both solutions are mixed a (benzene-soluble) quaternary ammonium compound is formed by the homogeneous reaction

$$(C_2H_5)_3N + C_2H_5Br = (C_2H_5)_4NBr \tag{3.17}$$

Experimentally it has been found that the rate is proportional to the concentration of both reactants, hence the rate equation is

$$\Re_{C_2H_5Br} = kC_{(C_2H_5)_3N}C_{C_2H_5Br} \tag{3.18}$$

In equation (3.18), k is called the "rate constant" for this reaction. It is independent of concentration (although it will vary with temperature). In the terminology of the chemical kineticist, reaction (3.17) is said to be "first order with respect to ethyl bromide" because the power of the ethyl bromide concentration appearing in equation (3.18) is one. Similarly the reaction is first order with respect to triethyl amine. In general we expect, for a homogeneous irreversible reaction (i.e., one with a high equilibrium constant, permitting us to ignore the back reaction),

$$aA + bB + cC + \ldots = \text{products} \tag{3.19}$$

$$\Re_A = kC_A^{n_A}C_B^{n_B}C_C^{n_C} \cdots \tag{3.20}$$

where n_A is the reaction order with respect to A, etc. Note that the reaction orders are usually small integers and are not necessarily equal to the stoichiometric coefficient appearing in reaction (3.19), that is, it is *not* necessary that $n_A = a$, and so on.

Chemical kineticists have determined reaction orders and rate constants (as a function of temperature) for a large number of reactions, particularly homogeneous reactions. However, this large number of reactions still represents only a fraction of all possible reactions and there are many common reactions, especially heterogeneous ones, for which rate equations have not yet been determined. This is unfortunate and contrasts strongly with the situation in thermodynamics. The scientist or engineer involved in materials production can reasonably expect to find thermodynamic data which will enable the prediction of the free energy of reaction, and therefore the nature of the chemical equilibrium, for most reactions of practical interest. It would be unusual, however, to encounter a heterogeneous reaction on which sufficient accurate kinetic data are available. Consequently, one may be obliged to perform one's own experiments on kinetics (or interpret the results of others) and a simplified account of how this is done now follows.

3.2.3 Determination of Rate Equations

Consider the simplest possible reaction where a species A reacts homogeneously to produce a second species B

$$A = B \qquad (3.21)$$

We will assume that the reaction is irreversible. From (3.20) we expect then

$$\Re_A = kC_A^{n_A} \qquad (3.22)$$

For convenience we will now drop the subscripts on $\Re$, C and n because their meaning is unambiguous.

If we carry out the reaction in a closed reactor of fixed volume, we can use equation (3.9):

$$\Re = -\frac{dC}{dt} = kC^n \qquad (3.23)$$

If the temperature is fixed, k is constant and we can rearrange (3.23) and integrate as follows:

$$\int k\,dt = -\int \frac{dC}{C^n} \qquad (3.24)$$

that is,

$$kt = \begin{cases} \dfrac{C^{1-n}}{n-1} + \text{constant} & \text{for } n \neq 1 \\[2em] -\ln C + \text{constant} & \text{for } n = 1 \end{cases} \qquad (3.25)$$

We can determine the constants of integration by examining the start of the reaction. Let C_o be the initial concentration of A in the reactor (at t=0). Hence

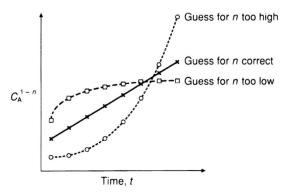

FIGURE 3.4 Determination of reaction order by trial plots of the left-hand side of equation (3.26).

$$C^{1-n} = k(n-1)t + C_0^{1-n} \quad \text{for } n \neq 1 \tag{3.26}$$

$$\ln C = \ln C_0 - kt \qquad \text{for } n = 1 \tag{3.27}$$

Suppose we measure C in our reactor (e.g., by inserting some kind of concentration-sensitive probe or by withdrawing small samples for analysis). The data which we must match with our assumed rate equation (3.22) then consists of values of C at various values of t. We can guess a value of n (other than 1) and calculate the left-hand side of equation (3.26) from the experimental data. If indeed our guess is correct then a plot of C^{1-n} against t will yield a straight line of slope $k(n-1)$ from which we get our rate constant. If the guess is incorrect a curve results and we must improve our guess. The procedure is illustrated in Fig. 3.4. Of course, it is also necessary to deal with the possibility that $n = 1$ and here the plot of $\ln C$ versus t shown in Fig. 3.5 is used. Again the rate constant is obtained from the slope.

To further illustrate the fitting of assumed rate equations to experimental data let us now do a worked example for a *heterogeneous* reaction.

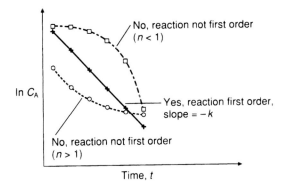

FIGURE 3.5 Plot to check whether or not reaction is first order, based on equation (3.27).

PROBLEM

In a laboratory equipment, 0.1 mole of pure hydrogen is contacted with two hundred non-porous 20mm diameter nickel oxide spheres at 300°C. The reaction

$$NiO + H_2 = Ni + H_2O$$

takes place. If the molecular weight of NiO is 74.7 and its density is 6700 kg/m³, show that there is a large excess of NiO. This being the case, it can be assumed that the area of the reaction interface changes negligibly during reaction. By analyzing minute samples of gas in the reactor the following results are obtained:

Time (s)	Mole fraction hydrogen in gas
0	1
20	0.64
40	0.41
60	0.26
100	0.108
150	0.036

The gas space in the reactor is 10^{-3} m³ and can be assumed to be at uniform composition. Assuming that the reaction is irreversible, show that it is first order with respect to hydrogen and determine the rate constant in mm/s.

Solution

$$\text{Volume NiO spheres} = (4/3) \, \pi \, 10^3 \times 200 \text{ mm}^3 \equiv 8.38 \times 10^{-4} \text{ m}^3$$

$$\text{Mass NiO spheres} = 8.38 \times 10^{-4} \times 6700 \text{ kg} \equiv 5.61 \times 10^3 \text{ g}$$

$$\text{Moles NiO} = 5.61 \times 10^3 / 74.7 = 75.14 \text{ mol}$$

Because the reduction of one mol of NiO requires one mol of hydrogen we see that, compared to the 0.1 mol of hydrogen, there is a great excess of NiO. In accordance with the assumption stated in the problem, the reaction interface area remains equal to the outside surface of the NiO spheres $= 4\pi \, 10^2 \times 200 \text{ mm}^2 = 2.51 \times 10^5 \text{ mm}^2$. Let x_{H2} be the mole fraction of H_2 in the gas, i.e., $0.1 \, x_{H2}$ is the moles of H_2 (because the consumption of one mole of H_2 yields one mole of H_2O and therefore the total number of moles of gas stays constant at 0.1). Hence the rate of decrease of the moles of H_2 is $-0.1 \, dx_{H2}/dt$ or the rate of reaction is

$$\mathfrak{R}_{H_2} = \frac{-0.1 \text{ mol}}{2.51 \times 10^5 \text{ mm}^2} \cdot \frac{dx_{H_2}}{dt} = kC_{H_2}^n$$

with k the rate constant and n the reaction order. But the concentration of H_2

$$C_{H_2} = \frac{\text{molH}_2 \text{ in reactor}}{\text{gas volume in reactor}} = \frac{0.1 x_{H_2}}{10^{-3} \times 10^9} \frac{\text{mol}}{\text{mm}^3} = 10^{-7} x_{H_2} \text{ mol/ mm}^3$$

Substitution yields

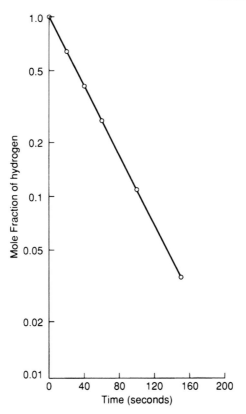

FIGURE 3.6 Data of the worked example plotted to show that reaction is first order with respect to hydrogen.

$$\frac{-0.1 \text{ mol}}{2.51 \times 10^5 \text{ mm}^2} \cdot \frac{dx_{H_2}}{dt} = k(10^{-7}x_{H_2} \text{ mol}/\text{mm}^3)^n$$

Trying $n = 1$,

$$\frac{dx_{H_2}}{dt} = -2.51 \times 10^{-1} \text{ mm}^{-1} kx_{H_2}$$

Integrating, $\ln x_{H2} = \text{constant} - 2.51 \times 10^{-1} \text{ mm}^{-1} kt$ with constant $= 0$ since $x_{H2} = 1$ at $t = 0$. A plot of $\log x_{H2}$ versus t appears as Fig. 3.6; clearly a straight line results (and therefore n does indeed equal 1). The slope of the line is $-0.0096 \text{ s}^{-1} = -2.51/2.303 \times 10^{-1} \text{ mm}^{-1}k$. Hence $k = .088 \text{ mm/s}$.

The examples above have been made deliberately simple. It is a more complicated procedure to derive rate equations from experimental data when several reactants participate in reaction or when more than one reaction may occur simultaneously. However, the principles involved are no different from those illustrated by means of the simple examples. Various graphical techniques have been developed for handling the more complicated cases and now computer-based curve fitting methods are widely used.

3.2.4 Rate Equations for Heterogeneous Reactions

The reader will have noted that in the worked example of section 3.2.3 there was nothing in the resulting rate equation corresponding to the "concentration" of the solid reactant (NiO). We shall now show that this is entirely natural because the reaction was a heterogeneous one with nickel oxide present in pure form.

From Chapter 2 it will be noted that the "driving force" for reaction is the chemical potential of the reactants (or more precisely, the chemical potential difference between reactants and products). It is therefore natural to expect that rate equations written in terms of chemical potentials or, more conveniently, activities

$$\Re_A = f(\text{activities}) \tag{3.28}$$

would be more "fundamental" than ones written in terms of concentrations (e.g. (3.20)). For homogeneous reactions this approach is not particularly fruitful but for heterogeneous reactions it leads immediately to the disappearance of pure solid or liquid reactants from the right-hand side of equation (3.28), because their activities equal unity.

By way of an example we now show that (3.28) is readily manipulated into a more familiar form. Consider

$$a\,A(g) + b\,B(s) = C(s)$$

which we shall regard as irreversible. Equating the activity of the gas to its partial pressure and setting the activity of B to unity (which assumes B and C do not form a solid solution) we expect (3.28) to become

$$\Re_A = kP_A^{n_A} \tag{3.29}$$

Since

$$P_A = C_A RT \tag{3.30}$$

from the ideal gas law, with R the gas constant and T absolute temperature, then

$$\Re_A = k(RT)^{n_A} C_A^{n_A} = k' C_A^{n_A} \tag{3.31}$$

is the more familiar form.

3.2.5 Rate Equations for Reversible Reactions

If the equilibrium constant for a reaction is not large then we must allow for back reaction in the rate equations. That is, as reaction proceeds and the concentrations of products start to increase, these products will start to react back, diminishing the net rate of forward reaction. As an example, for the reaction

$$FeO(s) + CO = Fe(s) + CO_2 \tag{3.32}$$

which occurs in the technologically important blast furnace used to produce iron, we might expect

$$\Re_{FeO} = k_f C_{CO}^{n_{CO}} - k_b C_{CO_2}^{n_{CO_2}}$$ (3.33)

where k_f and k_b are the *forward* and *backward* rate constants respectively. Note that the four parameters k_f, k_b, n_{CO} and n_{CO2} are not all independent since it is necessary that at equilibrium

$$0 = k_f C_{CO}^{*n_{CO}} - k_b C_{CO_2}^{*n_{CO_2}}$$ (3.34)

where the superscript * indicates equilibrium concentrations given by the ratio

$$\frac{C_{CO_2}^*}{C_{CO}^*} = \text{equilibrium constant}$$ (3.35)

3.2.6 Langmuir-Hinshelwood Kinetics

Sometimes, the simple "power law" correlations for reaction rates, such as equation (3.29), are not accurate for heterogeneous reactions. For example, if the reduction of nickel oxide by hydrogen is studied over a sufficiently wide range of hydrogen partial pressures, it is found that the experimental data are better matched to an equation of the form

$$\Re_{NiO} = \frac{k_1 C_{H_2}}{k_2 + C_{H_2}}$$ (3.36)

where k_1 and k_2 are constants. The reaction is described as "obeying Langmuir-Hinshelwood kinetics". A sketch of how the rate depends on the hydrogen concentration appears in Fig. 3.7. It is seen that the reaction is approximately first order at low hydrogen concentrations ($C_{H2} << k_2$ in equation (36)) and zeroth order at high concentrations ($C_{H2} >> k_2$). The reason for this behaviour is that the reaction is really the result of two consecutive steps, the first being adsorption of hydrogen on a finite number of

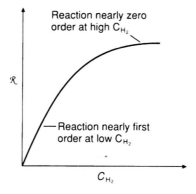

FIGURE 3.7 Dependence of rate on concentration of gaseous reactant for a reaction obeying Langmuir-Hinshelwood kinetics.

sites on the nickel oxide and the second the reaction of adsorbed hydrogen with the oxide. Competing with the first step is the desorption of the hydrogen. The net rate of adsorption R_{ads} is then given by

$$\mathfrak{R}_{ads} = k_{ads}C_{H_2}(1-\theta) - k_{desorb}\theta \tag{3.37}$$

where θ is the fraction of adsorption sites that are occupied by hydrogen. This equation assumes that the adsorption step is first order with respect to hydrogen and unoccupied sites while the desorption step is first order in the occupied sites. Assuming the reaction between the adsorbed hydrogen and nickel oxide is also first order then the rate of reaction is

$$\mathfrak{R}_{NiO} = k_{react}\theta \tag{3.38}$$

and is equal to the net rate of adsorption

$$\mathfrak{R}_{ads} = \mathfrak{R}_{NiO} \tag{3.39}$$

It is left as an exercise for the reader to show that the elimination of θ and R_{ADS} between equations (3.37)-(3.39) yields an equation of the form (3.36).

3.3 Temperature Effects

The rate "constants" appearing in rate equations typically show the following dependence on absolute temperature, T

$$k = A \exp\left(\frac{-E_A}{RT}\right) \tag{3.40}$$

which is Arrhenius's equation.

A is a constant for a particular reaction (the *pre-exponential constant*), as is E_A (the *activation energy*). Determination of the rate constant at two (or more) temperatures for a particular reaction enables the values of *A* and E_A to be determined. Frequently

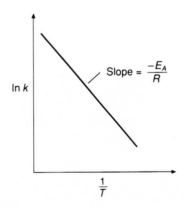

FIGURE 3.8 Arrhenius plot.

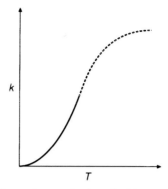

FIGURE 3.9 Plot of reaction rate constant against temperature.

this is done by plotting $\ln k$ versus reciprocal temperature, as in Fig. 3.8. The slope of the straight line on this *Arrhenius plot* is $-E_A/R$.

Fig. 3.9 presents a clearer picture of the effect of temperature on a rate constant. It is seen that at high temperatures the rate asymptotically approaches a constant value A. In fact, for reactions of practical interest this only happens at very high temperatures and in reality only the solid part of the curve in Fig. 3.9 need concern us. Along this solid part the rate increases greatly with increasing temperature as the following worked example shows.

PROBLEM

A reaction has an activation energy of 83.1 kJ/mol. By what factor does its rate constant increase on increasing the temperature from 400K to 440K? R = 8.31 J/ (mol•K).

Answer

$$\frac{k_{440}}{k_{400}} = \frac{A \exp\left[-83.1 \times 10^3 / (8.31 \times 440)\right]}{A \exp\left[-83.1 \times 10^3 / (8.31 \times 400)\right]} = \frac{1.35 \times 10^{-10}}{1.39 \times 10^{-11}} = 9.7$$

We see then that an increase in temperature of 40°C increases the rate by almost an order of magnitude. Reactions with higher activation energies than that of this worked problem are common and the effect of temperature on the rates of these reactions is even stronger.

Two things occur that effect the reaction rate when we raise the temperature. As seen above the rate constant increases greatly. The second effect is a diminution in concentrations due to thermal expansion. The latter effect is generally negligible compared to the former. Because in many cases we can write

$$\Re = kf(\text{concentrations}) \tag{3.41}$$

the result is

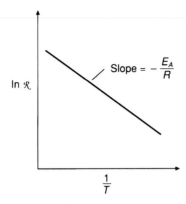

FIGURE 3.10 Arrhenius plot in which the reaction rate is used directly.

$$\Re = A \exp\left(\frac{-E_A}{RT}\right) f(\text{concentrations}) = \text{constant} \exp\left(\frac{-E_A}{RT}\right) \qquad (3.42)$$

and a plot of $\ln \Re$ versus reciprocal temperature (Fig. 3.10) is a straight line of slope -E_A/R. This is convenient because it enables us to determine the activation energy for reaction without going to the trouble to determine the rate equation. Note however that in order to employ this short-cut we must determine rates at several different temperatures but the same concentrations. Furthermore, this short-cut is only valid under circumstances where the rate of the reverse reaction is negligible (i.e., we can treat the

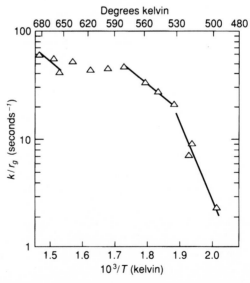

FIGURE 3.11 Arrhenius plot of experimental data on the kinetics of nickel oxide reduction by hydrogen. [Data from J. Szekely, C. I. Lin and H. Y. Sohn, *Chem Eng. Sci.* (1975) 1973.]

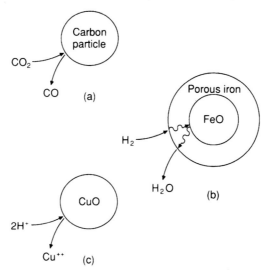

FIGURE 3.12 Mass transport in heterogeneous reactions: transport of CO_2 and CO in (a), of H_2 and H_2O in (b), and of H^+ and Cu^{++} in (c).

reaction as irreversible). Otherwise the activation energy is some (uncertain) average of those for the forward and reverse reactions.

Frequently, plots of $\ln k$ or $\ln \mathscr{R}$ versus reciprocal temperature are not simply one straight line. An example is provided in Fig. 3.11 which contains kinetic data for nickel oxide reduction. The change of slope at 530 K is attributed to a change (the disappearance of ferromagnetism) in the nature of NiO at this temperature. Such changes of slope are not uncommon in Arrhenius plots and are usually ascribed to changes in the mechanism of reaction. Another common occurrence (for heterogeneous reactions) is that at high temperatures the data tend to fall below the straight line which fits the data at lower temperatures. This is an effect of mass transport and the role of mass transport will now be considered.

3.4 Mass Transport and Heterogeneous Reactions

3.4.1 Diffusion

In heterogeneous reactions it is necessary for the reactants to meet at the interface between phases and this implies that one (or more) reactant must be transported to that interface. Some examples have already been discussed in this chapter and Fig. 3.12 illustrates the concepts.

One mechanism by which mass transport takes place is diffusion and Fig. 3.13 depicts an experiment in diffusion. A pipe with a transparent wall is separated into two compartments by a removable partition. One compartment is filled with the brownish-orange gas nitrogen dioxide while the other contains colorless argon. The gas molecules on both sides of the partition are in constant motion and, on removing the partition, NO_2 molecules start to move into the right-hand half of the pipe and argon mol-

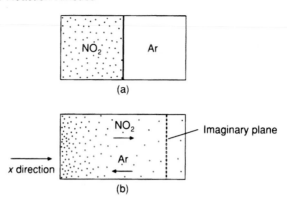

FIGURE 3.13 Diffusion illustrated by the interdiffusion of nitrogen dioxide and argon: (a) initially, (b) some time after partition is removed.

ecules into the left. This is visible as the movement of the brownish-orange coloration to the right and eventually the color is uniform throughout the container. In reality it is difficult to carry out this experiment (except perhaps in space) since the density difference between the two gases is sufficient to bring about a flow of the gases which mixes the gases by convection as well as diffusion. Nevertheless, the experiment provides a useful first description of the mechanism of diffusion. The diffusion (of NO_2 to the right and Ar to the left) we can think of as being "driven" by concentration gradients. Let us start a quantitative examination of this phenomenon by defining the *flux* of NO_2 molecules to the right. Suppose that in some way we could count the number of molecules of NO_2 passing in some brief time interval to the right through some imaginary plane perpendicular to the axis of the pipe and placed at an arbitrary position. If we divide the number by Avogadro's number, the length of the brief time interval and the cross-sectional area of the pipe we will have a number that is the "molar flux of NO_2" at some imaginary plane and for which we will use the symbol J_{NO2}. The reader will note that the flux has the dimensions of moles per unit area per unit time (i.e., the same dimensions as a heterogeneous reaction rate). As we have defined it, the flux is an intensive measure of the rate of diffusion at the point in question.

Fick's first law states that the flux of a diffusing species is proportional to the concentration gradient at the point. In mathematical symbols:

$$J_{NO_2 x} = -D_{NO_2-Ar} \frac{dC_{NO_2}}{dx} \qquad (3.43)$$

The subscripts on J indicate that it is a flux of NO_2 in the x direction that we are measuring. The minus sign appears because the flux is a positive number while the gradient dC_{NO2}/dx is negative. If the proportionality constant D_{NO2-Ar} is to be positive then the minus sign is necessary. D_{NO2-Ar} is the *diffusivity* or *diffusion coefficient* for the gas pair nitrogen dioxide-argon; it is independent of concentration gradient and (in this instance) of concentration itself, although dependent on temperature and total pressure.

We can use a similar equation to describe the diffusion of the argon:

$$J_{Ar\,x} = -D_{NO_2-Ar} \frac{dC_{Ar}}{dx} \tag{3.44}$$

In the experiment of Fig. 3.13 the concentration gradient of argon is positive which means that equation (3.44) will give us a negative flux for the argon. This is appropriate because we are treating diffusion in the x direction as positive and argon is diffusing in the opposite direction.

Diffusion occurs in liquids and solids (whenever there is a concentration gradient, or more precisely, a gradient of chemical potential); solid-state diffusion will be encountered in several later chapters. Generally diffusivities in liquids are a few orders of magnitude lower than in gases and those in solids are even lower. Furthermore the mechanisms of diffusion in solids are such that we must frequently use more than one diffusion coefficient even in a binary mixture. For example in the important iron-carbon alloys (steels) the diffusivity of carbon in the alloy is not the same as the diffusivity of the iron.

3.4.2 Convection

It is evident that a species in a mixture can be transported simply by moving the whole structure. For example, if we opened the ends of the pipe in Fig. 3.13 we could use a fan at the left end to blow the nitrogen dioxide-argon mixture along the pipe. If the fan is sufficiently powerful both gases are transported to the right. This mechanism of mass transport is called convection (or sometimes advection). Note that we can still count the molecules passing through the imaginary plane and therefore still define a flux but now it will be made up of a contribution from convection and a contribution from diffusion. The latter will still be given by the product of a diffusivity and a concentration gradient while the former is merely the product of the velocity and the concentration at the plane. Therefore, convection is dependent on velocity and concentration but independent of concentration *gradient* .

Gases and liquids are extremely mobile; quite small forces are sufficient to set them in motion. As a consequence, *within the bulk* of liquids or gases it is usual for convection to be the dominant mechanism of mass transport and diffusion can be ignored. Furthermore, convection is usually effective in bringing about mixing. The reader will recognize that the fan-forced convection that we have now introduced into the experiment of Fig. 3.13 tends to mix the two gases in the same way that smoke rising from a cigarette is dispersed in the air stream from a fan. Therefore it can be assumed frequently that the bulk of a liquid or a gas is "well mixed", i.e., at uniform composition. There are many circumstances where this assumption is not valid, e.g., the flame of a propane torch has sharp variations in composition. However, it may be reasonable to assume that the air in the room containing the torch is well mixed since combustion products are rapidly dispersed into the air by air currents within the room. Solids are not mobile and therefore diffusion dominates over convection (for stationary solids) in mass transport in solids. As a consequence the field of solid-state materials science is very much concerned with diffusion, which tends to be less important in other areas of science and engineering where fluids are the subject of interest.

The term *convection* is commonly (although imprecisely) used to describe a circumstance where both convection and diffusion play a role and which we will now treat.

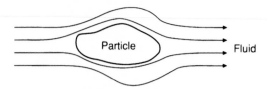

FIGURE 3.14 Solid particle within a flowing fluid.

3.4.3 Mass Transfer to Interfaces

Consider a solid that is immersed in a flowing fluid (gas or liquid), as illustrated in Fig. 3.14, and imagine that a species A in the fluid is being transported to the particle surface. For example the particle might be carbon in contact with a mixture of nitrogen and carbon dioxide. Carbon dioxide would be transported to the particle surface where it would react. We shall consider the bulk of the fluid to be well mixed and call the concentration of A within it C_{AB}, the second subscript indicating bulk conditions.

Imagine a miniature velocity probe that can be used to "map out" the velocity of the fluid in the vicinity of the particle surface. We should find that, due to the drag exerted by the particle on the fluid, the velocity near the particle surface is well below that in the bulk of the fluid, indeed it becomes zero right at the particle surface. We can therefore think of a nearly stagnant film of fluid, a "boundary layer" surrounding the particle. In the outer regions of this boundary layer, velocities are still sufficiently high that convection plays a dominant role in mass transport but, closer to the surface, diffusion becomes more important. Diffusion can only occur if there is a concentration gradient and the end result is the situation depicted in Fig. 3.15. The concentration departs from its bulk value more and more as we move down through the boundary layer, finally reaching some value adjacent to the surface that we will call C_{AS}. Suppose the flux J_A to the surface can be measured, i.e., the moles of A arriving at the surface per unit area per unit time is determined. Experimentally it is found that this flux is proportional to the concentration difference $C_{AB} - C_{AS}$. Mathematically

$$J_A = h(C_{AB} - C_{AS}) \qquad (3.45)$$

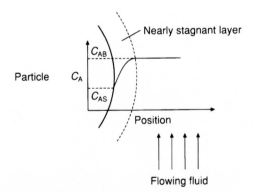

FIGURE 3.15 Concentration profile in vicinity of particle surface.

where the constant of proportionality, h, is called the *mass transfer coefficient*. Because mass transfer to the surface is made up of both convection and diffusive contributions, this coefficient will depend on the bulk fluid velocity, on parameters (fluid viscosity and density, and particle size) that affect the boundary layer thickness and on the diffusivity. It will come as no surprise that the mass transfer coefficient increases with increasing velocity of the fluid relative to the solid.

This approach, of expressing a flux as a product of a mass transfer coefficient and the concentration difference between the surface and a well mixed bulk, is equally applicable to fluid-fluid interfaces such as a gas-liquid interface. Indeed, it applies on either side of the interface, if boundary layers exist on both sides, but the mass transfer coefficients will, in general, be different for the two sides. Furthermore, it would be unusual for the surface concentrations to be the same on both sides of the interface even if chemical equilibrium exists there. For example, water in equilibrium with air does *not* contain 21 vol% dissolved oxygen.

Of course, circumstances arise where mass transfer is in the opposite direction, away from the surface. For example, in our oxidation of a carbon particle by carbon dioxide in nitrogen, the carbon monoxide produced undergoes mass transfer from the surface of the particle into the bulk of the gas. A worked example will now be used to illustrate this point further.

PROBLEM

Calculate the rate of evaporation of water from a raindrop at 27°C if the drop surface area is 10^{-5} m² and it is falling above the Mojave desert. The saturation vapor pressure of water at 27°C is 5.066 kPa and the velocity of the raindrop is such that the mass transfer coefficient is 0.02 m/s. The pressure of the atmosphere is 101.33 kPa.

Answer

We shall assume that the air above this desert is sufficiently dry that it contains negligible amounts of water vapor, i.e., C_{H2O-B} = zero. Furthermore, we shall assume that thermodynamic equilibrium exists locally at the droplet-air interface. This latter assumption is common in problems involving phase changes (evaporation, condensation, melting, etc.). We can then immediately calculate the water vapor mole fraction in the air next to the droplet surface

$$x_{H_2O\text{-}S} = \frac{5.066\text{kPa}}{101.33\text{kPa}} = 0.05$$

because the mole fraction for a gas constituent is the partial pressure (which by our assumption equals the saturation vapor pressure) over the total pressure. At 27°C and 101.33 kP there are 40.7 moles of an ideal gas mixture per cubic meter, from the ideal gas law. Hence

$$C_{H2O\text{-}S} = 0.05 \times 40.7 = 2.035 \text{ mol/m}^3$$

Therefore the flux of water vapor away from the droplet is

$$-J_{H_2O} = h(C_{H_2O-S} - C_{H_2O-B}) = 0.02\,\text{m/s} \ \times 2.035\,\text{mol/m}^3 = 0.0407\,\text{mol/m}^2/\text{s}$$

Multiplying by the droplet surface area gives the rate of evaporation:

$$40.7 \times 10^{-8}\ \text{mol water/s} \equiv 0.732 \times 10^{-5}\ \text{g/s}$$

It should be noted that this rate of evaporation is the instantaneous one pertaining to the moment when the conditions of the problem hold. As the drop continues to fall its evaporation rate can be expected to change; the reader is left to ponder why this change might occur.

3.4.4 Mass Transfer Accompanied by Chemical Reaction

Let us now consider the case where A, our chemical species in a fluid, undergoes reaction upon arrival at the surface of the particle immersed in the fluid.

$$A(g) + B(s) = \text{gaseous products} \tag{3.46}$$

We shall treat the reaction as irreversible. As in section 3.4.3, the flux of A to the surface is

$$J_A = h(C_{AB} - C_{AS}) \tag{3.47}$$

We expect the rate of the heterogeneous reaction at the surface to be given by

$$\Re_A = kC_{AS}^{n_A} \tag{3.48}$$

Note that it is the concentration in the gas next to the solid surface, not the bulk concentration, that appears in equation (3.48). An anthropomorphic explanation of this is that the reacting atoms and molecules can only "know" the concentration in their immediate vicinity, not the concentration a few millimeters away in the bulk of the gas. For mathematical convenience we shall now assume that the reaction is first order with respect to A (i.e., $n_A = 1$). A few moments' thought reveals that

$$J_A = \Re_A \tag{3.49}$$

because the net rate at which A arrives at the surface must equal its rate of consumption by reaction.[1] The alternative would be an accumulation (or depletion) at the surface and it is not possible to store significant amounts of a gas at a surface, in the present context.

Substituting (3.47) and (3.48) in (3.49) we can obtain an equation for the surface concentration:

$$C_{AS} = \frac{h}{h+k}C_{AB} \tag{3.50}$$

and substituting this in (3.48) gives us

---

[1] Note that this is *not* equivalent to assuming that every molecule reaching the surface reacts instantly. That assumption would be equivalent to putting C_{AS} = zero, which we do not do.

$$\mathfrak{R}_A = \frac{kh}{h+k}\left(C_{AB}\right) \text{ bulk} \tag{3.51}$$

Equation (3.51) has many virtues. One obvious one is that it expresses the rate in terms of the bulk concentration, C_{AB}, rather than the concentration next to the surface, C_{As}, which is probably impossible to measure. Another virtue is that it makes clear how chemical kinetics and mass transfer interact to set the pace of a heterogeneous reaction. To see this let us regard the concentration of A in the bulk of gas as the driving force for reaction. We can then rearrange (3.51) to

$$\mathfrak{R}_A = \frac{C_{AB}}{1/k + 1/h} \quad \text{total resistance} \tag{3.52}$$

and identify the quantity $(1/k + 1/h)$ as the *total resistance* to the progress of reaction that is the sum of two resistances in series

$$1/k \equiv \text{the chemical step resistance} \tag{3.53}$$

$$1/h \equiv \text{the mass transfer resistance} \tag{3.54}$$

Equation (3.52) is then of the form

$$k \gg h$$
$$C_{AS} = 0$$
$$R = hC_{AB}$$

$$\text{rate} = \frac{\text{driving force}}{\text{total resistance}} \tag{3.55}$$

that is encountered very often in science and engineering. The example of Ohm's law (rate = current, driving force = potential difference, total resistance = sum of electrical resistances in series) springs readily to mind.

3.4.5 Rate-controlling Step

Following the concepts of driving force and resistances described at the end of section 3.4.4, let us consider two extreme cases:

1. The case $k \gg h$ i.e., the chemical step resistance $\ll$ mass transfer resistance. From (3.50)

$$C_{AS} \cong \text{zero} \tag{3.56}$$

and the concentration profile across the boundary layer becomes as sketched in Fig. 3.16. From (3.51),

$$\mathfrak{R}_A \cong hC_{AB} \tag{3.57}$$

and we see that under this extreme condition the chemical step has no influence on the progress of reaction! In other words, changes in k (provided $k \gg h$ is maintained) have no effect on the rate. Changes in h *do* have an effect and we say that the rate is "controlled" by mass transfer, or that mass transfer is the *rate-controlling step*. From (3.53) and (3.54), total resistance $\simeq$ mass transfer resistance.

2. The case $k \ll h$, that is, chemical step resistance $\gg$ mass transfer resistance. From (3.50),

$$C_{AS} \cong C_{AB} \tag{3.58}$$

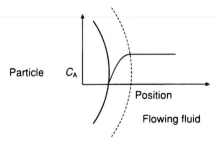

FIGURE 3.16 Concentration profile near particle surface for mass transfer control.

i.e., there are negligible changes in composition across the boundary layer (see Fig. 3.17). From (3.51),

$$\mathfrak{R}_A = kC_{AB} \tag{3.59}$$

and the mass transfer step has no effect on reaction rate (provided $k \ll h$ continues to hold). We say reaction is "controlled by the chemical kinetics" or that the chemical step is "rate-controlling". Note that the total resistance $\simeq$ chemical step resistance.

Between these two extremes the full equations (3.50) and (3.51) rather than their extreme forms (3.56) and (3.57) or (3.58) and (3.59) must be used. Both k and h influence the reaction rate, we describe reaction as proceeding under "mixed control", and both resistances are significant.

The rate-controlling step is sometimes referred to as the *rate-limiting step* or *rate-determining step*, which more clearly establishes the meaning of this concept which is important in science and engineering. The concept extends far beyond reaction kinetics; it is frequently possible to identify a rate-determining resistance in an electrical circuit, a rate-determining material in the conduction of heat through layers of different materials, a rate-determining orifice in the flow of liquids through pipe systems, or a rate-determining obstruction in the movement of vehicles along a road.

Unfortunately, in many texts, very poor semantics have been used in describing this concept. It is not uncommon to encounter words such as the "rate-controlling step is the slow step in the sequence of steps". This is at best confusing to the student and at

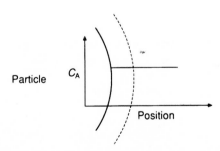

FIGURE 3.17 Concentration profile near particle surface for chemical step control.

the worst it is nonsensical. In the context of this section it would suggest that the rate of mass transfer is lower than that of reaction (or vice versa). The inaccuracy of this kind of statement is immediately obvious from equation (3.49); the two rates are equal, irrespective of which step is truly rate-controlling! Another example will serve to drive this point home. Consider an urban four-lane highway along a few miles of which road repairs have reduced the lanes to two. Suppose that sufficient time has elapsed that steady state has been reached (i.e., that whatever traffic jams are to happen have already built up), whereupon the number of vehicles per minute passing any point on the highway is everywhere the same. There is therefore no "slow" region of the highway in terms of traffic volume per minute. Nevertheless any reasonable person would identify the section undergoing repairs as being rate-determining in the sense of being the major cause of commuters being late for work. Amazingly, it is usually found that drivers accelerate on entering this section and we see that, as registered by speedometers, the rate-determining section is actually a "fast" one! The reader will be able to conceive of other examples where the rate-controlling step is poorly (or never) identified by the adjective "slow" and confusion is best avoided by stating that the rate-controlling step is the one that poses the major resistance to reaction (or current flow, or heat conduction, etc.), i.e., it supports the largest drop in the "driving force" (concentration difference, voltage, etc.).

PROBLEM

A carbon coating 20 μm thick is to be burned off a 2-mm-diameter sphere by air at atmospheric pressure and 1000 K. Calculate the time to do this, assuming that the product of reaction is carbon dioxide, that the mass transfer of oxygen from air to the carbon surface is the rate-controlling step, and that the mass transfer coefficient is 0.25 m/s. Density of carbon = 2250 kg/m^3. Air is 21% oxygen.

Solution

Because the coating thickness is only 1% of the sphere diameter (D), we will ignore any change in area as the carbon is burned off. Concentration of oxygen in air at 1 atm and 1000 K is 21% of P/RT. With appropriate values substituted this becomes 2.56 mol/m^3. Hence mass flux under mass transfer control (zero oxygen at carbon surface) is

$$2.56 \text{ mol/m}^3 \times 0.25 \text{ m/s} = 0.64 \text{ mol/m}^2 \cdot \text{s}$$

Surface area of sphere = $\pi \times D^2 = \pi \times 4 \times 10^{-6}$ m^2
Therefore,

$$\text{rate of transfer of oxygen to surface} = 0.64 \times \pi \times 4 \times 10^{-6}$$

$$= 8.04 \times 10^{-6} \text{ mol/s}$$

If carbon dioxide is the reaction product, 1 mol of carbon (i.e., 12 g) are consumed per mole of oxygen transferred to surface. Hence

$$\text{rate of consumption of carbon} = 12 \times 8.04 \times 10^{-6} = 9.64 \times 10^{-5} \text{ g/s}$$

Multiplying the sphere surface area by the coating thickness and then by the density of the carbon gives

$$\text{mass of carbon to be consumed} = \pi \times 4 \times 10^{-6} \times 20 \times 10^{-6} \text{ m}^3 \times 2250 \text{ kg/m}^3$$

$$= 5.65 \times 10^{-4} \text{ g}$$

Hence, time for removal of coating $= 5.65 \times 10^{-4}$ g/9.64 x 10^{-5} g/s $= 5.9$ s

3.4.6 Effect of Temperature on Reaction Involving Mass Transfer

Fig. 3.18 shows schematically the effect of temperature on rate constants and mass transfer coefficients. Although mass transfer coefficients increase with increasing temperature, they do so much more slowly than most rate constants increase. Also sketched as the solid line in this figure is the quantity $kh/(k + h)$ which, referring to equation (3.51), is the factor by which we multiply the bulk concentration to get the reaction rate. At lower temperatures where k is small this factor approaches k and we have chemical step control. At high temperatures the factor approaches h and we have mass transfer control. In between we have mixed control. There are important practical implications. At lower temperatures there would be no point in trying to increase the productivity of a reactor employing this reaction by increasing h (as can be achieved by flowing the fluid more rapidly past the solid, for example) but great increases in productivity might be achieved by moderate increases in temperature. The opposite would be true at high temperature. There are also important implications for the chemical kineticist. It would be difficult to measure the rate constant k at elevated temperatures since the rate is largely (or nearly entirely) dependent on h at these temperatures. Special precautions might have to be taken to increase h to the point where it becomes larger than k. Conversely, measuring h at low temperatures by measuring a chemical reaction rate will be fruitless unless the reaction has an unusually high k. Finally, Fig. 3.19 is an Arrhenius plot of the solid line of Fig. 3.18; the departure of the high temperature part of the curve from a straight line is evident.

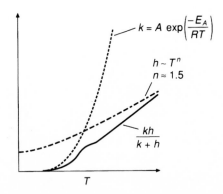

FIGURE 3.18 Effect of temperature on parameters determining reaction rate.

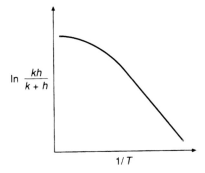

FIGURE 3.19 Arrhenius plot for a reaction where mass transfer is rate limiting at high temperatures.

3.5 The Case of the Disappearing Solid

We have been considering the reaction of a gas with a solid, according to equation (3.46), without considering what happens to the solid. What happens, of course, is that it shrinks away as it is consumed by reaction at its external surface. Let us develop a quantitative description of this disappearance by considering a spherical particle of B with initial radius R_o. We will assume that reaction takes place uniformly over the external surface of the solid so that it retains its spherical shape and we will call the radius at time t after reaction begins R. The volume of B is $4/3\pi R^3$ and the number of moles of B in the sphere is therefore $4/3\pi R^3 (\rho/M_B)$ where ρ is the density of B and M_B its molecular weight. Differentiating, we get the rate of increase of moles of B = $(4\pi\rho R^2/M_B)(dR/dt)$, which is actually a negative number since R is decreasing with time. The number of moles of B reacting per unit time is therefore $-(4\pi\rho R^2/M_B)(dR/dt)$, or dividing by the area of the reaction interface $(4\pi R^2)$ we get the rate of reaction in moles of B per unit area per unit time:

$$\Re_B = -\frac{\rho}{M_B}\frac{dR}{dt} = \Re_A \tag{3.60}$$

the last part of the equation arising from the stoichiometry of reaction (3.46). Now substituting equation (3.51) in (3.60) and rearranging

$$\frac{dR}{dt} = -\frac{M_B}{\rho}\frac{hk}{h+k}C_{AB} \tag{3.61}$$

where it will be recalled C_{AB} is the bulk concentration of A.

Let us regard the right-hand side of (3.61) as constant (which would be the case under chemical step control, i.e., when $h \gg k$). Then (3.61) can readily be integrated to give

$$R = R_0 - ct \tag{3.62}$$

where c is the right-hand side of (3.61). Let us look at the fraction of the weight initially

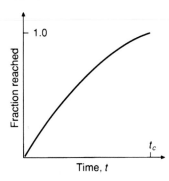

FIGURE 3.20 Plot of fraction of the solid that has reacted, versus time, for the case of a disappearing solid.

present that has been reacted away; that is,

$$\text{fraction reacted} = \frac{\text{initial sphere mass - present sphere mass}}{\text{initial sphere mass}}$$

$$= \frac{R_0^3 - R^3}{R_0^3} = 1 - \frac{(R_0 - ct)^3}{R_0^3} \tag{3.63}$$

A plot of the fraction against time appears in Fig. 3.20. The time for complete reaction, t_c, can readily be obtained from (3.62) by putting $R=0$:

$$t_c = \frac{R_0}{c}$$

We see from Fig. 3.20 that more than half of the solid has reacted away at $t = t_c/2$.

Now let us examine the case of mass transfer control. The mass transfer coefficient is inversely proportional to the particle size raised to a power between one half and one (see the problem on mass transfer coefficients at the end of this chapter). The time for complete reaction under these circumstances is approximately proportional to the particle size raised to a power between 1.5 and 2. These results are summarized in the upper half of Table 3.1 where the effect of temperature, discussed in section 3.4.6, and velocity (discernible through the effect of velocity on mass transfer coefficient) also are summarized.

3.6 The Case of the Shrinking Core

Finally we consider a more complicated case such as has already been illustrated in Fig. 3.2. This is the case where one of the products of reaction is a (porous) solid:

$$A(g) + B(s) = C(g) + D(s) \tag{3.64}$$

Also included in this case are examples where the gaseous product C is absent. We presume that B, unlike D, is not porous.

TABLE 3.1 Dependence of Fluid-Solid Reaction Kinetics on Particle Size, Temperature, and Relative Velocity

	Dependence of time for complete reaction, t_c, on:		
	Particle radius, R_0	Temperature T	Relative velocity of solid and fluid V
Disappearing solid			
Chemical step control	$t_c \sim R_o$	t_c decreases rapidly with increasing T	t_c independent of V
External mass transfer control	$t_c \sim R_o^n$ (approx., with $1.5 < n < 2$)	t_c decreases with increasing T	t_c decreases with increasing V
Shrinking core			
Chemical step control	$t_c \sim R_o$	t_c decreases rapidly with increasing T	t_c independent of V
Pore diffusion control	$t_c \sim R_o^2$	t_c decreases with increasing T	t_c independent of V

Figure 3.21 gives a more detailed picture of how this class of reaction proceeds. It is seen that our particle of solid (which here retains its external dimensions) contains a *shrinking core* or *kernel* of unreacted solid. Three steps must be considered (if the reaction is irreversible):

1. Mass transfer of A across the boundary layer external to the particle
2. Diffusion of A through the gas permeating the pores of the solid product D surrounding the core
3. The chemical step(s) at the interface between D and B.

If the reaction is reversible then two additional steps must be included:

4. diffusion of C(g) through a porous reaction layer, in a direction opposite to the flow of A
5. mass transfer of C(g) across the boundary layer, again in a direction opposite to the flow of A

The mathematics describing this situation is beyond the scope of this text [see the book by Szekely et al. (1976)], but it is worthwhile examining the results for each of the above steps being rate-controlling.

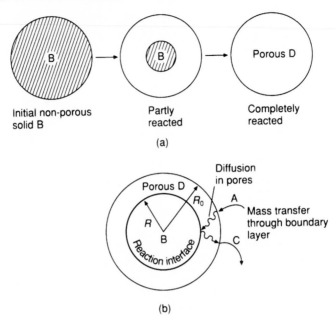

(a)

(b)

FIGURE 3.21 Reaction where a shrinking core of unreacted solid occurs:
(a) appearance of sectioned particles (b) diffusion of gaseous reactants and products.

Fig. 3.22 depicts the concentration profile when the chemical step is rate-controlling. The concentration at the reaction interface is C_{AB} and the rate of reaction is

$$\Re_A = kC_{AB} \tag{3.65}$$

(First order reaction is again assumed.) The radius of the shrinking core is given by

$$R = R_0 - \frac{M_B kC_{AB}}{\rho} t \tag{3.66}$$

and a plot of fraction reacted versus time appears in Fig. 3.20.

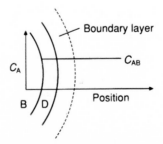

FIGURE 3.22 Case of the shrinking core; concentration profile for gaseous reactant under chemical step control.

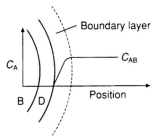

FIGURE 3.23 Shrinking core; external mass transfer control.

For external mass control the concentration profile appears as shown in Fig. 3.23 and the rate of reaction becomes

$$\Re_A = hC_{AB} \frac{R_0^2}{R^2} \tag{3.67}$$

As an exercise the reader may wish to show that this leads to a plot of fraction reacted versus time that is a straight line. In fact, external mass transfer control would normally be encountered only in the initial stages of reaction because diffusion through a thickening layer of porous solid product soon poses a greater resistance than diffusion through the boundary layer.

For pore diffusion control the concentration profile is shown in Fig. 3.24. The rate of reaction is now a complicated fraction of R, R_o and the effective diffusivity of the gas A in the pores of D (but not, of course, h or k). The lower half of Table 3.1 shows the effect of particle size, temperature and velocity on the time for complete reaction.

For the case of the shrinking core we expect chemical step control at lower temperatures, pore diffusion control at higher temperatures (particularly for large particles where a thick layer of solid product presents formidable barrier to reaction, or for low-porosity product) and external mass transfer control only for the initial stages of reaction.

Progress of reactions of this type is very much dependent on the porosity of the solid product. In the extreme, D may be non-porous. A classical example is the oxidation of aluminum

$$4Al + 3O_2 = 2Al_2O_3 \tag{3.68}$$

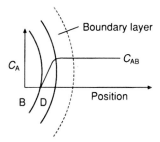

FIGURE 3.24 Shrinking core; porous diffusion control.

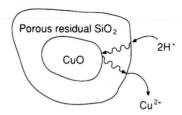

FIGURE 3.25 Shrinking core behavior in the leaching of a copper oxide ore with silica gangue.

This has a large negative free energy of reaction at all practical temperatures, i.e., thermodynamics predicts that aluminum is unstable in air. In fact, aluminum is a useful material because any Al_2O_3 formed by reaction (3.68) is non-porous and the aluminum is isolated from the air by a thin film of oxide which can be crossed only by (comparatively slow) solid state diffusion

3.7 Concluding Remarks

Materials are solids, or at least (in some instances) highly viscous liquids. It appears almost inevitable then that if a material is produced by chemical reactions at least one of those reactions will be heterogeneous. Indeed the authors can think of many materials which are formed by reactions that are exclusively heterogeneous. Because of this, and because the traditional emphasis in undergraduate chemistry texts is on homogeneous reactions, this chapter has been mainly concerned with heterogeneous reactions. Much of that treatment has focused on gas-solid reactions but the concepts advanced are applicable to other important heterogeneous reactions.

An example of a liquid-solid reaction wherein a shrinking core is encountered is the leaching of copper oxide ores by acids to extract copper into solution, as illustrated in Fig. 3.25. These ores typically consist of a copper oxide dispersed in a silica gangue that is insoluble in the acid. A porous layer of silica is formed enveloping a shrinking core of unleached ore.

An example of a gas-liquid reaction is found in steelmaking, the immensely important process of refining impure carbon-rich liquid iron to a useful material by oxidizing the carbon and other impurities. The carbon is actually removed from the melt by its mass transfer to a rising CO bubble as shown in Fig. 3.26. At the surface of the bubble

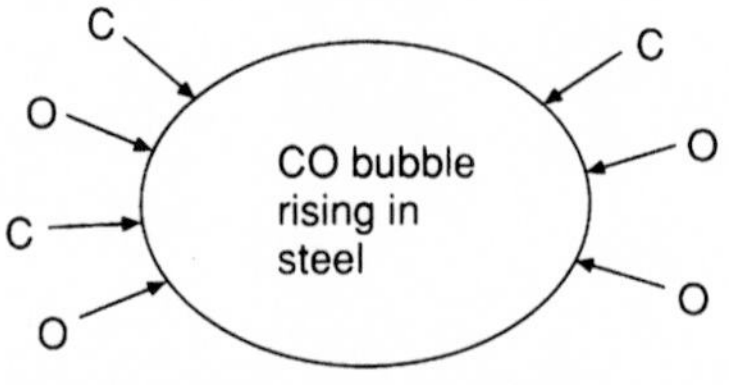

FIGURE 3.26 Mass transport of carbon and oxygen to a CO bubble in steelmaking.

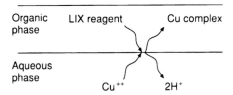

FIGURE 3.27 Mass transport to the liquid-liquid interface in a solvent extraction.

carbon combines with oxygen in the melt to form carbon monoxide that enters the bubble.

Finally, liquid-liquid reactions play an important role in materials production. As an example, in one method of producing copper an aqueous solution of a copper salt is purified and concentrated by *solvent extraction*. The first step in this is to transfer the copper ions from the aqueous phase to an interface with an organic phase. At that interface they react with an organic reagent (dissolved within the organic phase) that is specific to copper (i.e., will not react with impurities). This is shown in Fig. 3.27 for a particular commercial reagent (LIX). The copper-organic complex then enters the organic phase, the two phases are separated physically and the copper recovered from the organic phase by treating with an aqueous solution that reverses the reaction of the first step. Clearly, multiple mass transfer steps are involved in this example.

We conclude by emphasizing the practical value of the material presented in this chapter. The productivity of reactors where heterogeneous reaction is proceeding under mass transfer control can be increased by increasing the relative velocities of the two phases (e.g., by enhanced agitation). The productivity of reactors where the chemical step is rate-controlling can usually be increased markedly by a moderate increase in temperature. In both cases (or in the case of mixed control) the productivity can be increased by increasing the interfacial area available for reaction (e.g., using smaller particles or finer droplets).

LIST OF SYMBOLS

C	concentration
E_A	activation energy for reaction
h	mass transfer coefficient
J_i	flux of species i
k	chemical rate constant
M_i	molecular weight of species i
N_i	number of moles of species i
n, n_i	reaction order, with respect to species i
P	pressure
$\mathfrak{R}_i$	rate of chemical reaction expressed in terms of the disappearance of species i
R	gas constant (radius of sphere in Sections 3.5 and 3.6)
t	time
V	volume
x_i	mole fraction of species i

ρ density

Θ fraction of adsorption sites that are occupied

Subscripts

ads   for adsorbed species

B in the bulk of the gas or liquid

b backward

f forward

s at the surface

0 initially

PROBLEMS

3.1   What are the dimensions of the rate constant for
(a) a first order heterogeneous reaction
(b) a second order homogeneous reaction
(c) a zeroth order heterogeneous reaction?
A reaction has an activation energy of 150 kJ/mol. If the temperature is increased from 40°C to 50°C, by what factor does the reaction rate increase if everything else is held constant?

3.2   A sphere of nickel oxide is being reduced by hydrogen. Its diameter is 2 cm and initially it is pure, non-porous NiO. It is suspended in hydrogen that is flowing sufficiently rapidly past the sphere that we can assume that the gas in the reactor remains pure hydrogen. The temperature is 250°C and the pressure is atmospheric Calculate the time that it would take to completely reduce the sphere if external mass transfer were rate-controlling, with a mass transfer coefficient of 100 mm/s. If you carried out this experiment and the time to complete reduction was actually fifteen minutes, what would you conclude? Density of NiO = 6700 kg/m³, molecular weight = 74.7 g/mol.

3.3   A study is made of the kinetics of the dissolution of zinc in dilute hydrochloric acid. 10 g. of spherical zinc particles of diameter 1 mm are poured into an agitated reactor containing 10^{-2} m³ of hydrochloric acid of varying initial concentrations. Small samples of the solution are withdrawn at sixty-second intervals and analyzed for zinc; the results are as follows:

Experiment Number:	1	2	3
Initial HCl conc.:	0.05M	0.1M	0.2M
Time(s)		Zinc in solution (wt % Zn)	
60	0.7×10^{-4}	1.4×10^{-4}	2.8×10^{-4}
120	1.5×10^{-4}	2.9×10^{-4}	5.8×10^{-4}
180	2.2×10^{-4}	4.5×10^{-4}	9.1×10^{-4}
240	3.0×10^{-4}	6.1×10^{-4}	12.1×10^{-4}
300	3.7×10^{-4}	7.5×10^{-4}	15.0×10^{-4}

Show that after 300s the surface area of the zinc particles had decreased by a negligible amount and that negligible amounts of acid have been consumed. Determine the value of k and n in the heterogeneous rate equation:

$$\Re_{Zn} = k(HCl\ conc.)^n$$

At the concentrations under discussion, aqueous solutions can be assumed to have a density of 10^3 kg/m³. Density of zinc $= 7.14 \times 10^3$ kg/m³. Atomic weight of Zn $= 65.37$.

3.4 The following is a correlation for the mass transfer coefficient between a flowing fluid (velocity V) and a sphere (diameter d) immersed in it

$$\frac{hd}{D} = 2 + 0.6\,Re^{1/2}\,Sc^{1/3}$$

where

$$Re = \text{Reynolds number} \equiv \frac{dV\rho}{\mu}$$

$$Sc = \text{Schmidt number} \equiv \frac{\mu}{\rho D}$$

D = diffusivity in the fluid
ρ = fluid density
μ = fluid viscosity

Calculate the mass transfer coefficient between a solid sphere of diameter 10 mm. and an aqueous solution ($\rho = 10^3$ kg/m³, $\mu = 10^{-3}$ kg/(m•s) and $D = 10^{-9}$ m²/s) if $V = 0.1$ m/s.

3.5 A species A reacts with B in an irreversible homogeneous reaction:

$$A + 2B = AB_2$$

An experiment is carried out in which A and B are contacted in a closed, well-mixed reactor at fixed temperature. A probe inserted into the reactor measures C_A, the concentration of A as a function of time. The results are as follows:

Time (s)	C_A (mol/m³)
0	1.0
7	0.9
17	0.8
31	0.7
50	0.6
81	0.5
139	0.4
301	0.3

The initial concentration of B is 1.5 mol/m³. It is hypothesized that the reaction is first order in both A and B. Show that if this hypothesis is true then a plot of $\ln(0.75C_A/(C_A-0.25))$ against time should yield a straight line. Make such a plot of the data above and determine the reaction rate constant.

3.6 Electric power plants and metallurgical plants employing sulfide ores typically cause considerable atmospheric pollution by SO_2. One method for reducing sulfur dioxide emissions is to employ limestone (calcium carbonate). This would decompose to lime on exposure to hot SO_2-containing gases, according to

$$CaCO_2 \rightarrow CO_2 + CaO$$

and the sulfur dioxide would be trapped by

$$SO_2 + CaO + 1/2\ O_2 \rightarrow CaSO_4$$

This method of removing sulfur dioxide has been less successful than initially hoped for. In laboratory and pilot plant experiments, complete conversion of the solid to calcium sulfate seldom happens. The initial reaction rate on contacting sulfur dioxide, oxygen and lime is high, but the rate drops to zero after reaction has proceeded to some extent. Explain this drop in reaction in terms of what you know about gas-solid reactions. Hint:

	$CaCO_3$	CaO	$CaSO_4$
Densities (kg/m³)	2930	3300	2610
Molecular weights (g/mol)	88	56	136

3.7 For the data of problem 4, what is the time required to completely react the solid sphere if it consists of 4×10^4 mol/m³ of solid reactant and the solution is 0.1 molar in a species one mole of which reacts with two moles of the solid.

3.8 In each of the following cases list the reactions taking place (ignoring any chemical details that are unfamiliar to you) and state whether they are homogeneous or heterogeneous. Be as imaginative as possible, e.g., in (a) you may wish to include the respiratory and digestive system of the driver!
(a) an automobile passing along a highway
(b) a barbeque on which a steak is being broiled
(c) a percolator wherein coffee is being brewed.
What role do heat and mass transport play in the reactions you have listed? What methods have nature or mankind devised to enhance (or control) these transport phenomena in these cases?

3.9 In a laboratory experiment, measurements are made of the rate of reaction between a solid and a solution. Although the reaction is heterogeneous the results are reported as moles per unit time per unit volume of reactor. What information is needed to re-express this data in the form more appropriate for heterogeneous reactions?

3.10 The chemical species A undergoes two simultaneous irreversible reactions in a reactor:

(I) A (gas) + B (solid) → desired products

(II) A (gas) + C (gas) → undesired products

Both reactions are first order with respect to A; reaction II is homogeneous with C present in large excess. The temperature control on the reactor is sophisticated and the reactor temperature can be made to oscillate about some mean value. However, for reasons that are concerned with operational problems and reactor materials problems, the mean temperature cannot be changed from the value specified when the reactor was built. Assuming that the rate of reaction (I) depends on temperature as described in section 3.4.6, while reaction (II) obeys Arrhenius' equation, under what circumstances would such oscillation be beneficial in increasing the fraction of A converted to desired products? Be as quantitative as possible in answering this question.

3.11 A 5mm diameter sphere has a porosity (volume fraction of pores) of 0.2. Initially the pores are filled with water (which undergoes no chemical interaction with the solid). When held in dry air at atmospheric pressure flowing at 100 mm/s at 25°C the sphere takes five hours to dry completely. Using the correlation given in problem 4 and the necessary physical property data from handbooks, show that the drying does not take place under control by external mass transfer and that doubling the air flow rate would have negligible effect. What are other phenomena that could be controlling the drying rate?

REFERENCES AND FURTHER READINGS

BELTON, G.R. and W. L. WORRELL, (eds.), *Heterogeneous Kinetics at Elevated Temperatures*, Plenum Press, New York, 1970.

BOUDART, M., *Kinetics of Chemical Processes*, Prentice Hall, Englewood Cliffs, NJ, 1968.

O. LEVENSPIEL, *Chemical Reaction Engineering,* 3rd ed., Wiley, New York, 1998.

SHERWOOD, T. K., R. L. PIGFORD, and C. W. WILKE, *Mass Transfer*, McGraw-Hill, NY, 1975.

SZEKELY, J., J. W. EVANS, and H. Y. SOHN, Gas-*Solid Reactions*, Academic Press, NY, 1976.

POIRIER, D. R. and G. H. GEIGER, *Transport Phenomena in Materials Processing,* TMS, Warrendale, PA., 1994.

CHAPTER 4

Powders and Particles

4.1 Introduction

Nearly all practical ceramics and some metal products are fabricated from powders. Typically, these powders are mixed with a few percent of various organic and inorganic additives, compacted to form a porous "green" body[1], and heated (or "fired") to a high temperature. During the heating the porosity in the sample is reduced or eliminated. This phenomenon is called sintering. Care must be taken to distinguish this sintering *phenomenon* from the sintering *process* described in Chapter 1. The properties of the sintered ceramic body depend to a large extent on the chemical and microstructural imperfections that were not eliminated or that were produced as a result of the sintering. Many imperfections can be traced back to the structure of the green compact or to the nature of the starting powder; therefore, in this chapter we examine powders. Powders can have a complex structure. Sometimes in describing powders confusion may arise through the inconsistent use of terminology. The terminology proposed by Onoda and Hench (1978) has been adopted here.

4.1.1 Types of Particles

Primary Particles
Primary particles are low-porosity units that may either be single crystals, polycrystals, or glass. The pores are usually isolated. Attempts to break up such particles by, for example, ultrasonic agitation in suspension are unsuccessful because they are strong, whole particles rather than the more weakly bonded agglomerates. One may regard a primary particle as the smallest unit of powder with clearly identifiable surfaces.

Agglomerates
Agglomerates are clusters of primary particles that are internally bonded by surface forces or by liquid or solid bridges. Agglomerates that can easily be broken up (e.g., by ultrasonic agitation when in suspension) are called *soft agglomerates.* Those in which

---

[1] Some clay bodies are green in color when they are just shaped and dried. Upon heating in a kiln the color changes depending the presence of impurities such as iron or copper. The term "green body" has been extended to refer to all shaped and dried ceramic bodies before they are heated to develop the final ceramic properties.

the primary particles are bonded by solid bridges, such as may result from partial sintering, are difficult to break up and are called *hard agglomerates*. Often, it is the hard agglomerates that are responsible for defects in the ceramic after sintering. The pores in agglomerates are generally connected.

Particles
When no distinction between primary particles and agglomerates is made, the term *particles* is used. Particle size analysis would refer to such a mixture of primary particles and agglomerates.

Granules
Granules are intentionally formed agglomerates, usually between 0.1 and 1 mm in size. They can be formed by adding a granulating agent, usually a polymer, to the powder in small amounts, after which the powder is tumbled. Large, nearly spherical granules, resembling microscopic snowballs, are then formed which are useful for filling complex molds. This is similar to the technique of producing pellets or balls of concentrates described in Chapter 1, although granules would be much smaller.

Flocs
Flocs are often voluminous particle assemblies that form in liquid suspensions, and can immobilize a significant volume of liquid. They are usually held together by electrostatic forces. The can have extended structures that can be disrupted easily, e.g. by simple stirring, but then can spontaneously re-form. Changing the surface chemistry of the suspended particles, *e.g.* by modifying the pH of the suspension, may lead to spontaneous dispersion or to full sedimentation. If a sedimented floc is made to re-disperse one speaks of peptization.

Colloids
Colloids are very fine particles, often with a diameter of less than 10 nm, held in suspension for some time in a liquid by Brownian motion. If held for some time, colloidal particles may settle, form low concentration/high concentration suspension layers, or form flocs.

Aggregates
Aggregates are the coarse constituents in a mixture, such as the pebbles in concrete. The fine cement particles in this mixture are called the bond.
Some examples of particles are shown in Fig. 4.1.

4.1.2 Particle Size Distributions

It is frequently necessary to characterize the particle size distribution of a powder. It should first be recognized that the size of an irregular particle is an uncertain quantity. A simple definition would be the diameter of a spherical particle with the same volume as the particle. Of course, this begs the question because in many cases the volume of the particle is ill defined. For example, in the agglomerate depicted in Fig. 4.2, we would probably choose to include the volume at *A* in the particle volume and to ex-

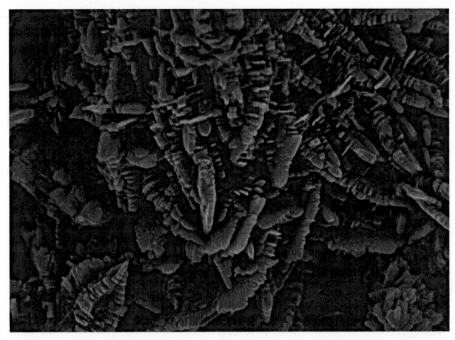

FIGURE 4.1a Scanning electron micrograph of a CuO powder compact.

FIGURE 4.1b $YBa_2Cu_3O_7$ powder mixture.

FIGURE 4.1c Transmission electron micrograph of Si_3N_4 powder.

clude the volume at B. But what about the volume at C? This problem of defining the particle size is more of a difficulty for flocs than for agglomerates and less of a difficulty for primary particles. Nevertheless, the uncertainty should be kept in mind. In fact, particle size is usually defined in a very *ad hoc* fashion in terms of the number generated by some size measurement devices such as those discussed in Section 4.2. Consequently, the same particle may yield a different value for its size when measured by different techniques. It will become clear after reading Section 4.2 that, for example, a needle-shaped particle will have a different size as measured by sieving (which tends to be governed by the largest dimension) than when measured by a Coulter counter (which tends to be more sensitive to the particle cross section), even when the two methods are working correctly.

Having arrived at a (somewhat vague, except for perfect spheres) definition of particle size, we can think of using it to describe the distribution of particle sizes in a batch

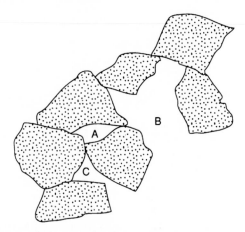

FIGURE 4.2 Floc consisting of seven primary particles.

of powder. If the total of n particles have sizes D_1, D_2, ..., D_i, ..., D_n we can calculate a mean diameter, $\bar{D}$:

$$\bar{D} = \frac{1}{n}\sum_{i=1}^{n} D_i \qquad (4.1)$$

and a standard deviation, σ:

$$\sigma = \left[\frac{1}{n}\sum_{i=1}^{n}\left(D_i - \bar{D}\right)^2\right]^{1/2} \qquad (4.2)$$

Most likely our size measurement device (such as a series of sieves) would sort the particles into a small number, k, (much smaller than n) of size categories, so that there would be N_1 particles of the size category centered about D_1, N_2 of size D_2, ... N_N of size D_N. Now the mean and standard deviation are

$$\bar{D} = \sum_{i=1}^{N} N_i D_i \div \sum_{i=1}^{N} N_i \qquad (4.3)$$

and

$$\bar{\sigma} = \left[\sum_{i=1}^{N} N_i\left(D_i - \bar{D}\right)^2 \div \sum_{i=1}^{N} N_i\right]^{1/2} \qquad (4.4)$$

Similar expressions can be used if the fractions are recorded as weights, W_i, rather than as the number count, N_i, for the diameter (range) D_i. Then, if the weight of the sum of all k fractions together is W

$$\bar{D} = \frac{1}{W}\sum_{i=1}^{k} W_i D_i$$

and

4.1 / Introduction   **115**

$$\sigma = \left[\frac{1}{W}\sum_{i=1}^{k} W_i\left(D_i - \overline{D}\right)^2\right]^{1/2}$$

The weight of a fraction could correspond for example to what is collected in a particular sieve after sieving with the sieves stacked in series, with the smaller size on the bottom.

There may be circumstances where we prefer to give a more complete description of the particle size distribution than is provided by merely the mean and standard deviation. One common way of doing this is to fit the measured size distribution to an expected size distribution, such as the normal distribution. For the normal (or Gaussian) distribution the expected number fraction $f(D)$ of particles with size D obeys the relationship

$$f(D) = \frac{1}{\sigma\sqrt{2\pi}}\exp\left[-\frac{1}{2}\left(\frac{D-\overline{D}}{\sigma}\right)^2\right] \tag{4.5}$$

For a limited data set of n measurements Eqns. 4.1 and 4.2 will still yield the mean and standard deviation. This relationship can be further fitted to the experimental data by adjusting the values of the standard deviation and the mean diameter until the best fit is obtained. The procedure typically uses some appropriate computer program. A problem involving such fitting appears at the end of this chapter. Since we are now regarding the particle size distribution as continuous, the forms of equations (4.1) and (4.2) that must be used to calculate means and standard deviations for particles obeying the normal distribution are

$$\overline{D} = \int_{-\infty}^{\infty} fDdD \div \int_{-\infty}^{\infty} fdD \tag{4.6}$$

and

$$\sigma = \left[\int_{-\infty}^{\infty} f.(D-\overline{D})^2 dD \div \int_{-\infty}^{\infty} fdD\right]^{1/2} \tag{4.7}$$

which can be further simplified by the normalization

$$\int_{-\infty}^{\infty} fdD = 1$$

Normalization fixes the amplitude of the distribution curve. The reliability of data is often cited with a confidence interval. When the standard deviation has been determined, a confidence interval of e.g. $\pm 2\sigma$ means that there is 95% confidence that the true value lies in this interval around the mean, or that 95% of an infinite number of measurements would lie within this confidence interval. Also, $\pm\sigma$ corresponds to 68.3% of the data, while $\pm 3\sigma$ corresponds to 99.7% of the expected data. It can be readily verified that these confidence intervals follow directly from an integration of Eqn. 4.5 between the appropriate limits.

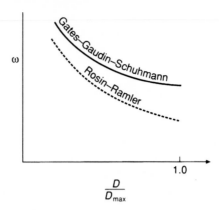

FIGURE 4.3 Sketch of Gates-Gaudin-Schuhmann distribution and Rosin-Rammler distribution for a= 0.7 and D_{max} = 1.

The integrals are carried out over the complete size range. A difficulty is immediately obvious: the normal distribution is symmetrically centered on $\bar{D}$ and predicts a nonzero number of particles of size less than zero! Therefore, real particle size distributions seldom fit the normal distribution at all closely. A more commonly observed distribution is the log-normal distribution

$$f(D) = \frac{1}{DA\sqrt{2\pi}} \exp\left[-\frac{1}{2}\left(\frac{\ln D - B}{A}\right)^2 \right] \qquad (4.8)$$

where B is now the mean of the natural logarithms of the particle sizes (i.e. the log of the geometric mean), A is the standard deviation of the logarithms of the particle sizes, and

$$f(D)\Delta D = \int_{D}^{D+\Delta D} f dD$$

the number fraction between D and $D+\Delta D$. Since the particles now cannot have negative sizes (only positive numbers have logarithms), we can expect a better fit to the measured distribution than from equation (4.5). The lower limits on the integrals of equations (4.6) and (4.7) can now be set to the more realistic zero. Other particle size distributions that are commonly encountered are the Gates-Gaudin-Schuhmann distribution,

$$\omega(D) = \alpha \frac{D^{\alpha-1}}{D_{max}^{\alpha}} \qquad (4.9)$$

where

$$\omega(D)\Delta D = \int_{D}^{D+\Delta D} \omega(D).dD$$

is the *weight* fraction of the particles between size D and $D + \Delta D$, α is an empirical factor known as the distribution modulus (typically between 0.4 and 1.0), and D_{max} is the maximum particle size, while the Rosin-Rammler distribution follows

$$\omega = \frac{\alpha D^{\alpha-1}}{D_m^\alpha} \exp\left[-\left(\frac{D}{D_m}\right)^\alpha\right]$$ (4.10)

where D_m is the Rosin-Rammler size modulus. These two distribution functions are sketched in Fig. 4.3. Unlike the log-normal distribution, these distributions have a finite size for the largest particle.

PROBLEM

A batch of particles obeys the Gates-Gaudin-Schuhmann size distribution with $\alpha = 0.8$. Calculate the total weight fraction of particles with a size less than one-tenth of the maximum size.

Solution

The weight fraction of particles with size between D and $D + dD$ is

$$\omega(D).dD = \int_D^{D+dD} \omega(D).dD$$

with $\omega(D)$ given by equation (4.9). Hence the weight fraction of particles of size between zero and $0.1D_{max}$ is

$$\int_0^{0.1D_{max}} \omega dD = \frac{\alpha}{D_{max}^\alpha} \int_0^{0.1D_{max}} D^{\alpha-1}dD = \frac{1}{D_{max}^\alpha}\left(D^\alpha\right)\Big|_0^{0.1D_{max}} = 0.1^\alpha = 0.1^{0.8} = 0.158$$

It should be recognized that the arithmetic mean particle size $\bar{D}$, defined by equation (4.3), or by (4.6) in the case of a continuous distribution, is not the only mean size that can be defined. For example, we may consider a geometric mean particle diameter D_g given by

$$\log D_g = \sum_{i=1}^{k} N_i \log D_i \div \sum_{i=1}^{k} N_i$$ (4.11)

or a harmonic mean diameter D_h given by

$$D_h^{-1} = \sum_{i=1}^{k} \frac{N_i}{D_i} \div \sum_{i=1}^{k} N_i$$ (4.12)

The harmonic mean is thought to be the appropriate choice when considering phenomena that are governed by the surface area of the particles because the surface area per unit mass of a particle is proportional to D_i^{-1}. Therefore, the surface area per unit mass of a number of particles is equal to the surface area of the same number of mono-sized particles with the harmonic mean diameter. It is noted that

$$D_h \leq D_g \leq \overline{D} \tag{4.13}$$

Note also that for skewed distributions, such as the log-normal one, the *mode* (which occurs at the maximum of $f(D)$), the *median* (for which

$$\int_0^{median} fdD = \int_{median}^{\infty} fdD,$$

and the *arithmetic mean* do not coincide. For the log-normal distribution we have: Mode < Median <Mean.

4.1.3 Particle Shape

Since the shapes of the particles that are produced by the traditional processing methods can be very varied and convoluted, the mathematical description of particle shapes is quite complex. It involves methods that are also encountered in pattern recognition. One commonly used descriptor of particle shape is the shape factor, defined as the particle surface area divided by the surface area of a spherical particle of the same volume. By definition the shape factor for a spherical particle is unity; the shape factor for all other shapes is greater than unity. For example, it is easy to show that a cube will have a shape factor of $(6/\pi)^{1/3} = 1.24$, while a right circular cylinder with a height equal to its diameter has a shape factor of $(1.5)^{1/3} = 1.14$. The modern trend in materials fabrication, however, is to avoid complex particle shapes and to try to approach mono-sized, spherical particles or particles with narrow size distribution for powders as much as technically and economically feasible. In general, it has not yet been possible to relate quantitatively the parameters needed to describe distributions of complex particles to the behavior of particle compacts during the fabrication of ceramic bodies. Instead, correlations are usually sought only in terms of average particle size, surface area, or less frequently, particle size distribution.

4.2 Characterization of Particles

Some of the commonly used techniques for particle size (distribution) measurement will now be described. Some of these techniques appear, together with other information, in Figure 4.4.

4.2.1 Sieving

Sieving is the oldest method, giving particle size distributions in rather coarse size intervals. Typically, sieves with decreasing mesh size are stacked with the largest mesh at the top. Mesh size is equal to the number of openings per linear inch in the sieve screen. Screens can have holes as small as 1 μm in special cases, with 5 μm the usual lower cutoff. Some sieve sets are pulsed by air, so that sieve clogging is diminished. Size distribution data obtained by sieving are only approximate, since it is too time consuming to sieve for a sufficiently long period to achieve the final distribution of the particles in the various sieves. Sieving can, however, be a quick method for product verification or product selection in technologies such as refractories manufacturing. For the fabrication of ceramics in more demanding applications, particle sizes are often in the micrometer range or below, and particle characterization or separation by sieving is not feasible.

4.2.2 Microscopy

Microscopy is perhaps the most straightforward technique for determining particle size. Optical microscopes permit measurement in a size range down to approximately 1 μm, while scanning or transmission electron microscopes can extend the range down to 0.1 nm. Measurements are made at the microscope itself (e.g., using a reticulated eye-piece in the case of an optical microscope) or from micrographs. Well-separated particles are necessary and it is common to measure two *diameters,* in orthogonal directions, on each particle. The technique is tedious in that a large number of particles must be measured, particularly if the size distribution is broad, to get a statistically valid set of data. This task may be facilitated by use of an image-analyzing computer or image analyzing software applied to digital photomicrographs of the particles.

4.2.3 Sedimentation (Stokes Diameter)

A particle falling through a fluid achieves (very rapidly) a terminal velocity, V, given by Stokes' equation:

$$D = \left[\frac{18\eta V}{\Phi g\left(\rho_s - \rho_f\right)} \right]^{1/2} \tag{4.14}$$

where

η = viscosity of the fluid
Φ = shape factor (usually assumed to be unity)
g = acceleration due to gravity
ρ_s = density of the solid
ρ_f = density of the fluid

Consequently, if we can measure the terminal velocity, equation (4.14) immediately gives us the particle size. Stokes' law has a limited range of validity because it only applies under laminar flow conditions (to Reynolds numbers, Re = $V\rho_f D/\eta$, of less than about 0.2[2]). Furthermore, at small particle sizes Brownian motion starts to interfere with the falling of particles. Also, electrostatic and other interactions between the particles would make equation (4.14) invalid. The practical consequence is that this method is reliable in the size range 2 to 50 μm when settling under gravity, in dilute suspensions.

If the particle size distribution is very narrow, it is simple to apply the technique. A dilute suspension of the particles is shaken, say, in a tall graduated cylinder and then allowed to settle. After a few seconds the suspension becomes stagnant and the par-

---

[2] For perfect spheres Stokes' law remains valid up to Re ≈1, but powders do not consist of perfect spheres.

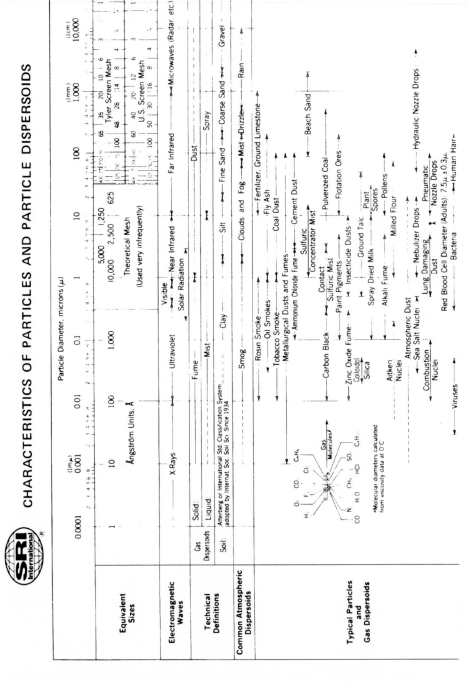

FIGURE 4.4 Chart showing characteristics of particles and methods for measurement. [C. E. Lapple, SRI Journal 5, 94 (Third Quarter, 1961). Reprinted with permission from SRI International]

Methods for Particle Size Analysis

- Ultramicroscope
- Electron Microscope
- Microscope
- Electroformed Sieves
- Sieving
- Centrifuge
- Ultracentrifuge
- Elutriation
- Sedimentation
- Turbidimetry
- X-Ray Diffraction
- Permeability
- Adsorption
- Light Scattering
- Nuclei Counter
- Electrical Conductivity
- Scanners
- Visible to Eye
- Machine Tools (Micrometers, Calipers, etc.)

Furnishes average particle diameter but no size distribution
Size distribution may be obtained by special calibration

Types of Gas Cleaning Equipment

- Ultrasonics (very limited industrial application)
- Settling Chambers
- Centrifugal Separators
- Liquid Scrubbers
- Cloth Collectors
- Packed Beds
- Common Air Filters
- Impingement Separators
- High Efficiency Air Filters
- Mechanical Separators
- Thermal Precipitation (used only for sampling)
- Electrical Precipitators

Particle Diameter, microns (μ)

0.0001   0.001 (1mμ)   0.01   0.1   1   10   100   1,000 (1mm)   10,000 (1cm)

Terminal Gravitational Settling [for spheres, sp. gr. 2.0]

In Air at 25 C, 1 atm — Reynolds Number, Settling Velocity cm/sec

In Water at 25 C — Reynolds Number, Settling Velocity cm/sec

Particle Diffusion Coefficient,* cm²/sec

In Air at 25 C, 1 atm
In Water at 25 C

*Stokes-Cunningham factor included in values given for air but not included for water

PREPARED BY C. C. LAPPLE

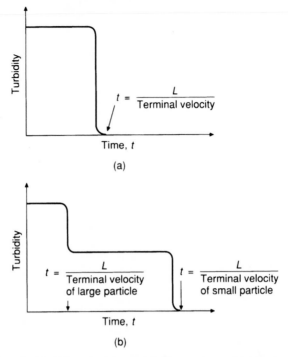

FIGURE 4.5 Results of sedimentation experiments using turbidity measurements for idealized particle size distributions: (a) single particle size; (b) two sizes of particle.

ticles start to settle at their terminal velocity. A layer of liquid clear of particles forms at the top of the cylinder and grows as particles continue to settle. The velocity of the downward movement of the interface between the clear liquid and suspension is the terminal velocity of the particles and can readily be obtained from a stopwatch and the cylinder graduations.

The technique becomes more complicated if there is a distribution of particle sizes. In that case it is usual to measure the particle concentration at some point in the fluid. A ready way of doing this is by measuring the *turbidity* of the fluid (its ability to absorb and scatter light) by some optical technique (e.g., using a light source, a slit, and a photodetector). Figure 4.5(a) is a sketch of the way the turbidity changes with time at a distance L below the top surface of the liquid for a dilute suspension of particles which are all the same size. Figure 4.5(b) depicts how the turbidity would change if the suspension contained roughly equal amounts of two sizes of particle. Clearly, if the particle size distribution is broad, the interpretation of the turbidity measurements is not simple. That interpretation is beyond the scope of this book, although a problem on this topic is included at the end of the chapter.

Differences in settling (terminal) velocity are exploited in separating particles by size (or density) in hydrocyclones, as mentioned in Chapter 2. Size measurement devices based on cyclones are also available.

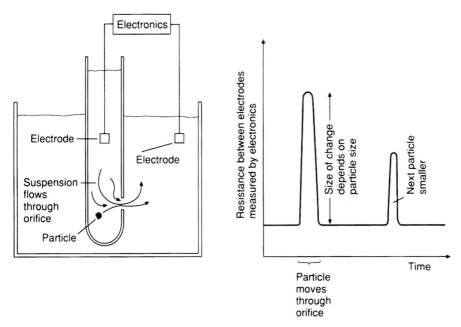

FIGURE 4.6 Principle of operation of the sensing zone technique (Coulter counter) for particle size determination.

4.2.4 Sensing Zone Techniques (Coulter Counter)

Consider a dilute suspension of electrically insulating particles in a conducting electrolyte (e.g., a salt solution) moving through an orifice on opposite sides of which electrodes are placed. An electric current is passed between these electrodes and through the orifice as shown schematically in Fig. 4.6. The resistance between the electrodes will change as a particle moves through the orifice and the magnitude of that change will be greater the greater is the particle size. Electronic circuits can count the number of changes and record them by magnitude, thereby obtaining a size distribution for the particles. The technique works best when the particles are comparable in size with the orifice, and therefore several orifices may be necessary if the particle size distribution is broad. This principle is used in the Coulter counter, which can conveniently measure particles in the size range 0.5 to 100 μm. Originally, the Coulter counter was used for determining the amount of red cells in blood, but it has now become a convenient and widely used tool for particle size measurement.

4.2.5 X-ray Line Broadening

A relatively simple way of measuring a mean particle size is by x-ray line broadening. The x-ray diffraction pattern produced by a large, perfect single crystal in a conventional diffractometer (with a slit source) is a series of sharp lines. These lines would become infinitesimal in width for well-collimated x-rays as the slit of the diffractometer is closed. For small crystals the diffraction lines broaden. A mean particle size can therefore be obtained by measuring the angular width of the diffraction lines. The crystal diameter, D, can be found from the Scherrer equation:

$$D = \frac{0.9\lambda}{\beta.\cos\theta}$$

where λ is the wavelength of the x-ray, β is the x-ray peak width (in radians) at half of its maximum intensity, and θ is the diffraction angle at which the line is measured. The technique is most suitable for particles in the range 0.005 to 0.1 µm. In practice, the technique is complicated by the need to apply a correction for the broadening due to the instrument itself (e.g., due to the finite width of source and detector slits). In principle, one can determine the instrument line broadening from the examination of a sample with know particle size. Furthermore, there is a more fundamental difficulty: it would not be unusual for primary particles to be made up of several grains, and under these circumstances the technique measures grain size rather than particle size.

4.2.6 Other Techniques

Very small particles (say less than 0.1 µm) display Brownian motion. This is an apparently random motion of a particle in suspension in a fluid that is due to the stochastic nature of the impacts of the molecules of the fluid on the small particle. The motion increases with temperature (as the momenta of the impacting molecules increase) and increases with decreasing particle size. The latter is a consequence of a much greater probability of the rate of molecular impact on one side of the particle not matching that on the other at small sizes. It is also a consequence of the much greater response of a smaller particle to such imbalance of impacts. Measurement of Brownian motion can therefore yield particle size information, and sophisticated instruments are now available that do this by measuring the Doppler shift of the frequency of laser light scattered from the particles.

Light scattering can be used in another way to measure particle size. Large particles tend to reflect impinging light back toward the source much as this page "backscatters" light toward the reader's eyes. As the particle size is diminished toward the wavelength of the light, there is an increasing tendency for the light to be forward-scattered (i.e., in roughly the same direction as the incident light). Measurement of the angle and the intensity of the forward scattered laser light can therefore be used to measure particle size.

Both of these last two techniques are expensive compared to the other techniques described in this chapter and are more likely to be encountered in specialized research labs, or in on-line process control where particle size is critical, than in general use. However, another frequently used technique, BET adsorption, is discussed in the following section since it applies equally well to agglomerates and compacts, as to particles.

4.3 Characterization of Agglomerates and Compacts

In this section we discuss three widely used techniques for arriving at a quantitative description of the structure of a porous solid, such as an agglomerate. The first of these is also widely used to obtain a measure of particle size for non-agglomerated powders.

4.3.1 Surface Area Measurement (BET Method)

The BET surface area method was devised by Brunauer, Emmett, and Teller (1938) and relies on the adsorption of gases onto a surface at low temperature. The mass of gas adsorbed on the solid surface is measured as a function of gas pressure at a fixed temperature, typically liquid nitrogen temperature. At low gas pressure the adsorbed gas is present as a partial monolayer. At higher pressures multilayer adsorption occurs, and at still higher pressures, the gas condenses on the powder surface to form a liquid film.

The theory of this physical adsorption is beyond the scope of this chapter, but the important result is the equation

$$\frac{1}{W(P_0/P - 1)} = \frac{1}{W_m C} + \frac{C-1}{W_m C}\frac{P}{P_0} \qquad (4.15)$$

where

W = mass of gas adsorbed at pressure P
P_0 = saturation vapor pressure of the adsorbing gas at the temperature of the experiment
W_m = mass of gas adsorbed at complete monolayer coverage
C = a constant

W is measured at several different pressures, usually by determining the mass of gas adsorbed from the gas pressure in a well-defined reservoir that contains the sample and a known amount of gas. A plot is then prepared of the left-hand side of equation (4.15) against P/P_0. That plot should be a straight line of slope s given by

$$s = \frac{C-1}{W_m C} \qquad (4.16)$$

and intercept

$$i = \frac{1}{W_m C} \qquad (4.17)$$

Hence

$$W_m = \frac{1}{s+i} \qquad (4.18)$$

But the mass of gas adsorbed at monolayer coverage is simply obtained from the area of the solid (a), the area occupied by each adsorbed gas molecule, Avogadro's number, and the molecular weight of the adsorbed gas. The final result is

$$a = \frac{A}{s+i} \qquad (4.19)$$

where A is $3.48 \times 10^3 m^2/g$ for nitrogen. The value of a obtained in this way is the accessible surface area of the solid (exclusive of occluded pores), and the specific surface area (surface area per unit mass) can be obtained by dividing by the mass of solid

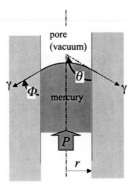

FIGURE 4.7 Penetration of mercury into a pore; q is the contact angle.

used in the experiment. The technique is useful for solids of surface area greater than a few m²/g when using nitrogen as the adsorbing gas. Solids of somewhat lower surface area can be measured by krypton adsorption.

If the specific surface area (surface area/g) of an non-agglomerated powder is measured, the particle size is readily estimated from the equation

$$D = \frac{6}{\rho_s \cdot \text{specific surface area}} \qquad (4.20)$$

where

$$\rho_s = \text{density of the solid}$$

The equation assumes that the particles are all spherical and of uniform diameter D; its derivation is left as an exercise. Alternatively, if the particle size is known from some other measurement (e.g., by Coulter counter), equation (4.20) can be used to estimate the specific surface area of the particles, assuming them to be spherical. If the actual specific surface area measured by BET is much greater than this estimated area, we have a clear indication that the particles are not simple spheres (e.g., they are porous, or they consist of agglomerates of smaller particles, or they are long needles).

4.3.2 Mercury Penetration Porosimetry

The pore size distribution in a compact or within particles can readily be measured by *mercury penetration porosimetry.* The material to be characterized is contacted with mercury under a vacuum. Because mercury does not wet most materials, the mercury does not initially enter the pores. Gradually increasing pressure is then applied to the mercury, forcing it into the pores, and the volume intruded at each pressure is measured.

Figure 4.7 is a sketch of mercury entering a pore, assumed here to be cylindrical and smooth. The pressure drop across the meniscus results in a force in the upward direction of $\pi r^2 P$, where r *is* the radius and P the pressure applied to the mercury. At equilibrium this must balance a downward force due to surface tension along the perimeter of the meniscus $2\pi r \gamma \cos\Phi$ where γ is the surface tension of mercury. Equating these two forces and recognizing that $\Phi = 180^0 - \theta$ (where θ is the contact angle), we

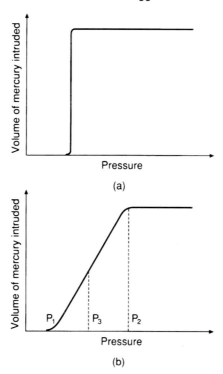

FIGURE 4.8 Mercury penetration porosimeter results: (a) idealized solid with cylindrical pores all of the same size; (b) real solid.

arrive at the result

$$r = 2g\,|\cos\theta|/P \tag{4.21}$$

Figure 4.8(a) depicts the experimental results we expect if all the pores are the same size and it is clear that we can easily obtain the pore radius from this plot and equation (4.21), along with the volume of pores in the sample. To do this we need the contact angle, θ, which is usually taken to be 140° for mercury, and the surface tension γ, which is typically 0.485N/m (at 25°C).

A sketch of results for a more realistic solid is depicted in Figure 4.8(b). Here the pores range in size from an upper limit [obtained by substituting P_1 in equation (4.21)] to a lower limit (obtained from P_2). One simple way of expressing the results is as an average pore size obtained from the pressure (P_3) needed to fill half the pores. A more precise expression of the results is in terms of a pore volume distribution function, $v(r)$, where $v(r)dr$ is the volume of pores with radius between r and $r + dr$ for $dr\rightarrow 0$, that is,

$$dv = v(r)dr \tag{4.22}$$

Let V_i be the volume of mercury intruded at pressure P into pores of radius r_i and above, with r_i given by equation (4.21):

$$r_i = 2\gamma\,|\cos\theta|/P \tag{4.23}$$

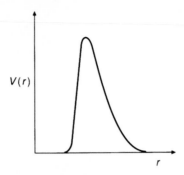

FIGURE 4.9 Sketch of pore volume distribution obtained from data of Fig. 4.8(b).

Consider a small increase dP in mercury pressure. The resulting increase in intruded volume dV_i is given by

$$dV_i = -v(r_i)dr_i \tag{4.24}$$

The minus sign appears in equation (4.24) because the penetrated radius *decreases* with increasing pressure (i.e., dr_i is itself negative). Hence

$$\frac{dV_i}{dP} = -v(r_i)\frac{dr_i}{dP} \tag{4.25}$$

Differentiating (4.23) with respect to pressure and recognizing that γ and θ are pressure independent, we have

$$\frac{dr_i}{dP} = -\frac{2\gamma \mid \cos\theta \mid}{P^2} \tag{4.26}$$

or substituting in (4.25),

$$\frac{dV_i}{dP}\frac{P^2}{2\gamma \mid \cos\theta \mid} = v(r_i) \tag{4.27}$$

Finally, we recognize that $v(r_i)$ is simply $v(r)$ and we get

$$\frac{dV_i}{dP}\frac{P^2}{2\gamma \mid \cos\theta \mid} = v(r) = \frac{dV_i}{dP}\frac{P}{r} \tag{4.28}$$

Hence we can use equation (4.27) and the mercury penetration results to obtain the pore volume distribution, $v(r)$. Figure 4.9 is a sketch of the result that would be obtained in this way from the data of Figure 4.8(b).

Care should be exercised in mercury porosimetry to allow sufficient time for the pressure-intrusion volume equilibrium to be reached because very fine pores take some time to be penetrated. Pores with an "ink bottle" shape pose a particular difficulty. These are pores accessible at only one end through a constriction. They fill at a pressure characteristic of the constriction size but empty at a pressure characteristic of the main body of the pore. Solids with this type of pore therefore show hysteresis in the plot of intrusion volume versus pressure.

4.3.3 Pycnometry

Both the BET method and mercury penetration porosimetry can only measure pores that are accessible (i.e., pores that are not completely isolated). One technique for measuring the volume of isolated pores is *pycnometry*. The technique consists of weighing a vessel completely filled with a liquid of known density that wets the solid. A known mass of the solid is then added and the vessel reweighed to determine the mass (hence volume) of liquid displaced. Usually, steps are taken to avoid errors due to air entrapped in pores (e.g., boiling the liquid). Let ρ_m be the density measured in this way and ρ_t the true density (i.e., density for completely pore-free solid). The volume fraction of closed pores is then $1-\rho_m/\rho_t$. More sophisticated versions of this apparatus employ a gas rather than a liquid, and the volume occupied by the solid is determined by the volume of gas displaced.

4.4 Concluding Remarks

In this chapter we have examined the way that particle size distributions are described and how they may be measured. The limitations of the measurement techniques have been discussed. The chapter closes with mention of techniques for characterizing porous solids. Particle technology is an important aspect of engineering that extends beyond the topic of materials production (into the fields of pigments and pollution control, for example). The coverage provided in this chapter is very much an introductory one and the interested reader is referred to the References. We have not examined the structure of the surfaces of particles in this chapter; that topic is treated in Chapter 5.

LIST OF SYMBOLS

a	surface area
$D, \bar{D}, D_g, D_h$	particle size, arithmetic mean particle size, geometric mean particle size, harmonic mean particle size
$f(D)$	particle size distribution function;

$$f(D)\Delta D = \int_{D}^{D+\Delta D} f\,dD$$

is the number fraction of particles between D and $D+\Delta D$

g	acceleration due to gravity
i	intercept in BET method
N_i	number of particles of size D_i
P, P_o	pressure, saturation vapor pressure in BET method
r	pore radius
s	slope in BET method

V	velocity
v, V_i	volume, intruded volume in mercury porosimetry;
	volume of pores with radius between r and $r + dr$ is vdr
W, W_m	mass of gas adsorbed, at monolayer coverage, in BET method
α	size distribution modulus in equations (4.9) and (4.10)
γ	surface tension
ϕ	shape factor
μ	viscosity
ρ	density
σ	standard deviation of particle size
θ	contact angle
$\omega(D)$	weight fraction distribution function; $\omega(D).\Delta D = \int_{D}^{D+\Delta D}\omega(D).dD$ *is* the *weight* fraction of the particles between D and $D + \Delta D$

Subscripts

f	fluid
max	maximum
s	solid

PROBLEMS

4.1 Use equations (4.6) and (4.7) to show that the constants $\bar{D}$ and s appearing in equation (4.5) are indeed the mean and standard deviation, respectively, of the normal distribution.

4.2 Consider the effect on $\bar{D}$, D_g, and D_h of adding one small particle to a large number of particles of much larger mean size. Consider the opposite (i.e., the addition of one very large particle). Deduce the relationship (4.13).

4.3 A particle size measurement device is used to determine the particle size distribution of a powder sample; the results appear in the table below.

(a) Calculate the mean particle size and the standard deviation.

(b) Prepare a plot of the fraction of particles smaller than size D versus D and draw a smooth curve through the data.

(c) Recognizing that the slope of your curve (which you can determine graphically) is the fraction of particles of size between D and $D + d D$, multiplied by dD, prepare a second plot of the latter fraction versus D. Compare the second plot with f given by equation (4.5); do this by plotting f on the same graph paper using $\bar{D}$ and s values you obtained in part (a).

(d) Repeat parts (a) and (c) using the natural logarithms of the particle size, instead of the particle size, to determine whether the log-normal distribution of equation (4.8) fits the data better or worse than the normal distribution.

(e) Calculate the geometric mean diameter and harmonic mean diameter and show that equation (4.13) holds true.

Particle size analysis results

Size range (μm)	Number of particles in size range
Less than 10	35
10-12	48
12-14	64
14-16	84
16-18	106
18-20	132
20-25	468
25-30	672
30-35	863
35-40	981
40-45	980
45-50	865
50-55	675
55-60	465
60-70	420
70-80	93
80-90	13
90-100	1
Greater than 100	4

4.4 Show that the width of the Gaussian distribution at half-maximum is 1.17 σ. Practically, the width at half maximum is easier to determine from a plot than the value at which σ occurs.

4.5 (a) A batch of particles is produced from a material of density 2500 kg/m³. The particles are of nearly equal size and shape and it is found that 10^9 particles weigh 5 kg. If the shape factor of a particle is 1.5, how much particle surface area will be contained in a cubic meter of space into which the particles are packed so that 35% of the volume is occupied by particles?

(b) An agglomerate consists of n spheres of uniform size in maximum point contact with each other. Calculate and plot the shape factor of the agglomerate as a function of n, in the range $n = 2$ to 50.

4.6 (a) Plot the shape factor for a cylinder as a function of its length-to-diameter ratio.

(b) Calculate the shape factor for a tetrahedron, octahedron, and dodecahedron.

4.7 Derive a general equation for the change of turbidity with time in a sedimentation experiment with a particle size distribution given by equation (4.5). Assume that the ability of a particle to scatter and absorb light is proportional to the square of the particle size (i.e., that the turbidity is dependent on the total particle surface area per unit volume at the position of the measurement). Assume also that the particles have the same density, and that the initial distribution is uniform.

4.8 Nitrogen adsorption experiments are carried out on 1 gram of an oxide sample at liquid nitrogen temperature. In the terminology of this chapter the results are

P (mm Hg)	100	130	160	190	210
W (mg)	0.98	1.13	1.27	1.41	1.50

For $P_0 = 760$mm Hg, determine the specific surface area of the oxide.

4.9 The following data were obtained in measurements on a porous oxide compact by mercury penetration porosimetry. Plot the data and then use equation (4.28) to obtain a plot of the pore volume distribution function.

Pressure (psi)	Mercury intruded (cm³/g solid)
2	0
5	0
10	0.01
20	0.02
50	0.07
100	0.15
200	0.25
500	0.35
1000	0.38
2000	0.39
5000	0.40
>5000	0.40

Use the calculated distribution to obtain the volume-averaged mean pore size

$$\bar{r} = \frac{\int rv(r)dr}{\int v(r)dr}$$

Compare this with the mean pore size calculated from the pressure required to half fill the pores.

4.10 A solid has a rather unusual size distribution; it contains large pores and small pores with none intermediate in size. The pore volume distribution function is as follows

$$v(r) = \begin{cases} 0 & \text{for } r < 0.1\mu m \\ 10^3 \text{cm}^2 / g & \text{for } 0.1\mu m < r < 1\mu m \\ 0 & \text{for } 1\mu m < r < 10\mu m \\ 10^2 \text{cm}^2 / g & \text{for } 10\mu m < r < 30\mu m \\ 0 & \text{for } > 30\mu m \end{cases}$$

Calculate and plot mercury intrusion volume versus pressure for mercury penetration porosimetry of this solid.

4.11 One stereological technique for determining the porosity (volume fraction of pores) of a solid, from a micrograph of the sectioned solid, is to select randomly many positions on the micrograph and to determine the fraction of these points that fall in pores. That fraction approximates the porosity if enough points are taken and certain other limitations, beyond the scope of this problem, are satisfied. Working with a map of the United States, use the technique to determine the fraction of the land area of the continental United States that is occupied by states with names beginning in the letter N. Select the random positions starting with a table of random numbers, generate them on a computer, or use the less sophisticated technique of throwing a pointer onto the map (without aiming). Plot the fraction you obtain as a function of the number of positions so that you can tell when you are approaching a statistically meaningful answer.

REFERENCES AND FURTHER READINGS

ALLEN, T., *Particle Size Measurement,* 2nd ed., Chapman & Hall, London, 1975.

BEDDOW, J. K., *Particulate Science and Technology,* Chemical Publishing Co., NY, 1980.

BRUNAUER, S. , EMMET, P.H., and TELLER, E., J. Amer. Chem. Soc., **60**, 309 (1938)

ONODA, G., and HENCH, L., *Ceramic Processing Before Firing,* Wiley, N.Y., 1978.

UNDERWOOD, E. E., *Quantitative Sterology,* Addison-Wesley, Reading, MA., 1970.

5

Surfaces and Colloids

5.1 Significance of Colloids in Materials Processing

Ceramic fabrication processes would often benefit from using powders with very small particles. However, such powders are notoriously difficult to handle when dry. They get airborne very easily, making everything look messy very quickly, and often pose significant health hazards. They readily form fluffy aggregates that make it difficult to pour the powder into dies and molds. They stick strongly to die walls and trap large amounts of air when pressed in dies. These processing problems led to the belief that there was an optimum particle size of about 1 μm. The optimum was supposedly determined by a trade-off between the benefit of improving sintering rates and the detriment of increasing particle handling problems with decreasing particle size. However, handling of fine particles as suspensions in liquids (or colloids) rather than as dry powders can facilitate the ceramic forming processes and thus extend downward the range of practically useful particle sizes significantly. Actually, particle suspensions have traditionally been used in tape casting and slip casting, as described in Chapter 12. More recently, success in producing submicrometer, nearly monodispersed powders (consisting of particles of all the same size and shape) has made possible vast improvements in the perfection of the ceramic bodies obtained by colloidal consolidation methods. One could think of colloidal consolidation methods as a glorified name for what are really just variants of the old mudpie baking. The practical aspects of these methods are described in Chapter 12.

The present chapter covers a large and complicated field, and we can only look at some of the more important phenomena that relate to practical issues of powder processing. First we look at the structure of surfaces and the experimental techniques that can be used to characterize surfaces. Then we see how surfaces act for particles suspended in liquids. After this we can proceed to look at how particles interact in colloidal suspensions, and a critical question will be whether or not the particles will come together and stick to form agglomerates or stay stably suspended. Later we will see how using polymers affects the stability and the flow properties of the suspension. Where possible, the practical implications of the fundamental background will be pointed out.

The properties of consolidated colloidal compacts will depend strongly on the nature of the suspension, which, in turn, is largely determined by the interactions of the particles with each other. The manipulation of the interactions between the suspended particles is then a key processing factor. While much of colloidal processing is still empirical, general principles can be formulated and provide a framework for experimental progress.

To understand the interactions in colloidal suspensions better, we will look at highly simplified surfaces in contact with an electrolyte,[1] a liquid that has positive and negative ions (in solution). We will find that suspended particles, if they have acquired a surface charge, will get surrounded by a diffuse cloud of ions from the electrolyte. This cloud has the effect of screening the surface charges. The spatial extent of the diffuse cloud is usually very small, perhaps a few tenths of a micrometer at most, but determines the interaction of the particles when they come close together. Two opposing forces will be at work: an attractive van der Waals force and a repulsive force that is due to the overlap of the charge-screening cloud. The balance of these two forces will determine whether the colloidal particles will stick together or repel each other as they move close together. The theory of the stability of colloidal systems in these terms is known as the DLVO theory, named after Derjaguin, Landau, Verwey, and Overbeek (see, e.g., Adamson, 1976). Nearly all other theories about colloidal stability start from this theory.

Later in this chapter we will discuss the behavior of colloidal suspensions that have been stabilized by polymers rather than by inorganic salt solutions. Although there are a number of similarities with the stabilization by ionic electrolytes, the interactions of the particles will now be much harder to quantify. However, practically useful trends can still be predicted. Polymeric stabilization of colloids is, in fact, the most widely used and preferred method in ceramic fabrication.

5.2 Surface Structure

The structure of clean, flat surfaces has been studied most. On an atomic scale the surfaces are not perfectly smooth because, at any temperature other than absolute zero,

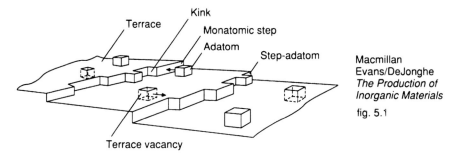

Macmillan
Evans/DeJonghe
*The Production of
Inorganic Materials*

fig. 5.1

FIGURE 5.1 Model of heterogeneous solid surface depicting different surface sites. These sites are distinguishable by their number of nearest neighbors. (Reprinted with permission from G. Somorjai, *Principles of Physical Chemistry*, Academic Press, New York, 1976).

[1] An electrolyte is a substance that produces positive and negative ions (dissociates) in a solvent such as water. If a high fraction of the electrolyte dissociates into positive and negative ions upon introduction in the solvent it is a strong electrolyte, if low concentrations of ions result it is a weak electrolyte.

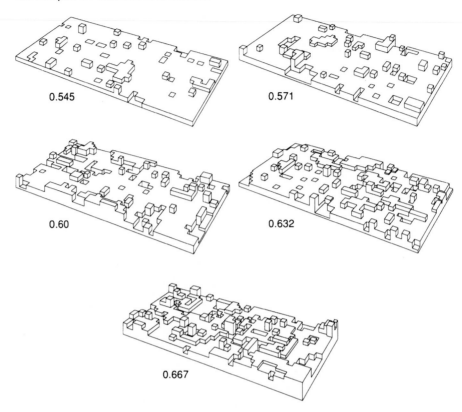

0.545   0.571

0.60   0.632

0.667

Figure 5.2 Typical configuration of the simple cubic (100) face generated by the Monte Carlo simulation for a crystal in equilibrium with its vapor. The numbers adjacent to the figures indicate the ratio kT/f. (Reprinted with permission from H. Leamy, G. Gilmer, and K. Jackson, "Statistical Thermodynamics of Clean Surfaces", in *Surface Physics of Materials*, Vol. I, J. Blakely, ed., Academic Press, New York, 1975, p. 184).

atoms will tend to pop up and make the surfaces more irregular. The somewhat disordered surfaces show features that can be described as terraces, steps or ledges, and kinks: TLK surfaces. A schematic summary of some possible surface disorders for solids is sketched in Fig. 5.1. The disorder develops as temperatures are increased; in other words, the surface entropy increases. An example of a computer-calculated surface disorder is shown in Fig. 5.2.

Real surfaces will be even more complicated than shown in Fig. 5.2. Surfaces of powders have complex shapes, departing considerably from the ideally smooth topology we assume in the DLVO calculations. Further, all sorts of chemical modifications are possible, making a detailed quantitative description of the surface chemistry quite difficult. Surfaces exposed to the environment will rapidly acquire carbon, water, and oxygen contamination. When submerged in liquids, a wide variety of impurities may be adsorbed. One of the challenges to modern colloid physics is to incorporate fully the structural and chemical aspects of real surfaces.

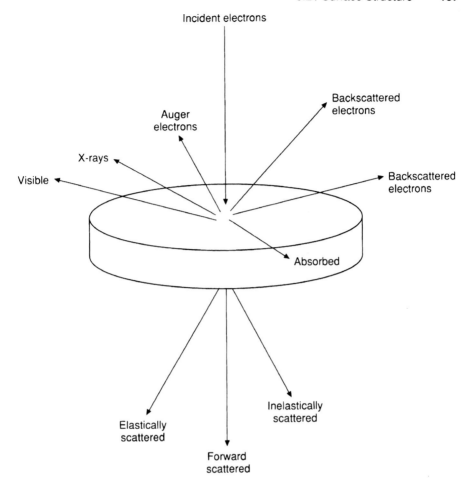

FIGURE 5.3 Various emissions as a result of incident electrons. Each of these emissions can be used to analyze the samples. (Courtesy of R. Gronsky.)

5.2.1 Surface Characterization Methods

Surface characterization methods could be classified in three main categories: microscopy methods, diffraction methods, and spectroscopic methods. Consider a beam of electrons incident on a sample as shown in Fig. 5.3. The reader is probably aware from courses in physics or materials characterization that a significant fraction of the electrons will pass through the sample, the fraction being dependent on the energy of the incident beam and the thickness of the specimen. These are forward-scattered electrons, and as shown schematically in Fig. 5.3, they fall into two groups: the elastically scattered electrons and the inelastically scattered electrons. The former have experienced interactions with the atoms of the sample that have been entirely elastic (i.e., have changed the direction of the electrons but not their energies). Consequently, provided that the incident electrons all have the same energy (are "monochromatic"), the elastically scattered electrons will all have the same energy (wavelength) and will form a diffraction pattern that can reveal the atomic arrangement of the solid. The inelastically scattered electrons have undergone more complex interactions with the atoms of the sample that have changed not only their direction but also their energy. Typically, the inelastically scattered electrons are much fewer in number than the elastically scattered ones. Of course, electrons can be scattered backward from the sample as well (see Fig. 5.3). Again the majority of these electrons will have undergone elastic collisions with the atoms of the sample and can yield a diffraction pattern. The penetration depth of the incident beam is strongly dependent on the electron energy and on the nature of the material. Typically, for electron energies below 1 keV the penetration depth is only one or two atomic layers; for 1 MeV the penetration depth is in the neighborhood of a micrometer. Surface analytic tools will therefore use only low-energy electrons. X-rays produce similar diffraction effects or can cause the ejection of electrons; they are therefore also used as incident beams.

Microscopy

Powders are not easily examined by conventional light microscopes, since the depth of focus at a useful magnification is very short. Additionally, most interesting features of technical powders fall below the resolution limit of optical microscopes. In a modern scanning electron microscope a much greater depth of field is available, revealing powder structure with a point-to-point resolution of around 2 nm. An example of a scanning electron microscope image of a powder is shown in Fig. 5.4. The agglomerates depicted are TiO_2 and are seen to consist of smaller, nearly monosized particles. Finer structural details of surfaces can be imaged with transmission electron microscopes

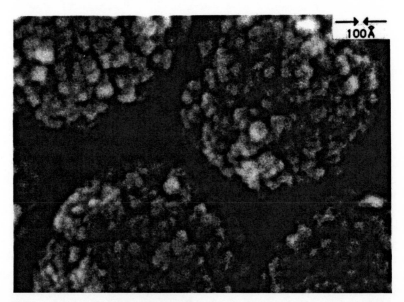

FIGURE 5.4 Structure of TiO_2 particles observed in the scanning electron microscope. The particles were prepared by chemical methods (sol-gel). (Courtesy of L. Edelson.

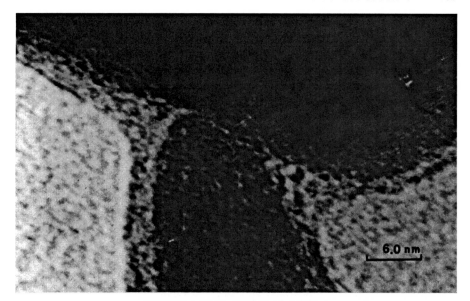

FIGURE 5.5 High-resolution electron micrograph of SiC powder particles showing an amorphous, fairly uniform silica layer approximately 3 to 5 nm thick.

(TEM), either on electron transparent thin foils or on particles if they are small enough. An example of a high-resolution TEM image is given in Fig. 5.5. Under favorable conditions surface features can be resolved that are only one atom step high.

A relatively new way of examining surfaces has been developed that permits the examination of surface topology on the near-atomic level. This method is called *scanning tunneling microscopy*. A very fine needle of tungsten or other refractory conductor comes within a few tens of angstroms of the surface of the solid, and a potential difference is applied between the two. The field will cause electron tunneling between the sample and the needle. The tunneling current turns out to be extremely sensitive to the tip-surface distance and also to the chemical nature and to the electronic properties of the solid. In this way, atomic surface roughness can be recorded directly. It is fascinating that this device can work if gases or even liquids are present, although with some loss of resolution. Researchers in this field have great hopes of now being able to examine surfaces in very complex situations. The principle of the method and an example of a surface topology are shown in Fig. 5.6. The examples shown are a rhenium surface and a sulfur overlayer; atomic or near-atomic features are clearly resolved.

Diffraction Methods

A narrow, parallel beam of x-rays or electrons is directed at the sample surface. The beam is scattered and reflected by the atoms that it encounters. The scattered beam will form an interference pattern that can be recorded on film or on a fluorescent screen. Analysis of the geometry of the pattern and the intensity of the pattern spots allows determination of the atom arrangement to within a fraction of a nanometer. This can be very useful structural information, although no direct images are obtained. In *low-energy electron diffraction* (LEED), a common method in surface structure analysis, a beam of monochromatic electrons, with an energy usually below 500 eV, is made to impinge on the sample. Such a low-energy beam usually penetrates only a few atom distances into the surface layer, and thus is sensitive to the surface structure only. It is possible to study the details of atom arrangements and adsorption patterns of foreign substances on crystal surfaces. In some cases, special phase relationships on the surface, differing

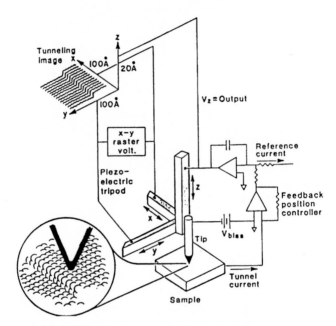

FIGURE 5.6a Sketch of the scanning tunneling microscope. (Courtesy of M. Salmeron.)

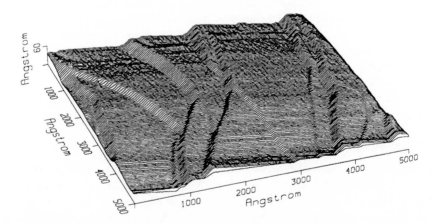

FIGURE 5.6b Low-resolution STM image of a stepped surface of rhenium with an absorbed layer of sulfur. (Courtesy of M. Salmeron.)

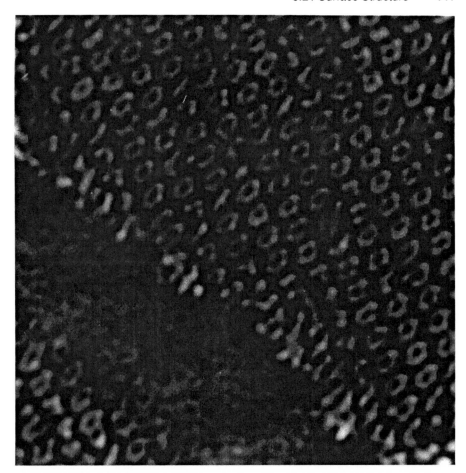

FIGURE 5.6c High-resolution image of a part of the surface shown in Figure 5.6b. (Courtesy of M. Salmeron.)

from those in the bulk, have been determined, illustrating the complexity of surface chemistry. A limitation of this method is that the low-energy electrons require an ultrahigh-vacuum environment so that only a small range of surface conditions can be explored. A sketch of a LEED apparatus, together with an example of a LEED diffraction pattern and the corresponding surface structure, is shown in Fig. 5.7

Spectroscopic Methods

Characteristic photon or electron emissions are excited in the first few surface layers of a solid, and the energy spectrum of the emitted particles is analyzed. Chemical information about the surface can be obtained in this way. A wide variety of spectroscopic methods are in use now. Most frequently used are *Auger spectroscopy, electron spectroscopy for chemical analysis* (ESCA) *or photoelectron spectroscopy,* and *secondary ion mass spectroscopy* (SIMS).

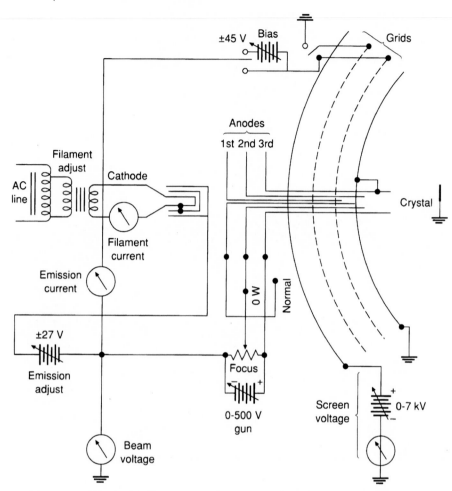

FIGURE 5.7a Sketch of a low-energy electron diffraction (LEED) apparatus.

Auger Spectroscopy: A beam of electrons with an energy of a few keV is directed at the sample surface in an ultrahigh-vacuum chamber (see Fig. 5.3). This beam will eject inner electrons of the atoms in the surface layer of the specimen. Electrons from outer orbits will fall into the free inner orbits, a transition that is accompanied by the emission of a characteristic x-ray (as described already) or the emission of a secondary electron to which the transition energy has been transferred. It is the energy spectrum of these emitted electrons that is recorded in Auger spectroscopy. This energy spectrum gives the identity of the emitting atom. The relative intensity of the various characteristic peaks in the electron spectrum can give a semiquantitative surface composition. Additionally, the surface of the specimen may also be bombarded with ions, sputtering off surface layers in a well-controlled manner. The evolution of the Auger spectrum as the surface layers are peeled off then gives information on the depth distribution of the elements in the near-surface region. An example of a result of a study where Auger

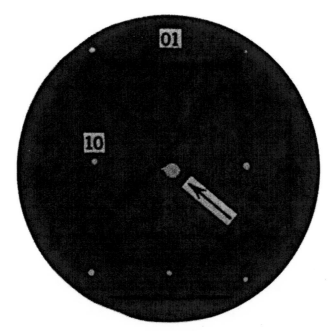

FIGURE 5.7b Diffraction pattern from a clean Ni(110), surface with 76-V electrons. The arrow indicates the position of the oo spot.

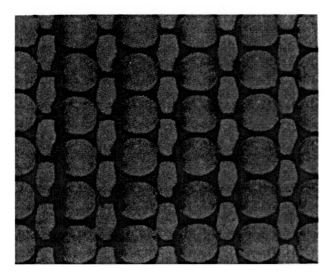

FIGURE 5.7c Model of the (110) surface of a face-centered crystal. (From L. H. Germer and A. U. MacRae, *J. Appl. Phys.* 33, 2923 (1962). (Reprinted with permission from Institute of Applied Physics.)

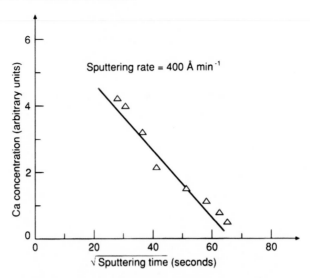

FIGURE 5.8 Relative Ca concentration as a function of sputtering time in the Auger analysis of the MgF_2–CaF_2 interdiffusion couple. The sample was annealed for 96 hours at 950°C. The mean Ca penetration depth is about 1 μm.

spectroscopy was used to determine a concentration profile in an investigation of the unsteady-state diffusion of calcium ions is shown in Fig. 5.8.

Photoelectron Spectroscopy (ESCA): A soft x-ray beam or an ultraviolet (UV) beam ejects electrons from atoms at the sample surface. The energy of the ejected electrons is recorded in a spectrometer. In this case the spectrum not only gives information on the chemical identity of the emitting atom, but also on its chemical bonding environment. Further. recording of the spectra as a function of the angle of incidence of the exciting beam can give information on the depth distribution of atoms in the near-surface region. Angle-resolved UV photo-electron spectroscopy has also given structural information. The use of the very intense and polarized radiation generated by synchrotrons has greatly improved the accuracy and information content of these surface analytical methods. An example of an ESCA characterization of a ceramic powder is shown in Fig. 5.9.

Secondary Ion Mass Spectroscopy: In this method a beam of electrons with an energy of about 3 to 5 keV bombards the specimen and ejects ions (sputters) from the surface. These ejected or secondary ions are analyzed in a mass spectrometer. An example of a SIMS spectrum of a ceramic powder is shown in Fig. 5.10.

The chemical identity of the ions can even be recorded as a function of depth as the surface is removed by sputtering. The method has good sensitivity for a wide range of ions and can yield a more quantitative comparison versus depth profile than Auger spectroscopy.

Because this is a book on materials production rather than materials characterization, this discussion of techniques for examining solid surfaces has necessarily been brief,

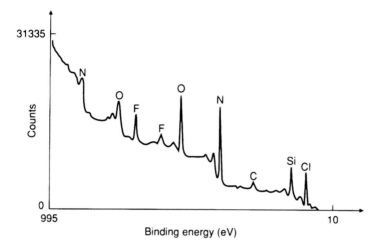

FIGURE 5.9 X-ray photoelectron spectrum of a SiC powder, obtained from a survey scan between 0 and 1 keV.

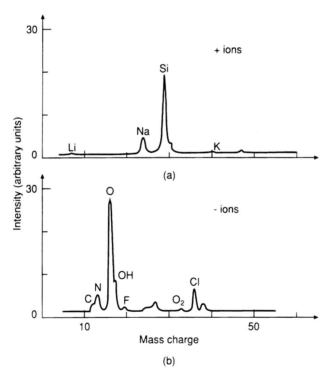

FIGURE 5.10 (a) Positive and (b) negative secondary ion mass spectra of the powder in Fig. 5.9.

and the reader seeking a more comprehensive treatment is referred to the literature (see, e.g., Vanselow, 1974-1982).

5.3 Colloid Interactions

Having briefly described the surfaces of solids and methods for their examination we now go on to discuss what happens when a solid is immersed in a liquid such as an electrolyte. Broadly, there are two classes of colloidal suspensions that one can consider. These are the *lyophillic*(solvent loving) and the *lyophobic* (solvent hating) one. A lyophillic colloid will spontaneously re-disperse when the solvent is added again after drying, since the particles like to tie up solvent; a lyophobic colloid will not. Most colloidal suspensions that we are dealing with fall in the lyophobic category, and their behavior is determined by surface charges and interactions with the solvent.

5.3.1 Diffuse Double Layer

To examine colloidal interactions of particles suspended in an electrolyte, we shall first consider the simplified geometry of a flat. smooth, semi-infinite solid in contact with a semi-infinite reservoir of electrolyte. We assume that for some reason (to be discussed later), the surface has acquired a surface charge; in other words, there is an excess (or deficiency) of electrons over the number needed to balance the protons of the surface atoms. This would, in turn, lead to a surface electrical potential, ψ_0, that we can define with respect to a zero reference potential at a position in the electrolyte an infinite distance from the surface. The electrical potential in the liquid will then be determined by how effectively the negative and positive ions in the liquid screen the surface charge. A schematic diagram of the situation is shown in Fig. 5.11.

Let us now consider an electrolyte. for example water with NaCl dissolved in it, in which we have two ions, A and B, of opposite charge $z+$ and z^-. We see a region (known as the *double layer*) where the normal balance (equal numbers) of positive and negative ions is disturbed by the charge on the surface. The chemical potential of these ions will depend on the magnitude of the local electrical potential, $\psi(x)$, in which they are located, where x is the distance from the surface. We again choose $x=\infty$ as our reference, with $\psi(\infty) = 0$.

If we were to use equation (2.60) to write down the chemical potential of A, we would have to add an extra term to allow for the fact that A is a charged species located in an electric field; we call the potential we get after adding this extra term the *electrochemical potential*. The extra term is simply the work done in bringing the ions from infinity to their location.[2] The work to bring one ion is $z|e|\psi(x)$, where $|e|$ is the absolute value of the charge on the electron. If we multiply by Avogadro's number N we obtain the work to bring a mole of ions from infinity, and in doing so, we can put $N|e| = F$, which is Faraday's constant (numerical value 96,484 C). Therefore, the electrochemical potential of A becomes

$$\mu_A = (x) = \mu_{A0} + RT \ln C_A(x) + zF\psi(x) \qquad (5.1)$$

---

[2] This is actually the definition of electrical potential if the charge were unity.

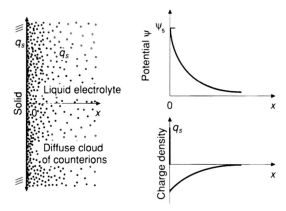

FIGURE 5.11 If a charge q_s per unit area is present on a solid surface in contact with a liquid, ions of opposite sign will be attracted from a solution and ions of the same sign will be repelled. These ions form a diffuse, rather than a discrete layer, extending for a small distance into the liquid. As a result, an electrical potential and a charge density distribution is present that falls off, approximately exponentially, with distance from the surface.

where, for convenience, we have assumed an activity coefficient of unity. [In what follows it is necessary to remember the distinction between the electric potential at a point $\psi(x)$, which is the work done in bringing a unit test charge to that point, and the electrochemical potential of species A, μ_A, which is different from that of species B at the same point and is a function of the activity of A.] At $x = \infty$ the electrochemical potential will be

$$\mu_A(\infty) = \mu_{A0} + RT \ln C_A(\infty) \tag{5.2}$$

Equilibrium requires that the electrochemical potentials of the ions be the same everywhere, and consequently,

$$\mu_A(x) = \mu_A(\infty) \tag{5.3}$$

which leads immediately to

$$C_A(x) = C_A(\infty) \exp\left[\frac{-zF\psi(x)}{RT}\right] \tag{5.4}$$

Similarly, for the negative ions, B, we would have

$$C_B(x) = C_B(\infty) \exp\left[\frac{+zF\psi(x)}{RT}\right] \tag{5.5}$$

The local charge density, $q(x)$, at any point x in the double layer follows from the difference between the concentration of positive and negative ions, so that

$$q(x) = zF[FC_A(x) - C_B(x)] = -2C(\infty)\sinh\frac{zF\psi(x)}{RT} \qquad (5.6)$$

where we have taken

$$z_A = z_B = z \text{ and } C_A(\infty) = C_B(\infty) = C(\infty) \qquad (5.7)$$

The local charge density, $q(x)$, can be related to the local potential gradient by one of the fundamental relations of electrostatics, Poisson's equation:

$$\nabla^2\psi(x) = -\frac{4\pi q(x)}{\varepsilon} \qquad (5.8)$$

where ε is the dielectric constant of the electrolyte.[3]
   Combination of equations (5.5) and (5.8) gives

$$\nabla^2\psi(x) = \frac{8\pi C(\infty)zF}{\varepsilon}\sinh\frac{zF\psi(x)}{RT} \qquad (5.9)$$

This equation can be written for the flat interface that we are considering here as

$$\frac{d^2x}{dx^2} = \kappa^2\sinh y \qquad (5.10)$$

if

$$\kappa = \left[\frac{8\pi C(\infty)z^2F^2}{\varepsilon RT}\right]^{1/2}$$

and

$$y = \frac{zF\psi(x)}{RT} \qquad (5.11)$$

The boundary conditions are $y = 0$ and $dy/dx = 0$ at $x = \infty$. A first integration can be done by putting $dy/dx = p$. Then $d^2y/dx^2 = dp/dx = (dp/dy)dy/dx = (dp/dy)p$. We can then separate the variables, so that the integration of equation (5.10) gives

$$\frac{dy}{dx} = 2\kappa\sinh\frac{y}{2} \qquad (5.12)$$

A second integration with $y = zF\psi_o/RT = y_o$ at $x = 0$, where ψ_o is the electric potential at the surface, then yields

$$\exp\left(\frac{y}{2}\right) = \frac{\exp(y_0/2) + [\exp(y_0/2) - 1]\exp(-\kappa x) + 1}{\exp(y_0/2) - [\exp(y_0/2) - 1]\exp(-\kappa x) + 1} \qquad (5.13)$$

---

[3] $\varepsilon = 8.85 \times 10^{-12}$ farad/m $= 8.854118 \times 10^{-12}$ Coulomb^2/Newton m^2 for empty space. For the electrolyte solution e needs to be determined. For dilute aqueous electrolyte solutions e is around 80x $?_0$ or larger.

If $y_o \ll 1$, then, after all this algebra, equation (5.13) can be reduced to a very simple expression:

$$\psi(x) = \psi_o \exp(-\kappa x) \tag{5.14}$$

This would be the case for a weakly charged colloid for which $\psi_o \ll RT/zF; (RT/zF \sim$ 25 mV at room temperature). Then the potential in the diffuse cloud of counter ions decays exponentially, with a characteristic distance $x_D = 1/\kappa$. x_D, called the Debye length, equals the distance from the surface at which the potential has fallen to ψ_o/e, where $e =$ 2.71828.... It is customary to take x_D as the effective thickness of the electrical double layer that surrounds the particle. Note that this thickness decreases as the inverse square root of the bulk concentration of ions in the electrolyte. For high concentrations of dissolved salts, the double layer will be very thin, perhaps only a nanometer or so. Interactions between particles will become increasingly long range the more dilute the electrolytes. As an example, for a solution containing 0.01 mol per liter of a salt such as NaCl, x_D would be about 3 nm.

Overall, the system of the particle plus the electrolyte must remain electrically neutral. Preservation of electroneutrality then requires that the total charge imbalance in the electrolyte must equal the opposite of the surface charge on the solid,[4] σ_s, so that

$$\sigma_s = -\int_0^x q(x)dx \tag{5.15}$$

We can go back to the Poisson equation, (5.8), and substitute for $q(x)$, equation (5.6). The evaluation of the surface charge density then becomes particularly simple if we restrict ourselves to the case of low surface potentials so that equation (5.14) applies. This yields immediately the following interrelationship between surface charge, surface potential, and Debye length:

$$\sigma_s = \varepsilon \kappa \psi_o \tag{5.16}$$

Thus, changing the Debye length by changing the electrolyte concentration, necessarily induces either a change in the surface charge or in the surface potential or both.

The treatment that led eventually to equation (5.14) can be further refined by taking a more detailed look at the structure of the diffuse double layer. One important modification is to limit the concentration of ions in the double layer, so as not to exceed what can be physically accommodated. This was first done in 1924 by Stern, who considered the first layer of counter ions to be at most a densely packed one. The Stern layer would be only a few ion layers thick. Beyond, the double-layer cloud is considered to be properly described by the treatment that led to equation (5.14). but with ψ_o now having the meaning of the potential at the interface between the Stern layer and the rest of the double layer. The difficulty is then how to take various chemical and electrostatic interactions into account to describe accurately the ion arrangement and the potential in the Stern layer. These matters are still subjects for research.

---

[4] This implies that the missing charges in the electrolyte solution have been tightly bound in an infinitesimally thin layer on the surface of the colloidal particles.

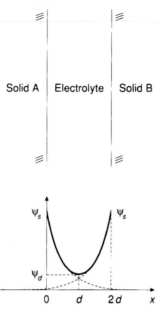

FIGURE 5.12 An electrolyte trapped between two solid surfaces a distance $2d$ apart will develop a distribution of counterions that is symmetrical around $x=d$. A good approximation to the electrical potential distribution is obtained by summing the individual double-layer potentials as if these did not affect each other. At $x=d$, $d\psi/dx = 0$, and the mutual repulsion between solid A and solid B follows from the difference in ion concentration at $x=d$ and far away in the electrolyte reservoir; the effect is an osmotic pressure.

5.3.2 Repulsion Between Overlapping Double Layers

Suppose that we now look at a situation in which two flat, submerged surfaces would come sufficiently close together so that the diffuse double layers overlap, as in Fig. 5.12. We wish to calculate the force acting on the solid surfaces (i.e., the interaction forces). Consider the midplane between the two surfaces (at $x = d$). The force exerted by the fluid to the left of this midplane ($x < d$) must balance that exerted by the fluid to the right, which, in turn, must balance the force exerted by the solid on the fluid between $x = d$ and $x = 2d$. Hence the force exerted by the fluid to the left of the midplane on the midplane equals the interaction force we are seeking. In the present case the fluid on the left can exert a force on the midplane by two possible mechanisms: because of its hydrostatic pressure and because the fluid contains ions and is in the electric field of the double layer. However, the latter is proportional to the electric field, which is (minus) the gradient of the potential. At the midplane, from symmetry, the potential gradient is zero. Therefore, the second contribution to the force on the midplane is zero and the interaction force is just the hydrostatic pressure at the midplane between the surfaces. However, we must recognize that the hydrostatic pressure will be different from the normal hydrostatic pressure between two surfaces that do not have surface charges. That difference arises because the ionic concentration will be different between the two charged surfaces than between two uncharged surfaces. In fact, it is this difference that leads to the net repulsive force (per unit area) between the charged surfaces. This pressure difference due to a concentration difference is an osmotic pressure and we can use van't Hoff's relationship for osmotic pressure, Π_{os}, that gives Π_{os} as RT multiplied by the excess concentration. For the concentration of positive and negative ions at the midpoint when the surface charge is present, we can use equations (5.4) and (5.5). The concentrations in the absence of surface charge would simply be $C_A(\infty)$ and $C_B(\infty)$ (which are equal for a symmetrical electrolyte, hence we drop the subscript). Subtracting gives us the excess concentration of each ion, and adding the two excesses gives us the total excess concentration. Using van't Hoff's relationship gives

$$\Pi_{OS} = 2C(\infty)\left\{\exp\left[\frac{zF\psi(d)}{RT}\right] + \exp\left[\frac{-zF\psi(d)}{RT}\right] - 2\right\}RT$$

$$= 2C(\infty)RT\left[\cosh\frac{zF\psi(d)}{RT} - 1\right]$$

(5.17)

Of course, this equation gives us the interaction force in terms of the potential midway between the plates and is not very useful as such. However, if the plates are sufficiently far apart (greater than three Debye lengths), their double layers are not greatly perturbed by the presence of the other double layer; consequently, we can approximate the potential using equation (5.14). Substituting in equation (5.17) and carrying out some algebra that we leave as an exercise for the reader, we arrive at the following results.

For $zF\psi(d)$ large, compared to RT, the interaction force is proportional to $1/d^2$. For $zF\psi(d)$ small, compared to RT (i.e., for small surface charge), the repulsive force is proportional to $e^{-\kappa/D}$ where κ is the reciprocal of the Debye length and is given by equation (5.11).

For spheres, the relationships discussed above become more complex, although the principles that are involved are the same.

5.3.3 Van der Waals Attraction

Neutral atoms can exert a mutual force on each other which arises from the interaction of temporary dipoles generated by the oscillations of the negative electrons around the positive nuclei. For identical substances, this interaction is always attractive, and is known as the *London-Van der Waals* or *dispersion force*. For two atoms, a distance x apart, the interaction energy, $E_d(x)$ may be expressed to a first approximation as

$$E_d(x) = \frac{-C_1}{x^6}$$

(5.18)

C_1 is a constant on the order of 10^{-77} J•m^6 but differs for different materials. Although this force is of only very short range for an isolated pair of atoms, it can lead to long-range interactions for solid bodies suspended in a fluid. This can be illustrated when one tries to find the interactions of two bodies; we must perform the integration over the volumes of the solids, and for each integration the power of x in the denominator of $E_d(x)$ is reduced. For the interaction of an atom with a slab of the same material, one has, for example, found (Adamson, 1976).

$$E_{\text{atom-slab}}(x) = -\frac{\pi}{6} NC_1 x^{-3} \tag{5.19}$$

where N is the number of atoms per unit volume in the slab. For the interaction between two slabs it was found that

$$E_{\text{slab-slab}}(x) = -\frac{\pi}{12} \cdot \frac{N^2 C_1}{x^2} = -\frac{1}{12\pi} \cdot \frac{H}{x^2} \tag{5.20}$$

where H is the Hamaker constant and $E_{\text{slab-slab}}$ is the interaction energy per unit area of surface. H is usually between 10^{-19} and 10^{-20} J. The interaction energy for other shapes of solids may be expressed in a similar manner. For example, for two spheres of radii a_1 and a_2, the interaction energy, $E_{1\text{-}2}$, may be expressed as

$$E_{1\text{-}2}(r) = -Hg(a_1, a_2, r) \tag{5.21}$$

where r is the center-to-center distance between the spheres and g is a function of the indicated variables. Tables for the detailed form of $g(a_1, a_2, r)$ can be found in the literature (see, e.g., Israelachvili, 1985).

5.3.4 Interaction Potentials

The interaction of two bodies in an electrolyte may now be considered as the sum of the double-layer repulsion and the van der Waals attraction. For flat slabs, separated by a distance x, the total interaction energy per unit area of surface is

$$E_{\text{tot}} = \frac{AC(\infty)RT}{\kappa} h\left(\frac{zF\psi_o}{RT}\right) \exp(-2\kappa x) - \frac{1}{12\pi}\frac{H}{x^2} \tag{5.22}$$

where $h(zF\psi_o/RT)$ means a function of the parameter $zF\psi_o/RT$. For a given electrolyte, with ψ_o and T fixed, h will be a constant. A is a numerical constant.

Figure 5.13 shows a few calculated pairwise *interaction potential* curves for spheres. Although equation (5.22) suggests that E_{tot} goes to minus infinity at zero separation, there is in fact a finite minimum (called the primary minimum in Fig. 5.13) which results from the very rapidly rising repulsion that sets in when the electronic orbitals of the atoms in the particles start to interpenetrate. This repulsion was not included in equation (5.22). The details of the close range repulsion are not important here; rather, the height of the potential energy barrier that particles have to surmount before they get trapped in the strong primary minimum is what is important for interactions of particles in colloidal suspension. The effect of electrolyte concentration is strong and can be used to manipulate the stability of the suspension. With increasing electrolyte concentration (larger values of K), the double layer becomes compressed, as follows from equation (5.14), and the energy barrier that prevents the two particles from coming together becomes smaller. For a small range of conditions, a secondary minimum may occur, as indicated in Fig. 5.13, although it is usually only a few kT energy units deep, and is smaller for smaller particles. For particles of 1 μm and below, this secondary

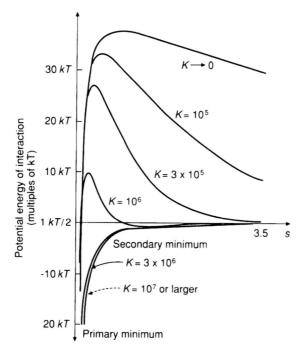

FIGURE 5.13 Effect of relative electrolyte concentration on the interaction potential energy between two spheres separated by a distance s. (Reprinted with permission from E. J. W. Verwwey and J. Th. G. Overbeek, *Theory of the Stability of Lyophobic Colloids*, Elsevier Science Publishers, B.V., Amsterdam, 1948.)

minimum has little significance. For large particles it may be more important and may lead to the formation of ordered suspensions. However, it must be noted that a secondary minimum is not required to produce ordering of particles in a suspension. For purely repulsive particles in a finite box (i.e., a finite amount of liquid), ordering of the particle arrangement will also occur. since this helps to minimize the total free energy of the system. The particle spacing in that case can be much larger than that corresponding to ordering in a secondary minimum. Ordered arrangements of particles of the same size can give interesting optical diffraction effects, (just as seen in gem opals, which are amorphous or microcrystalline silica-containing ordered domains of small hydrated silica spheres), and can be used to obtain information about the strength of the particle interactions.

In principle, all colloidal suspensions would have to settle eventually, since for every potential energy barrier described here there is a finite probability that some particle pair will eventually go over it and combine in the stable primary minimum. The consideration of such probability is, however, not as straightforward as in the case of energy barriers for chemical reactions, since now the spatial extent of the pair potential comes into play. Flocculation, the rapid clustering of particles to form agglomerates that settle on the bottom of the vessel containing the colloidal suspension, will only occur readily if there is a monotonic decrease of potential energy as the particles approach each other. As Fig. 5.13 indicates, this may be achieved by manipulating the concentration of ions in the electrolyte.

5.3.5 Zeta Potential

The description of the diffuse double layer as a cloud of ions that shield the surface charge was based on a perfectly immobile particle in an infinite amount of still liquid. In actuality, the colloidal particles are moving around continuously due to thermal motion, and as a result, they lose part of the double layer. Relative motion of the particles with respect to the electrolyte prevents the double layer from being established fully. Instead, the charges on the particles will end up being only partially screened by that part of the ion cloud that is fairly firmly bound. For a particle moving relative to the liquid, the counter ions that remain with the particle are only the first few ion layers of the diffuse double layer. Beyond this thin layer the electrolyte behaves like a normal liquid, while inside the layer the ion and solvent molecules are essentially immobile. The result is that now the colloidal particles, when in motion, will behave as if possessing an effective charge, q_{eff}, that is only a fraction of the surface charge, q_s (the surface area multiplied by σ_s).

An interesting observation is that if an electrical field is applied across a colloidal suspension the particles start to move. This phenomenon is called electrophoresis. By measuring the particle velocities as a function of applied electric field, one can obtain a quantity, the *zeta potential,* that is commonly used as one of the parameters that characterizes colloids.

To understand how the applied field relates to the movement of the particle, let us consider a spherical particle as in Fig. 5.14. This particle is assumed to move through the stationary electrolyte with a velocity V. The ions immediately adjacent to the particle surface are considered to be stuck firmly to it, so that the particle plus the affixed fraction of the double layer has an effective charge, q_{eff}, that is lower than the surface

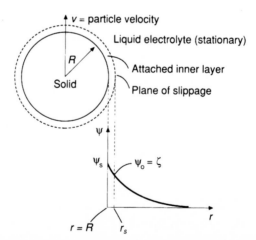

FIGURE 5.14 A colloidal particle of radius r moves through the stationary liquid electrolytes with a velocity v. The inner layer remains attached to the particle so that a surface of slip develops between the stationary electrolytes and the attached layer. The potential $\psi_o = z$ is assumed to be equal to the potential at the plane of slip position for a stationary particle. In motion, the particle loses the double layer beyond r_s, acquiring a charge e_{eff}.

charge of the particle alone, q_s. The interface between the attached double-layer part and the remainder of the mobile liquid is referred to as the *plane of slippage*.[5] The difficulty is now to find the effective charge of the particle with its thin layer of attached counter ions. We can get some idea of what this should be by considering how much charge was contained in the stationary cloud of ions that was stripped as a consequence of the motion. The charge contained in that part should be equal, but opposite in sign, to the effective charge, q_{eff}, on the particle. We can find this charge by considering the solution to the Poisson equation for a spherical particle that has a potential zeta (ζ) at the plane of slippage. Without needing to worry about exactly how the charges are distributed in the sphere that the plane of slippage encloses, we can then use the expression for the potential in a stationary double layer for a surface potential ζ.

If the effective charge on the particle of radius R is fairly small, then, as before for the derivation of equation (5.14), it will be possible to find an approximate analytical expression for the solution to the Poisson equation, equation (5.8), this time in spherical coordinates which for small values of the potential can be written as[6]

$$\frac{1}{r^2}\frac{\partial}{\partial r}\left(r^2\frac{\partial \psi}{\partial r}\right) = \kappa^2\psi,$$

using the expansion

$$e^{-\kappa} \approx 1 - \kappa$$

The answer is

$$\psi(r - r_s) = \zeta\left(\frac{r_s}{r}\right)\exp\left[-\kappa(r - r_s)\right] \tag{5.23}$$

where r is radial position and r_s is the radius of the interface of slippage. The effective charge on the particle, q_{eff}, will then be equal to

$$q_{eff} = -4\pi\int_{r_s}^{\infty}r^2q(r)\ dr \tag{5.24}$$

where we mean the lower integration limit, r_s, to correspond to the point where $\psi=\zeta$. Thus the simplified Poisson equation is

$$\frac{1}{r^2}\cdot\frac{\partial}{\partial r}\left(r^2\frac{\partial \psi}{\partial r}\right) = -\frac{4\pi q(r)}{\varepsilon} = \kappa^2\psi \tag{5.25}$$

The first part of this equation follows from equation (5.8) and the second part comes

---

[5] for the particle considered here, we are actually dealing with a surface of slippage, not a plane.

[6] we substitute $y=u/r$ in Eqn.5.9 which can then be reduced to $du^2/dr^2=k^2u$, for which a general solution is $u=Ae^{kr}+Be^{-kr}$, leading eventually to Eqn.5.25.

from the direct differentiation of $\psi(r-R)$, equation (5.23). With equations (5.24) and (5.25), we can find

$$q_{eff} = \varepsilon \kappa^2 \int_{r_s}^{\infty} r^2 \psi(r) dr \qquad (5.26)$$

so that

$$q_{eff} = \varepsilon \zeta r_s \left(\kappa r_s + 1 \right) \qquad (5.27)$$

In the special case that $r_s \kappa \ll 1$, we would have an effective charge on the moving colloidal particle that is proportional to the zeta potential ζ, and to the radius r_s. Taking that simple case, we can now find the velocity V with which the charged particle would move through a liquid of viscosity η. The layer that is affixed to the particle surface is usually only a few ion diameters thick, and thus is negligible compared to the radius of the colloidal particle, R. The viscous drag force on the particle can be found from the Stokes equation

$$f_{visc} = -6\pi\eta RV \qquad (5.28)$$

while the electrical force exerted by the field E is

$$f_{elec} = E q_{eff} \qquad (5.29)$$

Putting the net force on the particle equal to zero. as required by the steady-state condition. we get the electrophoretic velocity V in terms of the zeta potential ζ, the applied electric field E. and the viscosity of the liquid η:

$$V = \frac{\varepsilon \zeta E}{6\pi\eta} \qquad (5.30)$$

Note that the velocity is independent of the particle size in equation (5.30), so we do not have to measure the particle size to determine the zeta potential; measurement of the velocity under a known electric field and knowledge of the physical properties ε and η is sufficient.

We must expect that the electrophoretic velocity V, which is, in general, proportional to the effective charge on the particle rather than to ζ, will depend on the electrolyte concentration and surface charge conditions of the colloidal particle. One can add substances to the liquid that do change the surface charge q_s or the surface potential of the particle. For oxide particles and many other kinds of colloids, acids or bases, even in low concentrations. affect the surface charge of the particle. Then one is usually able, by adjusting the pH of the liquid, to find a condition in which the surface charge, and hence the effective charge, is zero. The colloidal particle will then not move in an applied field. This point is called the isoelectric point. Figure 5.15 shows how electrophoretic mobility V/E, which is proportional to the zeta potential, according to equation (5.30), might vary as a function of pH. Typical values for the zeta potential might be between +100 and -100 mV.

The zeta potential is frequently correlated with colloidal properties. It is. for ex-

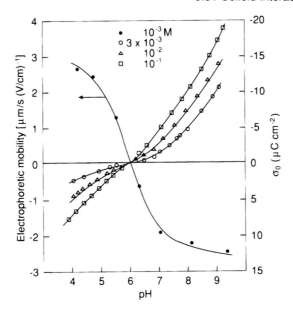

FIGURE 5.15 Surface charge density an electrophoretic mobility of rutile as a function of pH in aqueous solutions of potassium nitrate. (Reprinted with permission from H. M. Jang and D. W. Fuerstenau, *Colloids Surf. 21*, 238 (1986). Elsevier Science Publishers, B. V., Amsterdam.)

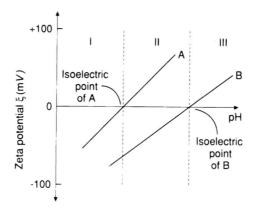

FIGURE 5.16 Schematic representation of the dependence of the zeta potential, z, on pH, of two different colloidal materials, A and B, simultaneously present in a liquid. Since the isoelectric points occur in this example at different pH values, the A and B colloids will have similar charge signs in regions I and III, and opposite charge signs in region II. Bringing the pH suddenly from a stable region (I or III) to a value in region II will cause electrostatic attraction between A and B to form neutral complex that precipitate. This process is an example of coprecipitation.

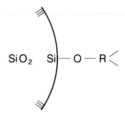

FIGURE 5.17 Schematic representation of a polymer anchored onto an SiO_2 particle surface.

ample. a measure of the strength of the interactions between colloidal particles, and hence relates to colloid stability. At the isoelectric point the colloidal suspension will flocculate; however, flocculation may occur at zeta potentials that are somewhat different from zero as well, depending on the height of the energy barrier in the pairwise interaction potential (Fig. 5.13). The farther away from the isoelectric point, the slower we should expect flocculation to proceed, since the potential energy barrier gets progressively higher.

Different materials will have different dependences of the zeta potential on the pH, so that there can be a pH range for which different colloidal particles that have opposite zeta potentials will coprecipitate, since they attract each other. This can be practically exploited when preparing ceramics powders, containing more than one type of material, that need to be intimately mixed (Fig. 5.16).

5.3.6 Stabilization by Polymers

Much of the theory describing colloidal behavior has been developed for inorganic electrolytes. In practice, colloid stabilization by additions of polymer to a slurry of particles in water is the oldest and most widely used method. Stable suspensions can be produced that can have a high solids content while still having a low viscosity. A well-known example of antiquity is ink, in which soot particles are suspended with a natural polymer such as egg white or gum arable. Solvents may be water or organic compounds, depending on the solubility of the polymer of choice. Thus a very wide range of possibilities is at our disposition to produce the desired suspension.

The polymers may contain ionizable groups, such as acetates, and thus produce ions that work in a manner analogous to the inorganic salt electrolytes that we have discussed so far. However, the most interesting and beneficial aspects of polymers as stabilizing agents derive from the interactions of the polymer chains themselves, when they have been attached to the particle surface somehow.

The polymers that are most frequently used are long-chain copolymers of which one half (or *moiety,* after the French for "half") reacts with the particle surface and anchors the polymer, and the other half sticks out into the solvent in which the particle is suspended. Indifferent polymers—those that do not contain reactive, anchoring moieties—can also be used, but they are not as effective. An example of a possible polymer anchored on a silica surface is depicted in Fig. 5.17.

The mechanism by which flocculation is prevented in polymerically stabilized col-

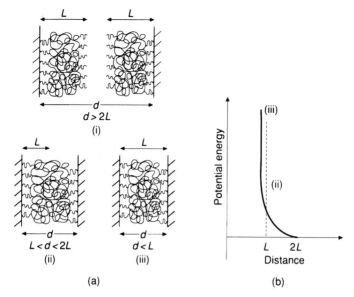

FIGURE 5.18 (a) Three domains of close approach of sterically stabilized flat plates: (i) noninterpenetration; (ii) interpenetration; (iii) interpenetration plus compression. (Reprinted with permission from the publisher, R. Hunter, *Foundations of Colloid Science,* Oxford University Press, London, 1987.) (b) Corresponding energy diagram.

loids involves the entanglement of the polymers upon close approach. These mechanisms fall in the category of steric stabilization. Free polymers can also affect the interaction of two approaching particles, depending on whether or not they are caught in between.

5.3.7 Steric Stabilization

Figure 5.18 shows conceptually what happens when two particles, on which polymers have been anchored, approach one another. The stretched-out polymer chains will get in each other's way and limit the number of possible spatial configurations that they can assume. This lowers the degree of disorder and thus decreases the entropy of the system. Because the entropy is decreased, the free energy of interaction is less negative; in other words, there is less tendency for interaction. The entropic effect opposes flocculation, providing what amounts to a repulsive interaction. Other effects upon overlap of the polymer layers on the particles can contribute to the repulsive interaction as well. They include the increase in local polymer concentration in the overlap zone, causing an osmotic pressure rise, as in the case of inorganic salt electrolytes discussed in Section 5.3.2. At the same time, there will also be an enthalpy of mixing in this layer-overlap region as well as other chemical interaction effects between solvent and polymer that give enthalpic contributions. The enthalpic contributions to the free energy are likely to be fairly temperature independent in the absence of any phase changes at the surface and may be either positive or negative. It is then possible to classify the action of polymers in colloidal stabilization according to the magnitude and sign of the thermodynamic factors that are involved. If we put the total free energy

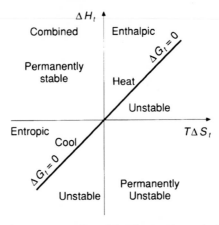

FIGURE 5.19 Schematic representation of the thermodynamic factors controlling steric stabilization. [D. H. Napper in *Colloidal Dispersions*, J. Goodwin, ed., Royal Society of Chemistry (1981). Reprinted with permission from the publisher.]

of interaction upon close approach, ΔG_t, equal to

$$\Delta G_t = \Delta H_t - T\Delta S_t \tag{5.31}$$

where ΔH_t refers to all enthalpic and ΔS_t to all entropic effects, the classification can now be depicted as in Fig. 5.19. It is still difficult to predict the sign or magnitude of the various thermodynamic factors. However, the thermodynamic description provides a rationalization of the behavior of the polymeric stabilized colloids with respect to temperature. For example, heating a polymerically stabilized colloidal suspension can give rise to rapid flocculation when the condition for which $\Delta G_t = 0$ is crossed, as shown in Fig. 5.19. The temperature at which the stability of the colloidal suspension decreases dramatically, as measured by the decrease in time it takes for the colloid to flocculate, is called the *critical flocculation temperature (CFT)*. Similar effects can be produced by changing the pressure over the colloidal suspension. Further, flocculation may also be induced by changing the character of the solvent (e.g., by mixing in another solvent in which the stabilizing polymer moiety is not very soluble).

In certain cases, the polymers will not be able to stabilize the suspension. This can happen when at low polymer concentration, the colloidal particle has usable anchoring sites left on the surface that are not yet occupied. Then polymers from a neighboring particle may attach themselves, and now the colloidal particles are connected by polymer bridges and flocculate. This is called *bridging flocculation*.

The effect of free polymers on the colloid stability is more subtle. One can imagine the particles so close together that now only small solvent molecules can fit in between, and the macromolecular polymers are excluded because they are too big. Then we have actually created a negative osmotic pressure that pushes the particles together. This is *depletion flocculation*. On the other hand, if we had started out at some distance and had trapped some polymer in between, we would have to do extra work to make the polymer leave the interparticle gap before we could make the particles approach one

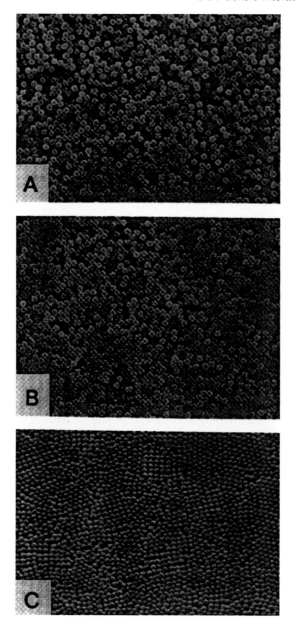

FIGURE 5.20 Microstructures of particle domains formed by centrifugal consolidation of SiO_2 *colloidal* suspensions at (a) $z = 0$ mV, (b) $z = 68$ *mV*, and (c) $z = 110$ mV. Average particle diameter is 0.7 m. [I. Aksay, in *Advances in Ceramics*, Vol. 9, J. Mangels, ed. (1984), p. 96. Reprinted with permission from the American Ceramic Society.]

another any closer. This would provide a repulsive interaction between the particles. In short, we must have some sort of potential energy barrier to overcome before particles can flocculate, even when only free, noninteracting polymer macromolecules are present in the solvent.

An important difference in the pair-potential relationships between ion- and polymer-stabilized colloids is that for the latter, the particles do not fall into the strong primary energy minimum (see Fig. 5.13) as long as the polymers remain firmly attached. Thus flocced colloids in polymerically stabilized systems will be easier to disrupt or to redisperse. The relative ease with which the colloids are redispersed will have significant beneficial effects for keeping the viscosity of such suspensions low, even at high solids content. It is also possible to mix separately prepared colloids that have been stabilized with different polymers. If these polymers are mutually compatible and tend to coprecipitate, the colloidal dispersions will also coprecipitate on mixing. If, however, the polymers are not compatible, stability is likely to be maintained in mixing. This general principle is useful in the preparation and manipulation of mixed ceramic powders. An example of where this might be useful is the fabrication of particulate ceramic composites. Such composites consist of a ceramic matrix of one type of material, say aluminum oxide, in which a second phase, say of fine particles of zirconium oxide, has been dispersed. The homogeneity of the dispersions of the second phase affects very strongly the mechanical properties of particulate composites. The polymeric coprecipitation method would help in achieving homogeneously mixed starting powders for such ceramics.

In the case of polymeric stabilization we will, in general, have a combination of all the charge and steric effects, making the situation very difficult to predict in detail. This is one of the reasons why polymeric stabilization, while being the most effective and desirable way to stabilize concentrated colloidal suspension, is still somewhat of an art.

5.3.8 Structure of Consolidated Colloids

To make a practical powder compact, the colloidal suspension must be made to settle either by itself (under gravity) or by forcing it to settle by filtration or by centrifuging. The structure of the deposit that is formed in each case will be the starting point for the powder that is to be turned into a useful ceramic. The quality of the finished ceramic will depend strongly on the structure of the powder compact. and therefore we need to understand how we can manipulate the colloidal suspension to achieve the best powder compact.

The settling of the colloidal suspension by itself was discussed in previous sections of this chapter, and depended on how we could manipulate the zeta potential of the particles to cause flocculation. The resultant floc structures are quite loose and have a highly irregular pore structure. One can envision the structure as coming about in two steps: (1) sticking an arriving particle on another particle or cluster of particles, and (2) rearrangement of the captured particle or particle cluster into a lower-energy position. The second step depends on how tightly the arriving particle will stick. If the arriving particle sticks completely, without being able to move any further, a low-density floc will result. The further agglomeration of small floc clusters (sometimes called domains)

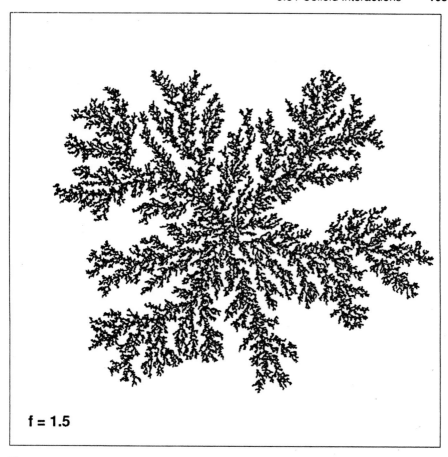

f = 1.5

425 DIAMETERS

FIGURE 5.21 A 25,000 particle aggregate generated using an off-lattice aggregation model in which particles travel along a Levy walk trajectory with a fractal dimensionality of 1.5 (Courtesy of P. Meakin.).

into larger ones will then make the overall structure even looser, with pores between the clusters. In fact, we could envision that repetitive agglomeration would result in the formation of a hierarchical structure in which progressively larger pores are present between progressively larger agglomerates. For strongly attractive forces between particles, we expect highly irregular clusters or domains for which rearrangement is difficult, and we end up with a low-density deposit consisting of small domains with considerable interdomain spacing. The weaker the attraction between the particles, and the more opportunity for rearrangement, the denser the deposit will be, with progressively larger and denser domains and smaller interdomain pores. These principles can readily be incorporated in computer modeling of floc structures, and Fig. 5.21 gives an example of a computer-generated floc cluster. In the computer calculations we could include finer points, such as whether or not the particles will firmly stick when they collide with another particle or cluster due to some chemical bonding reaction probability. The degree or rate of such bonding reaction can then be the variable that determines to what extent the particle cluster can rearrange. Actual flocs have been examined by microscopy, and their structure can be remarkably similar to that shown in Fig. 5.21. Obviously, the manipulation of the zeta potentials of the particles is one way to manipulate the structure of the deposit. Control of the surface chemistry to affect the rate at which bonding reactions between particles occurs would be another. One drawback of letting the deposits form by themselves rather than forcing them to settle by filtration or centrifuging is that these deposits usually have too low a density. Forced settling is therefore the method of choice.

In forced settling. such as filtration or centrifuging, we can make the particles come together, even if they have repulsive interactions. Again, the magnitude of the repulsive interactions can be tailored by changing the zeta potential. The principles discussed for free flocculation can be invoked to understand the structural trends in the deposits that form. If the particles are strongly repulsive, they will certainly stick less well together, and thus more rearrangement can occur and denser deposits will result. In the case of highly repulsive interactions for particles that are all the same size. with lots of rearrangement possible, we may even generate deposits in which particles are arranged very regularly, almost like atoms in crystals. The principles are very well demonstrated by Fig. 5.20 for nearly monosized colloidal silica.

5.3.9 Rheology of Colloidal Suspensions

An important factor in the processing of colloidal suspensions, especially concentrated suspensions, is the ease with which the particle/solvent mixture flows. For simple, pure fluids, we would expect to find Newton's law obeyed. This law puts the shear stress τ on a liquid layer proportional to its shear rate $\dot{\gamma}$. The proportionality factor is the viscosity η. Newtonian fluid flow is then described by

$$\tau = \eta \dot{\gamma} \tag{5.32}$$

Colloidal suspensions should not be expected to follow such a simple relationship. During flow, particles will collide with one another, move past each other, or possibly, stick together. This must affect the viscosity of the suspension. Two types of effects may be distinguished: (1) hydrodynamic effects, which are solely the result of the forces that the fluid exerts on the particles as they alter their trajectories during collision or flow-by, and (2) effects due to the interactions of the colloid double layers on close approach.

The possible responses of the colloidal suspensions to increasing shear rates are shown in Fig. 5.22. For Newtonian viscous fluids, η is independent of shear rate. If the viscosity becomes dependent on the shear rate, we have to define it by taking the derivative of the shear stress-shear rate relation

$$\eta(\dot{\gamma}) = \frac{d\tau}{d\dot{\gamma}} \tag{5.33}$$

When the viscosity increases with increasing shear rate, we have what is called shear thickening. The case where the viscosity decreases with increasing shear rate is called

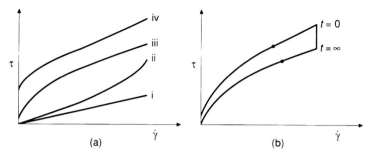

FIGURE 5.22 Typical rheological behavior of colloidal dispersions. (a) (i) Newtonian; (ii) shear thickening; (iii) pseudoplastic; (iv) plastic (b) thixotropic. [J. Goodwin in *Colloidal Dispersions*, J. Goodwin, ed., p. 166, Royal Society of Chemistry (1981). Reprinted with permission from the publisher.]

pseudoplasticity. When the viscosity decreases with increasing shear rate but some initial threshold stress needs to be surmounted, we have plasticity. Below this threshold stress the suspension behaves like an elastic material. Another effect can be that the viscosity is time dependent. When we have plasticity or pseudoplasticity with a viscosity that depends on the time of stirring, as depicted in Fig. 5.22, one speaks of thixotropy. It is this phenomenon that is exploited in some no-drip paints. Thixotropic effects occur for partially flocced, anisotropic particle suspensions. When left alone, a floc network will gradually develop that increases the viscosity of the suspension and that may even exhibit a yield point. Stirring, as when moving paint with a brush, will disrupt the network and cause a drop in the viscosity.

For a dilute dispersion of hard, noninteracting spheres, Einstein derived a relationship between the viscosity of the dispersion, η, the viscosity of the parent fluid, η_0, and the volume fraction of the dispersed particles, f, by analyzing the hydrodynamic forces that were acting on particles during flow:

$$\eta = \eta_0(1 + 2.5f) \tag{5.34}$$

This is quite a good approximation up to a few volume percent of inert particles. For higher-volume fractions, higher-order terms in f need to be included and rigorous derivations are not yet possible. It has been found useful to define a quantity known as the intrinsic viscosity, $[\eta]$, by

$$[\eta] = \frac{1}{\eta_0}\left(\frac{\partial \eta}{\partial f}\right)\bigg|_{f=0} \tag{5.35}$$

evaluated at the low particle concentration limit of $f \to 0$. A common, semiempirical expression for the case of high concentration of inert solid particles then reads

$$\frac{\eta}{\eta_0} = (1 - k_p)^{-\frac{[\eta]}{k_p}} \tag{5.36}$$

where the parameter k_p stands for the inverse volume fraction, f^{-1}, at which the viscosity

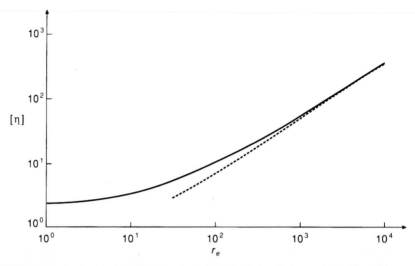

FIGURE 5.23 Intrinsic viscosity as a function of axial ratio for prolate ellipsoids [J. Goodwin in *Colloidal Dispersions*, J. Goodwin, ed., p. 178, Royal Society of Chemistry (1981). Reprinted with permission from the publisher.]

becomes practically infinite. For noninteracting spheres that critical volume fraction is around 0.65. In practice, particulate suspensions tend toward high viscosities for volume fractions of solids as low as 0.4 to 0.5. This is because particles interact with each other and are not inert spheres. The particle shape can play an important role in the viscosity of the suspension, as is evident from considering Fig. 5.23. This figure shows the values of calculated intrinsic viscosities, as defined by equation (5.35), for various values of the aspect ratio of hypothetical, monosized particles.

For charged particles, mutual interactions will be stronger. This is intuitively clear if one regards the sphere of particle interaction to include the double layer as well. Then the effective hydrodynamic radius would be larger, compared to particles without a double layer. The interactions can be strong, and viscosities can be many times larger for charged colloids with an extended double layer than for the same uncharged ones. The higher the zeta potential, the larger one should expect the effective radius to be, with a concomitant increase in viscosity. Thus adjustment of the zeta potential provides a practical means of manipulating the viscosities of colloidal suspensions. At the same time, effects such as shear thinning occur in these suspensions, making this subject rather difficult to handle with theory. The important effects of the zeta potential remain, however, qualitatively predictable.

During the rapid flow of electrostatically stabilized colloidal suspensions, collisions between the particles could be vigorous enough to propel them over a potential energy barrier that would otherwise prevent flocculation. Whether or not this will occur will depend on the strength of the repulsive interactions (i.e., the magnitude of the zeta potential) and the shear rate. At sufficiently high shear rates flocculation may occur due to the particle collisions. This is known in the literature as orthokinetic flocculation. The higher the zeta potential, the higher the shear rates that will be necessary to cause such flocculation. At extremely high shear rates, however, the hydrodynamic forces acting on particles and particle groups may be sufficient to redisperse agglomerates. Very high shear rates are required for this because the potential barrier to flocculation in the primary minimum is not symmetrical with respect to interparticle distance, as is evident in Fig. 5.13. For polymerically stabilized systems, orthokinetic flocculation is less likely to happen since, as we discussed. particles do not tend to cluster in their primary minimum except under extreme compression. This is yet another argument in favor of stabilization of practical suspensions by polymers.

Polymeric additives may change the viscosities of slurries rather dramatically. Particularly effective are the polyelectrolytes, which are polymers that contain ionizable groups. Adding as little as 0.5 wt % of such polymer to a slurry may permit solids contents of 50 or 60% while maintaining relatively low viscosities. The effectiveness of such polyelectrolytes can be affected still further by control of the pH of the slurry. One class of the recently developed polyelectrolytes are water-soluble naphthalene sulfonate formaldehyde copolymers that can be obtained as sodium or potassium salts. as ammonium salts, or as acids. Many other dispersants now on the market have been shown to be effective in reducing the viscosities of slips, such as methylcellulose esters, ethylene oxide ethers of alkyl benzene, and so on.

5.4 Concluding Remarks

In this chapter we have provided an introduction to the complicated field of colloid behavior. The necessity of this introduction follows from the increasing importance that colloidal suspensions play in the production of high-performance ceramics. Many aspects of colloid behavior cannot. at present. be treated systematically; for example. the full incorporation of the surface chemistry remains quite difficult. Another difficulty is that most colloid theories lose their quantitative applicability for highly concentrated slurries, the very thing the ceramist is interested in. Nevertheless, the predictions of colloid theory are, at the very least, qualitatively correct and form the foundation of a rational approach to colloidal processing. In this, control of the zeta potential and judicious use of polymeric stabilizers are key ingredients for success.

LIST OF SYMBOLS

C_A	concentration of A
E	applied electric field
$E_{1\text{-}2}$	interaction energy
$\lvert e \rvert$	absolute value of the charge of one electron
f	force
H	Hamaker constant
k	Boltzmann constant = gas constant/Avogadro's number
q_{eff}	effective charge on a particle
$q(x)$	charge density at x
R	radial distance
T	absolute temperature (Kelvin)
V	electrophoretic velocity

V/E	electrophoretic mobility
x_D	Debye length
z	number of charges on an ionic species, in units of $|e|$
ε	dielectric constant
κ	1/Debye length
μ_A	electrochemical potential of A
$\psi(x)$	electrical potential at position x
η	viscosity
∇^2	Laplacian operator (spatial second derivative)
Π_{OS}	osmotic pressure
σ_s	surface charge
ζ	zeta potential

PROBLEMS

5.1 **(a)** A solid with a flat surface is submerged in a 0.01 molar (M) solution of NaCl. It has a surface potential of +25 mV. Plot the electrical potential as a function of distance from the surface and determine the Debye length, assuming that the dielectric constant of the solution is 78.

   (b) If NaCl is replaced by $CaCl_2$, what would be the concentration of $CaCl_2$ at which the Debye length is 0.01 μm?

5.2 For the conditions of Problem l(a), find the surface charge density.

5.3 For two surfaces and a solution as in Problem l(a), determine the force distance relationship. [*Hint:* Consider equation (5.20).]

5.4 Estimate the probability of a particle-pair collision leading to agglomeration in the primary minimum for a potential energy barrier of 10 kT. Assume that the particles are agitated as a result of Brownian motion.

5.5 Derive equation (5.l6).

5.6 Show that when $zF\psi$ *(d) is* small compared to *RT*, the potential energy of repulsion would be proportional to $\exp(-2\kappa d)$, where κ is the inverse of the Debye length and d is the separation between two flat plates immersed in an electrolyte.

5.7 Electrophoretic separation is a technique in which colloidal particles are deposited on an electrically conductive substrate (e.g., a metal tube) that acts as the negative or the positive electrode, depending on the sign of the charge on the particles. For a ceramic powder with a particle size of 0.3 μm, suspended in an electrolyte with a dielectric constant of 10 and a viscosity of 1.3 cP, an electrophoretic velocity of 2.3 μm/s was observed at an applied field of 10 V/cm. The concentration of the monovalent salt in the electrolyte is 5×10^{-6} mol/L. Calculate what the electrophoretic deposition, as the deposited layer thickness as a function of time, would be at a current density on the electrode of 25 mA/cm², assuming that the deposited layer has a relative volumetric density of 0.45. [*Hint:* You will have to find the effective charge on the particles.]

REFERENCES AND FURTHER READINGS

Monographs

ADAMSON, A. W., *Physical Chemistry of Surfaces,* Wiley, New York, 1976.

HEIMIT, P., *Principles of Colloid and Surface Chemistry,* Marcel Dekker, New York. 1986.

ISRAELACHVILI, J., *Intermolecular and Surface Force,* Academic Press, Orlando, FL, 1985.

SHAW, D. J., *Introduction to Colloid and Surface Chemistry,* Butterworth, London, 1985.

Articles and Article Collections

AKSAY, I., "Microstructure Control Through Colloidal Consolidation," in *Forming of Ceramics* (Advances in Ceramics, Vol. 9), J. Mangels, ed., American Ceramic Society, Columbus, OH, 1984.

FAMILY, F., and D. R LANDAU, eds., *Kinetics of Aggregation and Gelation,* Elsevier, New York, 1984.

GOODWIN, J. W., ed., *Colloid Dispersions,* Royal Society of Chemistry, London, 1982.

HU, S. C., and L. C. DE JONGHE, Pre-eutective Densification in MgF_2-CaF_2, *Ceram. Int. 9 (1983).*

JANG, H. M., and D. W. FUERSTENAU, The Specific Adsorption of Alkaline-Earth Cations at the Rutile/Water Interface, *Colloids Surf. 21* (1986).

KEEFER, K., "Growth and Structure of Fractally Rough Silica Colloids," in *Better Ceramics Through Chemistry 11,* C. J. Brinker, D. Clar, and D. Ulrich, eds., Materials Research Society, Pittsburgh, PA, 1986.

LEAMY, H., G. GILMER, and K. JACKSON, Statistical Thermodynamics of Clean Surfaces," in *Surface Physics of Materials, Vol. I,* J. M. Blakely, ed., Academic Press, New York, 1975.

OVERBEEK, J. T., Monodisperse Colloidal Systems, Fascinating and Useful, *Adv. Colloid Interface Sci. 15 (* 1982).

RAHAMAN, M. N., Y. BOITEUX, and L. C. DE JONGHE, Surface Characterization of Silicon Nitride and Silicon Carbide Powders, *Ceram. Bull. 65* (1986).

VANSELOW, R., *Chemistry and Physics of Solid Surfaces,* Vols. I-III, CRC Press, Boca Raton, FL, 1974-1982.

CHAPTER 6

Fundamentals of Heat Treatment and Sintering

6.1 Introduction

So far in this book we have had little interest at the microscopic level in the structure of the solids that we have encountered. Exceptions have been in Chapter 3, where we recognized that a reacting solid might be porous (and the pores affect the rate of reaction), and in Chapter 4, which was concerned with the shape and size of the small particles that we frequently encounter in ceramic or powder metallurgical processing.

We must now recognize that the microstructure of a material is, along with composition, the fundamental determinant of its properties. In producing and processing materials considerable attention must be paid to the microstructure that results. In what follows we assume that the reader has already become aware that the majority of inorganic materials consist of grains, typically containing defects, that are separated by grain boundaries. The grains need not be of the same crystal structure (polyphase materials) and there may be pores present. In materials processing the microstructure is manipulated by heat treatment and sintering.

The heat treatment and sintering phenomena involve a number of fundamental concepts governing equilibrium phase relationships and movement of matter. During heat treatment particular microstructures are developed by second-phase formation or grain growth, and powder compacts densify so that pores are eliminated. The interaction of all these processes can make the development of a desired microstructure a rather complex and difficult task. The fundamental principles, however, are general, and can be treated separately. In this chapter we deal first with atom movement, then with phase relationships, and last with some application to the elementary aspects of densification.

6.2 Transport of Matter

6.2.1 Elementary View of Diffusion

We have already touched on this topic in Chapter 3, where it was necessary to discuss the role that diffusion and convective mass transport play in heterogeneous reaction kinetics. Now we shall make a more detailed examination of diffusion, particularly in the solid state.

Usually concentration differences, or more precisely concentration gradients, are regarded as the *driving forces* for the movement of atoms (or molecules) that we call diffusion. On first examination this concept may seem strange in that it implies that the diffusing atoms somehow "know" how many of their fellow atoms are nearby, and furthermore, they can distinguish like atoms (contributing to the concentration of their species) from unlike atoms (not contributing). Let us start with a few calculations which show that no such flight of fancy is necessary and that, to a first approximation, diffusion is simply a consequence of elementary statistics (this is best illustrated by the intermixing of gases by diffusion).

Consider the system sketched in Fig. 6.1. It consists of a box divided into two chambers, each containing a mixture of two gases. The total pressures in the two chambers are the same (so that no hydrodynamic-pressure driven flow occurs), but the partial pressure of one of the gases in the left-hand chamber is P_1 and in the right-hand chamber is P_2. For the sake of this discussion we will put $P_1 > P_2$. Henceforth we shall consider only this gas and ignore the other gas component. If the gas is ideal and the contents of each chamber are well mixed, we can write

$$C_1 = \frac{P_1}{RT}$$

(6.1)

where C_1 is the concentration (moles per unit volume) in chamber 1. The corresponding equation for C_2 is obvious. The kinetic theory of gases tells us that the rate at which molecules strike the hole connecting the chambers from the left is $P_1 A(2\pi k T m)^{-1/2}$, where k is the Boltzmann constant, T the absolute temperature, and m the mass of the molecule. These molecules therefore pass from the lefthand chamber to the right. Correspondingly, molecules pass from the right-hand chamber to the left at a rate of $P_2 A(2\pi k T m)^{-1/2}$ and the net rate of transfer is $(P_1 - P_2)A(2\pi k T m)^{-1/2}$. We will find it convenient to work in terms of a flux, J, which we obtain by dividing by the area of the hole A:

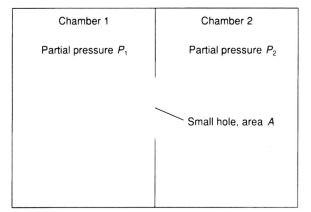

FIGURE 6.1 Schematic diagram of an apparatus in which diffusion of a gas component occurs through a hole in the partition dividing the apparatus into two chambers.

$$J = \left(P_1 - P_2\right)\left(2\pi kTm\right)^{-1/2} \tag{6.2}$$

This flux would be in units of molecules per unit area per unit time and frequently we will prefer to express the flux in moles per unit area per unit time. We obtain the flux in these new units by dividing by Avogadro's number N,

$$J = \left(P_1 - P_2\right)\left(2\pi RTm\right)^{-1/2} \tag{6.3}$$

where M is the molecular weight and we have used the relationships

$$Nk = R \tag{6.4}$$

and

$$Nm = M \tag{6.5}$$

Now substituting for the partial pressures gives

$$J = \left(\frac{RT}{2\pi M}\right)^{1/2}\left(C_1 - C_2\right) \tag{6.6}$$

We see then that diffusion through the hole is simply the consequence of molecules striking the hole from the left more frequently than from the right, just because there are more molecules on the left (of the type we are considering). The driving force for diffusion is therefore no mysterious knowledge possessed by the molecules, but just a consequence of the way their numbers are distributed. When the number per unit volume in the left-hand chamber equals that in the right-hand chamber there will no longer be a net diffusion.

Let us now extend these concepts to a long series of chambers shown in Fig. 6.2(a).

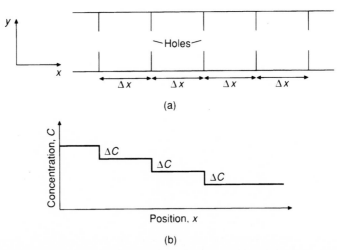

FIGURE 6.2 Series of well-mixed chambers between which diffusion is taking place through small holes and the corresponding concentration profile.

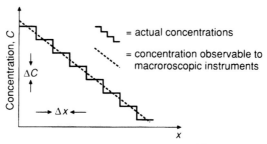

FIGURE 6.3 Concentration profile when the chambers of Fig. 6.2 become small.

Each chamber is of dimension Δx, measured in the x direction. Note that Δx is the distance the "average" molecule travels to hop from one box to the next. For the sake of this discussion the concentration differences between adjacent boxes will all be the same ΔC, as shown in Fig. 6.2(b). The flux from left to right (through each of the holes) is still given by equation (6.6), which we choose to rewrite as

$$J_x = \left(\frac{RT}{2\pi M} \right)^{1/2} \Delta x \frac{\Delta C}{\Delta x} \qquad (6.7)$$

where a subscript has been added to the J to indicate that we are discussing a flux in the x direction.

Now consider what happens as we shrink Δx so that the chambers become very small. To be precise, we suppose that they become smaller than can be resolved by conventional (macroscopic) gas sampling or measuring devices. The consequence is shown in Fig. 6.3; the actual concentration curve is a series of steps, but to our instruments that curve appears to be a smooth one. An analogy may be useful here. At a distance of several kilometers the Egyptian pyramids appear to be smoothsided but from a distance of less than I km it is clear that they are made up of large blocks of stone. But for the laws wisely laid down by the authorities, it would be possible to ascend the pyramids by the "staircase" formed by these blocks.

The slope of the dashed line in Fig. 6.3 is the concentration gradient observable to our macroscopic instruments, dC/dx, and for Δx small,

$$\frac{dC}{dx} = - \frac{\Delta C}{\Delta x} \qquad (6.8)$$

Substituting in (6.7) gives

$$J_x = - \left(\frac{RT}{2\pi M} \right)^{1/2} \Delta x \frac{dC}{dx} \qquad (6.9)$$

Note that this equation does not contain the area of the holes, A, and therefore ought to apply for any size of hole, including the case where the hole occupies the whole side of a chamber. Of course now all the chambers have merged into one and we have diffusion down a long tube. This poses a problem, however; how are we to interpret Δx if we do away with the chambers? Recall that Δx was the average distance that a molecule traveled to hop from one chamber to the next.

If we regard diffusion as a sequence of hops, Δx is the average distance traveled on each hop. We shall emphasize this by replacing Δx by the symbol L_H henceforth. Therefore,

$$J_x = -\left(\frac{RT}{2\pi M}\right)^{1/2} L_H \frac{dC}{dx} \tag{6.10}$$

The question is then: What is the average distance a molecule hops in the absence of the chambers? In a gas, the likely answer is the mean free path of the gas molecule, λ. This is the average distance a gas molecule travels before striking another gas molecule. From the kinetic theory of gases it is known that

$$\lambda = \frac{RT}{\sqrt{2}\pi d^2 NP} \tag{6.11}$$

where d is the diameter of the gas molecule (assumed spherical in this simple treatment) and P is the total pressure. Substituting λ for L_H in equation (6.10) gives

$$J_x = -\left[\frac{1}{2}\left(\frac{RT}{\pi M}\right)^{1/2} \frac{RT}{\pi d^2 NP}\right] \frac{dC}{dx} \tag{6.12}$$

For a given gas at a given temperature and pressure the quantity in brackets is a constant and we can compare equation (6.12) with Fick's first law,

$$J_x = -D\frac{dC}{dx} \tag{6.13}$$

where D is the diffusivity. In fact, the quantity in brackets in equation (6.12) gives us an estimate of the diffusivity of gases. That estimate is an approximate one because of the simplifying assumptions we have made, but it does correctly indicate that diffusivities of gases increase with increasing temperature and decrease with increasing pressure or molecular weight.

Let us reexamine equation (6.10). The quantity $(RT/2\pi M)^{1/2}$ is related to the average speed of a gas molecule, $\overline{V}$, which is given by the kinetic theory of gases,

$$\overline{V} = \left(\frac{8RT}{\pi M}\right)^{1/2} \tag{6.14}$$

and we can therefore write (6.10) as

$$J_x = -\frac{\overline{V}}{4} L_H \frac{dC}{dx} \tag{6.15}$$

Now recognize that the time taken to travel the distance L_H is $L_H/\overline{V}$, or, in other words, the molecule hops $\overline{V}/L_H$ times per second, or it hops with a frequency f given by

$$f = \frac{\overline{V}}{L_H} \tag{6.16}$$

Substituting in (6.15) yields

$$J_x = -\frac{f}{4} L_H^2 \frac{dC}{dx} \tag{6.17}$$

Again comparing with Fick's first law, we have

$$D = -\frac{f}{4} L_H^2 \tag{6.18}$$

Equation (6.18) is informative. Diffusion in liquids and solids can be thought of as hopping from one "chamber" to the next by an atom. In these cases the chamber size (i.e., L_H) is no longer defined in terms of a mean free path but rather as the distance between an atom and an adjacent site into which it can hop (e.g., a vacancy). Furthermore, the frequency is no longer given by simple kinetic theory of gases. Nevertheless, equation (6.18) can be regarded as a semiquantitative one predicting (correctly) that factors, such as increasing temperature, that increase the hopping frequency should markedly increase the diffusivity.

6.2.2 Fick's Laws of Diffusion

We encountered Fick's first law of diffusion in the preceding section and earlier in Chapter 3. It states that the flux of a diffusing species is proportional to the concentration gradient and is in the direction of decreasing concentration. The proportionality constant is called the *diffusivity* (or *diffusion coefficient*). We should recognize that concentration gradients may exist in more than one direction, and the more general form of Fick's equation is then

$$J = -D\nabla C \tag{6.19}$$

with J a vector with components (in a rectangular coordinate system) J_x, J_y and J_z. In other words,

$$J_x = -D\frac{\partial C}{\partial x} \tag{6.20}$$

$$J_y = -D\frac{\partial C}{\partial y} \tag{6.21}$$

$$J_z = -D\frac{\partial C}{\partial z} \tag{6.22}$$

It should be noted that equation (6.19) and the equivalents (6.20)-(6.22) assume that the diffusivity is the same in all directions (i.e., that it is isotropic). These equations contain partial derivatives because the concentration may be a function of all three spatial coordinates. It can also be a function of time, in which case Fick's first law still holds, but we can also write

$$\frac{\partial C}{\partial t} = D\nabla^2 C \qquad (6.23)$$

which in longhand, for rectangular coordinates, is

$$\frac{\partial C}{\partial t} = D\left(\frac{\partial^2 C}{\partial x^2} + \frac{\partial^2 C}{\partial y^2} + \frac{\partial^2 C}{\partial z^2}\right) \qquad (6.24)$$

This is Fick's second law. It assumes that the diffusing species is not consumed or created by homogeneous reaction, that no (macroscopic) motion is taking place with respect to the coordinate system, that the diffusion is isotropic, and that the diffusivity is independent of position. Fick's second law can be derived from his first law and the principle of conservation of matter; history has therefore been kind to Fick in according it the status of a law.

The differential operator ∇^2 appearing on the right-hand side of (6.23) is known as the Laplacian operator. For its form in cylindrical or spherical coordinates, and for equations appropriate when the assumptions listed above are not applicable, the reader is referred to texts on transport phenomena (e.g., Bird et al., 1960).

6.2.3 Mechanisms of Diffusion in Solids

In crystals, the lattice points are occupied by atoms, ions, or ion groups. However, they do so imperfectly, leading to missing atoms or *vacancies*; atoms inserted in spaces that would ideally be empty: *interstitials*; or atoms sitting on the sites that would ideally be occupied by another type of atom: *substitutionals*. These *point defects* are illustrated in Fig. 6.4. Additional deviations from an ideally periodic lattice population may also occur for chemical reasons leading, for example, to deviations from the "law" of multiple proportion (i.e., nonstoichiometry). All of these deviations from the complete, ideal lattice population have strong effects on the rate at which atoms get transported

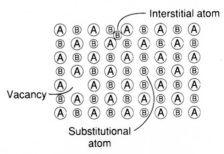

FIGURE 6.4 Point defects encountered in crystal lattices.

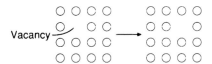

FIGURE 6.5 Vacancy mechanism of diffusion.

through solids. In fact, if it were not for disorder, atoms would hardly be able to change places in solids. We thus must consider how these imperfections affect diffusion and how we can possibly manipulate them. A description of exactly how atoms move about to accomplish matter transport is called the *transport mechanism*.

Vacancy Mechanism

An atom exchanges place with a neighboring vacant site (Fig 6.5). When one tracks the movement of the vacancy rather than of the atoms, one speaks of vacancy diffusion. A flow of vacancies must be compensated by an equal but opposite flow of matter; otherwise, vacancies would accumulate and produce pores. Pore formation can actually happen during interdiffusion of two species with greatly differing diffusion coefficients.

Interstitial Diffusion

Atoms populate and move over lattice sites that are not occupied in the ideal crystal (Fig 6.6). The octahedral and tetrahedrally coordinated spaces in face-centered cubic lattices are, for example, common interstitial sites. The diffusion of carbon in iron alloys, important in steel processing, is an example of interstitial diffusion.

Interstitialcy Diffusion

A lattice atom exchanges places with a neighboring interstitial atom (Fig 6.7). The two need not necessarily be the same species.

Direct Exchange or Ring Mechanism

Atoms exchange places directly, without involving defects (Fig 6.8). In principle, several atoms could participate in a simultaneous exchange. The significant momentary lattice distortions that such an exchange requires makes this a very unlikely mechanism.

Vacancy and interstitial diffusion are by far the most common atom transport mechanisms. The concentration of these defects is determined by temperature in pure, one-component systems. Composition, as for example modified by alloying or by

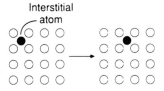

FIGURE 6.6 Interstitial diffusion mechanism.

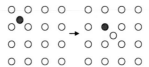

FIGURE 6.7 Interstitialcy diffusion.

oxidizing or reducing atmospheres, can also be a dominant factor in establishing particular defect types and concentrations. The latter is especially of significance for transport in oxide ceramics.

Recall that equation (6.18) shows us that the diffusivity is proportional to the frequency with which an atom jumps from one site to the next. We can now perceive that this jump is over some energy barrier separating one site from another. For example, to pass from one interstitial site to another, the lattice must be distorted momentarily so that the diffusing atom can squeeze between the lattice atoms standing in its way. We therefore expect the diffusivity to depend on temperature in the same way that other activated processes (e.g., chemical kinetics) depend on temperature:

$$D = D_0 \exp\left(\frac{-E_A}{RT}\right) \tag{6.25}$$

where D_0 and E_A are constants for any given chemical system. E_A is the activation energy for diffusion and ranges greatly from around 10 kJ/mol to a few hundred kJ/mol, where large lattice distortions accompany each jump. The consequence is that solid-state diffusivities are very dependent on both the chemical nature of the system and temperature. They typically range downward from $10^{-8} m^2/s$, for small atoms diffusing through lattices at high temperatures, to below $10^{-17} m^2/s$, for large atoms in lattices at low temperatures.

The reader will be aware that the grain boundaries separating grains are highly defective regions and we can therefore anticipate that the flux in a grain boundary would be higher, for the same concentration gradient, than in adjacent grains. This implies that the rate of transport averaged across a region of the solid containing several grains should be somewhat dependent on grain size because the fraction of the solid that is grain boundary will increase with decreasing grain size.

Another type of disordered region is the free surface of a solid. The diffusion of atoms (either of the solid or an adsorbed species) by this surface diffusion mechanism is therefore usually rapid compared to the bulk diffusion.

Finally, the ultimate in disorder, short of a gas, is the liquid state and we anticipate, correctly, that diffusion in liquids is usually faster than in solids. Diffusivities in liquids are typically of the order of $10^{-9} m^2/s$ and weakly dependent on temperature compared to those for solids.

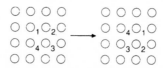

FIGURE 6.8 Direct exchange or ring mechanism of diffusion.

6.2.4 Defect Chemistry

Anything in a crystal that is a departure from the perfect, ideal structure is called a *defect*. The defects may be either electronic, charged or neutral atoms, structural defects such as vacancies, interstitials, dislocations and stacking faults, or grain boundaries, and so on. In defect chemistry one considers mostly the changes in electronic and atomic populations as a function of temperature and composition. A broad distinction is made between intrinsic defects that occur naturally in the pure materials, and extrinsic defects that are caused by external agents such as impurities. For ionic crystals the intrinsic defect concentration such as vacancies are usually several orders of magnitude lower than for metals (e.g., 10^{-5} versus 10^{-3}cm^{-3} near the melting points would be representative numbers). The description of point defects commonly uses a notation developed by Kroger and Vink (*K-V*), although some other notations are sometimes used. In the *K-V* notation a crystal species is denoted with respect to the perfect, idealized structure. Thus M_X^c would mean an M atom on an X site, with an apparent charge of c. The apparent or effective charge is the difference between the valence of the atom on the X site in the ideal structure and that of the species that actually occupies that site. The convention is

$$\text{negative charge } c = {}'$$

$$\text{positive charge } c = {}^{\cdot}$$

$$\text{neutral charge } c = x$$

In some cases defects will be strongly attracted to each other, so that they form associated complexes. Some examples of defects in the *K-V* notation are given in Table 6. 1.

To deal simply with the complications of the equilibrium defect chemistry in crystals when charged species occur, we first consider the reaction equilibrium as in regular chemical equilibria, where reactants A_i lead to products B_i:

$$a_i A_i = b_i B_i + \Delta G \tag{6.26}$$

At equilibrium, at fixed temperature the mass action law must prevail so that we can write the usual equilibrium equation for the activities. For the present we shall regard the activities as equal to concentrations, hence

$$\frac{\prod_i \left[B_i \right]^{b_i}}{\prod_i \left[A_i \right]^{a_i}} = K \tag{6.27}$$

where the brackets denote concentration and the equilibrium constant,

$$K = \exp\left(\frac{-\Delta G}{RT} \right) \tag{6.28}$$

TABLE 6.1 Some defects in MX compound in K-V notation

M_x	An M atom on an X site; if the MX is an ionic compound consisting of M^{2+} and X^{2+}, this defect would have an effective positive charge of 4+, or M_x^{4+}
$(V_M V_X)'$	Associated vacancy complex for the compound ($M^{2+}X^{2-}$)
e'	Electron
$h^{\cdot}$	Missing electron or hole

Recall from Chapter 2 that ΔG is the free energy of reaction as written in the chemical equation from MX. In that case the free energies are usually given per gram atom of oxygen in the reaction.

When dealing with crystals in which charged species may arise, the following conservation rules must be observed in writing defect reactions:

1. *Mass conservation.* A mass balance must be made to ensure that no matter is created or disappears.
2. *Electroneutrality.* No *net* charges should be created. This does not exclude the possibility of charge separation, however, but the defect reaction does not include this information.
3. *Site ratio conservation.* This assures that no different crystal structure is created. It does not matter whether or not the sites are occupied; for example, some crystal may be destroyed as a consequence of a reduction of an oxide by a gas. The disappearance of the crystal must, however, occur in such a way as to preserve the structure of the remaining crystal.

The following examples illustrate some of these points.

Frenkel Disorder

One type of intrinsic disorder in ionic crystals, using (M^+X^-) as an example, may be written as

$$M_M + V_i = M_i^. + V_M' \qquad (6.29)$$

This means that in the ionic crystal, such as (M^+X^-), the positive, monovalent cations can leave their assigned M lattice site and occupy an interstitial site, denoted by the subscript i, where they will have an effective charge of $+1$, and leave a vacant M lattice site behind that must have an effective charge of -1, since now the normal positive charge is missing there.

Assuming that the Frenkel reaction has reached equilibrium we can use equations (6.27) and (6.28). Thus one should have[1]

$$\frac{\left[M_i^. \right]\left[V_M'\right]}{M_M} = \exp\left(\frac{-\Delta G}{RT}\right) \qquad (6.30)$$

but we can assume that the fraction of M_M atoms that go to interstitial positions is very small, so that $[M_M]$ is constant for all practical purposes. If ΔG is known, we would

---

[1] The concentrations would have to be expressed in site fractions. However, it is possible to express concentrations in other units, and this would change the right-hand side of equation (6.30) by the appropriate combination of conversion factors. One may consider this multiplicative factor as accommodated by a constant added to ΔG, so that the formalism remains unchanged and concentration of site fractions very nearly equal to unity again do not have to appear in equation (6.30). In choosing units for the concentrations, it is therefore very important to check if ΔG is given in a manner consistent with those units. The units of concentration chosen here are moles per unit volume.

readily find $[M'_i] = [V'_M] = \exp(-\Delta G / 2RT)$. We could write a Frenkel defect reaction for the anions

$$\left[X_X = X'_i + V^{\cdot}_x\right] \tag{6.31}$$

and similar considerations apply. Note that since cations and anion concentrations are not coupled through an electroneutrality requirement, the concentration of cation interstitials need not equal the concentration of anion interstitials.

Schottky Disorder

Another common type of intrinsic disorder in ionic crystals is one in which M and X leave their regularly assigned lattice sites but instead of producing interstitials, atoms are moved to the surface leaving vacancies behind, so that the crystal actually expands a little. Since we must avoid charging up the crystal, the defects created in the bulk of the crystal must obey the requirements of overall electroneutrality: The electrostatic energy that would otherwise arise would very quickly arrest charge buildup. The *Schottky disorder* can be written

$$0 \leftrightarrow V'_M + V^{\cdot}_X \tag{6.32}$$

Since atoms moved to the surface are only building extra perfect crystal, they do not have to appear in the reaction, although it is possible to write out the full reaction that would take them explicitly into account. The cation and anion vacancy concentrations are now coupled by the electroneutrality equation so that

$$\left[V'_M\right] = \left[V^{\cdot}_X\right] \tag{6.33}$$

Dissolving NX_2 in MX

Below the solubility limit NX_2 would completely dissolve in MX. In this case a dissolution reaction, causing the N cations to go into the $(M+X^-)$ lattice, for example in interstitial positions, can be written

$$NX_2 + V_i \overset{MX}{=} N^{\cdot}_i + 2V'_M + 2X_X \tag{6.34}$$

This reaction, one of the several that are possible in principle, states that when N+ ions occupy interstitial positions, charge compensation is accomplished by creating vacant M+ lattice sites at the same time. To keep track of the site balance, the vacant interstitial sites have been included in this example; usually, this is not necessary, since reaction (6.34) is sufficiently clear. The phase rule, treated later in this chapter, would tell us that this reaction cannot reach equilibrium until the solubility limit of NX_2 has been reached (and we would then have an extra phase). Below that limit one therefore *cannot* apply equations (6.27) and (6.28) to equation (6.34). While the reaction as described here has gone to completion, the other defect equilibria, such as Schottky or Frenkel disorder, are still present.

Suppose that we have Schottky disorder for the MX crystal. Then, since equations (6.27) and (6.28) must apply for the cations and anion vacancies, we have

$$\left[V_M'\right]\left[V_X^{\cdot}\right] = \exp\left(\frac{-\Delta G}{RT}\right) = K \tag{6.35}$$

while at the same time the stoichiometry of the dissolution reaction requires that

$$\left[NX_2\right] = \left[N_i^{\cdot}\right] = \frac{1}{2}\left[V_M'\right] \tag{6.36}$$

Combination of these equations yields

$$\left[V_X^{\cdot}\right] = \frac{K}{2\left[NX_2\right]} \tag{6.37}$$

Thus, by dissolving NX_2 in the ionic crystal MX, we have affected the anion vacancy concentration as well. It is clear that various impurities, dissolved in ionic crystals, can have a profound effect on all properties that involve movement of atoms, since these in turn depend on the concentration of imperfections.

Nonstoichiometry in Cobalt Oxide

Consider the possible defect response of CoO as a function of oxygen partial pressure. At any fixed temperature and composition, cobalt oxide must be in equilibrium with a specific oxygen pressure. If not, the crystal will give up or take up oxygen, thus changing the cation-to-anion ratio, until equilibrium has been reached. When cobalt oxide takes up oxygen, in an oxidation reaction, it will do so by building up the oxygen lattice and creating cation vacancies that take their required charge by combining with an electron, thus creating a missing electron or hole in the valence band. The overall reaction may be written as

$$\frac{1}{2}O_2(g) = V_{Co}'' + 2h^{\cdot} + O_0 \tag{6.38}$$

Site conservation is satisfied in this reaction since

$$\left[V_{Co}''\right] = \left[O_0\right] \tag{6.39}$$

where $[O_0]$ is the concentration of the new oxygen ions in the crystal. Charge neutrality is observed since

$$\left[V_{Co}''\right] = \frac{1}{2}\left[h^{\cdot}\right] \tag{6.40}$$

Note here that while the number of cation sites that are created equal the number of new anion sites, the cation sites are vacant; this is nonstoichiometry. We could define nonstoichiometry that way: cation/anion site ratio different from the cation/anion atom ratio. Nonstoichiometry thus is a departure from Dalton's law of multiple proportions (as are solid solutions).

If we call K the equilibrium constant associated with the reaction (6.38), mass action would give, again considering $[O_O]$ invariant,

$$\frac{\left[h^{\cdot}\right]\left[V''_{Co}\right]}{p_{O_2}^{1/2}} = K \tag{6.41}$$

or substituting from the charge neutrality relation (6.40),

$$4\left[V''_{Co}\right]^3 = Kp_{O_2}^{1/2} \tag{6.42}$$

or

$$\left[V''_{Co}\right] \sim p_{O_2}^{1/6} \tag{6.43}$$

Thus any property involving the transport of cations will be affected by the oxygen partial pressure that has been equilibrated with the cobalt oxide. In many cases, we could measure a physical property, such as electrical conductivity, and from its dependence on oxygen pressure infer what the proper defect reaction is for nonstoichiometry.

Having digressed to examine the mechanisms of diffusion and the nature of defects in solids, let us return to Fick's first law and more quantitative aspects of diffusion.

6.2.5 Tracer and Self-diffusion in Solids

From the discussion so far it is evident that atoms (and molecules) diffuse even in the absence of a concentration gradient. That is, they undergo random motions that sooner or later cause them to visit all regions of the body to which they are confined. In a sense the presence of a time-dependent concentration gradient is merely a manifestation of these random walks of the diffusing atoms and allows us to observe how species A spreads itself throughout a region formerly occupied only by B, and vice versa. An alternative way of revealing this diffusion is by using radioactive isotopes of the material under study; for example, we could examine the diffusion of radioactive gold in gold. A typical experiment coats a crystal with a thin layer of radioactive atoms, which then diffuse into the crystal, as in Fig. 6.9.

The diffusion of the radioactive gold can then be measured by cutting thin slices of gold along planes parallel to the surface and measuring the radioactivity in each slice.

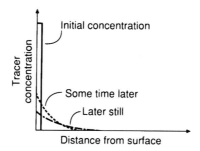

FIGURE 6.9 Diffusion of a tracer, initially deposited as a thin layer on the surface of a thick body into which it diffuses.

If the thickness of the layer of radioactive gold initially is small compared to the distance over which the gold diffuses, then the solution of equation (6.24) for this case is

$$C = \frac{R}{\left(\pi D * t\right)^{1/2}} \exp\left(\frac{-x^2}{4D * t}\right) \tag{6.44}$$

where R is the amount (moles per unit area) of radioactive gold deposited on unit area of the surface and $D*$ is the diffusivity of radioactive gold in gold. Consequently, a plot of the experimentally determined $\ln C$ versus x^2 should yield a straight line of slope $-1/(4D*t)$, enabling the diffusivity to be determined. The diffusivity measured in this or similar ways is termed the tracer diffusion coefficient. The self-diffusion coefficient D is related to the tracer diffusion coefficient by a factor f, so that $D = D*/f$, where f is a constant, depending on the crystal structure, somewhat less than 1. f, the correlation factor, can be calculated, but its determination is beyond the scope of this book. For now, we can take f to be between 0.6 and 1 (Shewmon, 1989).

Consider Fig. 6.9 or equation (6.44). The concentration of radioactive trace varies from its surface value of $R/(\pi D*t)^{1/2}$ to zero at $x=\infty$ (i.e., some atoms are close to the surfaces, whereas others have penetrated a considerable distance). The mean distance penetrated is given by the integral ratio

$$\int_0^\infty xC dx \div \int_0^\infty C dx$$

which can be shown to be equal to $(D*t)^{1/2}$, a result found by Einstein. We can therefore conclude that the "average" radioactive atom moves a net distance $(D*t)^{1/2}$ from its starting point in time t. Of course, this is not a result confined to a radioactive atom, and an average atom of a species having a diffusivity D will diffuse a distance $\sqrt{Dt}$ in time t.

6.2.6 More Precise View of Diffusion in Solids

In Section 6.2.1 we presented a simplified view of diffusion which treated the diffusing atoms or molecules as hard spheres that move independently of their neighbors (other than when they collide). This leads quite naturally to Fick's law and the concept that concentration gradients "drive" diffusion.

Let us now look more precisely at diffusion in solids. For example, we should allow for the possibility that an atom, particularly in a liquid or solid, is acted on by forces due to its neighbors. One way to do this is to regard chemical potential gradients, rather than concentration gradients, as driving diffusion. The gradient in chemical potential could then arise from gradients in pressure, or electrical potential, for example, as well as from concentration gradients. We then write, for the case where the species in question has gradients only in the x direction,

$$J_x = -CB\frac{d\mu}{dx} \tag{6.45}$$

where B is known as the thermodynamic mobility and μ is the chemical potential of the diffusing species. B is supposed to be independent of concentration (but dependent on such parameters as the size of the diffusing atom, the temperature, and perhaps the pressure). It has dimensions, in SI units, of $m^2 \cdot mol/(s \cdot J)$. Putting

$$\frac{d\mu}{dx} = \frac{d\mu}{dC} \cdot \frac{dC}{dx} \tag{6.46}$$

in equation (6.45) and comparing with Fick's first law, equation (6.13), we get

$$D = CB\frac{d\mu}{dC} \tag{6.47}$$

We now employ the usual relationship between chemical potential and concentration (see Chapter 2),

$$\mu = \mu_0 + RT \ln \gamma C \tag{6.48}$$

with γ the activity coefficient. Differentiating with respect to C and substituting in (6.47) gives

$$D = \frac{BRT}{\gamma} \frac{d(\gamma C)}{dC} = BRT\left(1 + \frac{d\ln\gamma}{d\ln C}\right) \tag{6.49}$$

For thermodynamically ideal systems (our earlier model for a diffusing gas made up of spherical noninteracting molecules is an example) the activity coefficient γ is independent of concentration. Consequently, $\ln \gamma$ does not vary with $\ln C$ and the second term in brackets on the right of equation (6.48) is zero.

We see then that the term $d \ln \gamma/d \ln C$ is a correction allowing for the departure of real materials from our earlier non-interacting sphere model. $(1 + d\ln\gamma/d\ln C)$ is therefore called the thermodynamic factor. If the activity coefficient is independent of concentration, this thermodynamic factor becomes 1.

An example of an ideal system is the diffusion of a radiotracer in a lattice of the same element. Because the interaction of a radioactive atom with its neighbors is not different from the interaction of a nonradioactive atom, the activity of the radioactive atoms is directly proportional to their concentration as we replace ordinary atoms with radioactive ones. In other words, the activity coefficient of the radioactive atoms is independent of their concentration. Consequently, the thermodynamic factor for tracer diffusion is 1 and

$$D = BRT \tag{6.50}$$

gives us the self-diffusion coefficient in terms of the mobility. Equation (6.50) will hold for any system where the activity coefficient is independent of concentration.

Most solid solutions depart from ideality and we must use the full form of equation (6.49), which substituting in Fick's first law leads to

$$J_x = -BRT\left(1 + \frac{d\ln\gamma}{d\ln C}\right)\frac{dC}{dx} \tag{6.51}$$

One virtue of this equation is its interesting prediction of "uphill" diffusion. We are used to thinking of diffusion as occurring down a concentration gradient. However, in rare circumstances diffusion may occur from a region of low concentration to one of high concentration. Manipulating (6.51) gives

$$J_x = -\mathcal{B}RT\frac{dC}{dx} - \mathcal{B}RT\frac{C}{\gamma}\frac{d\gamma}{dx} \tag{6.52}$$

The first term is always opposite in sign to dC/dx (i.e., downhill diffusion), but if $d\gamma/dx$ is opposite in sign to dC/dx and of sufficient magnitude, it will reverse the sign of J_x and diffusion will occur against the concentration gradient.

6.2.7 Binary and Multicomponent Diffusion: The Kirkendall Effect

So far in treating diffusion in solid solutions we have considered only one of the species present and neglected the other one(s). Usually, there will be at least one other species present and it will be diffusing with its own diffusion coefficient. For example, if only two species, A and B, are present (*binary diffusion*) and gradients exist in the *x* direction only, we would write Fick's first law (with obvious terminology)

$$J_{Ax} = -D_A\frac{dC_A}{dx} \tag{6.53}$$

$$J_{Bx} = -D_B\frac{dC_B}{dx} \tag{6.54}$$

The diffusivities defined in this way are *intrinsic diffusivities*. If the reader ponders the mechanisms of diffusion in solids, it will become clear that these intrinsic diffusivities will, in general, not be equal to each other nor to the selfdiffusivities of A or B.

The differences in intrinsic diffusivities in a binary mixture lead to an interesting effect known as the *Kirkendall effect*. Traditionally, this effect is described in terms of the interdiffusion of two metal bars (e.g., gold and nickel, as sketched in Fig. 6.10) that are joined to form a *diffusion couple*. *Markers* (such as fine tungsten wire) that do not dissolve in either metal are placed at the interface. Held at some elevated temperature the two metals diffuse into each other. Because the diffusivity of gold in gold-nickel alloys exceeds that of nickel, the flux of gold to the right past the markers in Fig. 6.10 exceeds the flux of nickel to the left. The net effect is a shift of the center of mass of the couple with respect to the markers. Indeed, if the couple were suspended by the markers (e.g., the tungsten marker wires extend outside the couple and the wires are used to suspend the couple), there would be a measurable movement to the right by the couple. In practice, it is more usual for the couple to be fixed with respect to the laboratory (e.g., by clamping one end of the couple, or merely resting the couple on a surface and relying on friction to immobilize it) whereupon the markers will move with respect to the laboratory.

This marker movement is more than a laboratory curiosity in that it can be used to identify transport mechanisms in the oxidation of metals. Figure 6.11 illustrates what frequently happens when markers (e.g., platinum wires) are attached to the surface of a

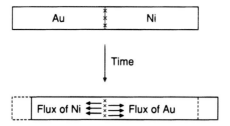

FIGURE 6.10 Kirkendall effect illustrated by the movement of inert markers initially at the interface between two interdiffusing metals.

metal which is then exposed to an oxidizing environment, and the metal is one of the many that form a dense "protective" film of oxide (rather than a porous one). After this protective film forms, further oxidation is slowed by the necessity for diffusion to occur throughout the oxide layer. Two possibilities will occur to the reader. First, as shown in Fig. 6.11, metal atoms could diffuse through the oxide film to meet oxygen with which they react at the outside surface of the oxide layer. The second possibility is that oxygen atoms could diffuse through the oxide film to encounter unreacted metal atoms at the metal-metal oxide interface, whereupon reaction takes place there. (There are two additional possibilities, namely the diffusion of metal vacancies from right to left in the oxide lattice or of oxygen vacancies from left to right, but these are equivalent to the first two possibilities and we will not discuss them further.) It should be clear that if, as shown in the figure, the first possible mechanism of diffusion is operative, the markers remain at the metal-metal oxide interface, while if the second possible mechanism is

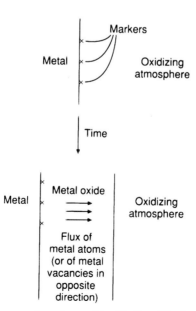

FIGURE 6.11 How marker experiments can identify the diffusing species (in this case metal atoms) during oxidation.

operative, the markers remain at the outside surface. Because the diffusivities of oxygen in metal oxide are typically small compared to metal diffusivities in the oxide lattice, the usual location of the markers is near the metal-metal oxide interface.

We see then that the Kirkendall effect can cause bulk movement of a solid and we must reexamine our definition of the flux, J, to be more precise about what it means. As used in this book, particularly in Fick's first law [equations (6.19)(6.22), (6.53), (6.54), etc.] the flux J is with respect to a reference frame that is fixed with respect to the *average* atom or molecule at that location. An example of such a reference frame would be one fixed with respect to the markers in Fig. 6.10, and we see that because of the Kirkendall effect, that frame of reference may move with respect to the solid as a whole, because the solid moves simply as a consequence of diffusion. Fick's first law can be modified so that it gives a flux with respect to a reference frame fixed to the laboratory and it incorporates the movement of the solid (due to the Kirkendall effect or any other cause), but the reader is referred to the texts on transport phenomena given at the end of this chapter for these equations.

A word of caution is necessary. In many texts (including some that we have listed at the end of the chapter), the distinction between D_A and D_B [see equations (6.53) and (6.54)] is sometimes ignored (i.e., D_A and D_B are made equal). Frequently, this is a very good approximation for diffusion in gases and liquids; however, the error in equating these diffusivities for solids may be substantial. For example, if the two diffusivities are equated, it becomes impossible to explain the Kirkendall effect. Also, Fick's first law, in the simple form presented here, is frequently used to calculate diffusion in solids with respect to a reference frame fixed to the laboratory, even when the Kirkendall effect is operative. The diffusivity used when these two imperfect approaches are combined is known as the "interdiffusivity." As a practical matter, the use of interdiffusivities is reasonable because the two intrinsic diffusivities are usually not known with sufficient precision to make their individual use meaningful.

Diffusion when more than two components are present in significant amounts is more complicated than the binary diffusion considered so far. We can glimpse this by examining equation (6.49). In a system where only two components are present (and the density is constant), fixing the concentration of component I fixes the concentration of component II. Therefore, even if component II affects the activity coefficient of component I, the thermodynamic factor can be expressed in terms of the concentration of I and the flux of I in terms of the concentration of I and its gradient. In a system with three or more components (I, II, III, . . .) fixing the concentration of I does *not* fix the concentration of the other components, and the thermodynamic factor then becomes a function of two concentrations (e.g., of I and II). The result is that the flux of I is a function of the concentration (gradient) of I and also of II, and Fick's first law cannot be used (unless II and III have no effect on the activity coefficient of I). These complications are beyond the scope of this book.

6.2.8 Diffusion of Charged Species

When mass transport involves the movement of species with a charge (e.g., ions), such that net charge is transported, we encounter some additional problems. A complication is immediately evident if some electric (potential gradient) is imposed on the system from the outside. The charged entities will tend to move (migrate) in this field. What may be less evident is that, even if an externally imposed electric field is absent, we must still recognize the charged nature of the transporting species since the separation of these charges may lead to an internally generated electric field.

Consider species i having a charge z_i (e.g., $z_i = +1$ for the hydrogen ion or -2 for an oxygen ion). The chemical potential of one mole of the ion at a place where the electric potential is ϕ, must be increased by an amount $Nz_i\phi q$, in other words,

$$\mu_i = \mu_{io} RT \ln \gamma C + Nz_i \phi q \qquad (6.55)$$

This equation is written in molar units. N is Avogadro's number and q is the absolute value of the charge on the electron. Employing equation (6.45), we obtain

$$J_{ix} = -C_i B_i \frac{d}{dx}\left(RT \ln \gamma_i C_i\right) - C_i B_i Nz_i q \frac{d\phi}{dx} \qquad (6.56)$$

where we have assumed (merely for convenience) that gradients exist only in the x direction. Putting

$$\frac{d}{dx}\left(RT \ln \gamma_i C_i\right) = \frac{d}{dC_i}\left(RT \ln \gamma_i C_i\right)\frac{dC_i}{dx} \qquad (6.57)$$

and assuming constant temperature,

$$J_{ix} = -\frac{B_i RT}{\gamma_i} \frac{d}{dC_i}\left(\gamma_i C_i\right)\frac{dC_i}{dx} - C_i B_i Nz_i q \frac{d\phi}{dx}$$

$$= -B_i RT\left(1 + \frac{d \ln \gamma_i}{d \ln C_i}\right)\frac{dC_i}{dx} - C_i B_i Nz_i q \frac{d\phi}{dx} \qquad (6.58)$$

Using the relationship between mobility and diffusivity [equation (6.49)] we get

$$J_{ix} = -D_i \frac{dC_i}{dx} - C_i B_i Nz_i q \frac{d\phi}{dx} \qquad (6.59)$$

The first term on the right is the familiar diffusion term, while the second term arises from the electric field and is called the migration term.

To treat an example, consider a system containing just two charged species. One will have charge z_+ (a positive number), the other z_- (a negative number). We will suppose that no net current flows through the system (i.e., no net transport of charge), which means that the flux of positive species J_+ and negative species J_- are related by

$$z_+ J_+ = -z_- J_- \qquad (6.60)$$

Hence, using equation (6.59), we have

$$-z_+ D_+ \frac{dC_+}{dx} - C_+ B_+ z_+^2 Nq \frac{d\phi}{dx} = z_- D_- \frac{dC_-}{dx} + C_- B_- z_-^2 Nq \frac{d\phi}{dx} \qquad (6.61)$$

Rearranging yields

$$\frac{d\phi}{dx} = \frac{-1}{\left(C_+ B_+ z_+^2 + C_- B_- z_-^2\right)Nq}\left(z_+ D_+ \frac{dC_+}{dx} + z_- D_- \frac{dC_-}{dx}\right) \tag{6.62}$$

Now let us make use of an electroneutrality condition. This is merely a statement that in the absence of other charged species besides the two we are considering, the positive charges must balance the negative ones, that is,

$$C_+ z_+ = -C_- z_- \tag{6.63}$$

or, differentiating and multiplying by D^+,

$$z_+ D_+ \frac{dC_+}{dx} = -z_- D_+ \frac{dC_-}{dx} \tag{6.64}$$

Substituting in (6.62) gives

$$\frac{d\phi}{dx} = \frac{-z_-}{\left(-B_+ z_+ C_- z_- + B_- z_- C_- z_-\right)Nq}\left(D_- - D_+\right)\frac{dC_-}{dx} \tag{6.65}$$

Substituting this equation back in equation (6.59), written for the flux of negative species,

$$J_- = -\frac{B_- z_- D_+ - B_+ z_+ D_-}{B_- z_- - B_+ z_+}\frac{dC_-}{dx} \tag{6.66}$$

If the thermodynamic factors are the same for both species, we can use (6.49) to arrive at

$$J_- = -\frac{D_- D_+(z_+ - z_-)}{D_+ z_+ - D_- z_-}\frac{dC_-}{dx} \tag{6.67}$$

and similarly [applying equations (6.60) and (6.63) to (6.66)],

$$J_+ = -\frac{D_- D_+(z_+ - z_-)}{D_+ z_+ - D_- z_-}\frac{dC_+}{dx} \tag{6.68}$$

That is, the combined diffusion and migration of these charged species can be described by an effective diffusion coefficient

$$D_{eff} = -\frac{D_- D_+(z_+ - z_-)}{D_+ z_+ - D_- z_-} \tag{6.69}$$

To examine a specific example, consider diffusion in some lattice where mobile defects arise as follows:

$$F_F = F_F^= + 2e' \tag{6.70}$$

In other words, an atom (or ion) on a lattice site undergoes ionization (or further ionization), producing electrons. Both the resulting ion and the electron are then free to move in the solid. We will assume that no other free electrons arise in this material from, say, intrinsic electric disorder. In this case $z_+ = 2$, $z_- = -1$, and applying equation (6.69), we obtain

$$D_{eff} = -\frac{3D_{F^-}D_{e'}}{2D_{F^-} + D_{e'}}$$
(6.71)

It is instructive to examine two extreme cases: First,

$$\text{if } D_{F^-} \gg D_{e'} \text{ then } D_{eff} = \frac{3}{2}D_e$$
(6.72)

Second,

$$\text{if } D_{e'} \gg D_{F^-} \text{ then } D_{eff} = 3D_{F^-}$$
(6.73)

In other words, the diffusion of the more slowly moving charged species determines the flux of the species F, although the effect of the other charged species is to accelerate the transport of the slower charged species. Physically, the phenomenon that is occurring here is that the more rapidly diffusing charged species, e', say, tends to reduce its concentration gradient far more rapidly than does the other species. Only a small amount of such diffusion is necessary before a large potential gradient with the same sign as the concentration gradient arises. Glancing at equation (6.59), it is seen that this potential gradient retards the transport of e' and enhances the transport of F_F [by the factor of 3 seen in (6.73)]. The potential gradient rapidly builds to the point where the fluxes are related by equation (6.60). Even very small differences in the numbers of charges are sufficient to do this, so that the electroneutrality condition holds true to a very high degree.

It should now be clear that the transport through the oxide film of Fig. 6.11 is not merely the transport of ions but must be accompanied by movement of electrons (or holes), and we see that the semiconducting properties of oxide films may be important in controlling their oxidation rates.

The coupled transport of charged species discussed in this section is known as ambipolar diffusion and is important in many processes, such as the operation of electrochemical cells (e.g., electrochemical sensors), oxide scale formation, and transport of matter during the sintering of powders.

6.3 Phase Equilibria

Many practical materials contain a number of different constituents. This leads to the possibility of having these constituents combine to form a variety of phases with different chemical composition or with different crystal structures. Most often, thermodynamic equilibrium is not achieved or desired in such multicomponent materials, since the nonequilibrium phase composition, distribution, and morphology add very importantly to the useful property of the material. To understand and manipulate multicomponent systems, it is necessary to consider the equilibrium phase relations before attempting a description of the kinetics of phase changes or the rate of morphological developments. Often, it is possible to predict how phase formation will proceed by combination of information on equilibrium phase relationships and on diffusion rates.

6.3.1 Phase Rule

In studying equilibrium phase relationships we essentially ask ourselves what would be present in a closed box containing fixed amounts of a variety of materials when we indefinitely keep the materials inside the box at a fixed temperature and overall pressure. The pressure would be established by the vapors of the materials contained in this box. The variables that we can control are thus composition, pressure, and temperature. The scale of the system is not relevant here. Before we proceed we must have a clear idea of the following concepts pertaining to equilibria of systems: component, phase, and degrees of freedom.

Component

The number of components of a system is the smallest number of independent substances that can constitute the system.

EXAMPLES

> If we consider NaCl and KNO_3 dissolved as an unsaturated solution in water, the system would have three components—water, NaCl and KNO_3—since any KCl and $NaNO_3$ that form in the solution can be generated by
>
> $$NaCl + KNO_3 = NaNO_3 + KCl \qquad (6.74)$$
>
> If we consider pure HI gas, we have only one component, even if at high temperature the HI would dissociate according to the equilibrium reaction
>
> $$2HI = H_2 + I_2 \qquad (6.75)$$
>
> If, however, extra H_2 were introduced, this relationship requiring equimolar amounts of H_2 and I_2 no longer exists, and the amounts of hydrogen and iodine become independent. Now the system has two components.

Phase

Phases are chemically and physically homogeneous parts of a system which, in principle, can be isolated mechanically from the other phases.

EXAMPLE

> A system of water and alcohol is composed of two phases: a homogeneous liquid phase and homogeneous gas phase. If we added NaCl below the solubility limit, we would still have two phases. However, if we added salt beyond the solubility of the system, we would have a three-phase system: solid, liquid, and gas.

Degrees of Freedom

Degrees of freedom represent the number of thermodynamic variables (pressure, temperature, and composition, i.e., concentrations) that we can manipulate independently and still maintain equilibrium. It is equivalent to the number of independent variables in a set of equations. From the requirement of thermodynamic equilibrium of the system, it follows that all the components have the same chemical potential in each of the phases, thus decreasing the number of independent variables. The relationship between the number of phases, n_P, the number of components, n_C, and the degrees of freedom, n_F, then follows and is set forth in the phase rule of Gibbs:

$$n_F = n_C - n_P + 2 \qquad (6.76)$$

This requirement restricts the number of independent variables (pressure, temperature, composition) that exists for a system in equilibrium. If we do change one of these variables, such as the pressure or the temperature for a system for which the degrees of freedom is zero, a phase change or a compositional change must occur if equilibrium is to be reestablished. If the system is not in equilibrium, the interrelation of the chemical potentials of the various components in the phases is not established and any number of phases may be present. In practical situations, full equilibrium is rarely reached in complex solid-state systems and nonequilibrium phases are common.

6.3.2 Equilibrium Phase Diagrams

One-component Systems

Figure 6.12 shows the various equilibrium phases of carbon as a function of temperature and pressure. Note that diamond is unstable at room temperature; however, the phase transformation to the equilibrium phase, graphite, does not occur unless one

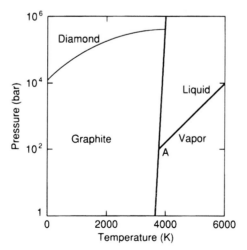

FIGURE 6.12 Phase diagram of carbon (Reprinted with permission from A.L. Ruoff, *Introduction to Materials Science,* Prentice-Hall, Englewood Cliffs, NJ, 1972.)

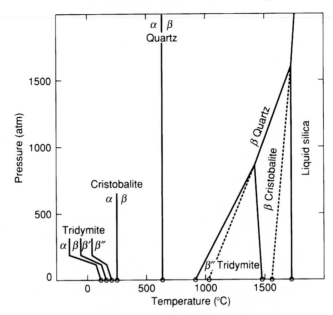

FIGURE 6.13 Phase diagram of SiO_2. (From *Phase Diagrams for Ceramists*, E. M. Levin, C. R. Robbins and H. F. McMurdies, eds., 1964, p. 84. Reprinted with permission from American Ceramic Society, Columbus, OH.)

raises the temperature quite a bit. Note also that three phases can be in equilibrium with each other, as at the point A, but this can only be at a fixed temperature. This is a consequence of the Gibbs phase rule.

A second example is the phase diagram of SiO_2, shown in Fig. 6.13. As the temperature is changed at constant pressure, quite a variety of crystalline phases, with somewhat different structures, will appear. In this particular case, however, some transformations are difficult, and the SiO_2 tends to form glass or nonequilibrium phases.

Two-component or Binary Systems

For two-component systems one usually plots the temperature vertically and the composition horizontally. Since only two components are involved, either A or B can be used for the compositional axis. The diagrams are always plotted with the understanding that the pressure is held constant for the system. This removes one degree of freedom, so that we would have no ability to change the composition, temperature, or number of phases present if, for some reason, the diagram would show three phases (plus the always present gas phase) coexisting. A simple, two-component system is shown in Fig. 6.14. This system exhibits a eutectic transformation (point E). A liquid of this composition transforms to a solid of the same (overall) composition at the unique temperature indicated. In the phase diagram we observed various lines that delineate phase fields. A phase field is a combination of temperature and composition for which all the possibly present phases change their relative amounts but not their nature or structure. Much information can be obtained by careful examination of phase diagrams.

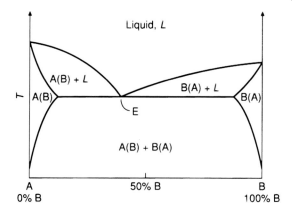

FIGURE 6.14 Simple binary eutectic. A(B), solid solution of B in A; B(A), solid solution of A in B.

We present a few examples to illustrate the use and meaning of the phase diagrams.

Isothermal Composition Change Consider the system at a fixed temperature T_1 (Fig. 6.15). In an isothermal section of the phase diagram we could start out with pure A at this temperature and describe the phases we find in the phase fields that we will traverse while increasing B. Following in Fig. 6.15, we first have a solid solution of B in A. This is also called the terminal solid solution. As we cross the first phase boundary at C_1, we will start to have a mixture of two phases: a liquid solution of composition C_2 and a solid solution of composition C_1. Note that the Gibbs phase rule requires that these compositions be constant. The change in the overall composition will then be accommodated by a change in the relative amounts of liquid C_2 and solid C_1. A mass balance will tell us exactly how much of the solid and the liquid solution is present. There is a simple geometrical construction in the diagram that gives the answer to the

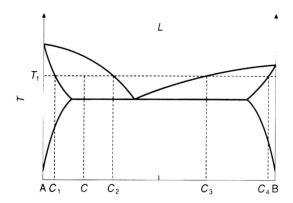

FIGURE 6.15 Binary eutectic isothermal section.

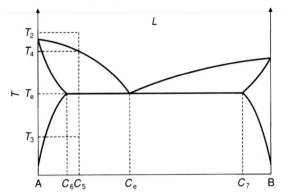

FIGURE 6.16 Temperature change at constant overall composition in a binary eutectic. L, liquid-phase field; T_e and C_e, eutectic temperature and eutectic composition.

mass balance. The construction is known as the lever rule. The lever rule will give

$$\%\text{liquid} = \frac{C - C_1}{C_2 - C_1} \times 100$$

$$\%\text{solid} = \frac{C_2 - C}{C_2 - C_1} \times 100$$

If the composition is in weight percent, so will be the amounts of liquid and solid. The lever rule is a very useful tool and applies in any system in which compositional axes are plotted linearly.

Increasing the amount of B so that the overall composition crosses the C_2 composition will result in all the solid disappearing, with only liquid solution present. Continuing the increase in B will make us enter the next two-phase field, where liquid of composition C_3 is in equilibrium with solid solution of composition C_4. Finally, we enter the B terminal solid solution field until we reach pure B.

Temperature Change at Constant Composition Consider the composition C_5 in Fig. 6.16, and let us decrease the temperature from T_2 to T_3. First, we have a liquid solution only. As we lower the temperature to T_4, we enter the two-phase field, forming the terminal solid solution of B in A. When we reach the temperature T_e, we still have two phases present simultaneously: a liquid of composition C_e and the terminal solid solution of composition C_6. Just below T_e the liquid of composition C_e will have been consumed in a phase change that forms the terminal solid solutions of B, at composition C_7, and more of C_6. This type of phase change is called a eutectic transformation. Note that at T_e we cannot have three phases present (plus the necessary gas phase) in arbitrary amounts, since this would be a violation of the Gibbs phase rule. The equilibrium of vapor plus the three phases (the eutectic liquid with the solid solutions of composition C_6 and C_7) can exist only for the eutectic overall composition. Thus, at the eutectic temperature, for an overall composition to the left of C_e we will have only solid C_6 and

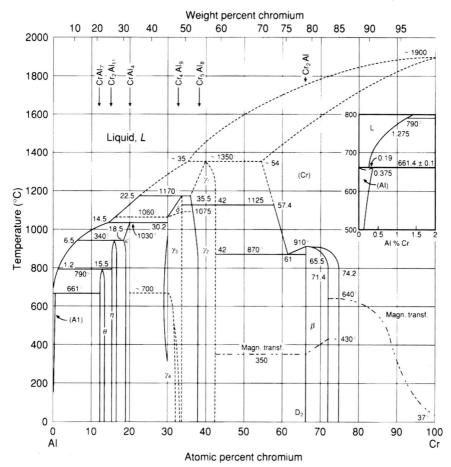

FIGURE 6.17 Aluminum-chromium binary-phase diagram. *L*, liquid; *D*, line compound; a, b, g, solid solutions. *D* Æ e + *L*, peritectic transformation, *D* Æ (Cr), congruent transformation; (Cr) Æ g + B, eutectoid transformation. (From M. Hansen, Constitution of Binary Alloys, 1954. Reprinted with permission from Genium Publishing Corp., Schenectady, NY.)

C_e, or solid C_6 and solid C_7, depending on whether we are infinitesimally above or below T_e. A more complex diagram, with some names of the types of transformations, is shown in Fig. 6.17.

Ternary Phase Diagrams Three-component systems can be represented at some fixed temperature in a triangular composition diagram. In many ceramic systems, the all-solid-phase situation can be particularly simple, because the mutual solubilities in ionic systems is usually very low. An all-solid equilibrium diagram can then be subdivided into compositional triangles as shown in Fig. 6.18. Note that for the compound D in this figure, we do not have five phases in equilibrium with each other; rather, we can only have the coexistence of at most three solid phases if the composition is to remain

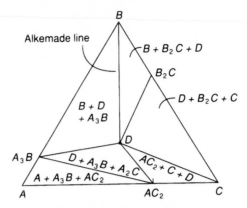

FIGURE 6.18 Hypothetical ternary diagram at the temperature where all phases are solid. Within each subtriangle the three phase constituting the corners of that triangle are present.

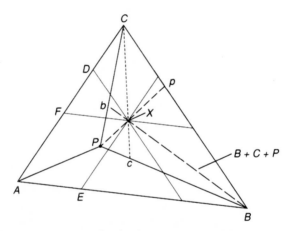

FIGURE 6.19 Composition determination in a ternary system.

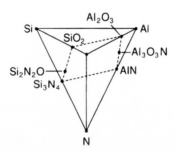

FIGURE 6.20a Represention of the Si-Al-O-N system as a tetrahedron with Si, O, Al, and N at the corners. (From I. J. McColm, *Ceramic Science for Materials Technologists*, 1983. Reprinted with permission from Blackie & Son, Ltd., Glasgow and London.)

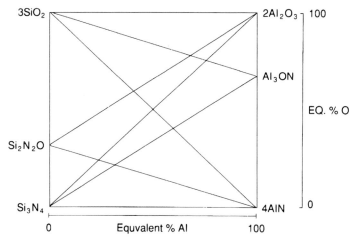

FIGURE 6.20b Representation of the Si-Al-O-section shown in Fig. 6.20(a). (From I. J. McColm, *Ceramic Science for Materials Technologists*, 1983. Reprinted with permission from Blackie & Son, Ltd., Glasgow and London.)

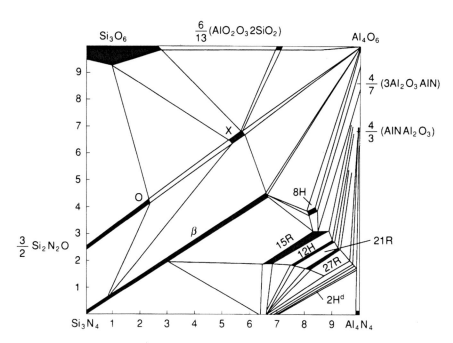

FIGURE 6.20c Si-Al-O-N systems at 1750°C. (From I. J. McColm, *Ceramic Science for Materials Technologists*, 1983. Reprinted with permission from Blackie & Son, Ltd., Glasgow and London.)

a variable. The connecting lines between the compounds of definite proportions are called the Alkemade lines. Any two compounds connected by an Alkemade line may be treated as a pseudo-binary-phase diagram; that is, we can plot temperatures of phase changes versus a single composition variable, as in Fig. 6.16, rather than needing a triangular diagram. Inside each compatibility triangle we can again use a construction equivalent to the lever rule to determine how much of each phase is present. An example is given in Fig. 6.19. The overall composition is determined by the indicated construction, so that for a composition X, the percentage of A = DC/AC, that of B = AE/AB, and that of C= AF/AC. Inside the triangle in which the overall composition X falls, we can determine the relative amounts of the compounds P, C, and B as follows: % P = pX/pP, % C = cX/cC, and % B = bX/bB each multiplied by one hundred. The liquidus lines of the binary-phase diagrams become surfaces for the ternary systems. The loci of points of the intersection of these liquidus surfaces are often superimposed on the all-solid phase relationships. A detailed analysis of such diagrams can give much information on exactly how the system crystallizes from a high-temperature melt and which crystallization path is followed.

Quaternary Phase Diagrams   Four-component systems would have to use three dimensions if they were to be represented in the same way as are three-component systems. Such a construction could be a regular tetrahedron where each of the vertices is assigned to a component. Instead, the four-component system is usually represented by an isothermal plane section of a compositional tetrahedron. An example for the Si-Al-O-N system is shown in Fig. 6.20. To be able to have a linear and self-consistent representation of the compositions in the quaternary phase diagram, one must use multiples of the molecular formulas. The composition can then be found in terms of equivalents. Traversing the diagram from the left to the right means replacing 3 Si by 4 Al, at constant O/N ratio. Thus the percentage Al can be figured from

$$\%\text{Al} = \frac{3\left[\text{Al}\right]}{4\left[\text{Si}\right] + 3\left[\text{Al}\right]} \times 100$$

Traversing the diagram from the top to the bottom means replacing 6 0 by 4 N, at a constant Si/Al ratio, so that

$$\%\text{O} = \frac{2[\text{O}]}{3[\text{N}] + 2[\text{O}]} \times 100$$

From these relationships one can derive the overall composition of an arbitrary point in the diagram. We read the percentage of Al and O from the horizontal and vertical axis and put these in the equations above. This will yield the Si/Al ratio and the N/O ratio. The total composition is then found by choosing the (Si + Al)/(N + O) ratio such that the total positive and negative charges balance.

   In isothermal sections we can find phase fields in quaternary diagrams that may be liquid or solid solutions, besides the regular compatibility triangles [Fig. 6.20(c)]. It should be obvious that the determination of reliable phase diagrams is a painstaking and time-consuming task. Nevertheless, their utility makes such labor a worthwhile enterprise. A large collection of these phase diagrams has been published by Levin et al. (1964).

6.4 Kinetics of Phase Changes

In this section we give a few examples of phase changes in which the rate of change is considered. The study of phase transformations and kinetics has now reached a fairly advanced stage, and more information can be found in other books (see, e.g., Christian, 1965).

6.4.1 Nucleation and Growth

\When a new phase B is formed from phase A, for example by changing the temperature, formation of B will have to start out somewhere. Complete, instantaneous changing from A to B does not occur. Even phase changes for which the only change would be a small lattice distortion still do not proceed all at once. If, for example, ice forms upon cooling water, we will first find small ice particles in the water. These usually have started to form on small bits of suspended dirt. If the water was totally clean, it actually becomes quite difficult to form ice, and water would have to supercool as much as 40°C below its normal freezing temperature before ice started to form. Then, suddenly, very many tiny ice particles would form throughout the undercooled water, and solidification would be quite rapid. The latter case is *called homogeneous nucleation*, since it does not require the assistance of some foreign phase. We will consider homogeneous nucleation for the moment.

As soon as rapid solidification starts, the temperature of the system would increase to the normal freezing temperature, with the necessary heat liberated by the heat of solidification of the water-to-ice transformation. Very small ice particles from which the solidification starts are called nuclei. We should note, however, that if we use the proper definition for what a nucleus is, it would be so small that we would need a very powerful microscope to detect it. We can imagine that upon cooling water to its freezing temperature, we would find the molecules tending to arrange themselves into ice crystals. When they do, they liberate an amount of energy proportional to their volume and to the heat of solidification. However, for each particle we have created some water-ice interface that requires energy to form: the interfacial energy. Whether or not we lower the overall free energy of the system by the incipient transformation will now depend on the surface-to-volume ratio and on the relative magnitude of the interface energy and the free energy of solidification. We could write

$$\Delta G = 4\pi r^2 \gamma_i + \Delta G_{sol}(4/3)\pi r^3 \tag{6.77}$$

where ΔG is the free-energy increase associated with the formation of a spherical nucleus, radius r, γ_i is the interfacial energy, and ΔG_{sol} is the free energy of solidification per unit volume (a negative number at temperatures below the freezing point). Graphically, the nucleation event could be depicted as in Fig. 6.21.

Therefore, the free energy of the system will start to decrease (i.e., solidification becomes spontaneous) only when a critical nucleus size, r^*, has been exceeded. Then, other molecules will add on rapidly, making the nucleus grow, thus letting the transformation proceed. After the nucleation step, the growth rate of a nucleus, and thus the local liquid-to-solid transformation rate, will be determined by how fast heat is rejected (a problem analogous to mass transport, where we substitute thermal diffusivity

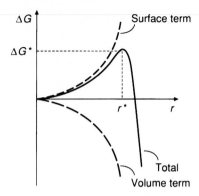

FIGURE 6.21 Free energy versus radius of nucleus. r^* is the critical radius.

for atom diffusion rate and temperature gradient for concentration gradient). For alloys, where there can be a change in composition associated with the phase transformation, heat flow is often not the limiting factor, especially in solid-solid transformations; rather, it is the transport of the atoms necessary to accommodate the composition change inherent in the phase change. For the critical nucleus size, r^*, the free-energy equation leads upon differentiation, and setting the differential to zero, to

$$r^* = \frac{-2\gamma_i}{\Delta G_{sol}} \qquad (6.78)$$

so that the free-energy change for the critical nucleus, ΔG^*, is

$$\Delta G^* = 16\pi \frac{\gamma_i^3}{3\left(\Delta G_{sol}\right)^2} \qquad (6.80)$$

ΔG_{sol} is dependent on temperature, since the Gibbs free energy is

$$\Delta G_{sol} = \Delta H - T\Delta S \qquad (6.80)$$

ΔH and ΔS are only weakly temperature dependent, so that the variation of ΔG_{sol} with degree of undercooling is due mainly to T in the last term. Thus if T_o, is the equilibrium transformation temperature, and T is the actual temperature,

$$\Delta G_{sol} = \Delta H \frac{T_o - T}{T_o} \qquad (6.81)$$

because $\Delta G_{sol} = 0$ at $T = T_o$. With these expressions we can get an approximate idea of how the nucleation rate depends on the degree of undercooling, $(T_o - T)/T_o$. Statistical thermodynamics would tell us that the number $n(\Delta G)$ of nuclei of a certain energy in a system in thermal equilibrium is exponentially dependent on that energy in the following manner:

$$n\left(\Delta G\right) = n_o \exp\left(\frac{-\Delta G}{RT}\right) \tag{6.82}$$

n_o is a proportionality constant, nearly equal to the number of atoms in the system, if $n(\Delta G)$ is small.

We now will assume that the rate of formation of stable nuclei, $\dot{N}$, is proportional to the number of nuclei of critical size, r^*, and to a rate constant, K, that describes how fast the critical nucleus forms a stable nucleus by adding on an infinitesimal amount of material. Thus

$$\dot{N} \sim n(\Delta G^*)K \tag{6.83}$$

and using the expression for ΔG^*, we have

$$\dot{N} \sim K \exp\left[\frac{-16\pi\gamma_i^3 T_o^2}{3RT\left(\Delta H\right)^2\left(T_o - T\right)^2}\right] \tag{6.84}$$

K itself involves a diffusion process and therefore has a similar temperature dependence as an atomic diffusion coefficient for a solid-solid transformation.

$$K \sim \exp\left(\frac{-E_A}{kt}\right) \tag{6.85}$$

Thus, whereas the nucleation rate would, on the one hand, sharply increase with increased undercooling, the diffusional step would become increasingly difficult and will eventually become nearly impossible, although the driving force for nucleation is quite substantial. If that occurs, the transformation is kinetically hindered and the phase that should form does not. This is actually not all that uncommon and is the reason why many silicates will cool to form glasses rather than equilibrium crystal phases.

In solid-solid transformations these concepts apply, but we may have to add a strain energy term to the free energy when the new phase does not occupy the same molar volume as the old one. Such strain energies could make the solid-solid transformation more difficult.

In heterogeneous nucleation, the new phase forms on a foreign object such as the container wall, the surface of the specimens, on grain boundaries or dislocations, and so on. For each of these defects there is an associated energy that may be reduced if it becomes coated with the new phase. Clearly, the geometry of how the new phase nucleates on the imperfection and the degree to which it lowers its energy can be included in the free-energy expression. Preferential nucleation on foreign objects will occur if the energy balance is favorable, and as a result, ΔG^* is usually significantly lowered. Inclusion of a particulate nucleation aid can shift the relationship between the nucleation rate and the growth rate. This possibility is exploited in the fabrication of glass-ceramics, in which a nucleating agent, such as submicrometer dispersed TiO_2 particles, lets a crystalline phase form from a glass at a temperature where growth hardly occurs otherwise. The growth of the crystalline phase is then carried out at a higher temperature. This process permits the formation of dense ceramic bodies from a molten glass, with the grain size determined by the concentration of the TiO_2 particles. A wide variety of

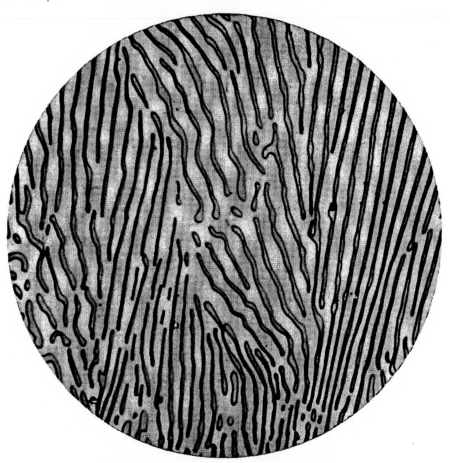

FIGURE 6.22 Typical eutectic structure in steel. (Reprinted with permission from E. C. Bain and H. W. Paxton, Alloying Elements in Steel, 2nd ed., American Society for Metals, Metals Park, OH, 1961.)

strong, crystalline, transparent household objects, such as coffee pots and cooking ware, are prepared this way from glasses in the system Al_2O_3-SiO_2-MgO-LiO_2.

6.4.2 Eutectic Transformations

When a liquid of eutectic composition in a binary system solidifies from the melt, two phases have to form simultaneously. In the simplest case, this would be the two terminal solid solutions. The transformation producing the two new phases occurs most often in a coupled manner, leading to the formation of the two phases side by side, so that lamellae or rods form with a composition that alternates between the two terminal solid solutions. The spacing, λ, of these lamellae or rods is quite regular when the solidification is made to occur at a constant solid-liquid interface velocity, v. In general, one finds that the spacing and the interface velocity are related by

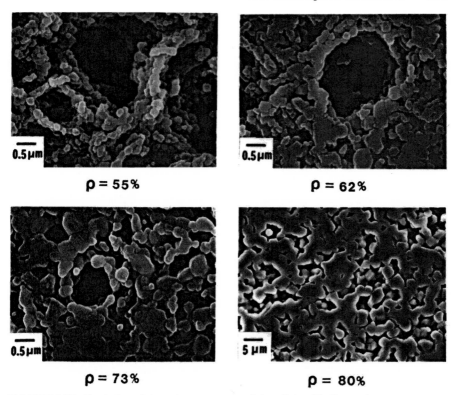

$\rho = 55\%$

$\rho = 62\%$

$\rho = 73\%$

$\rho = 80\%$

FIGURE 6.23 Evolution of the microstructure of densifying MgO powder compact.

$$\lambda^n v = \text{constant} \tag{6.86}$$

where n is a factor that may range from 2 to 4, depending on the details of the diffusion mechanism. The relative thickness of the lamella is determined by the eutectic composition and can be found from the phase diagram with the lever rule. An example of a typical eutectic microstructure is shown in Fig. 6.22. The microstructures that are formed will therefore depend on the rate of solidification. If a very fine distribution of the phases is desired, rapid interface movement will be required, but heat removal problems eventually limit the possibility of obtaining a reasonably uniform material.

Interesting composites can be made by the unidirectional solidification of eutectic systems. Van den Boomgaard et al. (1974) were, for example, able to produce composite ceramics in which cobalt ferrites were formed in barium titanate, thus producing a composite magnetoelectric material.

6.5 Sintering: An Introduction

Sintering is a process by which useful components are made from powder compacts through elimination of the inherent porosity in these compacts by heating. The driving force for the densification is the reduction of the free energy associated with the inter-

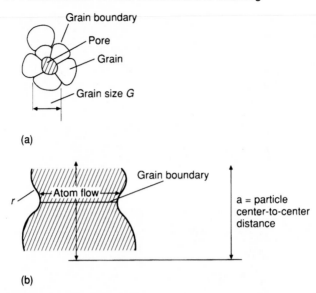

(a)

(b)

FIGURE 6.24 Geometry of sintering particles.

nal surfaces in the compact. Fine powder particles, typically with a size of 1 μm or less, "weld" together as atoms diffuse from their junction boundaries to fill in adjacent small pores. An example of the evolution of the microstructure of an MgO powder compact during sintering is shown in Fig. 6.23.

One of the difficulties in achieving the desired properties of a material prepared via powder processing is that usually not all of the porosity can be eliminated in sintering unless special techniques described in Chapter 15 are used. A simple picture of how mass transport, leading to densification, might occur for a set of a few particles is shown in Fig. 6.24. The negative curvature at the junction of the grain boundaries with the pore surface will cause the chemical potential of the atoms there to be lowered by an amount proportional to the curvature, as we examine in detail in section 13.4.1. This curvature is in turn determined by the particle size and the pore size. If atoms are transported away from the grain boundaries to the pore surface as shown in Fig. 6.24(b), we could determine how fast the particle centers would approach each other if we knew the diffusion coefficient. All that is required is the solution of Fick's law in the proper geometry and with the right driving forces. We consider the details of this later. In general, we would end up with an expression for the center-to-center approach rate, $\dot{a}$, of two particles, and thus the linear shrinkage rate, that is of the form

$$\dot{a} \sim D G^{-n} \Sigma \qquad (6.87)$$

where D is the diffusion coefficient, G the grain size, n a number that depends on the details of the transport mechanism, and is usually between 2 and 3. Σ is the driving force for the sintering, which in this model is determined primarily by the pore curvature at the grain boundary.

Practically, one often observes that the linear shrinkage during isothermal densification follows a simple time dependence for a large part of the sintering process:

$$\frac{\Delta L}{L_o} = Kt^m \tag{6.88}$$

where ΔL is the change in length of a compact of initial length L_o, t the time, and m a constant that is usually between 0.3 and 0.5. The proportionality factor K is often inversely proportional to the cube of the particle size. When the compact starts to approach full density, densification will start to slow down more rapidly than this expression would indicate, leading to a practical limit to the obtainable density.

It is obvious that the sintering rate will be proportional to the diffusion coefficient of the species that accomplishes removal of matter from between the particles. The manipulation of these diffusion coefficients by changing, for example, the oxygen pressure over a sintering oxide, or by adding certain compounds that affect the defect chemistry, is thus possible. Addition of some compounds may also cause the formation of liquid phases at the grain boundaries, which may accelerate densification very much. In that case the transport in the liquid determines densification rates.

In general, it is not possible to correlate the diffusion coefficients determined in a simple diffusion experiment with those obtained from sintering experiments. The difficulty is that, especially for ceramic systems, minor amounts of impurities may have drastic effects on transport rates and that the geometry of a sintering compact at the particle level is not sufficiently well known. In addition, powder compacts are almost never homogeneous at the microscopic scale of the particle size. The special heterogeneities, such as those due to the presence of powder agglomerates, will cause the macroscopically observed densification rates to be significantly lower than the ones that we would calculate on the basis of a two-particle geometry. In a subsequent chapter we shall deal with these complications at length.

6.6 Concluding Remarks

This chapter has served three purposes. First, a detailed description of diffusion, particularly in the solid state, has been provided. Second, a brief treatment of phase equilibria, phase diagrams, and phase transformations has been provided, the brevity being justified by the assumption that these topics are treated extensively in most undergraduate materials curricula and many texts. Finally, the practical topic of sintering was introduced, largely so that the reader may grasp, early in the present book, how solid-state diffusion and the formation of high-temperature phases play a vital role in the production of ceramics and other inorganic materials.

LIST OF SYMBOLS

$\dot{a}$ $-da/dt$ = center-to-center approach rate of two sintering particles
B thermodynamic mobility
C concentration
D diffusion coefficient
D^* radioactive tracer diffusion coefficient

f	correlation factor
G	grain size
ΔG	free energy of formation
J	atom flux
K	equilibrium constant
L	length
$\dot{N}$	nucleation rate (dN/dt, t = time)
q	absolute value of charge per electron ($\equiv |e|$ in Chapter 5)
R	gas constant
$r*$	critical radius of nucleus
v	interface velocity
z	valence
γ	thermodynamic activity coefficient
γ_i	interfacial energy
ϕ	electrical potential
λ	spacing of lamellae in eutectic
μ	chemical potential
Σ	driving force for sintering
$[\cdot]$	concentration

PROBLEMS

6.1 For the purposes of this calculation, we will regard the helium atom as a sphere of diameter 0.26 nm. Calculate the mean free path of helium molecules at:
(a) Atmospheric pressure and 25°C
(b) 10 torr and 1000 K

6.2 Using the approach of Section 6.2. 1, deduce Graham's law of diffusion.

6.3 You have been given the task of measuring the self-diffusion coefficient of a metal, by the tracer technique described in Section 6.2.5, at a temperature where the coefficient is thought to be 10^{-8} cm^2/s. You are able to deposit 0.005 mol of radioactive tracer per square centimeter on the surface of the metal initially and plan to determine the final distribution of radiotracer by machining off slices of metal 0.1 mm thick and measuring the radioactivity of each slice. You estimate that the lowest concentration of radiotracer you can measure in a slice is 10^{-3}mol/cm^3. It is Monday morning and you have just brought the sample up to temperature to start the diffusion. As bosses will do, your boss wants a result by the end of the week. You plan on letting the diffusion take place until Thursday morning. The rest of Thursday can be spent machining the slices and measuring their radioactivity. On Friday you will calibrate the apparatus (a Geiger counter) so that its measurements can be related to concentration. Late Friday you will plot the natural logarithm of concentration versus the square of the distance from the surface and determine the diffusion coefficient from the slope. Will the plot have a sufficient number of data points to be convincing (say, greater than 10)?

6.4   On Thursday morning, Vince Overhand, a colleague whose arrogance stems from an overly technical education, tells your boss that much time has been lost by your failure to calibrate the Geiger counter on Monday, Tuesday, or Wednesday, while the diffusion experiment was in progress. This, he declares, reveals your stupidity, laziness, or both. Clutching at straws you respond that the calibration can be dispensed with, that the results will be available that afternoon, and that the truly stupid person is Vince for not realizing that the calibration is unnecessary, provided that a certain condition is satisfied. Your boss then tells you that his budget has been suddenly cut back and that it will be necessary to fire one of you. His choice for firing is the more stupid one, as determined by the availability of valid results on Thursday afternoon. Who reports to the unemployment office on Friday? What is the condition referred to above?

6.5   Show that, as described in Section 6.5.1, the mean distance traveled by an atom in a diffusion experiment such as the one of problems 6.3 and 6.4 is $(D*t)^{1/2}$. [To do this requires good mathematical knowledge (e.g., concerning error functions) or a good table of integrals.]

6.6   From data provided in Chapter 2, calculate and plot the thermodynamic factor for the diffusion of carbon in iron at 1600°C as a function of the carbon concentration.

6.7   Show that the effective diffusivity appearing in equation 6.69 can never be greater than the greater of D_- and $D+$.

6.8   Two compartments containing argon gas at different pressures show a flux of 0.04 mol/s through an orifice in the separator. The orifice has a diameter of 500 μm. Calculate the pressure difference between the two compartments.

6.9   The self-diffusion coefficient of a metal is found to be 10^{-14}cm²/s at 1200°C. A very thin surface layer of a tracer of this metal is evaporated onto its surface. This layer is so thin that after 15 min of heating and interdiffusion, equation (6.44) holds well. Calculate the mean penetration depth of the tracer after 12 hours of heating. Also, plot the relative surface concentration as a function of time.

6.10 Calcium chloride is dissolved in sodium chloride. It is found that calcium substitutes for sodium. Using Kroger-Vink notation, write down the appropriate defect chemistry reaction.

6.11 Iron oxide (FeO) can be heated in an oxidizing environment to take up oxygen ions, with the residual iron ions taking up interstitial positions. Write the appropriate defect reaction and the dependence of the iron interstitial concentration on the oxygen pressure (in terms of the equilibrium constant), assuming that no electronic disorder is involved.

6.12 The diffusion coefficient of A in the metal oxide AMO has been found to be 10^{-16} cm²/s at 1450°C. Calculate the thermodynamic mobility, B, assuming that the activity of A can be expressed as $\gamma=0.03[A]$.

6.13 Derive the expression for the effective diffusion coefficient for ambipolar diffusion in Al_2O_3 assuming that Schottky disorder prevails.

6.14 Consider an alloy of 50 at % in the Al-Cr binary system (Fig. 6.17). Describe the sequence of events when cooling this alloy from interstitial melting temperature to room temperature.

6.15 Determine the compositional range of the X phase in sialon [use Fig. 6.20)].

6.16 Often, a critical nucleus consists of about 100 atoms or molecules. Assuming that the solid-liquid interface energy of a metal is about 1 J/m^2, calculate the free energy of solidification per unit volume for this metal if it has a molecular weight of 50 and a density of 6 g/cm^3. Is this number realistic? (Check with a data base.) Assuming that the enthalpy of solidification is 63 kJ/mol, what degree of undercooling would represent the critical nucleus?

REFERENCES AND FURTHER READINGS

BIRD, R. B., W. E. STEWART, and E. N. LIGHTFOOT, *Transport Phenomena*, Wiley, New York, 1960.

CHRISTIAN, J. W., *The Theory of Transformations in Metals in Alloys*, Pergamon Press, Elmsford, NY, 1965.

CRANK, J., *Mathematics of Diffusion*, Oxford University Press, London, 1956.

GEIGER, G. H., and D. R. POIRIER, *Transport Phenomena in Metallurgy*, AddisonWesley, Reading, MA., 1973.

GHEZ, R., *A Primer of Diffusion Problems*, Wiley, New York, 1988.

KROGER, F., *The Chemistry of Imperfect Crystals*, North-Holland, Amsterdam, 1964.

LEVIN, E., C. ROBBINS, and H. McMurdie, eds., *Phase Diagrams for Ceramists*, American Ceramic Society, Columbus, OH, 1964; and subsequent volumes.

MCCOLM, I. J., *Ceramic Science for Materials Technologists*, Leonard Hill, Glasgow, Scotland, 1983.

PFANN, W. G., *J. Met. 4*, 747 (1952).

SHEWMON, P. G., *Diffusion in Solids*, TMS, Warrendale, PA, 1989.

TILLER, W. A., K. A. JACKSON, J. W. RUTTER, and B. CHALMERS, *Acta Metall.* 1, 428 (1953).

VAN DEN BOOMGAARD, J., D. R. TERRELL, R. A. BORN, and H. F. GILLER, *J. Mater. Sci.* 9, 1705 (1974).

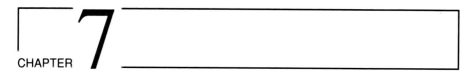

Process Engineering

7.1 Introduction

In this chapter we provide a brief introduction to the principles of process engineering. Process engineering is a large topic (much of it covered within chemical engineering) and a full introduction would require a book in itself. Our experience has been that, except for chemical engineering students, most students have had little or no exposure to process engineering, and therefore any text on how materials are produced must provide an introduction to the more important concepts. We begin with some definitions.

7.1.1 Unit Operations

The generally preferred way of describing a process is in terms of *unit operations*. These are the building blocks of the process and we illustrate the concept by means of Fig. 7.1, which is a simplified *block diagram* of the Bayer-Hall process used to produce aluminum. Each block represents a unit operation, and some of these operations (e.g., crushing and grinding) we have met already. The functions of other operations (e.g., solid-liquid separation) are obvious from their names. There are many simplified *flowsheets* similar to Fig. 7.1 in the remainder of this book. They are readily grasped descriptions of how unit operations are connected to form a process. The development of such simplified flowsheets would be one of the first steps in the translation of laboratory results into a full-scale commercial process for production of materials. Note that Fig. 7.1 conveys no impression of the equipment that is used in each unit operation; representing the unit operation as a simple block emphasizes the role that the unit operation plays in the entire process, rather than the details of each unit operation.

Treating processes in terms of unit operations is logical in that these unit operations are readily identifiable in any process; they are physically separated from each other, for example, sometimes being performed in different buildings within a plant. A more compelling advantage of the unit operations approach is that one can learn what is necessary of a given operation (precipitation for example) and then apply that knowledge in innumerable processes (both existing and potential). The remainder of this chapter contains little description of particular unit operations; that knowledge will be acquired by the reader in later chapters. Rather, this chapter is concerned with principles that are fundamental to all unit operations.

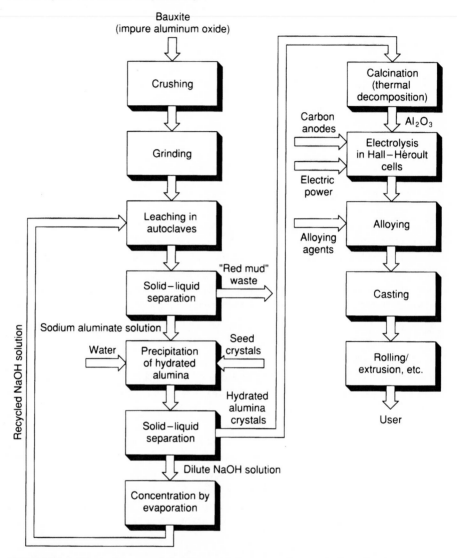

FIGURE 7.1 Block diagram of the Bayer-Hall process for producing aluminum.

7.1.2 Classification of Unit Operations

A first categorization of unit operations is by whether they are batch, continuous, or in-between. In a batch operation all the raw materials are fed to the unit, then the unit is operated, and then the products are discharged. A familiar example is that of taking a bath. The tub is filled with water, a person enters, bathing takes place, the person climbs out, and the water is drained.

In continuous operations the raw materials flow continuously into a unit that is operating continuously, and the products are discharged continuously. Again stretching

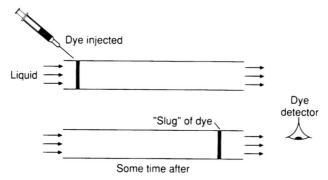

FIGURE 7.2 Movement of a tracer (in this case a dye) through a plug flow unit.

our imagination beyond materials processing, we can point to a jet engine in flight as a continuous operation. Fuel flows continuously into the engine, as does air, and the engine continuously rotates, producing an exhaust stream and thrust.

Between these extremes are operations labeled semibatch (or semicontinuous). Taking a shower would be so described; the water passes through continuously, the person passes through batchwise. Note also that some precision is called for in applying this label. An automobile engine might correctly be labeled batch or continuous, depending on whether we apply the label to individual cylinders or to the engine as a whole.

A second categorization of unit operations is by whether they are steady state or unsteady state. Unsteady-state operations are those in which one or more important variables change with time. What constitutes an important variable may vary from operation to operation, but typically they are easily identified. The mass of the contents, temperature, and composition are frequently encountered important process variables.

Batch operations are inherently unsteady state; continuous operations may be steady state or unsteady state. Consider, for example, what happens to a jet engine as the aircraft taxies to the end of the runway and then accelerates to takeoff.

Typically, laboratory experiments are batch (or at least semibatch) because such experiments are usually easier to conduct and understand. Full-scale processes are frequently made up of continuous operations, usually designed to operate at steady state. Such processes are typically easier to control and have other advantages, such as a more uniform product.

A final categorization of continuous processes is in terms of whether they are *backmixed* or *plug flow*. The concept is illustrated in Fig. 7.2 in terms of a liquid flowing along a pipe. Under the right conditions (high-enough Reynolds number) the liquid moves down the pipe much as a piston moves down a cylinder. If we color part of the liquid with a dye and place a detector for this dye at the outlet of the pipe, we will get the response shown in Fig. 7.3. We say that the liquid moves through the pipe *in plug flow* or *piston flow*. The time, t_R, between the instant of dye injection and the detection at the output is the *residence time* of the dye in the system. The dye is acting here as a tracer to trace the movement of the contents through the unit. If a tracer is well chosen, it moves through the unit in the same way as the rest of the contents, and therefore the residence time of all the contents is t_R. The reader is asked to ponder whether lead

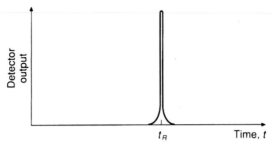

FIGURE 7.3 Signal from tracer detector at outlet of plug flow unit.

particles would have been suitable as a tracer in this case.

The residence time is easy to calculate:

$$t_R = \frac{\text{length of pipe}}{\text{speed of liquid}} \tag{7.1}$$

If we multiply the numerator and denominator of this equation by the pipe cross-sectional area, the numerator becomes the volume of liquid in the pipe (or the *holdup* of liquid in the pipe). The denominator becomes the volumetric flow rate of liquid in the pipe, hence

$$t_R = \frac{\text{holdup}}{\text{flow rate}} \tag{7.2}$$

Note that equation (7.2) still holds if holdup is expressed as mass of liquid in the pipe and the flow rate in mass of liquid per unit time. This can be seen by multiplying top and bottom of equation (7.1) by the product of area and liquid density. Equation (7.2) is a general equation for plug flow units.

Many examples of plug flow can be found in everyday life. A checkout line at a supermarket is in plug flow. Note that at any instant the person leaving is the one who has been in line longest and that, to a first approximation, everyone in line has to wait the same time, given by dividing the holdup (number of persons in line) by the flow rate (number of persons checked out per hour).

Now imagine our dye tracer injected into a well-agitated tank through which a liquid flows continuously. As sketched in Fig. 7.4, the detector observes dye in the tank output at virtually the instant it is injected. The quantity of dye in the output then diminishes as the dye is washed out of the tank by the flowing liquid. We call this a *backmixed* operation; if the agitation is sufficiently vigorous that the contents of the tank become uniform, a more descriptive label is *perfect mixing*. In this type of operation, the dye molecules (and the liquid molecules) spend varying times in the unit, in contrast to the plug flow unit, where everything had a residence time of t_R. There is a distribution of residence times; the average (arithmetic mean) residence time, $\bar{t}_R$, is given by

$$\bar{t}_R = \frac{\text{holdup}}{\text{flow rate}} \tag{7.3}$$

and again any consistent units can be used to express the holdup and flow rate.

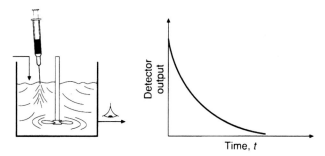

FIGURE 7.4 Movement of dye tracer through a backmixed unit.

An example of a unit operation that approximates a backmixed process is the ball mill that we met in Chapter 1. The churning of the mill contents, as the mill rotates and the balls tumble over each other, results in nearly perfectly mixed behavior.

Note that in perfect mixing each molecule of the contents of a unit has the same probability of exiting in the next interval of time; that is, there is no correlation between the probability of leaving and the length of time the molecule has been in the unit. This is in contrast to the plug flow unit where the molecules leaving at any instant are those that have been in the unit for the longest time.

The way in which materials move through a unit is important in terms of process performance and control. Consider, for example, whether you would prefer plug flow

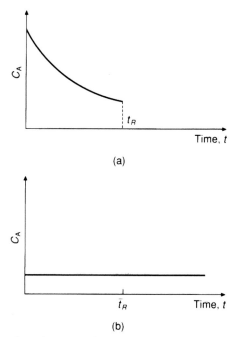

(a)

(b)

FIGURE 7.5 Concentration of reactant A "seen" by a molecule moving through a reactor: (a) plug flow; (b) backmixed.

or backmixed behavior in a continuous oven for baking pies. Would a plug flow or backmixed operation be better in a unit designed to be fed materials of rapidly varying composition but produce a product of uniform composition? Consider finally the concentration C_A of a reactant A "seen" by a molecule of B in a continuous reactor, where these two species are reacting to form products. These concentrations are sketched in Fig. 7.5 and it should be clear that reaction rates can be quite different in the two types of reactor. Indeed, in the plug flow reactor, the reaction rate in a packet of fluid moving through the reactor will change as the packet moves from a region of high concentration at the inlet to lower concentration at the outlet. In a backmixed reactor the rate is uniform throughout. Usually, for reactors where the (mean) residence times are equal, the plug flow reactor will yield a higher conversion to products than that of the backmix reactor.

The tracer experiments described above are characterization experiments that are frequently carried out on actual unit operations. Such experiments yield information on whether the operation is working fully as well as characterizing the flow through the system. For example, consider a well-agitated tank of 1000 m³ capacity through which liquid is flowing at 100 m³/min. From equation (7.3) we expect a mean residence time for a tracer of 10 minutes. If a tracer experiment measures an actual mean residence time of 7 minutes, the tank is only 70% full, or alternatively, 30% of the liquid in the tank is "dead space" that is cut off from the rest of the liquid.

PROBLEM

A 100-m³ tank through which water is flowing continuously is to be studied by a tracer method. A concentrated salt solution is poured into the inlet (along with the inflowing water) in a period of a few seconds. An electrical conductivity meter is used to measure the salt concentration at the tank outlet versus time since the salt was introduced at the inlet.

Time (s)	Salt concentration at outlet (mol/m³)
0	0
200	0.69
400	1.28
600	1.39
800	1.35
1,000	1.25
1,500	0.99
2,000	0.77
2,500	0.60
3,000	0.47
4,000	0.28
5,000	0.17
7,000	0.06
10,000	0.01

Show that the salt concentration, C, fits the equation

$$C = 2\{\exp[-0.0005(t-100)] - \exp[-0.005(t-100)]\}$$

and use it to determine the mean residence time for the tank.

Solution

Demonstrating that the equation fits the data is left as an exercise for the reader. The mean residence time is

$$\bar{t}_R = \int_0^\infty tCdt \div \int_0^\infty Cdt$$

The numerator can be split into two integrals, both of the form

$$\text{const.} \times \int_0^\infty te^{-at}dt = \text{const.}\left[\frac{e^{-at}}{a^2}(-at-1)\right]_0^\infty = \frac{\text{const.}}{a^2}$$

The denominator can similarly be split and the integration is straightforward. The result is

$$\bar{t}_R = \left[\frac{\exp(0.05)}{25\times10^{-8}} - \frac{\exp(0.5)}{25\times10^{-6}}\right] \div \left[\frac{\exp(0.05)}{5\times10^{-4}} - \frac{\exp(0.5)}{5\times10^{-3}}\right] = 2000s$$

7.2 Material and Enthalpy Balances

Material and enthalpy balances are among the major "tools" used by process engineers.

7.2.1 Control Volumes and a General Balance Equation

The first task in drawing up such a balance is to define a "control volume" over which the balance is to be carried out. A control volume might contain a unit operation, part of one, an entire plant, or may even be microscopic in size. Conceivably, a control volume might change in its dimensions; a mass balance on the falling/evaporating droplet of Chapter 3 would use as its control volume the (shrinking) surface of the droplet.

Having defined a control volume, we can then invoke the truism: Whatever enters a control volume, plus whatever is generated in the control volume, must either leave it, be destroyed within it or stay within it. In a more symbolic form,

input of X + net generation of X = output of X + accumulation of X   (7.4)

where by net generation we mean generation minus destruction. X can be any one of the following (besides a few other things):

1. Mass:
   (a) Total mass (*overall mass balance*)
   (b) Mass of a particular *chemical* species
   (c) Mass of all compounds of a particular element

    (d) Mass of an elemental form of an element
2. Heat (enthalpy)
3. Total energy
4. Mechanical energy
5. Momentum
6. Number

Note that in (la), leaving aside nuclear reactions where significant amounts of mass can be destroyed, the net generation is zero; we cannot destroy mass in chemical reactions but only convert it from one chemical form to another. Note also in (2) that it is necessary to specify a reference temperature since enthalpy can only be defined with respect to such a temperature. In this book we confine our attention to balances of types (1) and (2).

7.2.2 Basis of a Balance

Clearly, the amount of X input to a control volume is likely to increase with the time interval under consideration, as are the other terms in equation (7.4). It is therefore essential not only to define the control volume used in a particular balance but also to state the time interval used. If, as is usual, that interval is a unit of time (a second, an hour, a week, etc.), the terms in equation (7.4) become rates of input, accumulation, and so on. The interval chosen is called the *basis* of the balance. In performing balances on batch operations a basis of one batch is sometimes chosen. Sometimes, unit mass of a feed or product is chosen as the basis; for example, a material balance could be drawn up for the Bayer-Hall process of Fig. 7.1 to determine the inputs (bauxite, carbon anodes, etc.) to produce a ton of aluminum.

7.2.3 Stoichiometry

In doing mass balances on units where chemical reactions take place, the mass balances are usually insufficient in themselves to permit calculations to be completed, and we need, in addition, to exploit the stoichiometry of the reactions.

    Consider, for example, a continuous reactor where ferric oxide is being reduced by carbon to produce iron and carbon monoxide:

$$Fe_2O_3 + 3C = 2Fe + 3CO$$

We will select, as a control volume, the entire reactor and, as a basis, 1 hour of operation. The reactor is fed 10 metric tons per hour of Fe_2O_3 and the problem is to calculate the input of carbon necessary to reduce all this Fe_2O_3, together with the output of iron and CO. We shall need the atomic weights, which (to a precision sufficient for our purposes) are

$$Fe : 55.8 \qquad O : 16.0 \qquad C : 12.0$$

meaning that molecular weights are Fe_2O_3: 159.6 and CO: 28.0. We shall assume that the reactor is at steady state; this has the consequence that it makes the accumulation term in equation (7.4) zero. Any nonzero accumulation would mean that an important process variable is changing with time and therefore the assumption of steady state could not hold.

Let us start with a balance on metallic iron. The input is zero (the oxide, not the metal, is fed to the reactor) and, with zero accumulation, we get

net generation of metallic iron (tons) = output of metallic iron (tons)   (7.5)

This (rather trivial) balance is useless unless we can relate the left-hand side to the input of iron oxide; we can do this by considering the stoichiometry of the reaction, namely that 2 mol of iron is generated by reaction for every mole of Fe_2O_3 consumed. If there is zero accumulation in the reactor and all Fe_2O_3 fed is reduced (implying no output of Fe_2O_3), then

$$\text{consumption of } Fe_2O_3 \text{ in the reactor } = \frac{10 \times 10^6}{159.6} \text{ mol}$$

that is,

$$\text{generation of iron } = \frac{2 \times 10 \times 10^6}{159.6} \text{ mol}$$

$$= \frac{2 \times 10 \times 10^6}{159.6} \times \frac{55.8}{10^6} \text{ tons } = 6.99 \text{tons}$$

We have now used the stoichiometry of the reaction, in conjunction with mass balances, to calculate the output of metal. The reader should carry out similar balances on carbon to show that the input of carbon must be 2.26 tons and the output of carbon monoxide 5.26 tons. As a check, an overall mass balance yields

| input | input | output | output |
10 tons (as Fe_2O_3) + 2.26 tons (as C) = 6.99 tons (as Fe) + 5.26 tons (as CO)

to the precision of our calculations. Note that in the overall material balance the net generation is zero (since we cannot create or destroy matter but only convert it from one chemical to another), and in this case, the accumulation term is also zero, since we have assumed steady state. We defer considering heat balances to Section 7.5 in order to apply equation (7.4) to some common but more complex situations.

7.3 Recycles

In Fig. 7.1 a stream (sodium hydroxide solution) leaving one of the operations is recycled to a point earlier in the process. Recycle streams are common in process engineering and they complicate the calculation of mass balances. Let us consider the simple situation depicted in Fig. 7.6, where part of the output from a continuous steady-state unit is recycled directly back to the unit. This would make no sense for a perfectly mixed unit since such recycle is already taking place within the unit itself. However, there are some units where simple recycles of this kind are used. The ball mill coupled to the size classification device sketched in Fig. 1.14 is an example. In that instance the recycle stream is different from the output stream in that it has a different size distribution of particles. In the present example we consider the simpler case where the recycle stream is identical in composition to the output stream. Let us suppose that the unit is a reactor that converts a chemical species A to a chemical product B and that the conver-

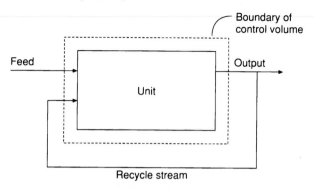

FIGURE 7.6 Control volume for a unit operation with a recycle.

sion *in the reactor* is 75 %. Let the feed rate of A (fresh feed, not including the recycle stream) be F volume units per unit time with a concentration of 0.5 mass unit per unit volume. The volumetric flow rate of the recycle stream we will set at half the flow rate of the fresh feed. We shall assume that the reaction does not significantly affect the volume of the material flowing through the reactor, so that the volumetric flow rate out of the reactor is one and a half times that of the fresh feed. The problem is to calculate the concentration of A in the output; let this be y.

We select as a control volume that shown as a dashed line in Fig. 7.6 and a unit of time as a basis. We then perform a mass balance on species A. The net generation of A in this case is negative and 0.75 times the input in magnitude. Since the unit is at steady state, the accumulation is zero. The balance becomes

$$\underset{\text{(fresh feed)}}{0.5\ F} + \underset{\text{(recycle)}}{0.5y\ F} - \underset{\text{(net generation)}}{0.75(0.5\ F + 0.5y\ F)} = \underset{\text{(output)}}{1.5y\ F} + \underset{\text{(accumulation)}}{\text{zero}}$$

that is

$$0.125\ F = y(1.5 - 0.125)\ F \qquad \text{or} \qquad y = 0.0909$$

Without employing algebra we can immediately calculate that a reactor with 75 % conversion, no recycle, and the same fresh feed would have an output concentration of $0.5 \times 0.25 = 0.125$, and we see that in this instance the recycle has been beneficial in increasing the fraction of the feed stream that has been reacted to B. This comparison is perhaps unfair, as the reader will see if he or she examines the reactor size (holdup) necessary for the two configurations by, for example, making the assumption that the two configurations require the same (mean) residence time to achieve the same 75% conversion.

The recycle stream forced us to use algebra in setting up the mass balance on the unit of Fig. 7.6. (The alternative would have been an iterative calculation of the "dog chasing its tail" variety.) The use of algebra is central to calculations on practical processes. Symbols are given for unknown compositions and flow rates, and equations containing these symbols are obtained from mass balances, stoichiometry, and (frequently) equilibrium relationships (when equilibrium holds). The task is completed when the number of equations equals the number of unknowns; the rest is merely the solution of a large set of equations. Usually, that solution is obtained with the aid of standard software packages on a computer. The equations and their solution may be complicated, but obtaining the equations is the simple application of the truism of equation (7.4) coupled with stoichiometry and equilibrium calculations. In some cases the number of equations does not become equal to the number of unknowns. This indicates that additional assumptions or measurements on the unit will be necessary to complete the mathematical description.

7.4 Staged Operations

We have already encountered unit operations of this type in Chapter I when we considered groups of flotation cells arranged together, with a slurry flowing from one cell to the next. To illustrate this concept further, we examine a different kind of staged operation known as solvent extraction.

7.4.1 Solvent Extraction

Solvent extraction has proved to be a useful tool in the production of copper (and other metals) by the hydrometallurgical route and will be encountered again in Chapter 9. Solutions of certain reagents in an organic liquid (e.g., an inexpensive liquid such as kerosene) are capable of selectively removing metal ions from aqueous solutions into the organic solution. A laboratory-scale experiment where a valuable metal in aqueous solution is extracted into an organic phase is sketched in Fig. 7.7.

Frequently, the ability of the organic phase to extract the desired metal ion is dependent on the pH of the aqueous phase, for example, as illustrated in Fig. 7.8. This is convenient since (by maintaining the aqueous pH at the point indicated by the arrow by adding suitable acids or alkalies) much of the valuable metal can be removed into the organic phase, leaving behind the undesirable impurities in the aqueous phase. Furthermore, if we now contact the separated organic phase with a fresh strongly acid aqueous solution, the valuable metal will enter the aqueous solution in accordance with Fig. 7.8. Transfer of the metal ion to the organic phase is known as *loading* and its recovery into a fresh aqueous solution as *stripping*. After stripping, the organic phase can be recycled to process a fresh batch of impure aqueous phase. The net result of the technology is the separation of an impure aqueous stream into two aqueous streams, one containing most of the impurities and the other most of the valuable metal ion.

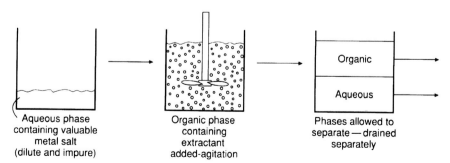

FIGURE 7.7 Experiment in which a metal salt is transferred from an aqueous phase to an organic phase.

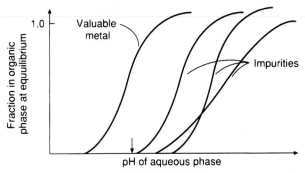

FIGURE 7.8 Sketch of how the fraction of the species reporting to the organic phase in solvent extraction might depend on pH.

A close examination of Fig. 7.8 shows that operation at the point indicated by the arrow loads most, *but not all*, of the valuable metal into the organic phase. If it is desirable to extract the remaining valuable metal from the aqueous phase, we could employ a second batch of organic (or a third, etc.), to make the loading a staged operation. Similarly, it may be desirable to make the stripping a staged operation.

There is a second incentive to use stage operations in solvent extraction. In practice the stages are carried out continuously, frequently in devices called *mixer-settlers*, one of which is sketched in Fig. 7.9. Usually, these mixer-settlers are large, approximating the size of a swimming pool, and expensive to build. The size and expense can be kept to reasonable values by not insisting that equilibrium be achieved. Use of a mixer-settler for loading would then result in the organic phase containing less valuable metal than predicted by equilibrium experiments and a second (or third, etc.) mixer-settler might be necessary to achieve the desired degree of loading.

If multiple mixer-settlers are used, various schemes can be designed for connecting them. The scheme depicted in Fig. 7.10 contains four mixer-settlers (now shown as blocks), two for loading and two for stripping. The loading mixer-settlers are operated in *countercurrent* flow (as are the stripping units). Note that the aqueous phase flows from 1 to 2 while the organic phase flows from 2 to 1. An alternative is the *cocurrent* operation depicted in Fig. 7.11. The reader will note that nothing is achieved in mixer-settler 2 if equilibrium is reached in 1; generally speaking, cocurrent operations are less

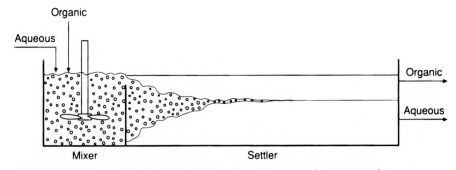

FIGURE 7.9 Mixer-settler for industrial-scale continuous solvent extraction.

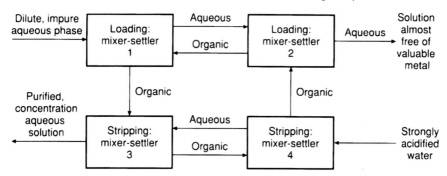

FIGURE 7.10 Arrangement of mixer-settlers into a solvent extraction scheme. Units are in countercurrent operation.

efficient than countercurrent operations. A third possibility is the *cross-current* operation illustrated in Fig. 7.12.

7.4.2 Example of Solvent Extraction

We conclude our treatment of staged operations by using mass balances in an example to illustrate the superiority of staged countercurrent operations over a single mixer-settler unit. We use the concept of a distribution coefficient, K.

$$K = \frac{\text{concentration in organic phase leaving a stage}}{\text{concentration in aqueous phase leaving a stage}} \quad (7.6)$$

If equilibrium is achieved (or approached) in a stage, this quantity can readily be measured (or approximated) in the laboratory for the particular chemical system of interest. Usually, K will be pH dependent, so that it will be different in the loading section than in the stripping section. For the present problem K = 4.

Consider first the loading section consisting of one mixer-settler as shown in Fig. 7.13. The aqueous input is a stream containing 5000 parts per million of copper (as the Cu^{2+} ion). The other input stream is an organic phase from stripping, presumed here to be free of copper. At steady state we can perform a mass balance on all of the copper to get

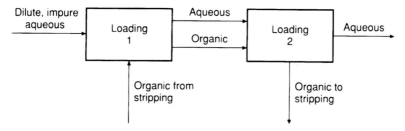

FIGURE 7.11 Loading section of the extraction scheme in cocurrent operation.

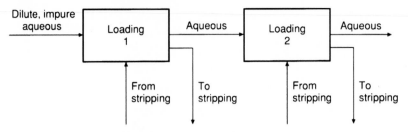

FIGURE 7.12 Loading section in cross-current operation.

$$
\begin{array}{ccc}
10^4 \times \dfrac{5000}{10^6} & \text{zero} & \text{zero} \\[2mm]
\text{(input in} & \text{(input in} & \text{(net generation = zero} \\
\text{aqueous stream)} & \text{organic)} & \text{since copper not generated)}
\end{array}
$$

$$
= \quad
\begin{array}{ccc}
10^4 \times \dfrac{x}{10^6} & 5 \times \dfrac{10^3\,y}{10^6} & \text{zero} \\[2mm]
\text{(output in} & \text{(output in} & \text{(no accumulation} \\
\text{aqueous)} & \text{organic)} & \text{because steady state)}
\end{array}
$$

where kilograms have been used as the mass unit, the basis is 1 hour, and x is the concentration of copper in the exit aqueous stream (in ppm) and y that in the organic stream. We now have one equation containing two unknowns and we use (7.6) as the necessary additional equation:

$$
\frac{y}{x} = 4
$$

Substitution yields

$$
50 = 3 \times 10^4 \, \frac{x}{10^6}
$$

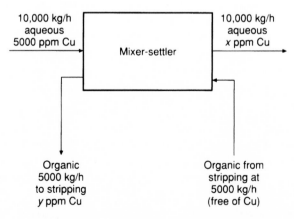

FIGURE 7.13 One mixer-settler of the loading section in the example.

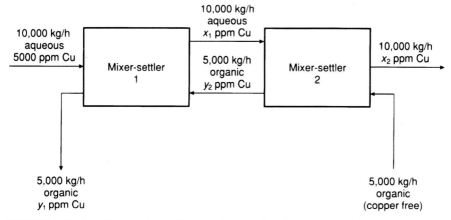

FIGURE 7.14 Loading section with two mixer-settlers in the example.

or $x = 1667$ and $y = 6667$.

Now consider the staged countercurrent operation depicted in Fig. 7.14. Note that x_i = the concentration of Cu in the aqueous phase leaving unit i, and similarly for y_i. There are now two mass balances on the copper, resulting from the two obvious choices for control volumes. Again the basis is 1 hour. The mass balance for mixer-settler 1 is

$$10^4 \times \frac{5000}{10^6} + 5 \times 10^3 \times \frac{y_2}{10^6} = 10^4 \times \frac{x_1}{10^6} + 5 \times 10^3 \times \frac{y_1}{10^6}$$

or

$$50 + 5 \times 10^{-3} y_2 = 10^{-2} x_1 + 5 \times 10^{-3} y_1$$

For mixer-settler 2 we get

$$10^{-2} x_1 = 10^{-2} x_2 + 5 \times 10^{-3} y_2$$

The necessary additional two equations are obtained by invoking (7.6):

$$\frac{y_1}{x_1} = 4 \qquad \frac{y_2}{x_2} = 4$$

The solution of this set of four equations is straightforward and yields

$$x_2 = 714 \qquad y_2 = 286 \qquad x_1 = 2143 \qquad y_1 = 8571$$

Note that the copper concentration leaving the first mixer-settler in the aqueous phase is much higher than in Fig. 7.13 but that overall the scheme of Fig. 7.14 is much better at transferring the copper from the aqueous phase to the organic phase. It is left as an exercise for the reader to show that the scheme of Fig. 7.14 is also better than a cross-current scheme where there are two fresh (copper-free) organic streams, each of 2500 kg/h, each fed separately to a mixer-settler.

7.5 Heat Balances

In performing enthalpy (heat) balances on a control volume, equation (7.4) becomes

$$\begin{matrix} \text{input of} \\ \text{enthalpy} \end{matrix} + \begin{matrix} \text{net generation} \\ \text{of enthalpy} \end{matrix} = \begin{matrix} \text{output of} \\ \text{enthalpy} \end{matrix} + \begin{matrix} \text{accumulation} \\ \text{of enthalpy} \end{matrix} \qquad (7.7)$$

Not only must we define the control volume and the basis (as in mass balances), but we must also specify the reference temperature that we are selecting for the enthalpy. The net generation term may include heat generated (or consumed) by chemical reaction. Each such reaction will make a contribution to this term given by minus $\Delta \tilde{H}_{R0}$ multiplied by the amount of reaction for the selected basis. The quantity $\Delta \tilde{H}_{R0}$ is the standard heat of reaction *at the selected reference temperature*, given by equation (2.57). The minus sign occurs because exothermic reactions (for which $\Delta \tilde{H}_{R0}$ are negative) generate heat (i.e., make a positive contribution to heat generation). Enthalpy can enter (or leave) a control volume by three possible mechanisms:

1. Convection (sometimes called *advection*), with the material flowing into and out of the control volume. This contribution is frequently called the *sensible heat* of the entering (or leaving) materials.
2. Conduction through the boundaries of the control volume.
3. Radiation through the boundaries of the control volume.

This is shown in Fig. 7.15. Note that the heat accumulation can be nonzero because the temperature in the control volume changes with time, because the contents are not at the reference temperature and their mass changes with time, or both. The implication of the remarks above is that it is usually necessary to perform a mass balance before a heat balance can be attempted. This is a consequence of the dependence of the terms in equation (7.7) on mass input, mass accumulation, amount of reaction, nature and mass flow rate of outputs, and so on. Next we illustrate the principles of heat balance by considering how the energy requirements for a unit operation might be met.

7.6 Supply of Heat to Unit Operations

Many operations in materials production require a supply of heat. For example, the sintering of ceramic parts is done in furnaces that are supplied with electricity or a combustible fuel. In many cases this supply of energy is the principal operating cost of the unit and it is therefore appropriate to examine heat flows in such units.

7.6.1 Fuel Combustion Units

Let us start by considering units, such as furnaces, where a fuel (oil, natural gas, etc.) is burned with air. A schematic diagram appears as Fig. 7.16 and a control volume has been drawn (with a dashed line) to enclose that space where the combustion takes place and to *exclude* whatever the furnace is designed to heat. The last might be the ceramic parts already mentioned, a pile of steel scrap we wish to melt, or even the water-filled tubes of a boiler that we are using to generate steam. We will draw a distinction between the heat transferred (by conduction and radiation) through the boundary of the control

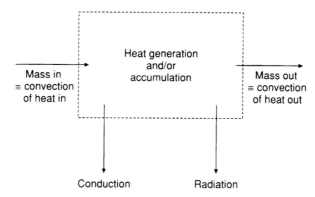

FIGURE 7.15 Control volume for a heat balance.

volume into the "load" that we are trying to heat, and the heat lost through the walls, roof, and bottom of the furnace and thence into the surroundings. We call the former *available heat* and the latter *heat losses*. There is a third heat output in the form of the enthalpy of the hot combustion products (*flue gases*) leaving the control volume.

To simplify what follows, let us treat a unit operating continuously at steady state. We will assume that *within the control volume* there is no generation of heat other than the combustion of the fuel. Equation (7.7) can then be rewritten

$$\begin{matrix} \text{sensible heat of} \\ \text{input fuel} \end{matrix} + \begin{matrix} \text{sensible heat of} \\ \text{input air} \end{matrix} + \begin{matrix} \text{heat generated by} \\ \text{combustion} \end{matrix}$$

$$= \text{available heat} + \text{heat losses} + \text{sensible heat of flue gases} \qquad (7.8)$$

Let us choose as a basis unit mass of fuel. Rearranging gives

$$\begin{matrix} \text{available heat} \\ \text{(heat units per unit} \\ \text{mass of fuel)} \end{matrix} = \begin{matrix} \text{sensible heat of} \\ \text{input fuel} \\ \text{and air} \end{matrix} + \begin{matrix} \text{Heat generated} \\ \text{by combustion} \end{matrix}$$

$$- \text{heat losses} - \text{sensible heat of flue gases} \qquad (7.9)$$

which makes it clear that the available heat will be increased if the flue gases leave at a lower temperature. Usually, there will be a lower limit on this temperature. In many combustion units it is possible to identify a minimum operating temperature (henceforth $T_{\min}$) for the unit. For example if the purpose is to boil water at atmospheric pres-

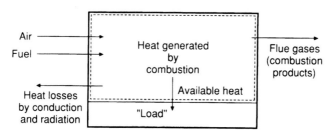

FIGURE 7.16 Schematic diagram of a fuel-fired furnace.

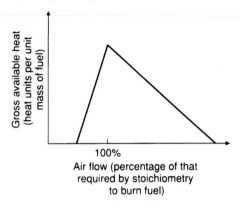

FIGURE 7.17 Gross available heat for a fuel-fired furnace as a function of air flow rate.

sure, 100°C would be the minimum. T_{min} would be approximately 660°C (the melting point of aluminum) in a furnace designed to melt and homogenize aluminum scrap. T_{min} (sometimes called the *critical process temperature*) frequently sets a lower limit on the temperature at which the flue gases leave. Heat is transferred from the gases in the control volume to the material that is to be heated, and if that material is at T_{min} the gases can be no cooler.[1]

Taking an ideal case where the flue gases leave at T_{min} and the heat losses are neglected, we have

$$\begin{matrix} \text{available} \\ \text{heat} \end{matrix} = \begin{matrix} \text{sensible heat} \\ \text{of input fuel} \\ \text{and air} \end{matrix} + \begin{matrix} \text{heat generated} \\ \text{by combustion} \end{matrix} - \begin{matrix} \text{sensible heat} \\ \text{of flue gases} \\ \text{at } T_{min} \end{matrix} \qquad (7.10)$$

The available heat calculated in this way is known as *the gross available heat*. Note that it can readily be calculated once the temperature of the inlet fuel and air, the ratio of their flow rates, and T_{min} are specified, provided it can be assumed that the combustion reaction proceeds to completion (the consumption of all the fuel or air). That is because a mass balance and the stoichiometry of the combustion yield the amount of all the flue gas constituents (including nitrogen entering in the air) per unit mass of fuel burned. Enthalpy tables or specific heats then yield the sensible heat of the flue gases at T_{min} as well as the sensible heat in the input fuel and air (with respect to the reference temperature selected). The heat generated by combustion of unit mass of fuel (at the *reference* temperature) can be determined from equation (2.57) and standard heats of formation. It is seen, then, that the gross available heat is independent of the furnace details (e.g., of what is being heated) and is therefore a useful means for comparing fuels.

---

[1] There are important exceptions to this generalization. For example, we might preheat aluminum scrap in a separate region of the furnace where the scrap is brought from room temperature to near its melting point by contacting it in countercurrent flow with hot flue gases. The flue gases might leave this section at well below 660°C. In this case, where there is a distribution of temperature in the material being heated, the simple analysis presented here falls down, although its conclusions are qualitatively correct.

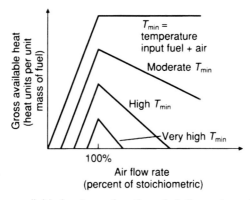

FIGURE 7.18 Gross available heat as a function of air flow rate and minimum furnace temperature.

Suppose that a calculation of gross available heat is made as a function of the ratio of air to fuel flow rates. (A problem of this type appears at the end of this chapter.) The results would appear as sketched in Fig. 7.17, where the air flow rate is expressed as a percentage of that required by stoichiometry to just completely burn the fuel. In this calculation T_{min} and the sensible heat of the inlet fuel and air are fixed. For air flow rates less than 100% the gross available heat is less than at 100% because now only a fraction of the fuel is burned. Above 100% the gross available heat is also less than at 100% because the sensible heat of the flue gas is increased by the excess air in that gas. Now let us think of the calculations repeated at different minimum operating temperatures. The result is shown in Fig. 7.18. In all cases the maximum gross available heat is obtained at an air flow rate set equal to that required by stoichiometry to just burn the fuel. Two other facts are obvious:

1. The gross available heat is small at high T_{min}; indeed, at very high values of T_{min} it becomes zero, and the particular fuel under consideration cannot be used where still higher temperatures are to be achieved unless radical changes are made .[2]
2. Operations at high T_{min} would require very precise control of the air-to-fuel ratio if the unit is to yield the maximum gross available heat possible.

The concept of gross available heat provides a ready means for a first comparison of fuels. It should be clear that the real cost of a fuel is not the dollars per kilogram, or the dollars per unit of heat generated by combustion but (to a first approximation) the dollars per unit of gross available heat. That cost will escalate rapidly as the minimum operating temperature is raised.

A more thorough analysis would recognize that the actual available heat would be still lower than the gross available heat due to (1) heat losses, (2) flue gases leaving at well above T_{min} and (3) because the combustion of the fuel may not be complete because

---

[2] Such radical changes might include using oxygen for combustion rather than air, or preheating the incoming air. The reader is asked to ponder the effect of such changes in terms of equation (7.10).

of kinetic limitations. Figure 7.19 is a sketch of how the actual available heat falls short of the gross available heat and has a maximum (typically at an air flow rate somewhat higher than required by stoichiometry).

7.6.2 Units with Electrical Generation of Heat

The rate of electrical generation of heat in a unit is simply the product of the current (I) through the unit and the voltage drop (V) across it. This heat is generated without the combustion of any fuel and therefore with the generation of no (or negligible) flue gases. Consequently,

$$\text{gross available heat (heat units per unit time)} = IV$$

and is *independent* of the minimum temperature of operation. For example, a furnace operating with a voltage of 220 volts and a current of 100 amperes has a gross available heat of 79.2 MJ/h, irrespective of whether it is used to boil water or melt steel. Contrasting this with Fig. 7.18, we see that electric furnaces will be more costly to operate than a fuel-fired furnace if T_{min} is low (e.g., in boiling water) but would tend to be less costly than fuel-fired furnaces at higher temperatures (e.g., the very high temperatures needed to reduce silicon dioxide to silicon). The "break-even temperature" would depend on the local cost of fuels and electricity as well as other factors, such as the capital cost of the furnace, the costs of pollution control equipment (which would be high for the fuel-fired furnace because of the large volumes of flue gases involved), and the relative ease of operation of the two types of furnace.

A more careful examination would recognize that some recovery of heat from the flue gases of a fuel-fired furnace may be possible. For example, the fuelfired reverbatory furnaces used in smelting copper concentrates (see Chapter 8) frequently have waste heat boilers that generate steam from the heat in the hot flue gases. Such heat recovery can make a significant contribution to the competitiveness of fuel-fired furnaces.

7.7 Process Control

No discussion of process engineering, however brief, could be considered complete without a treatment of process control. This topic is increasingly important in modern processes for producing materials. In older materials-producing technology, control (in the feedback sense described below) was rarely practiced. Today, most units in a process for producing materials would be equipped with some sort of control system. In the high-tech materials area, very sophisticated controls are used. For example, growth of the large single crystals employed in the semiconductor industry requires very precise temperature control, frequently under the guidance of a computer.

7.7.1 Typical Feedback Control System

Consider, as an example, the task of controlling the electric furnace depicted as a block in Fig. 7.20. The solid arrows in this figure do not (as in process block diagrams) indicate a flow of materials but rather a flow of information in the form of an electrical signal. Thus the thermocouple (shown without its reference junction) transmits the

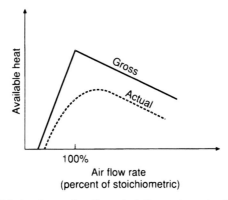

FIGURE 7.19 Available heat as a function of air flow rate; actual compared to gross.

furnace temperature to the electronic device labeled *controller*; also input into this device is the desired temperature (*set point*). This is usually achieved by means of a second electrical signal (perhaps provided by a supervising computer) or merely by the operator turning a knob on the controller. The controller has the function of determining the error, ε, given by

$$\varepsilon = \text{set-point temperature - actual furnace temperature} \qquad (7.11)$$

The controller then sends a signal (usually a voltage of a few volts, at a current in the milliampere range) to the power supply. This control signal is dependent on the error in the manner described below. The power supply is designed to supply power to the furnace (perhaps as much as a few hundred volts, at several tens or even hundreds of amperes) at a rate dependent on the control signal. This is exemplary of *feedback control*; the power to the furnace (i.e., the heat generated within it) is adjusted according to the difference between the set-point temperature and the actual temperature. The in-

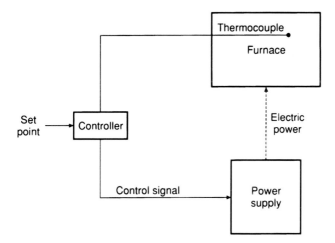

FIGURE 7.20 Feedback control scheme for a furnace.

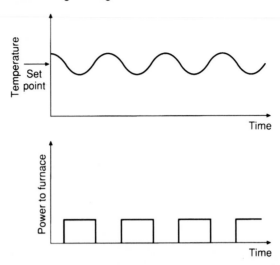

FIGURE 7.21 On-off control showing temperature variation as power to furnace turned on and off.

ventive reader will have no difficulty in conceiving a feedback control scheme for controlling a gas-fired furnace, the pH in a reactor containing an aqueous solution, the speed of an electric motor, or the heading of a torpedo. These schemes all have a *sensor* (e.g., the thermocouple), a controller, and a *final control element* (e.g., the power supply).

7.7.2 On-off Control

One very simple controller used frequently in the laboratory and for controlling home heating systems is the *on-off controller*. In the context of Fig. 7.20, such a controller would have no output unless ε exceeded zero (i.e., the actual furnace temperature dropped below the set point). The controller would then send a signal to the power supply so as to supply full power to the furnace. As the furnace heated up, ε would drop and finally reach a point at which the controller turns off its signal to the power supply, thereby turning off electric power to the furnace, which begins to cool, and so on. The resulting temperature history is sketched in Fig. 7.21. The amplitude of the fluctuations about the set point are dependent on the *thermal inertia* of the furnace, the amount of power supplied when the power supply is switched on, and the rate of heat loss. Under some circumstances a controller of this nature will switch the power supply off and on too frequently. One modification that can avoid this difficulty allows the temperature to drift within a narrow band about the set point without switching taking place. Some fluctuation in the temperature is inherent in on-off controllers, so they are generally not used for precise control.

7.7.3 Proportional Controllers

Proportional controllers give an output signal to the final control unit that is proportional to the error:

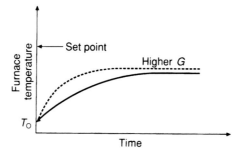

FIGURE 7.22 Typical response of a furnace under proportional control showing the effect of higher controller gain.

$$\text{control signal} = G\varepsilon \qquad (7.12)$$

with G a constant known as the gain. Usually, the gain would be adjustable by means of a knob on the controller. Sometimes the reciprocal of G (known as the *bandwidth or proportional bandwidth*) is marked on that knob. If the power supply is designed so that its output to the furnace is proportional to its input control signal, the power to the furnace will be proportional to ε. Let us consider a furnace where the rate of increase in temperature is proportional to the power input less heat losses (and any heat load imposed on the furnace, e.g., by cold material being fed into the furnace), and the latter are proportional to temperature difference between the furnace and the surroundings at T_{sur}:

$$\frac{dT}{dt} = K_1 G(T_{set\ point} - T) - K_2(T - T_{sur}) \qquad (7.13)$$

with T the furnace temperature and K_1 and K_2 constants. The solution of this ordinary differential equation with an initial temperature of T_{sur} is

$$T = \frac{K_1 G T_{set\ point} + K_2 T_{sur}}{K_1 G + K_2} - \left(\frac{K_1 G T_{set\ point} + K_2 T_{sur}}{K_1 G + K_2} - T_{sur} \right) e^{-(K_1 G + K_2)t} \qquad (7.14)$$

This result is sketched in Fig. 7.22. It is seen that on heating the furnace from some initial temperature, the temperature asymptotically approaches a steady-state value. The dashed line indicates what happens if the gain of the controller is increased. The behavior is "better" in the sense that the furnace more rapidly approaches the set-point temperature. Furthermore, manipulation of equation (7.14) shows that as t $\rightarrow\infty$

$$\left(T_{set\ point} - T \right) \rightarrow \left(T_{set\ point} - T_{sur} \right) \frac{K_2}{K_1 G + K_2} \qquad (7.15)$$

that is, that the final temperature is *offset* from the set-point temperature by a small amount that diminishes as the gain is increased. The need for this offset is seen from equation (7.12); unless the temperature falls a little short of the set point, the control signal will be zero and therefore the power to the furnace will be zero.

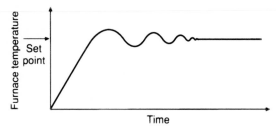

FIGURE 7.23 Response of a furnace when the gain on the controller is high.

Unfortunately our simple assumptions about the dynamic behavior of a furnace, embodied in equation (7.13), are not completely accurate for real furnaces. That dynamic behavior is such that at too high a gain, the furnace temperature will overshoot the set point, as sketched in Fig. 7.23. Such overshoot, followed by decaying oscillations, is not usually a difficulty. However. under some conditions, the oscillations grow, rather than are damped, and under these conditions the gain must be sharply reduced to restore stability.

7.7.4 Controllers with Multiple Action

A method of avoiding offset is to incorporate *integral action* in the controller. Hence

$$\text{control signal} = G\varepsilon + G_I \int_{t_0}^{t} \varepsilon \, dt \tag{7.16}$$

where G_I is a second constant. The knob on the controller that sets the value of G_I is usually labeled "reset." In a sense the integral on the right results in the controller "remembering" what the error was over the interval from t_0 to the present; in this way the present error can sink to zero without the power to the furnace becoming zero as it does with proportional action alone. Usually, the controller circuitry is such that t_0 is set (i.e., integral action switched on) only when the temperature approaches the set point; otherwise, the control signal would be dominated by what may have happened hours ago. While integral action permits offset to be avoided, excessive integral action (high G_I) can induce oscillation and instability.

Finally, we mention *derivative action*, which is sometimes called *anticipatory action*. A controller employing all three modes of control would have an output control signal given by

$$\text{control signal} = G\varepsilon + G_I \int_{t_0}^{t} \varepsilon \, dt + G_D \frac{d\varepsilon}{dt} \tag{7.17}$$

The function of the derivative term can be seen by examining a furnace where the error is positive but diminishing rapidly (i.e., the furnace temperature is well below the set point but climbing toward it rapidly). A prudent human controller might anticipate that the furnace temperature would soon reach the desired value and start to cut back on the power to avoid overshoot. This is what is attempted by our controller since $d\varepsilon/dt$ is negative and therefore tends to diminish the power. Unfortunately, the dynamics of furnaces (and other process units) are such that the effect is usually the opposite of that desired. Derivative action therefore frequently contributes to oscillation and instability and is best used with caution.

7.7.5 Other Controller Features

Many other features appear on modern controllers; examples include the opportunity to manually override the controller to set the output signal, the ability of the controller to detect a failed sensor and generate an appropriate signal, alarms that sound if the controlled variable experiences excursions outside prescribed limits, and limits that can be set on the output signal so that (for example) excessive power is not supplied to a furnace during heat-up.

7.7.6 Adjustment of Controllers

The behavior of a furnace (or other unit we wish to control) is dependent on the furnace, the controller settings, the power supply, and the heat load placed on the furnace. In an ideal situation, quantitative relationships would exist describing the transient response of the furnace temperature to a specified change in power input and heat load. Such relationships would have to be obtained experimentally or from mathematical models. Given those relationships, it would be possible to predict the transient response of the furnace under various controller settings, much as we obtained equation (7.14), and thereby to select the optimum settings. Seldom are these relationships available, and in reality it is usual to adjust the settings of a controller in a very empirical way. Such "tuning" of a controller to its application is described in detail in controller manuals. Usually, it consists of making small adjustments in the set point and observing how the temperature responds. Typically, the gain is adjusted first (with the integral and derivative action turned off) until the furnace temperature rapidly approaches a new set point with acceptable overshoot. Integral action is then adjusted to eliminate offset without introducing instability, and finally, if both necessary and possible, fine tuning is obtained by incorporation of derivative action.

7.8 Concluding Remarks

This chapter has provided an introduction to the field of process engineering. Certain important topics (e.g., process simulation) have been omitted altogether, and others (e.g., process economics) have only been hinted at. Nevertheless, it has been possible to convey the essence of a number of important concepts (unit operations, classification of units, residence times, tracer experiments, material and enthalpy balances, control volumes, role of stoichiometry, recycles, staged operations, supply of available heat to units, electrical generation of heat, and feedback, on-off, proportional, integral, and derivative control). The reader is now equipped to grasp, in the remainder of the book, why the technology for producing a given material is as it is, to glimpse the possibility of better technology, and to begin, conceptually, to design processes for new materials.

LIST OF SYMBOLS

G, G_I, G_D	controller gain, constant for integral, and derivative action of controller (see Section 7.7)
I	current (in electric furnace)
K	distribution coefficient in solvent extraction
$T, T_{min}, T_{setpoint}, T_{sur}$	temperature, minimum temperature for unit to operate, set-point temperature, temperature of surroundings
V	voltage (electric furnace)
x_i	concentration in aqueous phase leaving mixer-settler i
y_j	concentration in organic phase leaving mixer-settler i
ε	difference between set-point value and actual value

PROBLEMS

1.1 You are a plant engineer for Globall Corp. and responsible for the process that applies a tough day-glow orange coating to golf balls in order to make them more easily locatable by players. Due to financial shortsightedness on the part of Globall, budgets for development of this process at Globall were insufficient, and ultimately to maintain its competitive position in the world golf ball market and to avoid contributing to the trade imbalance problem, Globall was forced to purchase the technology from Honest Ed's Engineering Ltd. (HEEL). As it arrived from HEEL, the process (a continuous one) consisted of a large box 10 m by 20 m by 30 m which processes 1000 golf balls per minute. The terms of the contract with HEEL prevent you from opening the box to look inside. In order to learn something about the unit, you decide to perform tracer experiments. Using a surplus reactor from the nuclear engineering department of a nearby university, you irradiate several hundred golf balls. In your experiment you are assisted by Vince Overhand, a summer student. Equipped with a Geiger counter, a lead apron, and a false impression of the safety of the experiment, Vince pours several hundred irradiated balls onto the conveyer belt feeding the process (along with its usual load of normal balls) and counts the number of radioactive balls leaving the unit in each 1-minute interval. The results are as follows:

Time since radioactive balls entered unit (min)	Number of radioactive balls leaving in previous minute
1	105
2	130
3	92
4	66
5	49
6	33
7	24
8	18
9	12

10	9
11	7
12	4
13	4
14	2
15	2

Vince is able to reconfirm these numbers (with slight variation) in a second experiment two weeks later just before the company president shuts down the process following the death by radiation sickness of her pet Pekinese, which swallowed a golf ball.

(a) Calculate a mean residence time for the golf balls in the unit and then the number of balls in the unit at any time. What assumption(s) are you making when you do this?

(b) A plot of number of radioactive balls versus time suggests to you that the unit is approximately backmixed. Develop a mathematical equation for the fraction of radioactive balls in the stream leaving the unit, f, assuming perfect mixing. (Hint: If the unit is perfectly mixed, the fraction of the balls left in the unit that are radioactive will also be f.) Show that, to a good approximation, your equation fits the data.

(c) Might it be possible to avoid being fired by persuading the company that Honest Ed's has not lived up to its name?

7.2 Derive equation (7.3) for a unit of volume V and flow rate F (volume units per unit time). Use C for the concentration of tracer in the unit at time t after injection of tracer into the unit. The rate of output of tracer from the unit will then   equal $-V \, dC/dt$ and another value can be obtained from the flow rate and concentration of the output.

7.3

A → 10 m³/s	Plug flow Volume = 500 m³	→	Perfectly mixed 300 m³	B →

(a) Sketch the tracer response observed at point B above when a "pulse" of tracer is injected at point A.

(b) Would the tracer response be different if the perfectly mixed unit were up stream from the plug flow unit? Explain your answer.

(c) Sketch the tracer response for the system shown below.

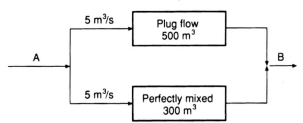

7.4 You are a plant manager for Grubstake Mining Company, a company that mines gold ores and extracts the gold from them by a cyanide leaching process. The leaching plant is your responsibility. Recently, you have become suspicious of Mr. Johnson, the foreman of the night shift, who has bought a Lear Jet, and his brother, the security guard, who is driving to work in a Ferrari Testarossa. Secretly you gather data for a material balance on the gold. Over a period of one week, 15,291 short tons of ore enter the plant and the average assay is 0.12 troy ounce of gold per ton of ore. During this week there is no change in the inventory of gold bullion and 1500 troy ounces of gold bullion are shipped. You measure and analyze the various solids and solutions in the plant throughout the week and find no significant change, except for one large tank containing (throughout the week) 10,000 U.S. gallons of solution. At the start of the week the solution analyzes 0.050 wt % $NaAu(CN)_2$, but at the end of the week this has risen to 0.088 wt %; the density of the solution is 1 g/cm^3. Are your suspicions justified, or did Johnson & Johnson inherit a fortune from a rich relative in the baby powder business, as they claim?

7.5 Calculate how much steel automobile scrap and how much direct reduced iron (DRI) should be added to an electric furnace to produce a batch of 100 metric tons of finished steel for a customer who sets the following upper limits on impurities in the steel: C less than 0.5%, Cu less than 0.3%, and Zn less than 0.5%. The steel scrap is 92% Fe, 3% C, 3% Zn, and 1% Cu. The DRI is 88% Fe, 7% O and 5% Si; the oxygen being present in the form of oxides of both silicon and iron with all the silicon being oxide (SiO_2). Previous experience shows that, during melting, all the SiO_2 in the DRI forms a separate liquid phase that can be poured separately out of the furnace, while half of the zinc in the material charged to the furnace will vaporize. During melting, oxygen that is present in the DRI other than as SiO_2 can react with carbon from the scrap to produce carbon monoxide, which leaves the melt. The steel scrap is less expensive than the DRI, so it is desired to minimize use of the latter. All compositions are in weight percent.

7.6 The system shown is a solvent extraction scheme for producing a concentrated and purified copper sulfate solution, starting with one containing 2 g/L of copper (as copper sulfate). Each block below represents a mixer-settler unit and it can be assumed that the streams leaving each unit are in equilibrium.

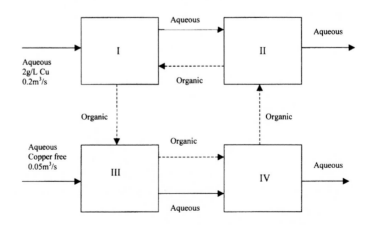

Units I and II are loading units and have a distribution coefficient of 4. Units III and IV are stripping units with distribution coefficients of 0.2 (in reality, this can be achieved by adjusting the pH of the aqueous streams). Assuming steady state and that the flow rates of each stream remain unchanged in passing through each unit, calculate the concentration of copper in the aqueous streams leaving the system. Show how the performance of the system can be significantly improved (without changing flow rates of the streams or the distribution coefficients).

7.7 Polycrystalline silicon is produced in a continuous reactor at the rate of 10 kg/h by reacting trichlorosilane ($HSiCl_3$) and hydrogen on the surface of an electrically heated silicon rod at 1300 K. For the purposes of this problem the reaction can be assumed to go to completion and the products can be regarded as HCl and pure silicon, both at 1300 K. If the hydrogen and trichlorosilane are both fed to the reactor at 300 K, calculate the minimum electric power requirements (i.e., neglecting heat losses from the reactor other than in the exit gas). Obtain the necessary thermodynamic data from the JANAF tables or similar sources. [The assumption in this problem, concerning the reaction going to completion and HCl & Si being the only products, is intended to make the calculation as simple as possible. Reality is quite different and is treated in section 16.3. Avoiding this assumption makes the calculation much more difficult but the CD supplied with this book shows how this is done using the software THERBAL.]

7.8 In a process, hydrogen is burned as a fuel with dry air. Calculate and plot the gross available heat (in kilowatthours per standard cubic meter of hydrogen) as a function of the air flow rate (expressed as a percentage of that required to just completely burn the hydrogen). The minimum operating temperature is 600° C, and for the purposes of this calculation, specific heats can be assumed to be constant with the following values: H_2, 7.1 cal/K. mol; N_2, 7.1 cal/K. mol; O_2, 8.4 cal/K. mol; and H_2O(g), 8.3 cal/K. mol. Dry air is 79% N_2, 21% O_2. The standard heat of formation of water (g) at 25° C is -57,800 cal/mol. Use 25° C as a reference temperature and assume that the hydrogen and air enter at this temperature. What is the effect of raising the inlet air temperature to 200° C at an air flow rate 50% greater than required for complete combustion?

7.9 A robot arm positions a piece of metal in a forging press by moving it in one horizontal direction. Friction can be neglected so that the acceleration of the piece, multiplied by its mass M, equals the force F applied by the robot. Let S be the distance of the piece of metal from its starting point (where its velocity is zero). The intention is to bring the piece to a halt at distance S_F, where the forging dies will strike it. Sensors detect the position of the piece as it approaches S_F so that the present error, $\varepsilon \equiv S_F - S$, is known to the control system, which has both proportional and derivative action [see equation (7.17)]. The output from the robot's motors and therefore F are proportional to the signal from the control system (F = Ks, where s is the control signal). Derive a differential equation for ε together with two initial conditions at t = 0. Show that the solution is

$$\varepsilon = e^{-at} \left[\frac{S_F KG_D}{2Mk} \sin kt + S_F \cos kt \right]$$

and identify the constants a and k. Sketch the solutions for G much bigger than KG_D^2/M, and vice versa.

7.10 **(a)** Foolishly, a plant manager leaves a summer student, Vince Overhand, in charge of a furnace in a continuous operation intended to be at steady state. Vince is a rather ungainly person and one day he accidentally lurches into the furnace control panel, turning off the controller that regulates its temperature. In a panic, Vince immediately switches the controller back on again, guessing (wrongly) that nobody will be any the wiser. Unfortu nately for Vince the controller has both proportional and integral action, and the consequence of its being switched off momentarily and then back on is that t_0, [see equation (7.16)] is reset to the present instant. Assuming that the heat load on the furnace is unchanged, sketch the plot of furnace temperature versus time subsequent to the mishap.

(b) Write and run a computer program to calculate the temperature of part (a) as a function of time. Plot the result. Start with the assumptions of the sentence preceding equation (7.13) and derive an equation similar to (7.13) except that now the control signal will be given by equation (7.16) rather than (7.12). Consequently, the differential equation that you derive will have an integral on the right-hand side. You will need an algorithm for solving ordinary differential equations (such as the Runge-Kutta method). Pick values for the various parameters so that your computed results are reasonable. Write the program so that the set-point temperature can be varied with time. Deter-mine whether or not Vince might have been able to keep a constant temperature (and thereby retain his job) by manipulating the set point following the mishap.

REFERENCES AND FURTHER READINGS

DAVIES, C., *Calculations in Furnace Technology*, Pergamon Press, Oxford, 1970.

FELDER, R. M., and R. W. ROUSSEAU, *Elementary Principles of Chemical Processes*, Wiley, New York, 1978.

FINE, H. A., and G. H. GEIGER, *Handbook of Material and Energy Balance Calcula-tions in Metallurgiical Processes*, Metallurgical Society. Warrendale. PA. 1979.

MCCABE, W. L., and J. C. SMITH, *Unit Operations of Chemical Engineering*, 3rd ed., McGraw-Hill, New York, 1976.

RAO, Y. K., *Stoichiometry and Thermodynamics of Metallurgical Processes*, Cambridge University Press, Cambridge, 1985.

SMITH, C. A., and A. B. CORRIPIO, *Principles and Practice of Automatic Process Control*, Wiley, New York, 1985.

CHAPTER

8

High-Temperature Processes for the Production of Metals and Glass

Roasting
Reduction of oxides
Smelting
Steelmaking
leaching
Electrometallurgy

8.1 Introduction

The production of metals has long been associated with high temperatures. Man moved beyond the use of naturally occurring materials, such as wood and stone, in approximately 6000 B.C. It is probable that the discovery that subjecting certain rocks to high temperatures resulted in metals was made accidentally in a cooking fire or, more likely, a potter's kiln. Since that time the major part of our metal production has been carried out in processes that employ temperatures in the range of 300 to 1600°C. This technology, known as *pyrometallurgy*, is treated first in this chapter. Alternative or complementary technologies — *hydrometallurgy* (employing aqueous solutions and comparatively low temperatures, say, less than 200°C) and *electrometallurgy* (exploiting direct current to bring about electrochemical reactions) — are treated in Chapter 9. At the end of the present chapter we include a brief treatment of the production of glass. Glass production bears a resemblance to pyrometallurgical operations in that high-temperature liquids are handled and ultimately cast into solid form.

8.2 Thermodynamic Considerations

From Chapter 1 the reader will be aware that many ores or concentrates are oxides. To produce metals from these oxides it is only necessary to reduce them. Usually, such reduction will produce an impure metal that must subsequently be refined, but the essential chemistry involved in producing the metal is one step: reduction. For economic reasons the reduction would typically employ one of the inexpensive reducing agents, hydrogen or carbon. In Chapter 2, in connection with the Ellingham diagram for oxides (Fig. 2.12), we saw that these reducing agents are effective for a number of oxides, but that more expensive reducing agents, such as aluminum, may be necessary to reduce oxides far down in the Ellingham diagram.

The other large class of compounds occurring in nature and thereby forming our ores and concentrates is that of the sulfides. If we compare Fig. 8.1, an Ellingham diagram for sulfides, with Fig. 2.12, we can immediately reach a technologically important conclusion: The inexpensive reducing agents carbon and hydrogen, which serve us well in the case of oxides, are nearly useless for producing metals by reduction of sulfides. The free energies of formation of both hydrogen sulfide and carbon disulfide

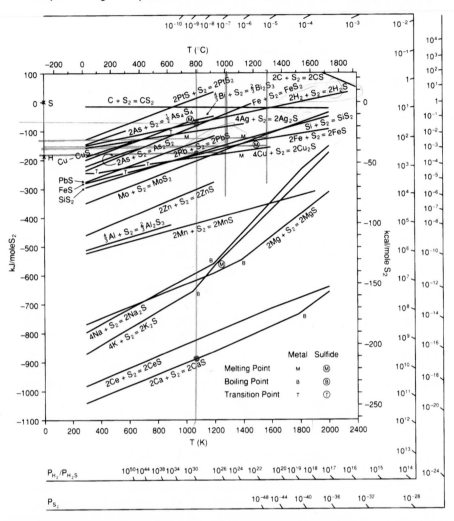

FIGURE 8.1 Ellingham diagram for sulfides. A larger version of this figure appears as an endpaper at the front of this book.

are insufficiently negative for hydrogen or carbon to be useful reagents in reducing sulfide directly to metal. We could, of course, use a more expensive reducing agent, such as aluminum, but two alternative strategies are possible.

8.2.1 Alternatives for the Treatment of Sulfides

One possibility is to convert a sulfide to an oxide by reaction with air or oxygen: for example,

$$2PbS + 3O_2 = 2PbO + 2SO_2 \tag{8.1}$$

following which the oxide can be reduced to metal.

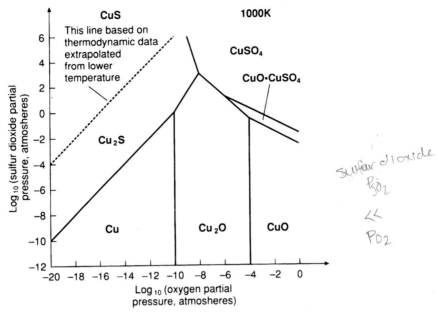

FIGURE 8.2 Predominance diagram for the copper-oxygen-sulfur system at 1000 K.

Figure 8.2 suggests another strategy. This is the predominance diagram for the copper-oxygen-sulfur system. For this system the area of stability of Cu is particularly large (contrast, e.g., Fig. 8.2 with Fig. 2.17 for Ni-S-O, where the metal is stable only at very low pressures of sulfur dioxide and oxygen). At sulfur dioxide partial pressures only a little below atmospheric (but much lower oxygen partial pressures), copper metal is the stable species. These conditions are practically[1] achievable, and copper sulfides can be converted to metal by oxidation:

$$CuS + O_2 = Cu + SO_2 \tag{8.2}$$

Note that the differences between reactions (8.1) and (8.2) (aside from the differences in the affinity for oxygen between Pb and Cu) is that in (8.2) the oxygen partial pressure must be kept low in order to produce SO_2 without forming copper oxide simultaneously. These alternative chemical schemes for the treatment of sulfides are shown in Fig. 8.3 together with a scheme for treating oxides.

8.3 Kinetic Considerations and Selectivity

One reason why high-temperature pyrometallurgical processes have been so widely used in the past is that the chemical steps involved in the processes pose little resistance to the progress of reaction at these elevated temperatures. It is common for mass trans-

---

[1] In reality, oxidation would be carried out at higher temperatures than the 1000 K of Fig. 8.2, in order that the products be molten.

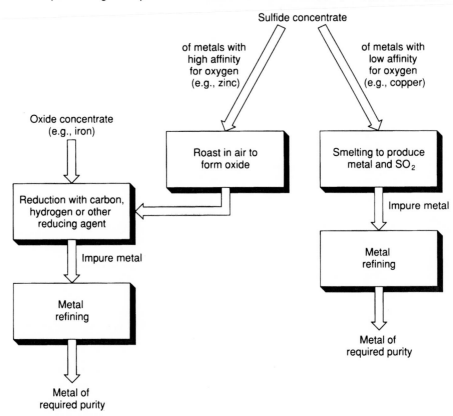

FIGURE 8.3 Alternative schemes for the treatment of concentrates.

port to be the rate-determining step in heterogeneous reactions in pyrometallurgical operations. In the past few decades significant improvements have been achieved in some pyrometallurgical operations, which can be ascribed to either increased interfacial area available for reaction within the unit or increased agitation, both of which would enhance the progress of reaction, as discussed at greater length in Chapter 3.

One important consideration is that we would typically like to carry out our reactions *selectively*, that is, to have the reactions yielding the desired product proceed without other reactions proceeding. This minimizes the contamination of the desired reaction product by the products of other reactions. Furthermore, energy or reagents are not wasted on undesirable reactions.

For example, most iron concentrates (pellets or sinter) fed to an iron blast furnace contain, besides iron oxides, the (gangue) oxides of silicon and manganese. The furnace is designed to reduce the iron oxides to metal, but unfortunately, some oxides of manganese and silicon are also reduced, resulting in contamination of the iron with these elements and extra consumption of the reducing agent (carbon).

Because chemical kinetics tend not to influence the progress of reaction at the temperatures encountered in pyrometallurgy, the selectivity of a unit operation can frequently be estimated from thermodynamic considerations, at least in a semiquantitative way.

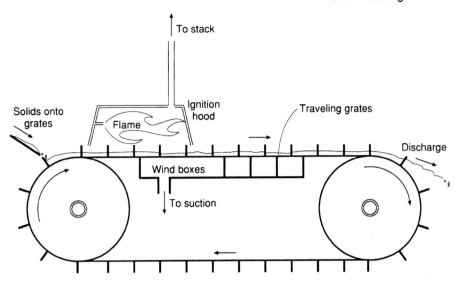

FIGURE 8.4 Traveling grate furnace for roasting of sulfides.

With the foregoing generalizations as a framework, let us examine some technologically important pyrometallurgical operations, starting with the oxidation of sulfides to oxides.

8.4 Roasting

The oxidation of sulfide ores or concentrates by heating them in the presence of air to produce solid products is known as *roasting.* Reaction (8.1) is a typical example. These oxidation reactions are typically highly exothermic and the heat generated is frequently sufficient to bring the air and solids fed to the reactor to the reaction temperature, saving on expensive fuel. The four common pieces of equipment in which roasting is carried out are the traveling grate furnace, the multiple-hearth furnace, the rotary kiln, and the fluidized bed roaster.

8.4.1 Traveling Grate Furnace

The *traveling grate furnace* is sketched in Fig. 8.4. This is the furnace that was mentioned in Chapter I as being used in one method of agglomerating concentrate particles (the sintering process) and is sometimes known as a *sinter strand.* As it is carried along on a grate moving along the top of the furnace, the sulfide passes first under an ignition hood (where a small amount of fuel is burned to ignite the sulfide) and then over a section of the furnace where *wind boxes* beneath the traveling grate draw air down through the material on the grate or force air upward through the material. This flow of air brings about oxidation of the solid deep within the material on the grate and raises the temperature. If fluxing agents (such as lime if the remaining gangue in the concen-

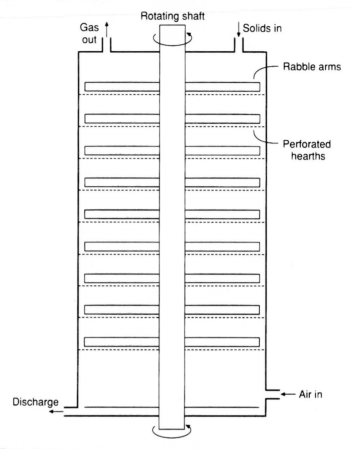

FIGURE 8.5 Multiple-hearth furnace for roasting.

trate is siliceous) are added with the concentrate, partial fusion of the material on the grate may occur. Toward the end of the machine, oxidation of the sulfide will be complete, and now air passed through the mass serves to cool (and solidify) it.

Depending on the next step in the processing sequence, the roasting product may be entirely oxide or contain a significant remainder of its original sulfur and may be either a powder or large porous lumps (*sinter*). For example, in treating lead sulfide concentrates, it is usual to attempt to remove all the sulfur and produce a sinter that can readily be reduced to lead in the lead blast furnace. In roasting copper sulfide concentrates (which typically contain a great deal of iron sulfide as well as oxide gangue compounds), only part of the sulfur is removed, leaving the rest available as "fuel" for the next unit operations, and a powder, which can be readily handled, is produced. In a traveling grate furnace the degree of oxidation is controlled by the time that concentrate spends on the grate at high temperature, the flow rate of air drawn down through the wind boxes, and the amount of combined sulfur in the concentrate. If partial fusion of the concentrate is to be avoided, fluxing agents would not be added, or would be added after roasting.

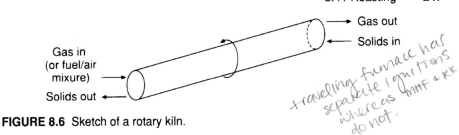

Gas out

Solids in

Gas in
(or fuel/air
mixture)

Solids out

travelling furnace has separate ignitions whereas MHF & RK do not.

FIGURE 8.6 Sketch of a rotary kiln.

8.4.2 Multiple-hearth Furnace

The *multiple-hearth furnace* is sketched in Fig. 8.5. The solid feed enters the top hearth and is moved across this hearth by rabble arms mounted on a rotating vertical shaft. Holes at the periphery of the hearth or at the central shaft permit the concentrate to fall from one hearth to the next lower one, passing inward across one hearth and outward across the next, against a countercurrent flow of gas that enters (as air) at the bottom of the furnace. The solids are heated as they move downward through the hot gases, finally reaching a temperature where the exothermic oxidation of sulfides starts. The roasted solids pass out of the bottom of the furnace while hot waste gases pass out of the top. Such furnaces might have four to 12 hearths and be 3 to 7.5 m in diameter and 7 to 15 m tall. The countercurrent flow of gas and solid means that there is no need for separate ignition of the solids (as in the traveling grate), except during startup. Again the degree of oxidation is controlled by the residence time of the solid (determined by the speed of the rabble arms), the air flow rate, and the amount of sulfur (as sulfide) in the concentrate.

8.4.3 Rotary Kiln

The *rotary kiln* is probably best known for its use in Portland cement production but is occasionally used for roasting of sulfides. The kiln (see Fig. 8.6) is a refractory-lined steel cylinder mounted so that it rotates about an axis inclined a few degrees to the horizontal (sloping toward the solids discharge end). A kiln might be 2 to 3 m in diameter and 10 to 20 m long (although even larger ones are used in cement production). Solids are fed into the higher end and usually move countercurrent to a flow of gas from the other end. Usually, the oxidation of the sulfides generates sufficient heat that no fuel need be burned, and the solids, which are tumbled through the hot gas as they descend, require no ignition, except at startup.

8.4.4 Fluidized Bed Roaster

The roasting of sulfides is one of the most successful applications of fluidized bed technology. *Fluidized beds* are beds of particles that are maintained in a state of agitated suspension (*fluidization*) by an upward flowing stream of gas (or liquid). The upflow of gas is set between a lower limit, where the particles are stationary, and an upper limit, where they are blown out of the bed (*elutriated*). The churning bed of particles shows much of the behavior of a liquid (hence its name), and there is good heat and mass transfer between the particles and gas. The rapid mixing ensures uniformity of the bed.

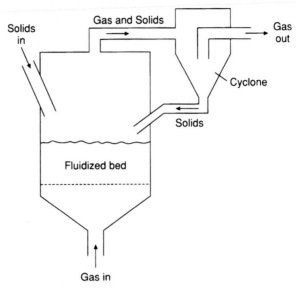

FIGURE 8.7 Fluidized bed roaster.

Figure 8.7 is a sketch of a fluidized bed roaster. The gas leaving the fluidized bed contains a small amount of solid; these are the finer particles fed to the roaster or particles formed in the roaster. They are collected by means of a cyclone and recycled to the bed. As in the other roasting units, degree of oxidation is dependent on the amount of sulfide in the concentrate, residence time, and air flow rate.

8.4.5 Comparison of Roasting Units

The four units described above differ significantly; all are continuous, but little else is similar. Solids on the traveling grate and within the multiple-hearth furnace or rotary kiln are approximately in plug flow, while the solids in a fluidized bed are nearly perfectly mixed. This might suggest that the first three would give greater conversion than the last for equipment of comparable residence time; however, mass transfer is so rapid in fluidized beds that at present the fluidized bed roaster is the usual choice for roasting sulfides. The multiple-hearth furnace and the rotary kiln generally have countercurrent flow of gas and solid, whereas flow is crosscurrent in the case of the traveling grate furnace. The traveling grate furnace is particularly suitable for producing sinter; partial fusion of the solid would usually lead to operating problems (accretions, poor fluidization, etc.) in the other units. All units generate a great deal of hot waste gas, and it is usual for part of that heat to be recovered by passing the waste gas through a boiler to raise steam.

8.4.6 Unusual Roasting Operations

The roasting of sulfides of mercury is unusual in that the metal, rather than the oxide, is formed:

$$HgS + O_2 = Hg + SO_2 \tag{8.3}$$

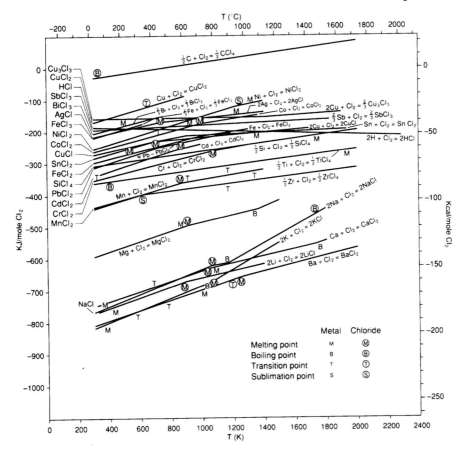

FIGURE 8.8 Ellingham diagram for chlorides. A larger version of this figure appears at the end of the book.

$\Delta \widetilde{G}_{F0}$ becomes positive for mercuric oxide above approximately 500°C. Hence at roasting temperatures, reaction (8.3) proceeds, generating mercury vapor that is condensed to metal as it leaves the roaster. This simple metal extraction, together with the distinctive pink color of cinnabar (the most common compound of mercury in the lithosphere) account for the fact that mercury has been known and extracted since ancient times. Multiple hearth furnaces are commonly used for roasting mercury ores.

Segregation roasting is another unusual technology; it has not found widespread use but is interesting from a thermodynamic viewpoint. It is a method of separating nickel, cobalt, or copper from gangue. For example, nickel oxide (either native to the ore or from roasting of the sulfide) can be separated from gangue by heating with carbon in the presence of a chloride (e.g., NaCl or $CaCl_2$). The probable reaction scheme is

$$NiO + CaCl_2 = NiCl_2 + CaO \qquad (8.4)$$

with the volatile $NiCl_2$ subsequently reacting with water vapor or oxygen on the surface of the carbon:

$$NiCl_2 + C + H_2O = Ni + 2HCl + CO \qquad (8.5)$$

In this way the carbon particles become encrusted with a nickel deposit and inexpensive physical techniques (e.g., sink-float or magnetic separation) can be used to separate these nickel-rich particles from residual gangue). Note that the reaction

$$2NiCl_2 + C = 2Ni + CCl_4 \qquad (8.6)$$

is unlikely. Examination of the Ellingham diagram for chlorides (Fig. 8.8) shows that carbon is an ineffective reducing agent for chlorides.

8.4.7 Calcining

Somewhat similar to roasting is calcining, by which is meant the thermal decomposition of solids. The technology gets its name from the production of lime ("calx" in old terminology) by the calcination of limestone:

$$CaCO_3 = CaO + CO_2 \qquad (8.7)$$

This (and other thermal decompositions producing gas) is highly endothermic; the reader may wish to ponder the thermodynamic reasons for this being so in terms of the entropy and free energy of reaction. Consequently, heat must be supplied in calcination, in marked contrast to roasting. This has made the rotary kiln, where fuel can readily be burned at the gas inlet end, a common unit for limestone calcining, although considerable amounts of lime are produced in updated versions of the older shaft-type furnaces, because of their lower fuel requirements.

8.4.8 Thermodynamics, Morphology, and Kinetics of Reaction

Figure 8.9 is a predominance diagram for the lead-oxygen-sulfur system. It is clear that roasting lead sulfide could readily yield the oxide [by reaction (8.1)] or the sulfate by the reaction

$$PbS + 2O_2 = PbSO_4 \qquad (8.8)$$

Usually, the sulfate is an undesirable product because, in this case, we wish to remove as much sulfur as possible, yielding an oxide that can then be reduced by carbon to metal. From Fig. 8.9 it is seen that sulfate formation should be minimized by avoiding high sulfur dioxide partial pressures in the unit. All four units described above permit us to do this by increasing the air flow rate. However, the sulfate formation reaction is more exothermic than the oxide formation reaction, and therefore we can deduce from thermodynamics that higher reaction temperatures should decrease sulfate formation. Consequently, excessive air flow (which would lower the reaction temperature) is to be avoided, and an optimum air flow rate (to minimize sulfate formation) should be expected. If sulfate formation is to be minimized (or indeed encouraged), variations of temperature and gas composition, both in time and location within the unit, should be avoided. The reader will be able to identify which of the four roasting units offers the greatest uniformity of composition and temperature.

Figure 8.10 shows idealized sketches of how individual particles of sulfide or calcium carbonate calcine would roast, displaying shrinking cores of unreacted solid. This suggests that transport of oxygen through the pores of the oxide layer might well be

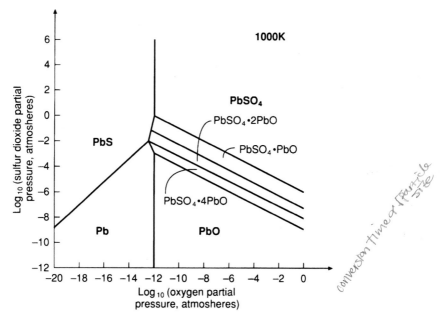

FIGURE 8.9 Predominance diagram for the lead-oxygen-sulfur system at 1000 K.

rate controlling in the case of sulfide roasting, whereas calcination might be controlled by transport of CO_2 through the oxide layer (or transport of heat through that layer). Recalling from Chapter 3 that under pore diffusion control, conversion times are proportional to the square of the particle size, and accepting (without proof) that this is also true for heat transport control, we see that it is desirable that the solid particles be small and of similar size. Satisfaction of the former condition, which is subject to the restraint that the particles are not so fine as to present operating problems (e.g., elutriation), will maximize the throughput possible. If the latter condition is not satisfied, some particles will be completely reacted after a short time in the unit (and thereafter be idle), whereas others will be incompletely reacted by the time they leave.

In reality, sulfide particles are usually mixed (e.g., a particle containing sulfides of

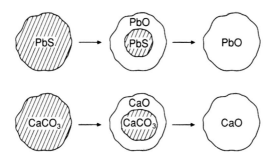

FIGURE 8.10 Shrinking cores shown by reacting solid particles during roasting or calcining.

nickel, copper, and iron) and they do not show the morphology of reaction idealized in Fig. 8.10. The sulfide with the greatest affinity for oxygen will tend to react first, and after its reaction the sectioned particle would show a patchwork morphology of oxidized and unoxidized grains. Nevertheless, the generalizations above concerning the effect of particle size are still valid.

We have now achieved the conversion of our sulfide into oxide; as depicted in Fig. 8.3, the next task is the reduction to metal, a task that is also common to the production of metal from oxide ores.

8.5 Reduction of Oxides

The Ellingham diagram for oxides (Fig. 2.12) illustrates the effectiveness of the inexpensive reducing agents hydrogen and carbon. The former can readily be generated from hydrocarbons such as natural gas by catalytic re-forming: for example,

$$CH_4 + 2H_2O = CO_2 + 4H_2 \tag{8.9}$$

carried out over a catalyst, with the carbon dioxide product being separated from the hydrogen by absorption in alkaline solutions. Carbon is usually employed in the form of coke; this is a porous form of carbon produced by heating coal in the absence of air:

$$coal = coke + volatile\ matter \tag{8.10}$$

Not all coals are suitable for producing coke. Coals low in mineral matter and sulfur are preferred for coke production because these yield low mineral matter and sulfur in the resulting coke. Furthermore, the mechanical strength of the coke is important because the coke must be transported to the unit operation where it is used and must withstand thermal or mechanical stresses in that unit; not all coals are capable of yielding cokes of sufficient strength.

Coke is produced in units called coke ovens which are heated externally and operated batchwise. Volatile organic compounds released during heating are condensed and used to produce organic chemicals. Coke ovens are deliberately kept small so that the externally supplied heat can readily be transferred through the bed of coal; economies of scale are therefore difficult to achieve in coke ovens. Furthermore, coke ovens have been a notorious source of pollution in the past, from leaking oven doors to pollution generated when the hot coke is quenched with water sprays as it is removed from the oven. These difficulties and the unsuitability of many coals have resulted in coking being an operation with significant capital and operating costs. Consequently, although coke remains the least expensive and most widely used reducing agent, its cost is frequently an important part of the economics of reduction processes that employ it. Let us now examine some of the technology of oxide reduction by carbon.

8.5.1 Zinc Production by Reduction of Zinc Oxide

Zinc has been used since ancient times, for example in the copper-zinc alloy brass, and the traditional method of production has been the batch reduction of zinc oxide (produced by roasting the naturally occurring zinc sulfide, sphalerite) using mixed in coke

FIGURE 8.11 Photograph of a zinc retort (probably dating from the fourteenth century) from Udaipur in India. Courtesy of D.W. Fuerstenau.

particles in fireclay retorts. Until coke production became widespread, the carbon was in the form of charcoal (produced by heating wood in the absence of air). Figure 8.11 is a photograph of an Indian fireclay retort found near what is now a plant of Hindustan Zinc Ltd. This technology was practiced with little change until the 1930s, when New Jersey Zinc Co. developed the vertical retort and St. Joe Zinc Co. the electrothermic process (described below). Figure 8.12 shows how a retort was placed in a furnace so that gases issuing from the neck of the retort passed through the furnace wall into a cooler *condenser* where zinc vapor would condense and solidify. A furnace might contain a few hundred retorts. The reduction reaction is

$$ZnO + C = Zn + CO \qquad (8.11)$$

The Ellingham diagram for oxides shows that this reaction proceeds readily at 950°C, which is above the boiling point of zinc. Consequently, both zinc vapor and carbon monoxide pass into the condenser, the former solidifying and the latter passing through to be burned at the mouth of the condenser. The reader will know, by now, enough process engineering to perceive the disadvantages of this technology:

1. Packing the retorts with oxide and carbon, loading them into the furnace, emp-

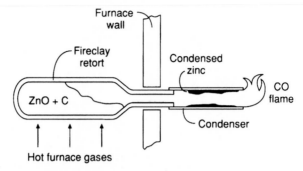

FIGURE 8.12 Retort in a furnace for the reduction of zinc oxide to zinc.

tying residues from the retorts, removing solidified zinc from the condensers, and so on, would be labor-intensive operations.
2. The large furnaces would take a long time to heat up and cool down, during which time their productive capacity would not be used.
3. The carbon monoxide is wastefully burned at the mouth of the condenser.
4. Thermal cycling of the retorts and their frequent handling results in their life-times being short, with significant replacement cost.

A modern technology that avoids many of these difficulties is the continuous *electrothermic* reduction of zinc oxide sketched in Fig. 8.13. Zinc oxide sinter (produced by roasting sulfide concentrates on a traveling grate) is preheated with coke in a rotary kiln using carbon monoxide (produced by the reduction reaction) as fuel. The hot oxide and coke then fall into a furnace, where further heat is generated by passing alternating electric current through the bed of solids. Current is fed into the bed through carbon *electrodes* projecting through the walls of the furnace. The reduction temperature of 950°C is achieved in this bed, producing carbon monoxide and zinc vapor, which are drawn to the center of the furnace by reduced pressure maintained in a *vapor ring* surrounding the furnace at this point. From the vapor ring the gases pass to a condenser maintained at a temperature where zinc condenses to a liquid (but does not solidify), while the carbon monoxide passes on to be used as fuel in the preheater. Residues (from gangue) leave the bottom of the furnace. Electrothermic reduction requires much less labor than the retort furnace, eliminates the cost of retorts, uses rather than wastes the carbon monoxide, produces liquid zinc that can be drawn periodically from the condenser and cast, and provides much less difficulty for control of air pollution. The main disadvantage is that the major part of the heat required is generated using expensive electrical energy.

Zinc oxide reduction poses an intriguing thermodynamic question. As zinc vapor and carbon monoxide pass into the condenser of either the retort or the electrothermic reduction, the temperature of this gas mixture drops below 950°C and we expect from thermodynamics that reaction (8.11) would be reversed to give only zinc oxide and carbon, which is what we started with. How is it possible for these technologies to produce metallic zinc? The answer lies in the field of kinetics; if the gas mixture is cooled rapidly (quenched), the reverse of reaction (8.11) becomes slow and insignifi-

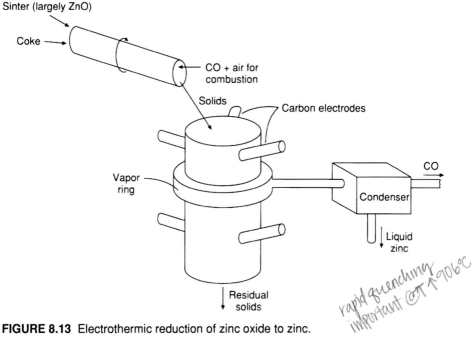

FIGURE 8.13 Electrothermic reduction of zinc oxide to zinc.

rapid quenching important @ T > 906°C

cant amounts of the zinc product are reoxidized. This rapid quenching is particularly important above 906°C (the boiling point of zinc). Below this temperature the zinc will be present mostly as liquid (or solid) and reoxidation is inhibited by the mass transfer necessary for the heterogeneous reoxidation. In the retort reduction of zinc oxide, the quenching is facilitated by the large condenser area presented by the few hundred tubular condensers. Quenching is not a problem in the electrothermic process; in both technologies only 2 mol of gas must be quenched per mole of zinc produced; this contrasts with the Imperial Smelting furnace, which we examine next.

The Imperial Smelting furnace (named after the company that developed it) avoids the principal disadvantage of electrothermic reduction (large electrical energy consumption) by using coke as a fuel to generate heat as well as as a reducing agent. The technology is sketched in Fig. 8.14. As in electrothermic reduction, the solids fed to the furnace are preheated by a rotary kiln and leave the kiln to pass into the furnace proper at close to reduction temperature. Now, however, the solids contain an excess of coke, above that required by the stoichiometry of reaction (8.11), and heat is generated by the combustion of this excess with air blown into the furnace through nozzles known as *tuyeres* (from the French word for "tubes"). That air is preheated to approximately 600°C to reduce the amount of heat it is necessary to generate within the furnace itself. The furnace has found its major application in the reduction of mixed oxides of zinc and lead, obtained by roasting of concentrates from mixed sulfide ores, and the liquid lead, having little vapor pressure at the reduction temperature, is drawn from the bottom of the furnace together with a molten slag formed by gangue constituents that are not reduced. Zinc is produced as a vapor, and the quenching of the vapor to avoid reoxidation has been a major technical challenge in the development of this technology. It should be noted that the task of quenching is much greater than in the electrothermic reduction because the gas stream contains not only the zinc vapor and carbon monoxide produced by reaction (8.11) but also carbon monoxide generated by combustion of the excess coke used as fuel and a large amount of nitrogen that entered the furnace in the air blown into the bottom. The solution is to spray molten lead (m.p. 327°C) into the gas stream, forming a lead-zinc alloy that is drawn from the condenser. On cooling, this alloy separates into two liquid phases; the zinc-rich phase is sent on for refining, and the lead-rich phase is recycled to the quenching condenser.

In the Imperial Smelting furnace, part of the coke is used as reductant and part as a fuel. This dual use of coke is also found in the case of iron oxide reduction in the blast furnace, to which we will turn our attention in the next section.

The pyrometallurgical route from zinc sulfide concentrates to zinc has lost ground in the twentieth century to the hydrometallurgical one of roasting, leaching, solution purification and electrowinning. The latter route is examined in the next chapter.

8.5.2 The Iron Blast Furnace

The iron blast furnace, one of the most important unit operations, is depicted in Fig. 8.15

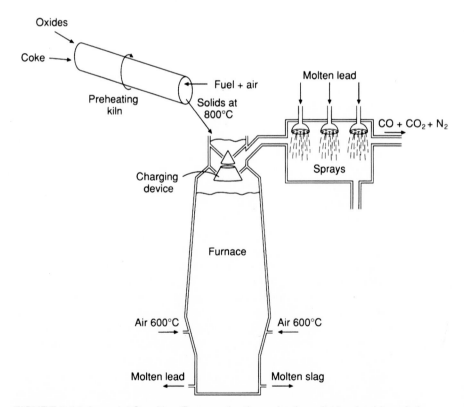

FIGURE 8.14 Imperial Smelting Furnace for the reduction of mixed oxides of zinc and lead.

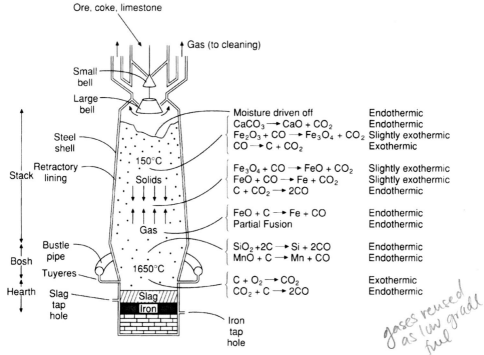

Ore, coke, limestone

Gas (to cleaning)

Small bell

Large bell

Steel shell

Retractory lining

Stack

150°C

Solids

Gas

Bustle pipe

Bosh

Tuyeres

1650°C

Hearth   Slag tap hole

Slag

Iron

Iron tap hole

Moisture driven off	Endothermic
$CaCO_3 \rightarrow CaO + CO_2$	Endothermic
$Fe_2O_3 + CO \rightarrow Fe_3O_4 + CO_2$	Slightly exothermic
$CO \rightarrow C + CO_2$	Exothermic
$Fe_3O_4 + CO \rightarrow FeO + CO_2$	Slightly exothermic
$FeO + CO \rightarrow Fe + CO_2$	Slightly exothermic
$C + CO_2 \rightarrow 2CO$	Endothermic
$FeO + C \rightarrow Fe + CO$	Endothermic
Partial Fusion	Endothermic
$SiO_2 + 2C \rightarrow Si + 2CO$	Endothermic
$MnO + C \rightarrow Mn + CO$	Endothermic
$C + O_2 \rightarrow CO_2$	Exothermic
$CO_2 + C \rightarrow 2CO$	Endothermic

gases reused as low grade fuel

FIGURE 8.15 Iron blast furnace.

together with the reactions that take place in the furnace. To give some idea of the scale of this unit, a typical furnace might be 12 m in inside diameter (toward the bottom of the furnace), 45 m high, and produce 3000 tons of molten iron per day. Recently constructed furnaces are even larger. As with many pyrometallurgical unit operations, the construction is of steel with a refractory lining. The lining (made of unreactive oxides or carbon in the hearth region of the furnace) serves to insulate the steel structure from the hot and hostile environment within the unit. This insulation reduces heat losses and minimizes both corrosion and mechanical problems (thermal expansion, creep, etc.) of the steel. The furnace is semicontinuous in operation in the sense that coke, ore, and limestone are fed into the top batchwise (but frequently) while preheated air (the "blast") is blown continuously into the bottom through tuyeres. Periodically, molten iron and molten slag are tapped from the bottom of the furnace. Most of the furnace is filled with a descending bed of solids; gases flow up through the interstices between the lumps of solids, making this a countercurrent operation. The gases exiting the top of the bed are nitrogen, carbon monoxide and carbon dioxide, water vapor, and hydrogen; they pass out of the top of the furnace and are frequently used as a low-grade fuel. Temperatures rise toward the bottom of the furnace, where coke, in excess of that required for reduction reactions, is burned with the incoming air:

$$2C + O_2 = 2CO \tag{8.12}$$

We see, then, the dual purpose of the coke: as a fuel and as a reducing agent, for example, for the reaction

$$C + FeO = Fe + CO \qquad (8.13)$$

Gangue (e.g., SiO_2) present in the lump ore, sinter, or pellets fed to the furnace forms a molten slag with calcium oxide, which in turn is formed by thermal decomposition of limestone [reaction (8.7)]. Limestone therefore serves to provide this CaO fluxing agent, which yields a silicate/oxide of melting point and viscosity low enough to flow readily from the furnace. However, some gangue constituents (notably the oxides of silicon, manganese, sulfur, and phosphorus) are partly reduced at the high temperature occurring at the bottom of the furnace. This can be seen by examining the Ellingham diagram for oxides (Fig. 2.12). The elemental silicon, manganese, sulfur, and phosphorus enter the iron as contaminants along with carbon, which has an ability to dissolve to the extent of approximately 4% by weight at the temperature of the molten iron. The molten iron tapped from the furnace is therefore quite impure; it is known as *pig iron* or *hot metal*. Most of the iron produced by blast furnaces is processed, while still in molten form, to remove most of the impurities and thereby produce steel. However, a significant amount is used to produce cast iron. Typically, the cast iron is made in a foundry, which may be at a different location from the blast furnace. In that case the pig iron is poured from a ladle into steel molds positioned on a conveyor belt of sufficient length that the pig iron has solidified by the time it reaches the end. The solidified "pigs" are then shipped to the foundry. The manufacture of cast iron from pigs is described in Section 8.5.4.

Significant improvements have taken place in blast furnace performance over the past few decades. These improvements have increased the productivity (tons of metal produced per day per unit of furnace volume) and reduced the coke rate (kilograms of coke consumed per ton of metal produced). Some of these improvements are higher blast temperature, oxygen enrichment of the blast, fuel injection, higher top pressure, and better feeding of the furnace.

The advantage of a higher blast temperature is that it increases the sensible heat of an input stream and therefore makes available more heat for endothermic reactions within the furnace. The productivity can therefore be increased and/or the coke rate decreased. A limit on this improvement is posed by the fact that increased blast temperatures tend to result in increased temperatures in the bosh region of the furnace. In this high-temperature region the highly endothermic reduction of silicon, manganese, phosphorus, and sulfur oxides occur. By Le Châtelier's principle, or the considerations of Section 2.5.7, these impurity-generating reactions are promoted by higher bosh temperatures. One way of circumventing this limitation is to increase the humidity of the blast (by injecting steam) along with an increase in blast temperature; excessive bosh temperatures are then prevented by the endothermic reaction

$$H_2O + C = H_2 + CO \qquad (8.14)$$

Hydrogen generated in this way acts as a reducing agent higher in the furnace. In any event, it is common practice to control the humidity of the blast so seasonal or diurnal variations in the humidity of the air do not result in irregular blast furnace behavior.

The nitrogen component of the blast does nothing but pass through the furnace, imposing a heat load that it would be best to avoid. This can be achieved, in part, by replacing part of the nitrogen with oxygen, that is, by using an oxygen-enriched airstream, rather than ordinary air, as the blast. Because the nitrogen imposes a heat load principally on the bosh region, oxygen enrichment tends to increase bosh temperatures, with subsequent increase in impurity levels in the iron produced. This can be avoided by increased blast humidity.

Carbon (coke) is consumed as both a fuel and a reductant in the blast furnace. An alternative source of such carbon, along with hydrogen, would be fuel oil, or other hydrocarbons, injected at the tuyeres. We can think of these hydrocarbons as undergoing *cracking* (thermal decomposition) in the bosh region:

$$C_nH_{2m} = nC + mH_2 \qquad (8.15)$$

Reaction (8.15) is highly endothermic and it is usual to accompany hydrocarbon injection with higher blast temperature, reduced blast humidity, or greater oxygen enrichment. Hydrocarbon injection is particularly effective in reducing the coke rate, although hydrocarbons are somewhat costly.

The term *higher top pressure* refers to throttling the gas flow out of the furnace so that the gas pressure at the top of the furnace rises. Of course, the pressure throughout the furnace rises, and it is at first sight paradoxical that this results in any benefits at all. The overall reactions can be represented as

$$\text{coke} + \text{iron oxides} = \text{iron} + \text{gases} \qquad (8.16)$$

and since there is a net generation of gas, we should expect from Section 2.6, or LeChâtelier's principle, that an increase in pressure would inhibit the overall reaction. However, if we examine the Ellingham diagram for oxides, we see that the oxide reduction reactions that generate gas [e.g., reaction (8.13)] have negative free energies of reaction of more than 100 kJ/mol at 1200°C; that is, they are essentially irreversible and increased pressure (by the 1 atm or so achievable in blast furnaces) has a negligible influence on equilibrium. The explanation is as follows.

The productivity of many reactors can be increased "simply" by supplying the feed materials to them more rapidly. This is particularly true in reactors, such as the blast furnace, where equilibrium is achieved or closely approached. In such reactors a negligible decrease in conversion of reactants to products occurs if we reduce the residence time by increasing the feed rates. Because the production rate is simply the feed rate times the fraction converted to products, there are big economic advantages to increasing the feed rates. In the case of the blast furnace, this means increasing the feed rate of ore, coke, and limestone (which is not difficult) and the feed rate of air. It is the latter that poses difficulty, because as we increase the volumetric flow rate of the blast, we increase the pressure drop necessary to drive this flow through the bed of solids filling the furnace. The electrical power (and cost) of driving the air compressor that does this is dependent on the product of volumetric flow rate and pressure drop. Consequently, beyond a certain point it becomes uneconomical to increase further the volumetric flow rate of the air.

Another difficulty is that with increased volumetric flow rate, the velocity with which the gases move upward through the interstices between solids in the furnace increases to the point where it interferes with downward flow of liquid iron and slag at the bottom of the furnace. If the point of maximum volumetric flow rate of the blast has been reached, there is only one way to increase the mass flow rate of the air, as we need to do to bring about further increases in productivity. That way is to increase the density of the air by increasing its pressure throughout the furnace. Hence higher top pressure enables a greater mass (mole) throughput of reactants, and therefore output of iron, without excessive penalties in terms of compressor costs and interference with the downflow of iron and slag.

Proper charging of ore, coke, and limestone to the top of the furnace is important in avoiding the formation of channels of low resistance to gas flow through which the rising gas can pass without contacting most of the bed. On modern furnaces, movable feed chutes, television cameras inside the furnace top, sophisticated computer control technology, and close sizing of the furnace feed are used to avoid this channeling.

EXAMPLE

An iron blast furnace produces 2×10^6 kg/day of hot metal, consuming in the process 0.8×10^6 kg of coke per day. The blast temperature is 1300 K and the flow rate of the blast is 3.3×10^6 kg/day. The reduction of the coke consumption that can be achieved by raising the blast temperature to 1400 K is to be estimated. It is assumed that this increase in blast temperature can be achieved without changing the flow rate, temperature, and composition of the hot metal and slag exiting the furnace; however, a reduction of blast flow rate is permitted. It is thought that the consequences of the change in blast temperature are determined by a heat balance on the lower part of the bosh region, where coke at 1950 K reacts with the incoming blast to produce carbon monoxide, which leaves this region at approximately 1950 K. The blast will be taken as dry air (21 mol % O_2, 79 mol % N_2) and the coke as pure carbon for the purposes of this calculation.

Data

Enthalpies in kJ/mol (with respect to 298 K):

T(K)	O_2	N_2
1300	33.31	31.47
1400	36.92	34.90
1950	57.24	53.73

Enthalpy of formation of CO at 1950 K $= -118.4$ kJ/mol. Mean molecular weight of air $= 28.84$.

Solution

As a basis we will take 1 kg of hot metal and 1950 K as a reference temperature. First consider operation with a 1300 K blast. For the basis selected, $3.3 \times 10^6/(2 \times 10^6$ kg) $\equiv 1650/28.84$ mol $= 57.21$ mol air used.

Part of coke is used to bring this air to 1950 K (the reference temperature). That amount of coke is obtained from the enthalpy difference of air between 1300 and 1950 K, divided by the enthalpy of the combustion reaction (= minus the enthalpy of formation of CO); that is, it is

$$\frac{57.2 \times 0.21 \times (57.24 - 33.31) + 57.21 \times 0.79 \times (53.73 - 31.47)}{118.4}$$

$$= 10.93 \text{ mol coke}$$

This is smaller than the amount of coke fed to the furnace (per kilogram of hot metal) because the rest is used as a reducing agent and to supply heat for other reactions, heat losses, and so on. We need not concern ourselves with this other part of the coke because it will remain the same (by the statement of the problem) when we raise the blast temperature to 1400 K.

Now consider operation at the higher blast temperature. Let X be the moles of 1400 K air blown into the furnace per kilogram of hot metal, and Y be the moles of coke consumed to bring this air to 1950 K. Proceeding as before yields

$$\frac{X \times 0.21 \times (57.24 - 36.92) + X \times 0.79 \times (53.73 - 34.90)}{118.4} = Y$$

We know that Y will be less than the 10.93 mol of coke used previously, and therefore less air is needed for combustion. The reduction in the amount of coke $(10.93 - Y)$ and the reduction in the amount of air $(57.21 - X)$ are related by simple stoichiometry (2 mol of carbon burned per mole of oxygen) and the composition of air:

$$\frac{10.93 - Y}{2 \times 0.21} = 57.21 - X$$

We now have two equations for the two unknowns, and simple algebra yields $Y = 8.2$ mol; that is, the reduction in carbon consumption is $10.93 - 8.2 = 2.73$ mol $= 32.76$ g for the selected basis of 1 kg of hot metal. Multiplying by the daily production yields a saving of 0.0655×10^6 kg/day, slightly over 8%, which would be a very worthwhile saving if the increase in blast temperature could be achieved inexpensively.

8.5.3 Direct Reduction Technology for Iron Oxides

In many countries with abundant supplies of hydrocarbons (e.g., Venezuela, Indonesia, and Canada), the use of hydrogen rather than coke as a reducing agent for iron oxides is practical or even preferable. The various processes that have been developed to achieve this are categorized as *direct reduction processes*. These processes make up a small but growing fraction of world iron production capacity. At present the two most successful technologies appear to be the HyL process (where packed beds of iron ore, pellets, or sinter are reduced batchwise) and the continuous Midrex process, where a downward-moving bed of solids is reduced by a countercurrent flow of hydrogen. Fluidized beds and kilns have also been used for direct reduction, at least through the pilot-plant stage. Most direct reduction processes produce a solid porous iron product known as *direct reduced iron* that is either subsequently melted and processed to steel or used as a substitute for scrap in oxygen steelmaking (see Section 8.7). In both the iron blast furnace and direct reduction units the reactions proceed as discussed in Section 8.5.2. While the kinetics of reaction are only of secondary importance in the iron blast furnace, the "reducibility" of an ore has a significant impact on the productivity of a direct reduction unit.

8.5.4 Manufacture of Cast Iron

Although not produced in the same quantities as steel, cast iron is still one of the most important metals. It is widely used for parts where hardness, rigidity, and relatively low cost are required, such as automobile engine blocks. Most cast iron is produced in foundries starting with iron pigs (produced as described in Section 8.5.2), steel scrap, and ferroalloys (such as ferrosilicon, the production of which is described in Section 8.5.6). The pig iron would usually contain approximately 4% carbon, 1 or 2% each of silicon and manganese, and between a few hundredths and a few tenths of a percent of phosphorus and sulfur. The steel scrap would typically contain much lower levels of these impurities, the exact composition depending on the degree of contamination inherent in the recycling process. Ferrosilicon would contain little carbon, manganese, phosphorus, or sulfur, but perhaps as much as 75% silicon. Because typical cast irons contain 3 to 3.5% carbon, 2 to 2.5% silicon, and a few tenths of a percent of sulfur, phosphorus, and manganese, it becomes clear that they can be produced by melting together the pigs, scrap, and ferrosilicon in appropriate proportions, with little need for composition adjustment by reaction.

The cupola is the most widely used device for achieving this melting. It resembles a small blast furnace in that it uses coke for fuel, and this is burned with air blown into the bottom of the furnace through tuyeres. However, the cupola is usually operated in batches and the iron-containing materials are pig iron, scrap, and ferroalloys rather than ore. Some change in impurities takes place by reaction, most notably the removal of manganese and sulfur from the metal to a slag formed from coke ash constituents, scrap contaminants, and fluxing agents, such as limestone, added with the charge. The metal and slag are removed from the furnace through separate tap holes.

Open arc electric furnaces have also been used to some extent for the melting operation. They have also been used as furnaces where molten metal from a cupola is adjusted to its final composition (e.g., by adding ferroalloys).

8.5.5 Lead Blast Furnace

The *lead blast furnace* is similar in concept to the iron blast furnace; coke is used both as a reductant and as a fuel to bring reactants to the reduction temperature and supply the endothermic heat of reaction. The differences, aside from the obvious one that lead oxide sinter is reduced, are mainly in details. For example, the furnace is rectangular in section rather than circular. The lower reduction temperatures make it unnecessary to preheat the blast.

8.5.6 Reduction of Silicon, Chromium, and Manganese Oxides

From the Ellingham diagram for oxides it is seen that comparatively high temperatures are necessary for the reduction: SiO_2, 1550°C; MnO, 1450°C; and Cr_2O_3, 1250°C. In fact, the reduction of chromium oxide is typically carried out at higher temperatures because it is desirable to produce liquid chromium (m.p. 1875°C). From the discussion

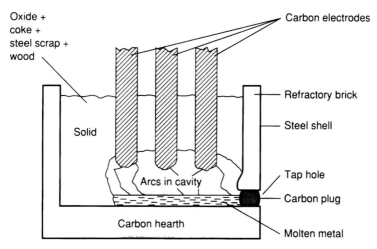

Oxide +
coke +
steel scrap +
wood

Carbon electrodes

Refractory brick

Steel shell

Solid

Arcs in cavity

Tap hole

Carbon plug

Carbon hearth

Molten metal

FIGURE 8.16 Submerged arc furnace.

of Section 7.6.2, it is seen that these temperatures are best achieved in an electric furnace, and the *submerged arc furnace* is widely used for this purpose. Such a furnace is sketched in Fig. 8.16; heat is generated within the furnace by the electric arcs between the carbon electrodes and the metal pool, as well as by resistive heating as current passes through the hot material in the furnace. Furnaces range in power from a few hundred kilowatts to a few tens of megawatts and are operated in a semicontinuous mode with intermittent tapping of metal and slag (if any). Coke and wood (added to the feed to provide porosity to the descending bed of solids as it reacts or burns away) are the reducing agents, and scrap steel is added along with the oxide when, as is common, the desired product is ferrosilicon, ferromanganese or ferrochromium, rather than silicon, manganese, or chromium. These ferroalloys are used in the production of special steels (e.g., stainless steels).

8.5.7 Reduction of Aluminum and Magnesium Oxides by Carbon

Aluminum and magnesium are produced in the electrolytic cells described in Chapter 9. These cells are expensive to build and operate and many attempts have been made to replace this electrolytic technology by the "carbothermic" reduction of the oxides, for example,

$$Al_2O_3 + 3C = 2Al + 3CO \qquad (8.17)$$

Examination of the Ellingham diagram for oxides shows that reaction (8.17) should occur at approximately 2100°C. Although such temperatures can be achieved in a submerged arc furnace, at 2100°C aluminum is close to its boiling point and there are significant aluminum losses by vaporization as well as by formation of a volatile suboxide Al_2O. Both the vaporized aluminum and the suboxide would tend to react with the CO on cooling the gas below 2100°C. What is possible is the reduction of the mixed oxides of silicon and aluminum (which are relatively abundant in nature) in submerged arc furnaces. An aluminum-silicon alloy is produced, with the silicon lowering the aluminum activity to the point where its vaporization is no longer a major difficulty. The principal problem is that there is no established technology for separating the silicon and aluminum economically. and the product therefore has limited use.

Vaporization is also a problem in the carbothermic reduction of magnesium oxide. From the Ellingham diagram this should occur at 1900°C, which is 793°C above magnesium's boiling point. The difficulty of quenching the gaseous mixture of magnesium and carbon monoxide produced by reaction (to prevent reoxidation of the metal) is therefore severe. Contrast this quenching difficulty with that for zinc oxide reduction. In the case of zinc, the reduction temperature is only 44°C above the boiling point of the metal. As an exercise, the reader is asked to determine why carbothermic reduction of magnesium oxide has proven technically infeasible, whereas reduction of this oxide by silicon has been quite successful, even though, at first sight, reduction with silicon appears to be precluded by thermodynamics. The reduction of magnesium oxide [or, more commonly, calcined dolomite (CaO MgO)] to metal by silicon or ferrosilicon was mentioned briefly in Section 2.6.2. It is exploited in the Pidgeon process for magnesium production. In the Pidgeon process calcined dolomite is reduced by ferrosilicon in externally heated retorts at approximately 1150°C and in a vacuum of about 10 kPa, under which conditions the reaction would proceed and the magnesium would boil out of the retort to be trapped in a condenser. A modern variant of this technology is the Magnetherm process, where the heat necessary for the reduction reaction and preheating the feed materials is generated by passage of electricity through a molten slag. The slag is formed by the CaO component of the calcined dolomite fed to the furnace (operated at approximately 1550°C and 5 kPa), by aluminum oxide added to lower the melting point, gangue oxides, and unreacted MgO. Again, the magnesium produced by the reaction boils out of the furnace and is collected in a condenser.

8.6 Smelting and Converting of Sulfides

Let us now examine the third route to metal (the second from sulfide ores) shown in Fig. 8.3, the "smelting" of sulfides to produce metal. Strictly, this usually entails two steps: a smelting step, in which the sulfide of the desired metal is separated from most of the gangue by forming two immiscible liquids, and a converting step, in which nearly all of the remaining gangue is separated and then the desired metal sulfide is converted to metal.

8.6.1 Reverberatory Smelting

Reverberatory smelters have been the traditional unit for carrying out the first step: (although now largely replaced by more modern technology described below): the formation of two liquids, one an oxide slag containing most of the gangue elements and the other a sulfide "matte" containing the valuable metal sulfide along with some gangue sulfides. The *reverb* is a furnace approximately 40 m long, 3 m high, and 10 m wide. Fuel is burned at one end (see Fig. 8.17) and the heat generated melts dry sulfide concentrates (in most cases after partial roasting) fed through ports in the roof of the furnace. The slag formed in this furnace is a waste product that is tapped separately from

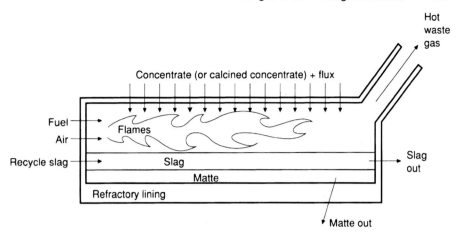

FIGURE 8.17 Reverberatory furnace smelting sulfide concentrate.

the matte and dumped. If the gangue is rich in silica, a lime or limestone flux would be added along with the concentrate to yield a slag with a sufficiently low melting point and viscosity. Conversely, if the gangue is high in calcite, sand would be used as a flux. In the case of smelting copper sulfide concentrates, the matte would consist largely of sulfides of copper and iron. Some oxidation of sulfides takes place in the furnace but the equilibrium for the reaction

$$Cu_2O + FeS = Cu_2S + FeO \qquad (8.18)$$

lies well to the right. Consequently, copper oxide in the slag phase is readily abstracted (as sulfide) into the matte phase by reaction (8.18). Matte is tapped from the furnace into ladles and conveyed to the converters.

8.6.2 Converting

The term *converting* is used to describe the batch treatment of matte to yield first an iron-rich slag that is recycled to the reverb and then an impure molten copper (or what-

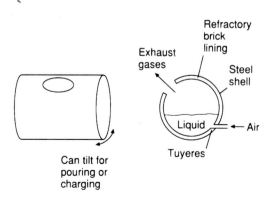

FIGURE 8.18 Copper converter: (a) sketch; (b) sectioned view.

ever the valuable metal is in the concentrate). A converter is a cylindrical vessel about 4 m in diameter and 10 m long that can be rotated about its horizontal axis for filling and emptying (see Fig. 8.18). As usual in pyrometallurgy, it is constructed of steel with a refractory lining. Tuyeres arranged in a row along the converter allow air to be blown into the contents.

The first stage in converting (of copper mattes, say) is the oxidation of iron sulfide in the matte to an oxide which forms a slag with silica (sand) added to the converter:

$$2FeS + 3O_2 = 2FeO + 2SO_2 \qquad (8.19)$$

Any copper sulfide oxidized tends to react with iron sulfide according to reaction (8.18), so that the transfer of copper into the slag is small. Nevertheless, the slags, which are poured out of the converter periodically as the processing of a batch continues, contain significant amounts of copper (3 to 5%) and are poured into the reverb for recovery of that copper (see Fig. 8.19). After nearly all the iron sulfide originally present in the concentrate has been removed as oxide in the slag, the oxidation of the copper sulfide begins:

$$Cu_2S + O_2 = 2Cu + SO_2 \qquad (8.20)$$

Any oxide formed at this stage tends to react according to

$$Cu_2S + 2Cu_2O = 6Cu + SO_2 \qquad (8.21)$$

The final product is impure copper metal with residual iron, precious metals (such as gold and silver present in the original ore), and oxygen as the chief impurities. The last

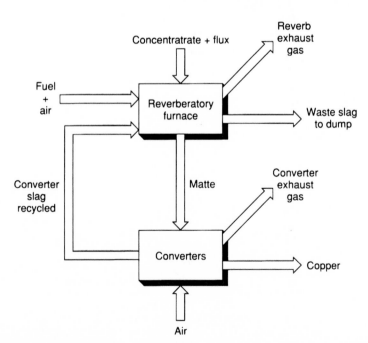

FIGURE 8.19 Arrangement of reverberatory furnaces and converters in the pyrometallurgical production of copper.

can be removed readily by contacting the molten metal with a reducing agent (e.g., natural gas) in a holding furnace prior to casting the metal. The metal is cast into anodes, which are slabs approximately a few centimeters thick and 1 m square. Pure copper cathodes are produced from the anodes by the electrolytic refining process discussed in Chapter 9.

8.6.3 Sulfur Dioxide Problem

A casual glance at the reactions listed above for smelting and converting sulfide concentrates to metal shows that they produce large amounts of sulfur dioxide. This must be recovered in some way if pollution of the environment is to be avoided; the need to avoid this pollution has had a major impact on the technology and economics of smelting in the last century. The most widely accepted method of trapping the SO_2 is conversion to sulfuric acid. SO_2 can readily be oxidized by air to sulfur trioxide over a platinum catalyst and the SO_3 absorbed in water to yield sulfuric acid, which can be sold or otherwise disposed of. However, the economics of conversion of SO_2 to sulfuric acid are most favorable if the gas stream containing SO_2 is both continuous and rich in SO_2. The SO_2 stream from a reverb is continuous but not rich in SO_2 because it contains carbon dioxide and water vapor, the products of the fuel combustion; nitrogen from the air, used in combustion of that fuel; and excess air. Conversely, the gas stream from the converters is rich in SO_2 (because fuel is not burned in the converters and most of the oxygen in the input air forms SO_2) but intermittent (because converters operate batchwise). These difficulties can be overcome, to a limited extent, by combining gas streams, but a more successful approach has been new technology for smelting. Some of this new technology will now be covered briefly.

8.6.4 Outokumpu and INCO Flash Smelters

The *Outokumpu flash smelter* (named after the Finnish corporation that developed it) represents an attempt to achieve a higher SO_2 level in the exit gas stream by minimizing the amount of gas generated by fuel combustion. In the furthest embodiment of this

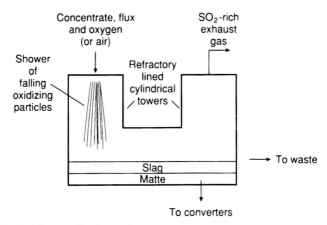

FIGURE 8.20 Outokumpu flash smelter.

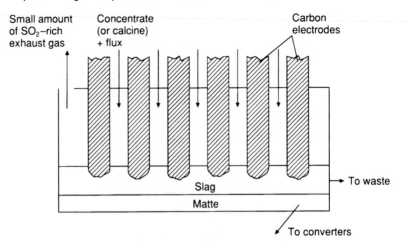

FIGURE 8.21 Electric furnace smelting of sulfide concentrates.

concept, fuel is dispensed with altogether and the heat required is generated by the flash combustion of part of the concentrate with oxygen. Figure 8.20 is a sketch of the Outokumpu smelter; it is seen that, as with the conventional reverb, matte and slag streams are produced, the former being tapped periodically and sent to conventional converters, with the latter being discarded. Another flash smelting unit is the INCO flash smelter which has been less widely adopted than the Outokumpu smelter. The INCO smelter is similar in concept, e.g., using oxygen for oxidation to avoid fuel costs) but differs in detail.

8.6.5 Other Technologies for Sulfide Smelting

On reading Section 8.6.3 the reader may have perceived that a continuous, SO_2 rich stream might be produced if the smelting and converting of the conventional technology could be combined in one continuous operation. One attempt to do this has been the Noranda process (developed by that Canadian corporation). However, in the present-day installation of this process, the units are used primarily as smelters, with converting done in separate conventional converters. A similar technology (the Teniente smelter) has been developed in Chile.

Another way of avoiding the generation of large volumes of gas low in SO_2 within a smelter is to use electricity rather than fuel for heat generation. In this way only small amounts of (SO_2-rich) gas are produced in the smelter. *Electric smelters* resemble the submerged arc furnaces described in Section 8.5.5, except that the heat is generated by resistive heating of the molten slag, into which the carbon electrodes dip, rather than within arcs. Figure 8.21 is a sketch of such a furnace.

As originally conceived, the Noranda process was an attempt to combine smelting and converting in one continuous operation. A major difficulty in doing this is to keep smelting and converting separate within the unit (e.g., to ensure that the slag leaving the unit is representative of a smelter slag—low in copper—rather than a high copper converter slag).

This difficulty is avoided in the *Mitsubishi continuous smelter*. which consists of three separate furnaces: smelting, slag cleaning. and converting with cocurrent flow (by gravity) of slag and matte from the first to the third. Dried concentrate, flux. and a granulated slag recycled from the converter are fed via overhead lances to the smelting furnace, where melting proceeds almost autogenously (i.e., heat generated by combustion of the sulfides is nearly sufficient to melt the feed). The small amount of supplementary heat needed is provided by burning oil, coal, or gaseous fuels. An SO_2-rich offgas is produced and the copper matte (approximately 65% Cu) flows along with slag to the slag-cleaning furnace, which is electrically heated by passage of alternating current through the slag and matte between immersed electrodes.

The *slag-cleaning furnace* is relatively quiescent, and by adding reducing agents (such as pyrite), the copper content of the slag is reduced to approximately 0.5%, a value sufficiently low that the slag can be discarded. Matte flows from the slag cleaning furnace to the converting furnace, where overhead lances oxidize it to copper, which flows from the converter to a holding furnace, whence it is transferred to anode casting. Limestone is injected into the converter through the oxygen lances and forms (after decomposing to CaO) a slag (15 to 18% Cu) which is granulated by a water stream as it flows from the furnace, then is dried and recycled to the smelting furnace.

8.7 Steelmaking: Preliminary Refining Technologies

In Section 8.5 we examined the iron blast furnace wherein impure iron is produced by the reduction of iron oxides with coke. The impurities are carbon (which dissolves in the molten iron to the extent of approximately 4% at the temperatures encountered by the molten iron in the furnace), manganese, silicon, phosphorus, and sulfur. Manganese and silicon are typically present in the molten pig iron from the blast furnace at the level of roughly 1%, with sulfur and phosphorus typically being in the range of 0.01 to 0.5%. They arise from gangue constituents (such as manganese oxide and silica) of the iron ore, as well as from mineral matter in the coal used to form the coke, and impurities in the limestone fed to the furnace along with the iron oxides and coke.

Most of these impurities are removed in the operation that we call *steelmaking*. The resulting iron-carbon alloy, which we call *steel*, and which is the most important structural metal, contains carbon and manganese at the level of a few tenths of a percent, sulfur and phosphorus at a few hundredths of a percent, and silicon at even lower levels. The exact composition would be determined by the specifications for the particular batch of steel to be produced, and in some cases further refining operations (discussed in Chapter 10) are necessary to meet those specifications.

Steelmaking is therefore a refining operation and might equally well have been treated in Chapter 10. However, steelmaking is closely linked to ironmaking and has usually been regarded as one of the most important pyrometallurgical unit operations; we have therefore chosen to cover it in this chapter.

The Ellingham diagram for oxides (Fig. 2.12) makes it clear that the manganese and silicon impurities in pig iron have a much greater affinity for oxygen (greater negative free energy of formation of the oxide) than does iron. This affinity is exploited in steelmaking. By contacting the pig iron with oxygen, we can oxidize the carbon, manganese, silicon, sulfur, and phosphorus with little loss of iron by oxidation. The oxides

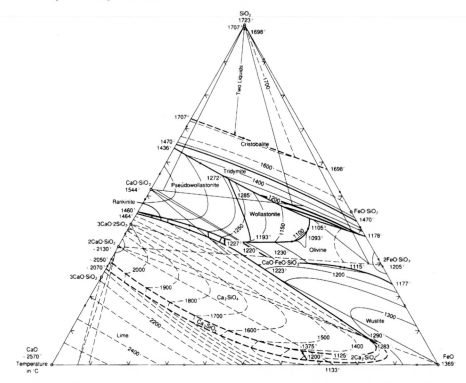

FIGURE 8.22 Phase diagram for the system CaO-"FeO"-SiO₂. Temperatures shown on contours are liquidus temperature. (From "Phase Diagrams for Ceramists", E. M. Levin, C. R. Robbins, and H. F. McMurdie, eds., Columbus, OH. 1964. Reprinted by permission of American Ceramic Society, Columbus, OH.)

produced by these highly exothermic reactions, for example,

$$2C + O_2 = 2CO \tag{8.22}$$

and

$$Si + O_2 = SiO_2 \tag{8.23}$$

will either escape as gases (CO) or become constituents of an oxide slag that floats readily to the surface of the metal. That oxide slag can be made to have a sufficiently low melting point and viscosity (that it can be readily poured from the steelmaking vessel) by adding lime (calcium oxide) to the vessel. Figure 8.22 shows the phase diagram for the CaO-FeO-SiO₂ system; in the region representative of steelmaking slags, melting temperatures are approximately 1500°C. The addition of lime also serves to make the slag basic, which implies that it is capable of absorbing the oxides of sulfur and phosphorus.

The oxygen used in steelmaking can be either in the form of "pure" oxygen, in the form of air, or as iron oxide (e.g., iron ore) that can react, for example, according to

$$2Si + Fe_3O_4 = 2SiO_2 + 3Fe \tag{8.24}$$

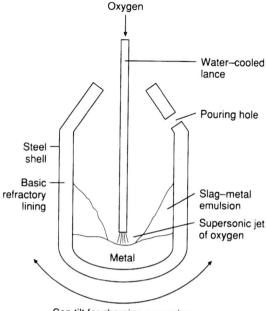

FIGURE 8.23 Basic oxygen furnace for producing steel.

Nowadays, oxidation by oxygen, known as oxygen steelmaking, is most important. Oxidation by air was employed in an obsolete technology known as open hearth steelmaking. The open hearth furnace was a large batch unit, akin to the reverb described in Section 8.6.1, wherein a few hundred tons of metal in the process of refining were covered by a layer of molten slag. Mass transport of oxygen through the slag to the slag metal interface and mass transport of the impurities to that interface limited the rate of reaction so that approximately 10 hours were required to process a batch. The rate of heat generation by the oxidation reactions was insufficient to compensate for heat losses, and fuel had to be burned to maintain the temperature. This obsolete technology is mentioned here only so that we can contrast it with oxygen steelmaking.

8.7.1 Basic Oxygen Furnace

The *basic oxygen furnace* (BOF), also known as the basic oxygen process (BOP) or Linz-Donawitz (LD) converter, is presently the most established oxygen steelmaking technology, although variants described in Section 8.7.2 are gradually superseding it. Figure 8.23 is a schematic diagram of a BOF (as usual in pyrometallurgy, a vessel constructed of steel with a refractory lining). A BOF would be approximately 5 m in diameter and 7 m high. The *lance* through which oxygen is blown (at supersonic velocities) onto the surface of the liquid metal is shown in place in the vessel. The sequence of operations involved in processing a batch of a few hundred tons is as follows:

1. The furnace is tilted to one side (the left in Fig. 8.23) so that molten pig iron can be poured in from a ladle, together with lime and scrap steel (or directly

reduced iron).

2. The lance is lowered into the furnace and the flow of oxygen started.
3. Upon completion of the oxidation, the metal is sampled for analysis and (the analysis falling within specifications) the lance is raised to permit the molten slag, then molten steel, to be poured out by tilting the vessel (to the right in Fig. 8.23).

Oxidation is typically completed in 20 to 25 minutes, and the entire operation is over in 35 to 40 minutes. The rate of heat generation by the exothermic oxidation reactions is sufficiently high that no fuel need be burned. Indeed, the steel scrap is added as a coolant to limit the temperature increase that occurs as oxidation proceeds. Final temperatures are approximately 1600°C and determined by the relative quantities of scrap and pig iron, as well as by the amount of impurities oxidized out of the latter during the steelmaking.

The reader may be tempted to suppose that the higher refining rates achieved in the BOF compared to the open hearth are due to the use of oxygen in the former, because pure oxygen contains five times as much oxygen as air, other things being equal. However, this is not a sufficient explanation, since refining rates are approximately twenty times greater in the BOF. The explanation is in terms of greater agitation (promoting mass transport) and interfacial area for heterogeneous reactions in the BOF. The agitation caused by the impacting oxygen jet (supplemented by rapid release of CO during refining) causes the formation of an "emulsion" of metal and slag. This has a very much greater interfacial area than the area of the horizontal slag-metal interface in the relatively quiescent open hearth furnace. We see, then, that the much improved technology of the BOF can be ascribed to the concepts of mass transfer and heterogeneous reaction at interfaces discussed in Chapter 3, although this understanding was achieved subsequent to the industrial adoption of the BOF, rather than as a step in its development.

Apart from the advantage of vastly greater productivity, the BOF avoids the cost associated with burning fuel in the open hearth, although energy costs are associated with the production of the oxygen used in the BOF. One major disadvantage of the BOF is that as no fuel is burned, there is little flexibility in the ratio of scrap to pig iron that can be used; that ratio is fixed by a heat balance. The furnace is also more difficult to control—to operate so that given specifications on the product composition and temperature can repeatedly and easily be met. This is because the rapid oxidation gives no time for the leisurely sampling and analysis practiced with the open hearth. Finally, dust and fumes are produced by the vigorous oxidation and must be trapped and disposed of.

8.7.2 Alternative Oxygen Steelmaking Technologies

It is feasible to blow oxygen into a steelmaking vessel through nozzles (tuyeres) located in the bottom of the vessel rather than via an overhead lance. Until recently this had proved difficult since the intense heat generated locally by oxidation of impurities had resulted in the lifetimes of the refractory bottoms of the unit being unacceptably short. This difficulty has been overcome by the development of better refractories and the use of double concentric tuyeres where oxygen can be blown through the inner nozzle and a "shroud gas" such as nitrogen or natural gas blown through the outer nozzle.

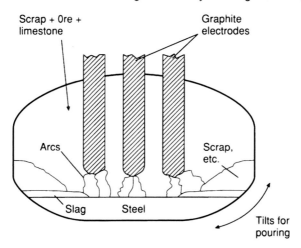

FIGURE 8.24 Open-arc furnace being used in steel production.

The most extensively used unit employing this *bottom-blown* technology is the *Q-BOP*, a vessel similar in shape, size, and operating procedure to the BOF but having many double tuyeres rather than an oxygen lance. One advantage is immediately clear; the high overhead clearance necessary for the oxygen lance is avoided and the relatively low buildings formerly housing open hearth furnaces can be more easily converted to Q-BOP operations than to BOF operations. Furthermore, although oxidation can be completed in a Q-BOP in approximately 20 minutes, implying that the Q-BOP matches the BOF in productivity, the fume and dust formation associated with the impact of a high-velocity oxygen jet on the metal surface is avoided. Additionally, by using natural gas as a shroud gas, the heat balance can be altered and the furnace has a much greater flexibility of scrap (or directly reduced iron) usage. In addition, argon degassing, discussed in Chapter 10, can readily be carried out in the Q-BOP.

Finally, mention is made of *combined blowing*, whereby oxygen (and other gases) are introduced both from above (through a lance) and below (through bottom tuyeres). In this way even more control is achieved over agitation, heat balance, and degassing.

8.7.3 Electric Furnace Steelmaking

A significant amount of steel is made in electric furnaces such as the *open arc fiurnace* sketched in Fig. 8.24. This furnace is distinguished from the submerged arc furnace (Fig. 8.16) by the fact that the arcs are readily visible on glancing into the furnace. Furthermore, the steelmaking furnace is operated batchwise, employs graphite rather than carbon electrodes, and is capable of being tilted to pour out the slag and finished steel; otherwise, the furnaces are similar in physical dimensions and electric power.

Open arc furnaces typically produce steel from scrap or direct reduced iron (DRI), although some use combinations of these and pig iron. Typically, in producing steel from scrap or DRI, much less oxidation is necessary than in producing steel from pig iron. Consequently, rust on the scrap, or residual oxygen in the DRI, are frequently sufficient to supply all the oxygen necessary. If additional oxygen is needed, it is commonly introduced by feeding iron ore to the furnace [which can react according to reactions such as (8.24)], although sometimes air or oxygen lances are used. As used in scrap processing, the productivity of the furnace is determined more by its ability to melt than by its ability to refine. Batches of approximately 100 tons can typically be processed in an hour or so. Compared to the BOF, the furnace produces little polluting fume or dust, can handle scrap, pig iron, or DRI in any proportion, and provides very easy control of the temperature of the finished steel. However, electric furnaces are expensive to build and operate, and electric furnace steelmaking has found its major use in areas where electricity is inexpensive and antipollution laws are strict. Along with the electric induction furnace (where heat is generated within the metal by currents induced by a coil that surrounds the metal and carries an alternating current) the open arc furnace has been widely used to produce special steel (e.g., stainless steels), where small batches are produced, usually from cold starting materials, and it is important to avoid contaminants that could be introduced by burning a fuel.

The open arc furnace and the continuous caster (see Chapter 10) have been central to the development of "mini-mills" in the twentieth century. These are plants producing small quantities of steel (compared to the blast furnaces – BOFs – of an "integrated steel mill") for local markets from scrap and DRI. Initially these mills produced a limited range of low-value products (such as reinforcing bar) but are now challenging the integrated steel mills in the range and value of their products.

8.8 Glassmaking

The reader is probably aware that a glass is a liquid that has been cooled, without crystallization, to the point where its viscosity has become enormous. For all practical purposes the glass is then rigid. Most (although not all) glasses are formed from SiO_2. The other principal oxides involved in the formation of glasses are Na_2O, B_2O_3, CaO, MgO, Al_2O_3, and PbO. Large numbers of glasses, each with a different composition tailored to its end use, are commercially available. Many of these are colored by the addition of Fe_2O_3, CuO or other oxides. Small amounts of other anions (notably fluoride and sulfate) are present in some glasses.

As a first approximation the manufacturing of glass can be regarded as simply the melting and mixing of various oxides (or compounds such as carbonates that form oxides on heating), followed by the cooling and shaping of the glass into finished forms such as flat glass or glass containers.

Silica is provided to the process in the form of sand. In most instances the sand is selected so as to be low in impurity content and is further purified by the physical separation techniques (mineral processing) described in Chapter 1. Sodium carbonate (soda ash) or trona ($Na_2CO_3 \cdot NaHCO_3 \cdot 2H_2O$) is the source of Na_2O, and limestone is the source of CaO. The oxides are mixed with recycled crushed glass, known as cullet, and pass into one end of the continuous melting furnace. Melting furnaces for the production of glass used in large quantities, such as flat glass or containers, are large, perhaps 1 to 1.3 m deep, 12 m wide, and 50 m long and are usually gas-fired furnaces where flames above the surface of the material in the furnace bring about the melting (see Fig. 8.25). Whereas the individual oxides typically have high melting points, the

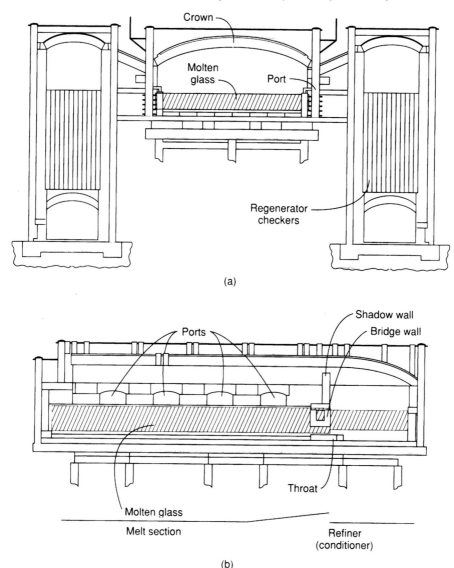

(a)

(b)

FIGURE 8.25 Glass furnace. (Reprinted with permission from Encyclopedia of Chemical Technology, Vol. 11, 3rd ed. Copyright 1980 John Wiley & Sons, Inc., New York.)

glass that forms would have a much lower melting point. The cullet therefore aids the melting by readily forming a liquid phase into which the oxides can dissolve. Melting and maintaining a uniform temperature is facilitated by the transparency of most glasses to infrared radiation. Radiant heat from the flames is therefore able to penetrate some distance into the molten glass. Temperatures in the hotter parts of the furnace might reach 1500°C or even higher, and extensive use is made of refractories resistant to glass at high temperatures (e.g., alumina-zirconia-silica refractories) to line the furnace. Typically, the furnace would produce several hundred tons per day of glass. The reader will note the similarity between this kind of furnace and the reverberatory smelter described in Section 8.6.1. Because most glasses have a significant electrical conductivity at high temperatures, it is also possible to heat them electrically. This is achieved by passing alternating current between molybdenum electrodes placed in the furnace. The electrical heating can be auxiliary to the gas-fired heating or can be alternative to it. The latter type of operation has the advantage that loss of any volatile constituents (e.g., boron compounds) from the melt is minimized. This is because the material in the gas-fired furnace is hottest at the surface and is exposed to a rapidly flowing gas stream, promoting mass transfer of volatiles. In the electric furnace heat generation is deep within the melt and the gas space above the melt is relatively quiescent.

The furnace is operated continuously and the material in the furnace passes gradually along its length and is discharged from the opposite end (the "refining end") to the charging end. The principal function of the refining section of the furnace is to provide a relatively calm region of the melt, where entrapped gas bubbles can escape by rising out of the liquid. The bubbles result from air entrapped between the grains of material fed to the furnace, from gas released by decomposition of carbonates, and so on. Many of these bubbles are small, particularly those resulting from air trapped between grains, and the glass being viscous, it is impossible for them to rise out of the melt in the residence time of the glass in the refining section. Larger bubbles rise more rapidly, however, and this fact is taken advantage of in glassmaking. *Fining agents* such as sulfates that release large volumes of large bubbles by thermal decomposition in the refining section are added to the furnace with the raw materials. As the large bubbles rise through the molten glass, they engulf the smaller bubbles, thereby removing them from the melt.

The refining end of the furnace is separated from the melting end by a *bridge wall* under which the molten glass flows. In this way surface scums and floating unmelted particles are prevented from reaching the refining section. In the production of flat glass the molten glass is drawn from the refining end by water-cooled rolls that serve to lower the glass temperature to below its softening temperature, producing a continuous ribbon of glass a few meters wide which passes into a continuous annealing furnace known as a *lehr*. This handling of the glass results in surface imperfections and distortion of the sheet that can be seen in most older window glass. Plate glass is produced by grinding and polishing both sides of the cooled ribbon so that these imperfections are removed. The grinding and polishing are expensive operations, and a less expensive way of obtaining a glass of comparable quality is the float glass process. In this technology the ribbon of glass from the melting furnace passes into a second continuous furnace, where it floats on a pool of molten tin. The furnace is designed so that the tin at the end where the glass enters is hot; this results in the soft glass becoming flat and smooth on both sides. At the other end of the furnace the tin is cooler and the glass becomes hard and rigid prior to leaving the furnace.

The manufacture of containers, light bulbs, and the many other items fashioned from glass, starting from the product of the melting furnace, is beyond the scope of this book.

8.9 Concluding Remarks

This chapter, one of the longest in the book, has dealt with the pyrometallurgical operations that are at the heart of the production of several important metals (iron and steel, copper, zinc, and lead) and (briefly) with the production of glass. We have emphasized the chemistry and engineering fundamentals that govern the behavior of these operations, rather than process detail.

These materials are very much the product of smokestack industries, and many undergraduate curricula in materials science and engineering neglect these metals in favor of more exotic materials such as optical fibers, compound semiconductors, and advanced ceramics. This is a mistake; the production of primary metals employs 654,000 Americans (mid 2001), and a greater portion of the population of Canada, and shipped $189 billion of products in 1999. These figures will exceed by far those for the high-technology materials through the next few decades, even assuming that the growth rate of the most optimistic projections is realized. The production of advanced materials is an important topic (treated elsewhere in this book) because those materials facilitate new applications (faster computer circuits, etc.). The production of traditional materials is important because of the sheer size of the industries involved and our dependence on their products for articles of everyday use (batteries, automobiles, airplanes, steel-framed buildings, windows, etc.) that are likely to be with us throughout this new century.

PROBLEMS

8.1   One important variable in fluidized bed roasters (or other types of fluidized beds) is the superficial gas velocity. This is the volumetric flow rate of the gas divided by the (empty) cross-sectional area of the bed. Dry air (20 mol % O_2, 80 mol % N_2) is fed to a roaster at a superficial gas velocity of 0.1 m/s (expressed at 25°C and 1 atm pressure). The roaster is 3 m in diameter. Calculate the maximum rate (kg/s) at which the roaster can process a concentrate (i.e., the rate achieved when the gases leaving the bed are in equilibrium with the solid). An elemental analysis of the concentrate shows 27% sulfur (by weight) and it is necessary to remove one-third of it by the reaction

$$MS + \frac{3}{2}O_2 = MO + SO_2$$

for which the equilibrium constant at the temperature in the roaster is 18.

8.2   One of the problems with the roasting referred to above is that it generates sulfur dioxide, a pollutant. A suggestion for avoiding this problem is to mix the sulfide concentrate with powdered lime and then to reduce by hydrogen, the idea being that coupled reactions such as

$$PbS + H_2 = Pb + H_2S$$

$$H_2S + CaO = H_2O + CaS$$

might take place, producing metal without evolution of sulfur dioxide. The solid reaction products would then be separated by some physical means (e.g., sink-float). Using the thermodynamic data contained in the Ellingham diagrams, evaluate the feasibility of carrying out these coupled reactions at 800°C.

8.3 A tin concentrate containing 15% wt SnO_2 and 25% wt Fe_3O_4, the rest inert gangue, is to be treated by reduction in a shaft furnace at 1000 K with a gas containing CO, CO_2, and 70 mol % N_2. The CO_2/CO ratio of the gas entering the furnace can be controlled. The reactor is operated in a semibatch manner, with all the concentrate being placed in the reactor and then reducing gas passed through until the reaction is complete. It is required to reduce the tin oxide to metal but not to do any reduction of Fe_3O_4. What is the minimum hourly input of gas (mol/h) to produce 1000 kg/h of tin? What are the compositions of the gases entering and leaving the furnace? Obtain from tables the necessary thermodynamic data for this problem. Why was it unnecessary to specify the furnace pressure? [In reality the reduction is complicated by the formation of an iron-tin alloy but ignore this for the present exercise.]

8.4 As presently practiced, steelmaking is a batch operation. There have been many suggestions over the years that a continuous steelmaking technology would be economically superior. Devise a scheme for making steel continuously, assuming a continuous supply of molten pig iron. What difficulties would you expect to encounter in operating your proposed scheme?

8.5 Calculate the rate of production of silicon from a 5-MW submerged arc furnace in which silica is reduced with coke. For the purpose of this calculation, assume that the silica is pure, that the coke is pure carbon, and that both are fed to the furnace at 25°C. Assume that the only gases leaving the furnace are carbon monoxide, at 700°C, and that reduction is complete, yielding silicon that leaves the furnace at 1550°C. Obtain the necessary thermodynamic data from the JANAF tables or a similar source.

8.6 A reverberatory furnace is used to smelt a copper concentrate at 1000°C. Estimate the copper content of the slag leaving the furnace if it is governed by the equilibrium

$$Cu_2O_{slag} + FeS_{matte} = Cu_2S_{matte} + FeO_{slag}$$

Assume that activities are equal to mole fractions and use an Ellingham diagram for oxides and sulfides to obtain the necessary thermodynamic data. The mole fraction of FeO in the slag can be taken as 0.5, and the matte is 40 wt % Cu_2S and 60 wt % FeS. Would the losses of copper to the slag be the same if the smelting temperature were increased to 1300°C?

8.7 One hundred thousand kilograms of molten pig iron (containing 4 wt % carbon, 1 wt % each of silicon and manganese, and 0.2 wt % each of phosphorus and sulfur) at 1300°C is to be processed in an oxygen steelmaking operation so as to yield molten steel at 1600°C. It is planned to achieve this final steel temperature by the addition of scrap steel at room temperature (25°C). Based on the following (somewhat crude) assumptions, provide a first estimate of the quantity of scrap needed.

(a) Oxygen is fed to the operation at 25°C and its utilization is 100%.

(b) Heat losses by conduction through the walls of the furnace are negligible.

(c) Essentially all the impurities are oxidized, along with 2% of the iron in the pig iron.

(d) The only gas leaving the furnace is CO at 1600°C.

(e) The heat capacities of the pig iron and steel are those of pure iron.

(f) The heat required to bring lime additions to the furnace up to temperature is negligible.

(g) The enthalpy of formation of the slag from oxidation products (FeO, MnO, P_4O_{10}, SiO_2, and SO_2) and CaO is negligible.

(h) Heats of solution of the impurities in molten iron are negligible. *Data*: Standard heats of formation at 1600°C (kJ/mol): FeO, –242; MnO, –408; P_4O_{10}, –3135; SiO_2, –914; SO_2, –59. Enthalpy of iron (relative to a zero for solid iron at 25°C): liquid pig iron at 1300°C, 63 kJ/mol; liquid iron at 1600°C, 77 kJ/mol. Enthalpy of oxygen at 1600°C (relative to zero at 25°C): 48 kJ/mol. What additional information would be required to carry out a more precise calculation of the amount of scrap that could be melted?

8.8 The gases leaving the top of the Imperial Smelting furnace contain zinc vapor, carbon monoxide, carbon dioxide, nitrogen, and traces of other gases. It is necessary to quench this gas stream as quickly as possible to prevent reoxidation of the zinc. This is achieved in a condenser by means of which molten lead is sprayed into the gas. A practical difficulty is that the condenser must be some distance from the furnace, and there is a tendency for the gas stream to cool in passing along the duct from the furnace to the condenser. Such cooling is relatively slow and leads to zinc reoxidation. One way this can be avoided is by injecting air into the gas leaving the furnace so as to oxidize carbon monoxide to carbon dioxide; the intention is that the heat evolved by this reaction is sufficient to raise the gas temperature to well above the temperature for reoxidation of zinc vapor by carbon dioxide. Carry out thermodynamic calculations to show that this air injection does indeed work. Assume that the gases leaving the top of the furnace are at 1000 degrees Celsius, suffer a heat loss of 5kJ/mol of gas passing along the duct and have a composition of 62 mol % N_2, 24 mol % CO, 6 mol % CO_2 and 8 mol % Zn vapor. Also assume that sufficient air (80 mol % N_2; 20 mol % O_2) is introduced to oxidize one-fourth of the CO to CO_2, and that the pressure at the furnace top is atmospheric. $\Delta \tilde{H}_{formation}$ (kJ/mol) at 800°C: CO_2, –395; CO, –112. Use the Ellingham diagram to obtain data on the free energies of formation of ZnO, and so on. Heat capacities in the temperature range of interest are (J/mol K) N_2, 32; CO, 35; CO_2, 58; and Zn vapor, 21.

8.9 It is necessary to produce a glass suitable for light bulbs; the composition (by weight) is to be 73% SiO_2, 17% Na_2O, 5% CaO, 4% MgO, and 1% Al_2O_3. Starting ingredients are sand (which can be assumed to be pure silica), trona, limestone, magnesia, and alumina. Calculate the weight ratios in which they should be fed to the glass furnace.

8.10 (a) A molten glass has a viscosity of 10^3 kg/(m • s) and a density of 2.5 x 10^3 kg/m^3. Calculate the time for the rise of a 5-mm-diameter gas bubble through 0.5 m of this glass if Stokes' law is obeyed and the bubble is spherical. By repeating the calculation for a 0.5-mm diameter bubble, show that the need for fining agents as described in Section 8.8 is a real one.

(b) An estimate is to be obtained of the quantity of gas that fining agents must release in the refining section of a glass furnace so that every point in the liquid glass is swept at least once by a bubble. Carry out this estimation based on the following simplifying assumptions:

1. Bubbles from the fining agents are of uniform size and are generated randomly across the bottom of the refining section.

2. The bubbles undergo negligible size change on passing up through the glass.

3. The bubbles are small compared to the dimensions of the refining section.

4. Your intuition tells you that the smaller the bubble diameter, the lower the volume of gas required; however, below a certain minimum diameter d_{min}, the bubbles will not rise fast enough, so you should take this as the bubble diameter.

5. The bubbles rise vertically.

Given the stochastic nature of the process, it could take an unreasonably large amount of gas to ensure that every point in the glass volume is swept. Calculate the volume required as a function of the percentage of the volume swept. Show that your intuition is correct, that the volume required is proportional to the bubble diameter. [*Hint.* The best way to solve this problem is probably by writing a computer program that generates circles (of constant diameter d_{min}) with centers randomly placed on a rectangle representing the refining section as seen from above. The location of each new circle is given by two coordinates obtained from the computer's random number generator. After a number of circles have been added, the remaining unswept area must be determined. The easiest way to do this is to use the stereological technique (again based on random numbers) for determining area fractions that is described in Chapter 4. This can be done by you on a drawing generated by your computer or, better yet, done within the computer itself with programming that tests whether or not randomly chosen points fall within circles.]

REFERENCES AND FURTHER READINGS

BISWAS, A. K., and W. G. DAVENPORT, *Extractive Metallurgy of Copper*, 3rd ed., Pergamon Press, Oxford, 1994).

Kirk-Othmer Encyclopedia of Chemical Technology, 3rd ed., Vol. 11, Glass Technology, Wiley, New York, 1980.

The Making, Shaping and Treating of Steel, 10th ed., Association of Iron and Steel Engineers, Pittsburgh, PA, 1984.

PEHLKE, R. D. *Unit Processes of Extractive Metallurgy*, American Elsevier, New York, 1973.

PEHLKE, R. D. et al., eds., *BOF Steelmaking*, Vols. I-V, Iron and Steel Society, New York, 1974-1977.

ROSENQUIST, T., *Principles of Extractive Metallurgy*, McGraw-Hill, New York, 1983.

STRASSBURGER, J. H., ed., *Blast Furnace—Theory and Practice*, Vols. I and II, Gordon and Breach, New York, 1969.

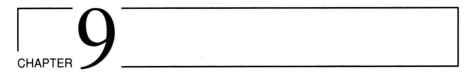

CHAPTER 9

Hydrometallurgy
and Electrometallurgy

9.1 Introduction

In Chapter 8 we examined some traditional techniques for the production of "tonnage" metals such as steel, aluminum, and copper. These techniques involved high temperature, molten metals, and slags and difficulties posed by the emission into the environment of particles, sulfur dioxide, and so on. In the present chapter we examine some alternative techniques that make use of aqueous chemistry to extract metals from ores and concentrates. This "hydrometallurgical" technology employs much lower temperatures than those encountered in Chapter 8. Nevertheless, hydrometallurgical techniques still entail the consumption of considerable quantities of energy, particularly when the energy costs associated with producing reagents are considered. However, hydrometallurgical processes typically pose no air pollution difficulties, but steps must usually be taken to avoid water pollution.

Figure 9.1 is a schematic diagram of a hydrometallurgical plant, in much simplified form. The "leaching" step, as implied by the name, is one where the desired metal compounds are abstracted into aqueous solution by suitable reagents ("lixiviants"). Typically, this step is imperfect in that undesirable compounds also find their way into solution. Solution purification usually follows the leaching, therefore. Frequently, this purification is accompanied by concentration of the solution. In this way, the volume of solution to be dealt with in subsequent steps is much reduced.

Finally, the desired metal (or in rare cases a compound) is recovered from solution. There are several ways to do this, but one widely practiced method is *electrowinning*. This term is applied to electrolysis of the solution, so that, for example, metal is deposited by the electrolytic reduction of the metal ion at a cathode. Electrowinning is a topic within *electrometallurgy*, the technology of using direct electric currents to obtain metals, and this chapter is therefore a convenient place to treat electrometallurgy. Of course, aqueous solutions are just one example of ionic conductors that can be electrolyzed between an anode and a cathode. Another example is a molten salt, and the electrowinning of aluminum and magnesium by electrolysis of molten salts is treated in this chapter. Finally, a second topic of electrometallurgy, *electrorefining*, wherein metals are purified (by passage into solution at an anode followed by reduction at a cathode), falls into the subject matter of this chapter.

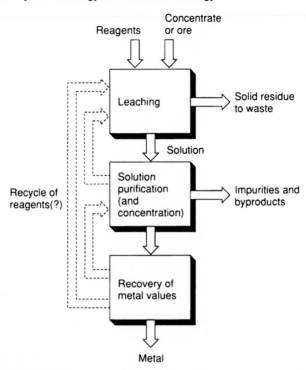

FIGURE 9.1 Simplified diagram of the production of metals from ores or concentrates by hydrometallurgical routes.

9.2 Leaching

Examples of the chemistry taking place in leaching operations are:

1. The dissolution of zinc oxide (produced by roasting of zinc sulfide concentrate):

$$ZnO + H_2SO_4 = ZnSO_4 + H_2O \qquad (9.1)$$

2. The dissolution of gold in alkaline cyanide solution:

$$4Au + 8NaCN + O_2 + 2H_2O = 4NaAu(CN)_2 + 4NaOH \qquad (9.2)$$

3. The dissolution of copper sulfide in water in the presence of oxygen:

$$CuS + 2O_2 = CuSO_4 \qquad (9.3)$$

The dissolution reactions can be specific to the mineral desired. For example, gold can be obtained economically from ores containing only a fraction of an ounce per ton of ore by reaction (9.2); the cyanide reagent is highly specific to gold (and silver); that is, it is inert to the vast majority of much less valuable minerals that make up the bulk of the ore. Economic extraction of gold from such lean ores would be impossible by pyrometallurgical techniques.

Equipment used in leaching ranges from the most rudimentary to sophisticated computer-controlled pressurized leaching vessels. Let us examine some of these methods.

9.2.1 Leaching Technology

One of the simplest technologies was mentioned in Chapter 1: solution mining, or in situ leaching as it is sometimes called. The technology has been practiced for the production of uranium from sedimentary deposits in Texas, using aqueous solutions of ammonium carbonate/bicarbonate with hydrogen peroxide as lixiviants.

[handwritten margin note: liquid medium in hydrometallurgy to selectively extract metal from ore]

$$UO_2 + 3CO_3^{2-} + H_2O_2 = [UO_2(CO_3)_3]^{4-} + 2OH^- \qquad (9.4)$$

The solution is continuously injected into the ore body through boreholes a few centimeters in diameter ("injection wells") drilled down to it from the surface, with uranium-containing solution pumped out continuously through similar boreholes (production wells). It is important that the ore body be permeable to solution and desirable that it be bounded by relatively impermeable strata (so as to avoid loss of solution). The similar importance of subsurface fluid flow in this technology and the oil industry has led to the application of many of the reservoir engineering techniques of the latter to solution mining. A solution mine might require years to complete the leaching of an ore body, during which time numerous boreholes would be drilled, used, and finally closed. *Monitor wells* would usually be drilled around the periphery of the ore body and used, during and after mining, to check on possible groundwater contamination by mine solutions.

There have been studies of the solution mining of impermeable rocks (which are hosts to a large fraction of our mineral resources) by first fracturing the rock with underground explosives, including nuclear explosives. A number of technical challenges still stand in the way of the realization of these rather aggressive concepts.

Mining and milling operations of previous decades were less efficient, typically, than are present-day ones. There are therefore numerous waste rock dumps containing significant amounts of metal value. One method of recovering such values is by *dump leaching*. Equipment that is little more than pumps, lawn sprinklers, and garden hoses is used to spray solutions onto the dump. For example, in leaching dumps from copper sulfide operations, water and air are sufficient, according to reaction (9.3), and as the solution trickles down through the dump, copper passes into solution as copper sulfate. The oxygen required for reaction passes into the dump from the outside through the interstices between particles. In reality, the solution sprayed over the dump is acid, the acid usually being self-generated within the dump by the oxidation of other sulfide minerals (such as pyrite). Bacteria play an important role in the leaching reactions. Ideally, the dump sits on a surface that is impervious to solution and slightly sloped so that copper-rich solution flows from one side. In some instances the dump might have to be moved to such a surface.

In some cases this method is used to leach low-grade ores directly (rather than the waste rock), whereupon it is called *heap leaching*. Now, however, the ore would be comminuted before leaching, and the cost of crushing (and perhaps grinding) the ore becomes significant, whereas there is no such cost in leaching a dump. Dumps or heaps leached in this way may be enormous (perhaps a few hundred feet in horizontal dimensions and a few tens of feet deep) and require many years to leach.

The next step upward in complexity is *vat leaching*. Here the crushed ore is loaded into vats that are typically made of concrete and are a few meters deep and several meters in horizontal dimensions. Usually, several vats are employed and leaching solution is pumped continuously from the bottom of one tank to the top of the next, as in Fig. 9.2. When leaching has been completed, the residual solids are dug out of the vat and replaced by fresh ore. Leaching of copper oxide ores by dilute sulfuric acid to produce aqueous copper sulfate solution is one example of the application of vat leaching. You will recall the advantages of countercurrent operation discussed in Chapter 7. This advantage is exploited in vat leaching by (in the example of acid leaching of copper oxides) contacting fresh incoming acid with the solid that has been in the leaching scheme the longest, while the last solid to be contacted with solution is fresh ore. This is achieved by switching the solution flows around between tanks as leaching proceeds.

Leaching reactions are heterogeneous, and as discussed in Chapter 3, considerable increases in productivity should result from (1) grinding the ore to a fine size, and (2) agitating the mixture of particles and solution (beyond the leisurely flow of solution typical of the leaching methods we have discussed so far) so as to enhance mass transfer between solution and particle.

These are the concepts behind *agitation leaching*, illustrated in Fig. 9.3. Crushed and ground ore particles (or concentrate particles) are maintained in suspension by either mechanical or air agitation. Typical tanks might be a few meters in diameter and of the order of 10 m high and be operated either continuously or batchwise. In the former case, several leaching tanks might be used in series to complete the leaching. Mechanically agitated tanks would employ one or more agitators mounted on a shaft driven by an electric motor to achieve the necessary circulation of the slurry. In air-agitated tanks, circulation is achieved by injecting air from a compressor into the bot-

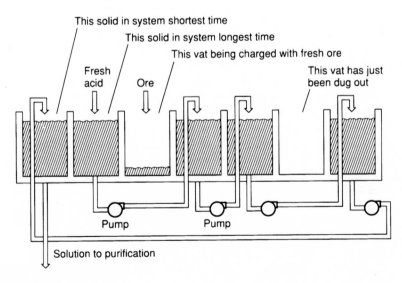

FIGURE 9.2 Vat leaching.

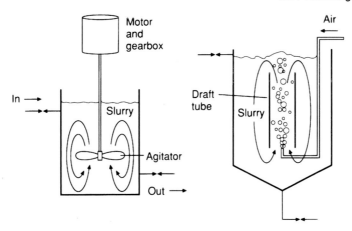

FIGURE 9.3 Leaching in mechanically agitated or air agitated tanks.

tom of the tank; the rising air bubbles (often confined by a *draft tube*, as shown in Fig. 9.3) cause an upflow of slurry in their vicinity, with a return downflow elsewhere in the tank. In either case, energy costs associated with agitation and the capital costs of the equipment are higher than for the leaching technologies discussed earlier. These costs would be justified by the higher productivity that is reflected in the shortened ore residence time (a few tens of minutes to a few hours for typical systems). Such residence times are sufficient because of the rapid reaction of the particles.

You will also discern that leaching might be accelerated by increasing the temperature or the concentration of lixiviant in solution. For equipment operating at atmospheric pressure, a natural limit to the former occurs in the vicinity of 100°C as the solution starts to boil. This limit can be circumvented if a pressure vessel is used, such

pressure vessel. used to increase limit of 100°C

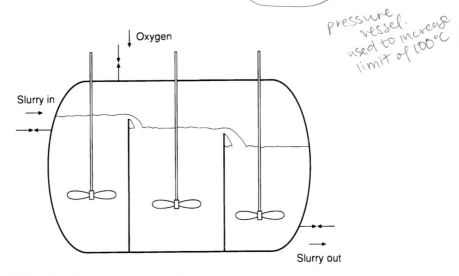

FIGURE 9.4 Leaching in a pressure vessel.

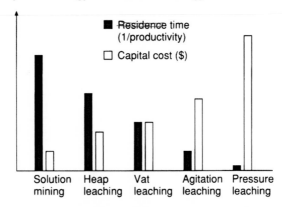

FIGURE 9.5 Residence time (1/productivity; solid bars) and capital cost (dollars; open bars) for various leaching technologies.

as that shown in Fig. 9.4. In the case of leaching of sulfides with oxygen [reaction (9.3), for example], the pressure vessel makes it possible to use oxygen under pressure, thereby increasing the dissolved oxygen in solution and the reaction rate. Of course, such pressure vessels are expensive to build, particularly if they are made of expensive materials such as titanium or stainless steels. Such expense may be justified by the rapid leaching, which might be completed in a few minutes of residence time. Ores or concentrates fed to such pressure leaching vessels are usually ground to a small particle size to maximize reaction rate and minimize the danger of particles settling out in the pressure vessel.

There is a progression, depicted schematically in Fig. 9.5, from solution mining (where leaching vessel costs are nonexistent and capital costs largely a matter of drilling wells) through vat leaching (where capital costs for the leaching units become significant) to pressure leaching (where they become large). Productivity (represented here as the inverse of the leaching time) increases sharply along this sequence, due to the faster leaching achievable in the more expensive equipment. The optimum technology would depend on aspects of economics, technology, and science that are beyond the scope of this book but which are now within the grasp of the reader. Leaching is an interesting example of the interplay among process economics, technology, and chemistry (kinetics).

In the case of agitation and pressure leaching, residual solids (gangue) leave the leaching reactor along with the solution, now containing the metal values. These solids must be removed from solution, and a preliminary step in this direction frequently is decantation carried out in *thickeners*. These are cylindrical vessels 10 to 15 m in diameter and perhaps half this height, into which the slurry is fed continuously. Particles settle to the bottom of the tank and flow out with a fraction of the solution. (Gentle agitation at the bottom of the tank prevents particles from accumulating on the bottom.) The major part of the solution overflows the top of the tank and the last traces of particles can be removed by filtration. The contact between solution and particles extends into the thickener, and sometimes a significant part (or the whole) of the leaching takes place in the thickener. The thickened slurry from the bottom of the thickener can be "washed" with fresh water in a second (or more) thickener(s), operated countercurrent to the first, prior to dumping.

9.2.2 Leaching Chemistry

The reader was introduced to predominance diagrams in Chapter 2. One form of predominance diagram that has been found particularly useful in hydrometallurgy is the Pourbaix diagram, of which Fig. 9.6 is an example. Originally developed by Pourbaix to treat the corrosion of metals, this type of diagram is now used by hydrometallurgists to determine the likely products of leaching reactions. The two independent variables are the pH and the Eh. The former will be familiar to the reader, but the latter requires more explanation. What is represented by this variable is whether conditions in solution are oxidizing or reducing. The reader will recall that reduction of an ion (e.g., as achieved electrochemically at a cathode) corresponds to adding an electron, whereas oxidation (e.g., at an anode) is the removal of an electron from an atom (or ion). The precise definition of Eh is therefore

$$Eh = -\frac{2.303RT}{F}\log_{10} a_\varepsilon = \frac{2.303RT}{F} pE \qquad (9.5)$$

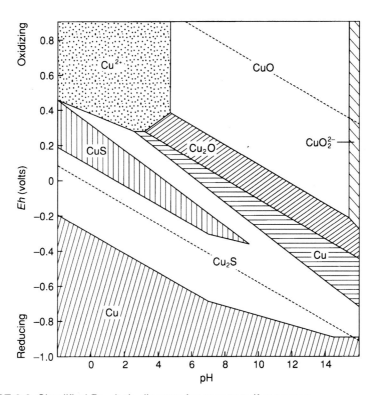

FIGURE 9.6 Simplified Pourbaix diagram for copper-sulfur-oxygen.

where a_ε is the electron activity and pE ($= -\log_{10} a_\varepsilon$) is to the electron activity what pH is to the hydrogen activity; R/F is a constant of value 8.614×10^{-5} V/K. This definition is not immediately helpful because the concept of electrons in solution seems unlikely. In fact, the concept of Eh really relies on the fact that reduction in aqueous solutions and aqueous suspensions entails the transfer of electrons, and we can therefore measure the driving force for such reactions by measuring the potential of an electrode in equilibrium with the solution. Adding a reducing agent to the solution (e.g., a solution of a ferrous salt in water) would therefore increase the activity of electrons in that equilibrium electrode. This is because the tendency of the electrode to add electrons to the species in solution must now match the considerable tendency of ferrous ions to donate electrons (by forming ferric ions). From equation (9.5) the electrode potential and therefore the Eh experiences a drop, and by implication the Eh of the solution can be thought of as having been lowered. That lowering of Eh will take place even if the electrode is not present, although a convenient way of measuring it is to have the electrode present. Indeed, this is how Eh is measured. For the present purposes it is sufficient to regard Eh as an indicator of the oxidizing nature of the environment into which we place our minerals. We shall return to the topic of Eh in Section 9.4 after a fuller treatment of electrochemistry.

Two lines on the Pourbaix diagram have a special significance. These are the sloping dashed lines at top and bottom of the figure representing the stability limits for water. At the upper line, oxidation of water proceeds according to

$$H_2O = 2H^+ + \frac{1}{2}O_2 + 2e^- \tag{9.6}$$

(with oxygen at 1 atm pressure), while at the lower line, reduction of water proceeds according to

$$2H_2O + 2e^- = 2OH^- + H_2 \tag{9.7}$$

(with hydrogen at 1 atm pressure). Thermodynamically, then, the key part of the diagram lies between these lines and the Eh in a solution would be: at the upper line if it were in equilibrium with pure oxygen at one atmosphere; at the lower line if the solution were in equilibrium with pure hydrogen at this pressure.

As an example of the use of Pourbaix diagrams, let us examine Fig. 9.6 in the context of a leaching operation where we wish to react copper sulfide minerals. The dotted area marked Cu^{2+} at the upper left represents the region of stability of cupric ion solutions (0.01 M or more concentrated) and we see that the sulfide minerals Cu_2S and CuS can only be leached at a pH value lower than 5 and an Eh exceeding 0.3 (higher Eh values if the pH is other than approximately 3). Attempts to leach with alkaline solutions would fail because the (solid) oxides of copper are formed under alkaline oxidizing conditions.

You will now perceive that the hydrometallurgist can, in many instances, shift leaching conditions to ones that are favorable for the desired reaction by adjusting Eh and pH. Furthermore, the areas of stability of the various species in a Pourbaix diagram shift as the temperature is changed, providing a third variable that may be manipulated to achieve the desired effect. Yet another variable that can be manipulated is the concentration of complexing agents (particularly CN^-, NH_3, and Cl^-) that can stabilize ions in solution. Pourbaix diagrams have now been obtained (by experiment or theoretical calculations or combinations of the two) for a large number of mineral systems and are invaluable in examining the chemistry of potential hydrometallurgical processes. However, it should be stressed that the diagrams are *thermodynamic* entities and therefore subject to the limitations discussed in Chapter 2: (1) they give no information on the kinetics of leaching reactions, and (2) they give no information on the state of the system if equilibrium is not achieved (or deliberately avoided).

Of course, the hydrometallurgist would be interested in the Pourbaix diagrams for gangue minerals, not just in the diagram for the mineral that is to be leached. In this way conditions can be selected that will minimize the dissolution of gangue minerals, reducing the contamination of the solution and the waste of reagent on the gangue. In reality, solutions produced by leaching are never pure, and next, we must examine their purification.

9.3 Solution Purification

The most common methods of solution purification are chemical precipitation, cementation, ion exchange, and solvent extraction (sometimes called liquid ion exchange).

9.3.1 Chemical Precipitation

The *chemical precipitation* of impurities by addition of chemical reagents at first appears an attractive means of removing impurities. For example, iron might be precipitated from an acidic solution from leaching by raising the pH:

$$Fe^{3+} + 3OH^- = Fe(OH)_3 \qquad (9.8)$$

Two difficulties present themselves, however. Reagent costs are associated with changing the pH from low values to high values. More important, the precipitation reaction is seldom as simple as represented in equation (9.8). Frequently, metal ions adsorb from solution onto the precipitate; if these ions are those of the metal we seek, a significant loss of product is entailed, or the oxide cannot readily be disposed of because of environmental problems posed by the adsorbed metals. Another difficulty is the occasional formation of mixed oxide precipitates with valuable metal ions, resulting in an unacceptable loss of the valuable metal; for example,

$$Ag^+ + 3Fe^{3+} + 2SO_4^{2-} + 6OH^- = AgFe_3(SO_4)_2(OH)_6 \qquad (9.9)$$

Chemical precipitation of a valuable metal compound, rather than the impurities, is more common. For example, in the Bayer process for the production of alumina from bauxite described in Chapter 7, the nearly pure hydrated oxide of aluminum is precipitated by dilution of an impure sodium aluminate solution, followed by seeding with a few alumina crystals.

As a second example, during processing of seawater (or other brines) to obtain magnesium, the hydroxide of magnesium is precipitated by raising the pH of a solution of magnesium salts (obtained following prior fractional crystallization of sodium chloride):

$$Mg^{2+} + 2OH^- = Mg(OH)_2 \qquad (9.10)$$

9.3.2 Cementation

Cementation is the reduction of an ion in solution using a metal that has a more negative electrode potential. An example is the purification of acidic zinc sulfate solutions formed by the leaching of zinc oxides with sulfuric acid. Such solutions typically contain other metal ions, such as nickel, cobalt, antimony, and arsenic. Although these metals are present only in small quantities, they are highly detrimental in the subsequent electrowinning step that is used to obtain the zinc product. By contacting the solution with zinc powder, these ions can be removed to a very low level by cementation reactions such as

$$Co^{2+} + Zn = Zn^{2+} + Co \tag{9.11}$$

The zinc particles thereby become coated with the impurity metals and are replaced periodically.

9.3.3 Ion Exchange

The reader is probably familiar with the *ion-exchange* resins that are routinely used for removing ions (such as the calcium ions causing "hardness") from water. Similar resins can be used to trap either the impurity or valuable metal ion streams from leaching. For example, in the solution mining of uranium discussed earlier, it is common to pass the solution from a production well through a bed of anion exchange resin. The uranium is trapped by the reaction

$$R(CO_3)_2 + \left[UO_2(CO_3)_3\right]^{4-} = RUO_2(CO_3)_3 + 2CO_3^{2-} \tag{9.12}$$

(where R indicates the resin), while the impurities pass through. When the resin approaches saturation, it can be regenerated by contacting it with strong alkaline carbonate solution, yielding a pure concentrate solution of carbonate complex, which is converted to ammonium uranate ("yellowcake") by first destroying the complex with hydrochloric acid and then precipitating with ammonia.

9.3.4 Solvent Extraction

Solvent extraction was described in Chapter 7, at least as far as the process technology is concerned. This technique has proven extremely valuable in recent years for the purification and concentration of solutions from leaching, particularly solutions produced by dump leaching of copper sulfide minerals.

A number of reagents have been developed by chemical manufacturers for this application. One such reagent is LIX 65N:

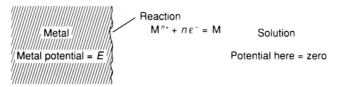

FIGURE 9.7 Piece of metal in contact with a solution containing its ions.

where R is C_9H_{19}. This reagent, other LIX reagents, and similar reagents from other manufacturers, owe their success to their ability, selectively but reversibly, to extract metal ions from aqueous solutions when dissolved in an inexpensive organic medium such as kerosene. Thus in copper extraction, copper ions are removed from the aqueous solution from leaching to the organic phase, leaving behind the impurities (such as iron salts) that are commonly encountered. The organic phase is regenerated by contacting it (*stripping*) with a strong aqueous solution of acid whereupon the copper enters the aqueous phase, yielding a concentrated pure solution of metal salt that is ideal for the subsequent electrowinning. The chemistry can thus be represented as

$$\underset{\text{(in }H_2O)}{Cu2+} + 2Org\text{-}H \underset{\text{stripping}}{\overset{\text{loading}}{\rightleftharpoons}} Org\text{-}Cu + \underset{\text{(in }H_2O)}{2H+} \tag{9.13}$$

The hydrogen ions added to the aqueous phase during loading render that solution sufficiently acid for its use in acid leaching (represented by one of the dashed recycle lines in Fig. 9.1).

9.4 Electrometallurgy

9.4.1 Metal ions in Solution

In Chapter 5 we examined the thermodynamics of ions in solution and concluded that an extra quantity must be added to the chemical potential of an ion to account for the extra free energy that the ion experiences in an electric field (i.e., when it is in an environment with an electrical potential):

$$\mu_i = \mu_i^0 + RT \ln a_i + n_i F E \tag{9.14}$$

Here μ_i is the chemical potential (more correctly now called the *electrochemical potential*) of ion i with charge n_i and activity a_i at electrical potential E. μ_i^0 is the chemical potential of i at standard state ($a_i = 1$). F is Faraday's constant, a conversion factor (23,066 cal/V•mol, for example, or 96,484 J/V mol) used to match the units of the last term to those in the rest of the equation. Now let us consider the case where i is a metal ion (M^{n+}) in solution in of the pure metal M, so that the *half-cell reaction*

$$M^{n+} + n\varepsilon = M \tag{9.15}$$

can occur at the interface between metal and solution (see Fig. 9.7). In general, a potential difference between metal and solution will exist even when reaction (9.15) is at equilibrium.

Let us calculate that equilibrium potential difference, taking as a zero of potential the solution a short distance away from the metal. Since metals are good conductors, they tend to have uniform potentials and we shall assume that our piece of the metal has a potential E right up to the interface with the solution. Now the electrochemical potential of an ion at the interface is given by equation (9.14):

$$\mu_i = \mu_i^0 + RT \ln a_i^s + n_i F E \tag{9.16}$$

where a superscript s has been placed on the activity of the ion to indicate that it is the value at the interface (which will differ from that in the bulk of the solution a_i)

Because the potential is zero in the bulk of the solution,

$$\mu_i = \mu_i^0 + RT \ln a_i \tag{9.17}$$

away from the interface. At equilibrium the electrochemical potential of an ion at the interface must equal that in the bulk and we can equate μ_i in (9.16) and (9.17) to yield

$$n_i FE = RT \ln a_i - RT \ln a_i^s \tag{9.18}$$

We can obtain a_i^s by examining reaction (9.15) at equilibrium:

$$\frac{a_m}{a_{M^{n+}}^s (a_{\varepsilon-})^n} = \exp\left(\frac{-\Delta \tilde{G}_{R0}}{RT}\right) = K_a \tag{9.19}$$

The metal being pure, its activity is 1. The activity of the electron is that in the metal at the metal-solution interface. With this substitution we can rearrange (9.19) to

$$a_{M^{n+}}^s = \frac{1}{\left(a_{\varepsilon-}\right)^n \exp\left(\dfrac{-\Delta \tilde{G}_{R0}}{RT}\right)} \tag{9.20}$$

or, taking natural logarithms,

$$RT \ln a_{M^{n+}}^s = -RTn \ln a_{\varepsilon-} + \Delta \tilde{G}_{R0} \tag{9.21}$$

and this can be substituted for $RT \ln a_i^s$ on the right of (9.18) to give

$$n_i FE = RT \ln a_i + RTn \ln a_{\varepsilon-} - \Delta \tilde{G}_{R0} \tag{9.22}$$

Let E_o be the electrode potential when $a_i = 1$ (i.e., when $\ln a_i = 0$); we can substitute this condition in (9.22) to obtain a value of $n_i FE_o$ for the last two terms on the right and then substitute in (9.22) to get (after minor rearrangement)

$$E = E_o + \frac{RT}{nF} \ln a_i \tag{9.23}$$

which is known as the Nernst equation. Note that E_o is the potential difference between the metal and solution when the solution contains unit activity of metal ions. We call it *the standard electrode potential* of the metal. The reader will recall from Chapter 2 that the choice of a standard state is an arbitrary one and that it is important to specify the standard state clearly when providing numerical values. The standard state used in this chapter is a 1 *molal*[1] solution of the relevant electrochemical species and 25°C. That standard state is only slightly different from a 1 *molar* solution, and generally we shall neglect the difference in this book. For the units to be correct, the activity appearing in the Nernst equation will then have units of molality (or molarity to within the precision required in this book). Mathematicians amongst our readers will be concerned at using a quantity with dimensions as the argument of a logarithm. Strictly speaking the argument should be the (dimensionless) ratio of activity to activity in the standard state. Because the latter has the numerical value of unity, it is common to exclude it from the Nernst equation. Equation (9.23) is an important one; it is the Nernst equation for this particular case. A more general form of the Nernst equation gives the potential of an electrode at which the reaction

$$aO + n\varepsilon^- = bR \tag{9.24}$$

takes place. (Here O indicates an oxidized species in solution and R the species produced by reduction at the electrode.) The more general form is

$$E = E_o + \frac{RT}{nF} \ln \frac{a_O^a}{a_R^b} \tag{9.25}$$

Again E_o is the standard electrode potential for reaction (9.24).

Actually measuring the potential difference between a piece of metal (an electrode) and a solution with which it is in contact poses a serious practical difficulty. To make a potential measurement using a voltmeter or similar device, two electrical connections must be made to the instrument: from the electrode and from the solution. The first connection is likely to involve a metal-metal contact (e.g., where a zinc electrode is connected to the voltmeter with a copper wire) at which a potential may arise to add error to our measurement. This first difficulty might be overcome (e.g., by making our voltmeter using only electrical parts of zinc), but a more substantial difficulty arises from the need to connect the solution to the voltmeter. This necessarily involves a contact between the solution and an electronic conductor (e.g., a copper wire), thereby producing a second metal-solution interface potential. The voltmeter can then measure only the *difference* between potential differences (i.e., the difference between electrode potentials). The measurement of absolute values of electrode potentials (and standard electrode potentials) therefore poses some practical difficulties (although indirect methods and calculations, both beyond the scope of this book, have been applied to this problem in recent years). Consequently, one particular electrode, the *standard hydrogen electrode*, has been arbitrarily assigned a standard electrode potential of zero and all other electrode potentials are expressed with reference to this. The standard hydrogen electrode consists of a platinum foil immersed in a solution with unit activity of hydrogen ions (pH = 0) and into which hydrogen at 1 atm is bubbled so that the bubbles contact the platinum, bringing about the reaction

$$H_2 = 2H^+ + 2\varepsilon^- \tag{9.26}$$

[1] A 1 molal (1 *m*) solution contains 1 mol of solute per kilogram *of solvent*. A 1 molar (1 *M*) solution contains 1 mol of solute per 1000 cm³ of *solution*. This definition of the standard state is sufficient for the purposes of this book. For a more precise definition the reader should consult books on thermodynamics, such as Pitzer and Brewer (1961).

The platinum foil serves to catalyze this reaction and permit an electrical connection (e.g., to a voltmeter). In this way the electrode potentials are "in error" by an amount equal to the unknown "real" value of the potential for the standard hydrogen electrode. All electrode potentials will have this same error, and since we are only concerned with differences in electrode potentials, there will be no practical effect. Because our electrode potentials are relative values, metal-metal contact potentials between electrodes and the voltmeter become immaterial.

Table 9.1 gives the standard electrode potentials for several metals, and more extensive tables are available in handbooks. These standard electrode potentials are *called standard reduction potentials,* for reasons that will become clear below. Remember that these are the potentials of electrodes with respect to the electrolyte, and it is best to think of them as metal potential minus solution potential at equilibrium, because then the direction of current flow becomes clear, as we shall see. Table 9.1 conforms to the convention adopted by the International Union of Pure and Applied Chemistry (IUPAC). Unfortunately, older American texts (and, regrettably, some newer ones) use an opposite

TABLE 9.1 Standard electrode potentials at 25°C, 1 atm
(Standard solution is 1 molal in appropriate ion)

	$E_o(V)$
$Au^{3+} + 3\varepsilon^- = Au$	1.50
$O_2 + 4H^+ + 4\varepsilon^- = 2H_2O$	1.23
$Hg^{2+} + 2\varepsilon^- = Hg$	0.80
$Ag^+ + \varepsilon^- = Ag$	0.80
$Cu^+ + \varepsilon^- = Cu$	0.52
$O_2 + 2H_2O + 4\varepsilon^- = 4OH^-$	0.40
$Cu^{2+} + 2\varepsilon^- = Cu$	0.34
$H^+ + \varepsilon^- = 1/2\ H_2$	0.000
$Pb^{2+} + 2\varepsilon^- = Pb$	−0.13
$Sn^{2+} + 2\varepsilon^- = Sn$	−0.14
$Ni^{2+} + 2\varepsilon^- = Ni$	−0.23
$Co^{2+} + 2\varepsilon^- = Co$	−0.27
$Cd^{2+} + 2\varepsilon^- = Cd$	−0.40
$Fe^{2+} + 2\varepsilon^- = Fe$	−0.44
$Cr^{3+} + 3\varepsilon^- = Cr$	−0.74
$Zn^{2+} + 2\varepsilon^- = Zn$	−0.76
$Cr^{2+} + 2\varepsilon^- = Cr$	−0.91
$Mn^{2+} + 2\varepsilon^- = Mn$	−1.05
$Al^{3+} + 3\varepsilon^- = Al$	−1.66
$Ti^{2+} + 2\varepsilon^- = Ti$	−1.75
$Mg^{2+} + 2\varepsilon^- = Mg$	−2.38
$Na^+ + \varepsilon^- = Na$	−2.71
$Ca^{2+} + 2\varepsilon^- = Ca$	−2.84
$K^+ + \varepsilon^- = K$	−2.92
$Li^+ + \varepsilon^- = Li$	−3.01

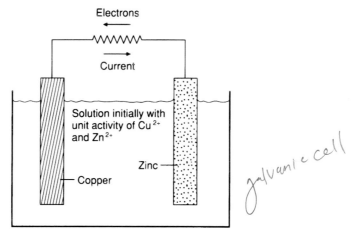

FIGURE 9.8 Two electrically connected dissimilar metals in contact with an aqueous solution (a galvanic cell).

convention in which the potentials are the negative of those in Table 9.1 and are referred to as *oxidation potentials*. It is hoped that the reader will not encounter these outdated potentials, but it is clearly wise to check when using data on electrode potentials. One simple way to do so is to look for Zn on the table. By the IUPAC convention this should be minus 0.76 V, and it is easy to remember this by the mnemonic Zn— zinc negative.

We are now in a position to examine what happens when two different metals in contact with the same solution are joined through an external conductor such as illustrated in Fig. 9.8. In this example the solution contains zinc and copper ions at unit activity, and the two pieces of metal are zinc and copper. With a sufficiently large resistance in the external circuit, any current flow will be small and departure from equilibrium values will be negligible. Consequently, the electrode potentials will be their standard values, from Table 9.1, and the potential in the solution will be uniform. From the table

$$\text{potential of Zn} - \text{potential of solution} = -0.76 \text{ V}$$

$$\text{potential of Cu} - \text{potential of solution} = +0.34 \text{ V}$$

Hence by subtraction,

$$\text{potential of Cu} - \text{potential of Zn} = +1.1 \text{ V}$$

We see therefore that current in the external circuit flows from the electrode with the more positive potential (Cu) to the one with the lower potential (Zn). Of course, we know that this corresponds to a flow of electrons in the opposite direction (see Fig. 9.8). Electrochemical reactions take place at both solution-metal interfaces:

$$\text{Zn} = \text{Zn}^{2+} + 2\varepsilon^- \tag{9.27}$$

generating electrons at the solution-zinc interface that flow through the external circuit to be consumed by

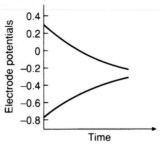

FIGURE 9.9 Expected change in the electrode potentials of the two metals of Fig. 9.8 with time due to flow of current.

$$Cu^{2+} + 2\varepsilon^- = Cu \tag{9.28}$$

at the solution-copper interface.

This is an example of a *galvanic cell* (one that produces electricity). Because copper ions are abstracted from solution, this might be a basis for a method of recovering copper from leaching solutions. The same reactions would occur if we dispensed with the external circuit and merely contacted the two metals. Indeed, copper is recovered commercially from solution by contacting with iron in the form of steel scrap. A glance at Table 4.1 shows that the reactions

$$Fe = Fe^{2+} + 2\varepsilon^- \tag{9.29}$$

and

$$Cu^{2+} + 2\varepsilon^- = Cu \tag{9.30}$$

should occur because iron has a lower standard electrode potential than copper, just as zinc does. The cementation reaction that is the sum of these two processes yields copper particles containing much residual iron which must be purified before use.

Returning to the example of Fig. 9.8, we see that as current continues to flow, the copper ion concentration diminishes while the zinc ion concentration increases, assuming no replenishment of solution. Consequently, we deduce [from equation (9.23)] that the electrode potential of the zinc increases while that of the copper diminishes, much as sketched in Fig. 9.9. Eventually, the two lines come together and the flow of current then ceases.[2]

Consider a slightly simpler example where the zinc ion concentration is held constant (at unit activity by some means that need not concern us) and the copper ion concentration allowed to diminish. What is the copper ion activity when current ceases to flow, that is, electrochemical equilibrium is achieved, at 25°C? The zinc electrode potential is fixed at $E_{zn} = E_{0Zn} = -0.76$ V. The copper electrode potential is

$$E_{Cu} = E_{oCu} + \frac{RT}{2F} \ln a_{Cu^{2+}} = 0.34 + \frac{RT}{2F} \ln a_{Cu^{2+}}$$

When these two electrode potentials become equal,

[2] In reality, this simple picture is complicated by the possibility of zinc ion concentration reaching some saturation value.

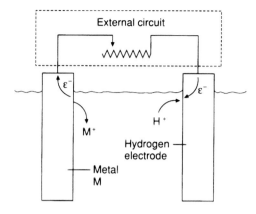

FIGURE 9.10 Metal electrode connected to a hydrogen electrode.

$$-0.76 = 0.34 + \frac{RT}{2F}\ln a_{Cu^{2+}}$$

that is,

$$a_{Cu^{2+}} = \exp\left[\frac{2F}{RT}(-0.76 - 0.34)\right]$$

$$= \exp\left(\frac{-2 \times 23066 \times 1.1}{1.98 \times 298}\right) = 3.7 \times 10^{-38}\, m$$

We see that copper is removed to extremely low concentrations. (At these low concentrations, activity would equal concentration, measured in molality.) Cementation reactions can be particularly effective in removing metal ions from solutions, especially if the two metals involved have widely different standard electrode potentials.

Finally let us consider another important example of the application of the galvanic cell: sensors for measuring concentration that are based on solid electrolytes. Solid electrolytes, as their name suggests, are solids that conduct electricity as a consequence of the movement of ions. One example is β-alumina which becomes conducting when sodium ions are introduced into its structure. Conduction is due to the movement of the sodium ions. Now consider two liquid alloys of aluminum-sodium that are separated by a wall of β-alumina containing sodium ions. If no current flows in the system then the activity of sodium ions will be the same everywhere in the β-alumina wall; we will call this activity a_{Na^+}. At the interfaces between the alloy and the solid electrolyte, potential differences develop that are just like those described earlier in this section and are given by the general form of Nernst's equation (9.25). For example if we call one of the alloys 1, then the potential difference between this alloy and the electrolyte is given by

$$E_1 = E_0 + \frac{RT}{F}\ln\left(\frac{a_{Na+}}{a_{Na1}}\right)$$

activity

where a_{Na1} is the activity of the sodium in alloy 1. [Note that n in the Nernst equation is 1 in this case because the equilibrium at the interface is established by the reaction

$$Na^+ + \varepsilon^- = Na$$

i.e. one electron is involved]. Writing down the similar equation for alloy 2 and subtracting we get the potential difference between the two alloys:

$$E_1 - E_2 = \frac{RT}{F} \ln\left(\frac{a_{Na2}}{a_{Na1}}\right)$$

If the activity coefficient is constant then this becomes

$$E_1 - E_2 = \frac{RT}{F} \ln\left(\frac{C_{Na2}}{C_{Na1}}\right)$$

Now concentrations of sodium in the alloy appear in the equation. Therefore, if the concentration of sodium in one alloy is known, then we can readily measure the potential difference and get the concentration in the other alloy.

This type of sensor, based on a "concentration cell" with a solid electrolyte, is now routinely used for measuring compositions in materials processing. For example the composition of steel is measured in this way during oxygen steelmaking as described in Chapter 8. These cells frequently measure oxygen potential and require a reference electrode of known oxygen potential, usually a mixture of a pure metal and its oxide. Another example, that might be closer to your experience, is the sensor that measures the gas composition in the exhaust system of most modern cars. The sensor feeds a signal to a microprocessor or some other controller that regulates the engine (for example the ignition timing) so as to reduce air pollution. This last example illustrates the fact that cells can be designed with solid electrolytes to determine *gas* composition and similar sensors are used for monitoring gas composition in furnaces and heat treating units common in materials processing. The CD with this book has an example on the use of these sensors.

9.4.2 Relationship Between Electrode Potential and Free Energy

Consider the cell depicted in Fig. 9.10, where an electrode of metal M is in the same solution as a hydrogen electrode. Let us suppose that the solution contains hydrogen ions and metal ions at unit activity and that there is a sufficiently large volume of electrolyte that we can carry out significant amounts of electrochemical reaction with negligible change in concentrations. Furthermore, let us suppose that the metal has a negative standard electrode potential, so there is a tendency for the electrochemical reactions shown in the figure to take place, driving electrons through the external circuit. Let us adjust the external circuit so that the current is infinitesimal and allow a sufficiently long time for 1 mol of metal ions to pass into solution. Because the current is infinitesimal the cell delivers the maximum energy to the external circuit, and if the pressure is constant, this energy will be the decrease in Gibbs free energy of the system, that is, the Gibbs free energy for the two reactions:

$$M = M^{n+} + n\varepsilon^- \qquad \Delta \tilde{G}_{RI} \tag{9.31}$$

TABLE 9.2 Standard free energies of formation
(kJ/mol)species in aqueous solution at 25°C

Cl^-	−131.
Co^{2+}	−51.
Cu^+	+50.
Cu^{2+}	+65.
F^-	+276.
H^+	0.
H_2O	−237.
H_2S	−27.3
HS^-	+12.6
S^{2-}	+92.4
I^-	−52.
Fe^{2+}	−85.
Fe^{3+}	−10.5
Pb^{2+}	−24.3
Mg^{2+}	−456.
Mn^{2+}	−223.
Ni^{2+}	−46.4
OH^-	−157.
Ag^+	+77.
SO_4^{2-}	−741.
HSO_4^-	−752.
U^{3+}	−520.
U^{4+}	−579.
Zn^{2+}	−147.

and

$$nH^+ + n\varepsilon^- = \frac{n}{2}H_2 \quad \Delta\tilde{G}_{RII} \tag{9.32}$$

As we shall see in Section 9.4.4, the amount of charge that is passed through the external circuit is nF ampere • seconds (where nF has a value of $n \times 96{,}484$ A • s per mol). Hence because this charge passes through a potential drop of E_o volts (the standard electrode potential for metal M), the electrical work done in the external circuit is nFE_o joules/mol, meaning that

$$nF E_o = \Delta\tilde{G}_{RI} + \Delta\tilde{G}_{RII} \tag{9.33}$$

By convention the standard free energy of reaction (9.32) is set to zero. Furthermore, $\Delta\tilde{G}_{RI}$ is simply minus the standard free energy of the reduction reaction [the opposite of reaction (9.31)]. Hence

$$nF E_o = -\Delta\tilde{G}_{R0} \tag{9.34}$$

which is a general result linking the standard electrode potential to the standard free energy of the electrochemical reduction reaction $\Delta \tilde{G}_{R0}$. It should be noted that to get the sign correct in using equation (9.34) it must be the standard free energy for the *reduction* reaction (electron addition to the ion) that is used; this yields the standard *reduction* potentials of Table 9.1, hence the name of these potentials and the convention used in the table of writing the reaction as reduction ones. Note that the standard free energy for an electrochemical reaction is the standard free energy that must be added to a system in order that, when the amount of reaction indicated takes place, the system be in the standard state before and after reaction. It is therefore just the standard free energy of reaction for chemical reaction that we met in Chapter 2 and can be calculated from the standard free energy of formation of reactants and products using equation (2.56). To do this we require information on the standard free energy of ions, and Table 9.2 provides some values. Note that the standard free energies of formation of elements appearing in electrochemical reactions are zero. It is left as an exercise for the reader to determine the (simple) connection between Tables 9.1 and 9.2.

We have arrived at equation (9.34) by envisioning a metal electrode connected to a hydrogen electrode, but the equation is really much more general than that. In its more general form,

$$nF\,E = -\Delta \tilde{G}_R \tag{9.35}$$

it provides the equilibrium potential E, given the sum of the free energies of the reactions taking place in the cell. For example, consider the zinc-air cell where the reactions are

$$Zn = Zn^{2+} + 2\varepsilon^- \tag{9.36}$$

and

$$\frac{1}{2}O_2 + H_2O + 2\varepsilon^- = 2OH^- \tag{9.37}$$

If the solutions contain ions at unit activity, the cell voltage is

$$E = -\frac{1}{2F}(\Delta \tilde{G}_{FZn^{2+}} + 2\Delta \tilde{G}_{FOH^-} - \Delta \tilde{G}_{FH_2O}) \tag{9.38}$$

greater than 1.1 V

9.4.3 Electrolytic Cells

Consider a modification to the experiment shown in Fig. 9.8; we shall introduce into the external circuit a device such as a dc power supply that will enable us to manipulate the potential difference between the two electrodes. The modified circuit is sketched in Fig. 9.11; the copper electrode is connected to the positive terminal of the power supply, and the zinc electrode is connected to the negative terminal. If (for solutions initially with unit activity of copper and zinc ions) we set the power supply so that the potential difference between its terminals is greater than 1.1 V (the voltage difference between the two metals in Fig. 9.8), then current flows in the direction *opposite* to that of Fig. 9.8. At the Cu solution interface the reaction

$$Cu = Cu^{2+} + 2\varepsilon^- \tag{9.39}$$

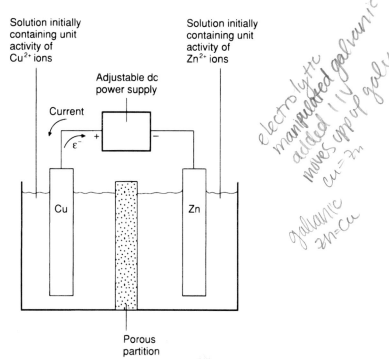

FIGURE 9.11 Electrolytic cell.

which is the opposite of reaction (9.28), occurs. Similarly, at the zinc electrode we have the opposite of reaction (9.27):

$$Zn^{2+} + 2\varepsilon^- = Zn \qquad (9.40)$$

We have also inserted a porous partition in the cell so as to avoid gross mixing of the two solutions on either side of the cell without preventing passage of ions through the partition sufficient to carry the cell current. In this way, at least initially, there will be negligible amounts of Cu^{2+} ion on the right-hand side of the cell and we can rule out the possibility of copper deposition, the reverse of reaction (9.39), occurring at the zinc electrode. The dc power supply serves to drive the cell away from electrochemical equilibrium. Initially, the dc power supply must impose a voltage greater than the difference between standard electrode potentials in order to do this. As an exercise the reader should decide what voltage is necessary for the power supply to impose after current has been passed for some time so as to change the solution composition.

The cell of Fig. 9.11 is an *electrolytic cell*. Such cells are commonly used to produce materials or chemicals (e.g., manganese dioxide or chlorine) or to store energy. In the last case the cell is an electrolytic one when it is being charged and a galvanic one on discharge (i.e., when the dc power supply is removed and the cell discharged through some external circuit, such as that of a flashlight or portable radio). Copper can be obtained from aqueous copper sulfate solutions, produced by leaching followed by solvent extraction: for example, by electrolysis between lead electrodes and copper

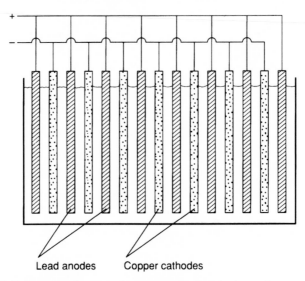

FIGURE 9.12 Schematic diagram of a copper electrowinning cell.

electrodes. The former are connected to the positive side of the power supply and are known as *anodes*; the latter are connected to the negative side and are known as *cathodes*. A typical cell is depicted in Fig. 9.12. The lead anodes are inert under the conditions prevailing in cells; consequently, the reaction at the anode solution interface is *not* the electrodissolution of the lead but the oxidation of water:

$$2H_2O = 4H^+ + O_2 + 4\varepsilon^- \tag{9.41}$$

The electrons produced by this reaction are driven through the external circuit by the dc power supply and react at the cathodes:

$$Cu^{2+} + 2\varepsilon^- = Cu \tag{9.42}$$

The copper cathodes, which start life as thin *starter sheets*, therefore grow by deposition of copper and are periodically removed and replaced by fresh starter sheets. For this cell to function, the dc power supply must supply a voltage that exceeds the difference of the electrode potentials corresponding to reactions (9.41) and (9.42). For example, if the solution contains unit activity of hydrogen and copper ions and the oxygen is released at atmospheric pressure, the voltage must exceed 1.23 - 0.34 V = 0.89 V.

Let us examine what happens if the solution is not replenished as electrolysis proceeds, so that the copper concentration drops. From equation (9.23) the potential of the copper electrode drops. Eventually, it reaches zero at a copper activity given by

$$0 = 0.34 + \frac{RT}{2F} \ln a_{Cu^{2+}}$$

that is, at 25°C,

$$a_{Cu^{2+}} = \exp\left(\frac{-0.34 \times 2 \times 23066}{1.98 \times 298}\right) = 2.1 \times 10^{-12} \, m$$

Although this concentration of copper is extremely low, it will eventually be reached and, in theory, the electrode potential is then zero, that is, exactly the value for the reduction of hydrogen by

$$2H^+ + 2\varepsilon^- = H_2 \tag{9.43}$$

(for unit activity of hydrogen ions in solution and hydrogen generated at 1 atm pressure). In reality, the electrode potential has to be made even lower than zero before significant amounts of hydrogen are generated. Nevertheless, we can still say that as we electrolytically strip metal ions from solution to lower and lower concentrations, there is an increasing tendency for reaction (9.43) to occur. Part of the electricity passed through the cell will then be "wasted" in the production of hydrogen rather than the production of metal. This leads to two important concepts to be examined in the following section.

9.4.4 Faraday's Law and Current Efficiency

If we examine the numerous electrochemical reactions so far presented in this chapter [(9.15),(9.24),(9.26–9.31)], we note a fixed relationship between the number of electrons involved in reaction (i.e., the number of electrons generated or consumed at the metal-solution interface) and the number of ions discharged or passed into solution. For example, in reaction (9.28), two electrons are required to discharge each copper ion because the charge on that ion is +2. In reaction (9.27), two electrons pass into the electrode as each zinc ion passes into solution. Hence we can write

$$\begin{matrix} \text{number of electrons consumed/} \\ \text{generated at electrode–} \\ \text{solution interface} \end{matrix} = \begin{matrix} \text{number of ions} \\ \text{discharged/created} \end{matrix} \times \begin{matrix} \text{charge} \\ \text{on ion} \end{matrix} \tag{9.44}$$

Now let us divide equation (9.44) by Avogadro's number N; this is the number of molecules in a mole, but it is also the number of atoms in a gram atom (a gram atom for any species being a mole of that species divided by the number of atoms in the molecule, that is, the number of grams numerically equal to the atomic weight). N is also, in our case, the number of ions in a gram atom. Hence we get

$$\frac{\text{number of electrons}}{N} = \begin{matrix} \text{number of gram atoms} \\ \text{in reaction} \end{matrix} \times \text{charge on ion} \tag{9.45}$$

We now divide and multiply the right-hand side by the atomic weight of the species in question. The first term on the right then becomes the mass of the species reacting:

$$\frac{\text{number of electrons}}{N} = \text{mass of species reacting} \times \frac{\text{charge on ion}}{\text{atomic weight}} \tag{9.46}$$

Now let us introduce *the gram equivalent weight*. In our case this is the mass of metal (or other chemical species that we are interested in) in grams that is numerically equal to the atomic weight divided by the charge on the ion. For example, for the copper ion, Cu^{2+}, the gram equivalent weight would be 63.54/2 = 31.77 g. Therefore, equation (9.46) becomes

$$\frac{\text{number of electrons}}{N} = \frac{\text{mass of species}}{\text{mass of 1 gram equivalent weight of species}} \quad (9.47)$$

or

$$\frac{\text{number of electrons}}{N} = \text{number of gram equivalents reacting} \quad (9.48)$$

Hence if we can count the number of electrons consumed or generated by reaction, we can use equation (9.48) to calculate the amount of electrochemical reaction taking place. The electrons flow to or from the electrode through the external circuit, and one way of "counting" them is to incorporate an ammeter in the circuit. An ammeter measures the *rate* of flow of electrons. In the simplest case when the current is constant, the number of electrons in equation (9.48) is

$$\frac{\text{current (amperes)} \times \text{time (seconds) for which current flows}}{\text{charge on electron [ampere} \cdot \text{seconds (coulombs)]}}$$

The charge on the electron is 1.6022×10^{-19} A•s. Substituting in (9.48) and using $N = 6.022 \times 10^{23}$, we get

$$\frac{\text{current} \times \text{time}}{96,484} = \text{number of gram equivalents reacting} \quad (9.49)$$

This is Faraday's law. The constant 96,484 (which has dimensions of ampere seconds per gram equivalent) is known as Faraday's constant and given the symbol F. It is the same constant as that appearing in the Nernst equation and equation (9.34), although care must be taken that the correct units are being employed. For example, 96,484 J/V gram equivalent can be used in the Nernst equation, when R is in J/mol K. If R is used in the older unit of cal/mol • K, F is 23,066 cal/V• mol. The reader is advised to check the units in employing F to ensure that the correct numerical value is being used.

Let us use Faraday's law to calculate the amount of copper deposited in a cell such as that of Fig. 9.12 in one day when the cell current is 10,000 A:

$$\text{number of gram equivalents} = \frac{10 \times 10^3 \text{ A} \times 24 \text{ h}}{96,484 \text{ A} \cdot \text{s} / \text{g equiv.}} \times \frac{60 \text{ min}}{\text{h}} \times \frac{60 \text{ s}}{\text{min}}$$

$$= 8955 \text{ gram equivalents}$$

Multiplying by the mass of 1 gram equivalent of copper, which we showed above is 31.77 g, we get

$$8955 \text{ x } 31.77 \text{ g of Cu} = 285.5 \text{ kg}$$

If we passed the same current for the same time through the cell of Fig. 9.11, we would expect 292.7 kg of zinc to be deposited, a result that the reader should verify. In reality, if we did this experiment, we should find that the quantity of zinc deposited was less than 292.7 kg, perhaps much less. Does this represent a failure of Faraday's law? In fact, it does not. The quantity of zinc falls short of expectations because of a second reaction, reaction (9.43), taking place at the cathode. If we measured the amount of hydrogen generated, converted it into gram equivalents, and added it to the gram equivalents of zinc deposited, we would obtain 8955, the number above. A clear statement of Faraday's law is therefore: *The amount of electrochemical reaction occurring at an electrode totals 1 gram equivalent for each 96,484 A s of electricity passed through the electrode.*

The word *total* serves to emphasize that any side reactions, as well as the principal one, must be allowed for in applying the law. This suggests the concept of *current efficiency.* For the example of zinc deposition,

$$\text{current efficiency} = \frac{\text{amount of zinc actually deposited} \times 100}{\text{amount calculated from Faraday's law}} \qquad (9.50)$$
$$\text{ignoring side reactions}$$

The current efficiency would be an important measure of the performance of any process for electrolytically producing a metal (or any other material). Typically, the current efficiency would depend on impurities in the electrolyte and on operating parameters (such as the cell current) as well as on the nature of the principal electrochemical reaction. For example, in the commercial electrodeposition of copper, there is little difficulty in running the cells so that the current efficiency is close to 100%. In the electrolytic production of zinc, the electrolyte must be purified and the cells carefully operated to achieve current efficiencies greater than 50%.

9.4.5 Kinetic Considerations

In Section 9.4.3 we examined the cell of Fig. 9.11 and reasoned that if the electrolyte were not replenished, then as electrolysis proceeds, the activity of copper would eventually drop to a value where hydrogen evolution would start according to equation (9.43). For an electrolyte containing unit activity of hydrogen ion, we calculated that this would occur at a copper activity of 2.1×10^{-12}. The reader is asked to calculate similarly the zinc ion activity in the cell of Fig. 9.11 at which hydrogen evolution should start if the hydrogen ion activity is unity and the temperature is 25°C. The answer is 7.0×10^{25} molality. Recognizing that we have picked as a standard state (in setting up Table 4.1) a 1 *m* solution of the ion (i.e., that an activity of 1 corresponds to a 1 *m* solution of zinc ion), we see that the calculated activity is enormous! The implication is that for all practical concentrations of zinc ions, reaction (9.40) should *not* take place at the cathode on electrolyzing the solution, but rather the evolution of hydrogen. What has gone wrong? We assure the reader that zinc can actually be deposited in the cell of Fig. 9.11. Indeed, zinc plants throughout the world carry out this electrochemistry daily at the rate of about 100 tons per plant.

The explanation lies in the fact that the Nernst equation is a *thermodynamic* equation that describes what will happen if the electrolysis is carried out so as to maintain *equilibrium. A* complete understanding of practical electrolysis can be achieved only if we also consider the *kinetics* of the reactions involved. Fortunately for the producers of zinc, the rate of hydrogen evolution in their cells, although not zero, is low; consequently, it is possible to electrodeposit zinc from acidic solutions. If catalysts for the hydrogen evolution reaction are present, hydrogen is indeed evolved in copious amounts and the deposition of zinc becomes impractical. The zinc ions in solution in electrolytic

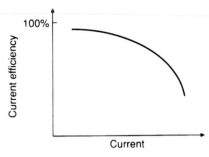

FIGURE 9.13 Current efficiency as a function of current for typical electrowinning cells.

zinc plants are produced by leaching zinc oxide calcine with sulfuric acid and impurities that are catalytic for hydrogen; notably, nickel, cobalt, antimony, and arsenic enter solution at this point. The solution is purified prior to electrolysis by cementation with zinc powder so as to bring these impurities to extremely low levels (parts per *billion* in the case of antimony and arsenic).

This is yet another example of thermodynamics yielding a result that, although correct, within the (equilibrium) limitations of thermodynamics, is misleading. Consideration of the kinetics involved gives a more practical result.

What is the difference between electrodeposition of zinc and copper from aqueous solution that makes the former prone to hydrogen evolution and loss of current efficiency? The calculations employing the Nernst equation tell us the answer. Thermodynamically, the onset of hydrogen evolution in a solution containing the metal ion and unit activity of hydrogen ions should occur when the metal ion activity drops to

$$\exp\left(\frac{-nFE_0}{RT}\right)$$

For copper, with $E_0 = +0.34$ V, this is a small number, whereas for zinc $(E_0 = -0.76$ V), this is an extremely large number, and it is only by the grace of electrochemical kinetics that zinc can be deposited at all from aqueous solution. Subject to possible modification by kinetic considerations, we can conclude that metals are most easily deposited from aqueous solutions if their standard electrode potentials are high. As we have seen, copper is readily deposited, as are lead and tin. Nickel and cobalt pose some difficulty, particularly if the solution is acid (i.e., the hydrogen ion activity is high), due not only to their standard electrode potentials but also to their catalysis of hydrogen evolution; hence kinetics is not helpful in the case of these metals. Cadmium, chromium, and iron can be electrodeposited, but when standard electrode potentials drop below -1 V, electrodeposition from aqueous solutions becomes impossible. Titanium, aluminum, magnesium, and sodium are examples of important metals that cannot be deposited from aqueous solution. However, as we shall see in the next section, these metals can be obtained by electrolysis of molten salts.

Of course, the kinetics of a reaction at an electrode will depend on the electrode potential. The net rate of reaction is zero at the equilibrium electrode potential given by the Nernst equation and increases rapidly as we move away from this potential by, for

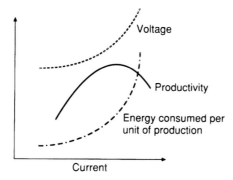

FIGURE 9.14 Performance of typical electrowinning cells.

example, imposing a different potential using the dc power supply of Fig. 9.11. Consequently, as we apply a greater and greater voltage across the cell in that figure by, for example, attempting to drive more current through the cell, the hydrogen evolution reaction will accelerate, eventually accounting for a significant fraction of the current and lowering the current efficiency. Many electrolytic cells therefore show a current efficiency that declines with increasing current, as sketched in Fig. 9.13.

Faraday's law tells us that the amount of metal deposited per unit time (i.e., the rate of production) is proportional to the current if there are no side reactions. In the presence of side reactions the production rate is proportional to the product of the current and current efficiency. The net result is that the productivity of a typical electrolytic cell behaves as shown schematically in Fig. 9.14.

The electrical energy consumed in electrolytic cells frequently represents a significant operating cost. The rate of electrical energy consumption is simply the product of current through the cell and voltage across it. The latter increases with increasing current (see Fig. 9.14), and therefore the energy cost increases sharply with current. An important index of cell performance is the energy cost per unit of production (e.g., kWh/kg of metal) and the reader will be able to deduce that this index is proportional to

$$\frac{\text{cell voltage} \times \text{current}}{\text{current} \times \text{current efficiency}} = \frac{\text{cell voltage}}{\text{current efficiency}}$$

This index of cell performance is also sketched in Fig. 9.14. As an exercise the reader is asked to consider what factors determine the optimum economic current for a cell with the characteristics of Fig. 9.14.

9.4.6 Technology

Electrolytic cells encountered in electrometallurgy are classified into electrowinning cells and electrorefining cells. *Electrowinning cells* obtain metal by the reduction of ions that enter the cell in solution (or by dissolving salts or oxides into the electrolyte within the cell). Fig. 9.12 is an example. A typical copper electrowinning cell would have an inflow of copper sulfate solution, perhaps from an upstream solvent extraction system, with an outflow (of a similar volume of solution but of lower copper content) to additional cells where more copper would be abstracted.

The cathodes are inserted into the cell as thin "starter sheets," usually produced by electrodeposition onto titanium cathodes in separate cells; the starter sheets can then be peeled from these titanium cathodes. Typically, these electrodes would be 1 to 2 m on edge and would be grown to a thickness of a few centimeters before removal from the cell. In some modern electrowinning plants "cathode blanks" are used instead of starter sheets. These cathodes are made of some metal (e.g. stainless steel) that is not corroded by the electrolyte. The metal to be electrowon is deposited on the blanks. When the deposited metal is thick enough the blank and metal are removed from the cell, the two separated by a mechanical device and the blank returned to the cell. The inert anodes, usually of a lead alloy, are interleaved between the cathodes, and oxygen evolution (by oxidation of water) is the usual anodic reaction.

Rather similar cells are used in the electrowinning of zinc and other metals that can be electrodeposited from aqueous solutions. Note that the cathodes in each cell are connected in parallel (as are the anodes), although it is usual to connect the cells in series; that is, the cathodes of one cell are connected to the anodes of its neighbor to one side, while the anodes (of the first cell) are connected to its neighbor on the other side. Cells of this type would have voltages of a few volts with currents of 10 to 20 kA. The *current density,* that is, the current per unit area of electrode, might range from 200 to 500 A/m². Electrolytes would usually be 1 to 5 molar in the salt of the metal to be deposited. Note that the anodic reaction (9.41) generates acid. Consequently, the solution becomes more acidic as it moves through the cells. This is advantageous in that a strongly acid solution is generated for recycle to an upstream unit such as solvent extraction, but disadvantageous in that for some metals (e.g., zinc) it promotes hydrogen evolution.

One of the most important electrowinning technologies is shown in Fig. 9.15. This is the Hall-Héroult cell, which is employed in the production of virtually all of the world's aluminum. This cell is a prodigious consumer of electrical energy, accounting for approximately 4% of the electricity generated in the United States.

The horizontal dimension of the cell shown in section in Fig. 9.15 is approximately 3 m; the other horizontal dimension would be roughly 10 m. The cell consists of a steel shell with a refractory lining within which is a thick lining of carbon blocks and baked carbon paste. This conductive carbon lining holds a pool of molten aluminum above which is a pool of a molten sodium-aluminum fluoride, both being at a temperature of 950 to 960°C. The molten salt, known as *bath* in the terminology of the aluminum industry, has blocks of carbon dipping into it. These carbon blocks, which serve as anodes, are connected by steel pins to aluminum *anode rods* which suspend them from a conductor *(anode bus)* passing over the top of the cell. A typical cell would have 10 to 30 anodes arranged in two rows.

Current flows down the anode rods, through the anodes, then molten salt, then molten aluminum into the carbon lining, whence it exits the cell via steel *collector bars* inserted into the carbon blocks. For a typical modern cell this current might be higher than 200 kA. The bath is capable of dissolving aluminum oxide, and this oxide (from the Bayer process discussed in Chapter 7) is fed intermittently onto the frozen "crust" of the molten bath which it enters when the crust is mechanically broken. For present purposes the anodic reaction can be thought of as

$$2O^{2-} + C = CO_2 + 4\varepsilon^- \qquad (9.51)$$

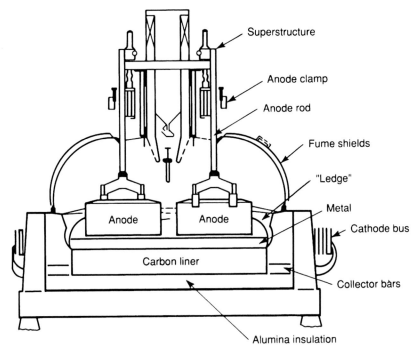

FIGURE 9.15 Cross section of the Hall-Héroult cell used in aluminum production. (Reprinted with permission from S. K. Banerjee and J. W. Evans. Measurements of Magnetic Fields and Melt Flow in a Physical Model of a Hall-Héroult Cell. Copyright 1990 *Metall. Trans., 21B*, 59-69 (1990).]

Hence the bottoms of the anodes are consumed as electrolysis proceeds and they must periodically be adjusted downward and replaced with new anodes every 10 to 20 days. An alternate anode is a *self-baking anode* (or *Soderberg anode),* which is a large anode (one per cell) formed within the cell itself by the baking of a carbon paste contained within a steel shell. Baking is brought about by heat from the cell itself and fresh paste is added periodically to the top of the anode. The anodes of Fig. 9.15 are formed by baking carbon pastes, coke, and other ingredients in a separate operation, and such a cell is therefore said to have *prebaked anodes.* Carbon dioxide generated by reaction (9.51) leaves the top of the cell and is vented to the air after passing through a scrubbing system to remove particles. The cathodic reaction can be thought of as

$$Al^{3+} + 3\varepsilon = Al \tag{9.52}$$

and occurs at the interface between the bath and molten metal. The latter therefore grows and some of the aluminum is siphoned out of the cells, typically at 2-day intervals. Cell voltages are usually between 4 and 4.5 V and a few hundred cells would be arranged in series to form a *potline.* A representative aluminum plant would have a few such potlines and produce perhaps a 100,000 tons of metal per year.

Hall-Héroult cells operate at current densities in the range 5 to 10 kA/m^2, that is, more than one order of magnitude greater than the cells encountered in electrowinning

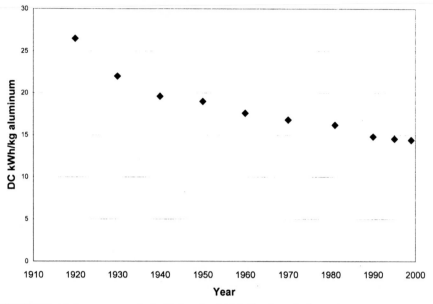

FIGURE 9.16 Improvements in Hall cells during the twentieth century.

from aqueous solutions. That is, the heterogeneous electrochemical reactions are proceeding very much faster in the former cells than in the latter and it is instructive to examine why. The reader will recall from Chapter 3 that rates of heterogeneous reactions are increased by (1) providing sufficient agitation to minimize the limitations of mass transport, and (2) operating at high temperatures so as to promote the chemical kinetics. The high temperature of the Hall-Héroult cell obviously satisfies (2), and within this cell, vigorous agitation is provided by the carbon dioxide bubbles evolved at the anodes, as well as by electromagnetic forces. The latter forces arise from the interaction between the current passing through the cell and the magnetic field generated by that current, in much the same way that forces are generated in an electric motor.

Present-day cells would have current efficiencies in the range 90 to 95%. The departure from 100% is *not* due to any side reaction at the cathode comparable, say, to hydrogen evolution in zinc electrowinning. Rather the inefficiency is due to a slight solubility of the aluminum product in the bath. One-tenth to one-twentieth of the aluminum produced therefore redissolves. The detailed chemistry is still uncertain, but this dissolved aluminum is then reoxidized, probably by carbon dioxide bubbles near the anode. The production of aluminum is therefore 5 to 10% less than expected from Faraday's law.

While present-day Hall-Héroult cells consume much electrical energy, they are superior in this regard to older cells, as shown by the historical data of Fig. 9.16.

EXAMPLE

A Hall-Héroult cell has a current of 280 kA, a current efficiency of 94.5%, and produces aluminum with a dc electrical energy consumption of 12.9 kWh per kilogram of aluminum. Determine (a) the cell voltage and (b) the heat losses from the

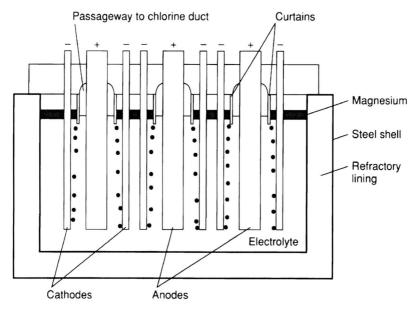

FIGURE 9.17 Magnesium electrowinning cell.

cell (other than in the hot gases leaving the electrolyte or in the aluminum).

Solution

(a) According to Faraday's law, the cell should produce

$$\frac{280 \times 10^3\,A \times 3600s\,/\,h \times 0.945}{96,484A \cdot s\,/\,g.equiv.}$$

gram equivalents of aluminum per hour. Aluminum has an atomic weight of 27, and three electrons are involved in the reduction reaction; hence

$$1 \text{ gram equivalent} = 9g = 9 \times 10^{-3} \text{ kg}$$

Hence

$$\text{production rate} = \frac{280 \times 10^3 \times 3.6 \times 10^3 \times 0.945 \times 9 \times 10^{-3}\,kg\,/\,h}{96.484 \times 10^3} = 88.85 \text{ kg / h}$$

Hence

$$88.85 \times 12.9 \text{ kW} = \text{electrical energy dissipated}$$

$$= 1146 \times 10^3 \text{ W} = \text{cell current} \times \text{cell voltage}$$

or

$$\text{cell voltage} = \frac{1146 \times 10^3}{280 \times 10^3} = 4.09V$$

(b) According to Grjotheim et al. (1977) the thermodynamic enthalpy requirement for the production of aluminum at 1250 K is

$$\frac{139.18}{x} + 477.24 \text{ kJ / mol aluminum}$$

where x is the current efficiency in percent/100. This includes enthalpy to bring reactants from a room temperature of 298 K, as well as the enthalpy of reaction. It also makes allowance for the back reaction of part of the aluminum (to aluminum oxide and carbon monoxide)—hence the appearance of the current efficiency in the expression.

The production rate from part (a) is

$$88.5\text{kg / h} \equiv \frac{88.5}{27} \times \frac{10^3 \text{molAl}}{h} \equiv \frac{88.5 \times 10^3}{27 \times 3.6 \times 10^3} \frac{\text{molAl}}{s}$$

Hence the thermodynamic requirement is

$$\left(\frac{139.18}{0.945} + 477.24\right) \times 10^3 \frac{W \cdot s}{\text{molAl}} \times \frac{88.5}{27 \times 3.6} \frac{\text{molAl}}{s} = 569 \times 10^3 W$$

From part (a) the actual energy usage is 1146 kW; therefore, energy loss from the cell is 577 kW. This is rather a substantial heat loss but is representative of modern cells. An additional amount of heat leaves in the form of the hot gases and aluminum exiting the cell.

The electrolytic production of magnesium provides us with another example of an electrowinning cell where the electrolyte is a molten salt, in this instance a mixture of molten chlorides. A major difference from the Hall cell is that magnesium is obtained by electrolysis of the chloride rather than the oxide. Figure 9.17 is a sketch of a cell that might carry a current in excess of 100 kA. Steel cathodes and graphite anodes are used. The intended overall reaction is

$$MgCl_2 = Mg + Cl_2 \tag{9.53}$$

and we see that, in contrast to the overall reaction of the Hall cell, the anodes are not consumed. This freedom from anode consumption can be achieved in cells fed with *anhydrous* magnesium chloride. With such a feed, anodes are consumed only *slowly* (by oxygen-containing impurities in the cell feed and by air leaking into the cell).

Unfortunately, the production of anhydrous magnesium chloride is not straightforward. Dehydrating the tetrahydrate, formed by crystallization from solution, yields oxides and oxychlorides as well as the chloride, the first two being formed in greater amounts during the later stages of dehydration. The usual method of producing the anhydrous chloride is by chlorination of the oxide in the presence of carbon:

$$MgO + Cl_2 + C = MgCl_2 + CO \tag{9.54}$$

This chlorination is carried out at temperatures above the melting point of $MgCl_2$ (perhaps as high as 1000°C), producing a liquid $MgCl_2$ that can be poured into the electro-

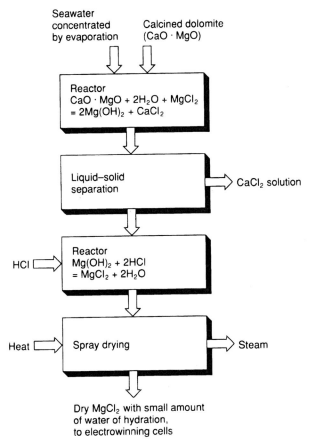

Seawater concentrated by evaporation

Calcined dolomite (CaO · MgO)

Reactor
$CaO \cdot MgO + 2H_2O + MgCl_2$
$= 2Mg(OH)_2 + CaCl_2$

Liquid–solid separation → $CaCl_2$ solution

HCl → Reactor
$Mg(OH)_2 + 2HCl$
$= MgCl_2 + 2H_2O$

Heat → Spray drying → Steam

Dry $MgCl_2$ with small amount of water of hydration, to electrowinning cells

FIGURE 9.18 Schematic diagram of the production of magnesium chloride from seawater.

lytic cell. The cost of energy, labor, and capital for production of the anhydrous chloride contributes significantly to the cost of production of magnesium by this route.

An alternative, practiced by the Dow Chemical Co. until recently, is to charge partly hydrated magnesium chloride (containing 1.25 to 1.5 waters of hydration) to the cell. Anode consumption is increased by about one order of magnitude due to reaction of the graphite with the oxygen of the hydrated chloride. Anodes must be adjusted daily and replaced periodically. Furthermore, magnesium oxide formed from the hydrated chloride has little solubility in the electrolyte; the oxide sinks to the bottom of the cell, where it forms a sludge that is removed daily. These difficulties of cell operation are compensated for by the fact that the partly hydrated chloride can be produced in a way that is relatively simple compared to the anhydrous chloride. Figure 9.18 is a block diagram of a scheme for producing the hydrated chloride from seawater.

Note that whether hydrated or anhydrous chloride is used as a feed, it is important to keep the magnesium produced at the cathode separate from the chlorine produced at the anode. The cell temperature is in excess of 700°C, and the molten magnesium pro-

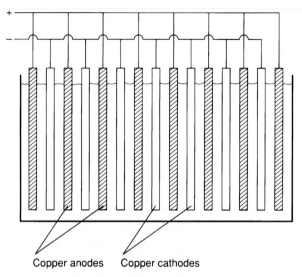

FIGURE 9.19 Sketch of copper electrorefining cell.

duced is less dense than the electrolyte and rises off the cathode. A *curtain is* interposed between the anode and cathode, dipping into the electrolyte and separating the two cell products. The molten magnesium is periodically ladled or siphoned from the cell while the chlorine is drawn off via a duct system. Electrical energy consumption ranges up to 20 kWh/kg of magnesium produced.

In *electrorefining cells,* the metal that ultimately becomes the cathode enters the cell in the form of an anode rather than in solution. Figure 9.19 is a sketch of a cell used extensively to refine copper. The impure copper, produced by the smelting operations described in Chapter 8, is cast into slablike anodes 1 to 2 m on edge and a few centimeters thick. These anodes are suspended in an acid electrolyte between cathodes that start life as thin starter sheets, just as in copper electrowinning cells. As in electrowinning cells, the anodes are connected to the positive side of a dc power supply (via other cells placed in series), and the cathodes are connected to the negative side. Here the similarity to electrowinning ends; the anodes, instead of being inert, now undergo electrodissolution according to

$$Cu = Cu^{2+} + 2\varepsilon^- \tag{9.55}$$

The copper ions thereby generated find their way to the cathode and are electrodeposited by the reaction

$$Cu^{2+} + 2\varepsilon^- = Cu \tag{9.56}$$

The anodes are replaced when they have become thin enough that there is danger of them falling apart, and the cathodes are replaced by fresh starter sheets when they have grown to a few centimeters' thickness. Principal impurities in the copper from the pyrometallurgical operations are iron and precious metals such as gold and silver. The latter do not anodically dissolve but fall as particles to the bottom of the cell, where they form an "anode slime" which is harvested from the cell periodically and pro-

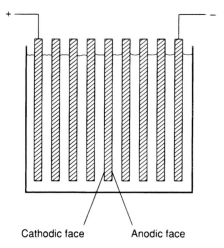

Cathodic face   Anodic face

FIGURE 9.20 Electrorefining cell using bipolar electrodes.

cessed to recover the precious metal values. Many copper plants derive a significant portion of their revenues from the anode slime. The iron impurities readily dissolve at the anode, but as a glance at Table 9.1 shows, copper is much more easily electrodeposited from aqueous solutions than iron. Consequently, the iron remains in the electrolyte and the cathodes generated are nearly pure copper.

An alternative design of electrorefining cell is shown in Fig. 9.20. This design employs *bipolar electrodes*. The flow of electricity in the cell is shown from left to right. Only the end electrodes are connected to the external circuit; the rest are insulated, apart from their contact with the electrolyte. Consider any one of the insulated electrodes. Current inflow through the left-hand face will result in deposition of pure copper on that face by cathodic reaction. On the right-hand face anodic dissolution of impure copper (and iron) will occur. These bipolar electrodes will thereby be converted to pure copper. An advantage over the cell of Fig. 9.19 is immediately apparent. Each 96,484 A s of electricity passed through that cell refines only 31.77 g of copper, while the same quantity passed through the cell of Fig. 9.20 will refine eight times as much (or even more if more bipolar electrodes are used). However, this advantage is more apparent than real. The current density on the bipolar electrodes must be similar to that on the electrodes of Fig. 9.19 (typically less than 300 A/m^2 to produce dense pure electrodes). Consequently, the current through the former cell must be one-eighth that of the latter and the productivities of the cells are equal. Furthermore, the voltage of the cell with the bipolar electrodes will be approximately eight times higher, leading to an electrical energy consumption rate (current x voltage) that is virtually identical in the two cells. Because the cost of electrical energy is the cost of kilowatt seconds, rather than ampere seconds, the cells are seen to have similar electrical energy costs. For these reasons and other reasons beyond the scope of this book, cells with bipolar electrodes have been less widely adopted than one might at first expect.

Electrorefining of other metals such as gold and silver is practiced. In some instances, such as electrorefining of nickel, the anode is nickel matte (impure nickel sulfide) rather than impure metal. The cell is then split into anode and cathode chambers by a diaphragm that will pass current (ions) but not solution, and impure *anolyte* (e.g., containing suspended sulfur produced during the anodic oxidation of the matte) is pumped from the anode chamber via a purification system to the cathode chamber.

Electrorefining using molten salts is also feasible. For example, in the Hoopes cell a layer of molten salt lies between a lower layer of impure molten aluminum (made dense by the addition of copper and connected to the positive side of the power supply) and an upper cathode of pure molten aluminum.

9.5 Concluding Remarks

This chapter started by examining how the chemistry of aqueous solutions can be exploited to transfer metal values from minerals into solution and how those solutions can be purified. In a sense this field of hydrometallurgy is in its infancy compared to the older field of pyrometallurgy, and significant advances can be expected in the years ahead. These advances are likely to be in novel chemistry rather than novel engineering, or at least in novel applications of knowledge from inorganic chemistry. An example is that of the newer solvent extraction reagents, which have had a marked impact on the production of copper.

Much of the chapter was devoted to the recovery of metal values from hydrometallurgical solutions, particularly by electrolysis (electrowinning). Other aspects of electrometallurgy were treated: electrolysis of molten salts to produce aluminum and magnesium and the refining of metals by electrolysis. We have found it necessary to provide an overview of electrochemistry, particularly of the central concepts of standard electrode potential, Nernst's equation, Faraday's law, and current efficiency. For heuristic reasons we have treated only simple electrochemical reactions in that overview (e.g., simple metal ions fully discharged at a cathode). Nevertheless, the reader should have little difficulty in handling more complicated situations: for example, the meaning of a gram equivalent when the ferric ion (Fe^{3+}) is reduced to ferrous ion (Fe^{2+}) rather than being fully discharged, or the meaning of a gram equivalent of hydroxide ion if it is discharged.

Certain aspects of electrometallurgy (e.g., the electrodeposition of coatings and aqueous corrosion) are beyond the scope of this book but may be glimpsed from the material presented here.

LIST OF SYMBOLS

a_i, a_i^s	activity of species i, at the metal-solution interface
a_ε	activity of electrons
E, E_o	electrode potential, standard electrode potential
Eh	oxidation potential
F	Faraday's constant
$\Delta \tilde{G}_{R0}$	standard free energy of reaction
N	Avogadro's number
n	number of electrons involved in reaction
n_i	charge on ion i
R	gas constant

T temperature
μ_i, μ_i^o electrochemical potential of species i, value at standard state

PROBLEMS

9.1   What are the advantages and disadvantages of air agitation over mechanical agitation? (See Fig. 9.3.)

9.2   In the text we discuss the difficulty of measuring an absolute value for an electrode potential; in any electrical circuit there must be at least two electrodes. Similarly, in any solution there must be at least two types of ions, and it is not possible to measure independently the free energy of formation of only one ion. Consequently (as indicated in Table 9.2), the standard free energy of formation of the hydrogen ion is arbitrarily set to zero. By considering the reaction taking place at the hydrogen electrode and equation (9.34), deduce the value for the standard free energy of formation of the electron. Use this value and equation (9.5) to calculate the equation of the line on the Pourbaix diagram (at 25°C) for

$$Cu_2S + 4H_2O = 2Cu^{2+} + SO^{4-} + 8H^+ + 10\varepsilon^-$$

that is, the line separating the predominance region for Cu_2S from that for Cu^{2+} (the activity of this ion being 0.01 M). Start by writing down an equation for the equilibrium of this reaction and recognizing that the activity of the electron appearing in this equation is linked to Eh by equation (9.5). Assume that in the pH range of interest the activity of the sulfate ion is 0.1 M. The standard free energy of formation of Cu_2S at 25°C is -86.1 kJ/mol.

9.3   A waste stream from a plant is an aqueous solution containing mainly sulfuric acid, and flows at 100 gallons (U.S.) per minute. It is to be neutralized by contacting it with lime before discharge into a nearby river. The calcium sulfate produced by reaction will be dumped outdoors. Plot the kilograms of lime required per day as a function of the pH of the stream discharged to the river over the pH range 3 to 7 for a waste stream pH of 2. Assume that all the lime reacts to form insoluble calcium sulfate and that the sulfuric acid in the waste stream is fully dissociated into hydrogen ions and sulfate ions. Plot a second curve for a waste stream of pH 1. Suppose it is found that the lime needed to neutralize the waste stream is more than that shown in your plot. Suggest causes for the excess lime consumption and experiments that you might carry out to test these hypotheses. The lime is a significant cost to the plant. One day the purchasing manager reports that he can get a great deal on calcined dolomite ($CaO \cdot MgO$). Would this be suitable for the neutralization?

9.4   A slurry from a cyanide leaching operation is to be washed with water in the countercurrent thickening operation sketched here. The slurry entering the operation is 3% solids (by volume) and the underflow stream from each thickener is 30% solids. The overflow streams are practically free of solids and at 10 L/s. The aqueous solution of the slurry entering the operation contains 300 ppm of sodium cyanide; the wash water is free of cyanide. Calculate the ppm of cyanide in the underflow stream leaving thickener II if the aqueous phases in both streams leaving a thickener are identical in composition.

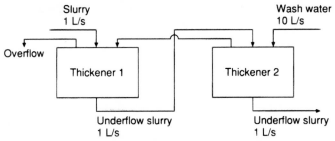

9.5 You are faced with the task of finding an extractant to be used in solvent extraction of rare earth salts from an aqueous solution containing a number of unwanted metal compounds. List the properties that you would look for in your search.

9.6 One inexpensive way of recovering (impure) copper from a solution by leaching is by contacting the solution with steel scrap, whereupon the cementation reaction

$$Cu^{2+} + Fe = Fe^{2+} + Cu$$

takes place. An aqueous solution containing 3 g of copper/liter is contacted with an excess of steel scrap (assumed here to be pure iron). Calculate the amount of copper and the amount of iron in solution when equilibrium is achieved at 25°C.

9.7 A Hall-Héroult cell operates at 4 V and a current of 160 kA; its current efficiency is 92%. It has 16 anodes, each of which has a bottom face measuring 100 cm × 150 cm. The horizontal dimensions of the aluminum pool are 9 m × 3.5 m. Calculate:

(a) The necessary feed rate of alumina in kg/day.

(b) The rate of increase in depth of the aluminum pool in mm/h.

(c) The rate of consumption of the carbon anodes (in mm/h) assuming that they are nonporous carbon with a density of 2.2 g/cm³, that consumption is only from their bottom faces, and that CO_2 is the product of the con sumption reaction.

(d) The minimum cell voltage at 960°C if the standard free energies of formation are as follows (at 960°C):

$$Al_2O_3 = -1287 \text{ kJ/mol} \quad CO_2 = -395 \text{ kJ/mol}$$

(e) The ratio of the energy actually consumed by the cell to that theoretically necessary to produce aluminum at 960°C (and 100% current efficiency) from alumina and carbon at room temperature. Standard heats of formation at 960°C are

$$Al_2O_3 = -1688 \text{ kJ/mol} \quad CO_2 = -394 \text{ kJ/mol}$$

and the specific heats can be assumed constant at

$$Al_2O_3 = 27.5 \text{ J/K} \cdot \text{mol} \quad C = 4.2 \text{ J/K} \cdot \text{mol}$$

9.8 Compare the theoretically necessary energy of Problem 9.7(e) with the number you obtain by multiplying the answer to Problem 9.7(d) by the cell current. What would happen to the temperature of a cell operated adiabatically at the minimum cell voltage of Problem 9.7(d)? What is the minimum cell voltage required to maintain temperature?

9.9 The following table gives some results from experiments in which the current efficiency of zinc deposition in an electrowinning operation was determined at various current densities (amperes per square meter of cathode surface area).

Current density (A/m^2)	Cell voltage (V)	Current efficiency (%)
300	1.5	92
400	2.3	88
500	3.0	82
600	3.9	72
700	4.8	52
800	5.8	10

Plot the energy consumed (in kWh/kg Zn) versus the production rate (kg Zn/day) if the cathode area in the plant is 10,000 m^2. If the cost of electrical energy is 2 cents/kWh and the other costs of the operation (labor, heating, building depreciation, etc.) are fixed at \$12,000 per day, what is the current density that minimizes the cost of electrowinning 1 kg of zinc?

9.10 At the startup of a copper electrorefining operation, copper anodes are placed with copper starter sheets (cathodes) in an aqueous electrolyte containing only sulfuric acid, and the current is turned on. At the moment the current starts to flow, will the current efficiency at the cathodes (for copper deposition) equal that at the anodes? If not, why not? What happens to the current efficiencies as time elapses, assuming negligible amounts of electrolyte removed from the cell?

REFERENCES AND FURTHER READINGS

BARD, A. J. and FAULKNER, L., *Electrochemical Methods: Fundamentals and Applications*, John Wiley and Sons, New York, 1980.

BOCKRIS, J. O M., B. E. CONWAY, E. YEAGER, and R. E. WHITE, eds., *Comprehensive Treatise of Electrochemistry, Vol. 2, Electrochemical Processing,* Plenum Press, New York, 1981.

FRAY, D. J., *The Use of Solid Electrolytes as Sensors for Applications in Molten Metals, Solid State Ionics, Vols. 86–88,* p 1045, 1996.

FRAY, D. J., *Potentiometric Gas Sensors for Use at High Temperatures, Materials Science and Technology, Vol. 16,* p 237, 2000.

GRJOTHEIM, K., C. KROHN, M. MALINOVSKY, K. MATIASOVSKY, and J. THONSTAD, Alu*minum Electrolysis: The Chemistry of the Hall-Héroult Process,* Aluminium Verlag GmbH, Dusseldorf, West Germany, 1977.

GRJOTHEIM, K., and B. WELCH, *Aluminum Smelter Technology,* Aluminium-Verlag

GmbH, Dusseldorf, West Germany, 1980.

HABASHI, F., *Principles of Extractive Metallurgy,* Vol. 2, *Hydrometallurgy,* Gordon and Breach, New York, 1970.

MINE, F., *Electrode Processes and Electrochemical Engineering,* Plenum Press, New York, 1985.

JACKSON, E., *Hydrometallurgical Extraction and Reduction,* Ellis Horwood, Chichester, England, 1986.

PITZER, K. S. and BREWER, L., *Revision of Thermodynamics* by G. N. Lewis and M. Randall, 2nd ed., McGraw Hill, New York, 1961.

PLETCHER, D., *Industrial Electrochemistry,* Chapman and Hall, London, 1982.

HABASHI, F., *Principles of Extractive Metallurgy; Amalgam and Electrometallurgy,* Métallurgie Extractive Québec, Sainte-Foy, Québec, 1998.

CHAPTER 10

Refining, Solidification, and Finishing of Metals

10.1 Refining and Alloying of Metals

10.1.1 Types of Impurities and Generalizations Concerning Their Removal

Consider a metal produced by one of the pyrometallurgical operations described in Chapter 8. The metal would be molten and in a few instances it may be of sufficient purity to cast (solidify) immediately. Usually, however, the metal will contain sufficient impurities that a further refining step must be carried out to remove these impurities. This point (prior to the first solidification) may also be a good one at which to add other metals to produce an alloy. Finally, this might be an appropriate point to add recycled metal, because any impurities (e.g., paint or corrosion products) may be removed along with those present in the virgin metal.

It will be useful to categorize impurities in three ways:

1. Particles or droplets of oxide, and so on, that might be separated by physical techniques such as filtration.
2. Dissolved gases that may also be removed by physical methods.
3. Metals (or nonmetallic species such as carbon or silicon) in elemental form that are *in solution* in the metal.

It is obvious that impurities in category (1) must be removed *prior* to solidification, their removal afterward being virtually impossible. Less obvious is the need to remove impurities in the other categories before solidification. After solidification the latter impurities could be removed only by solid-state diffusion, a relatively slow process. As we saw in Chapter 6, an estimate of the time required to bring about a significant adjustment in the composition of a solid by diffusion is L^2/D, where L is the distance over which the diffusion takes place and D is the diffusion coefficient of the impurity in the solid. L would be one-half of the smallest dimension of the solid (e.g., for a 2-cm-thick strip of metal, $L = 1$ cm. This would be roughly the lower limit for the casting of bulk metals by existing technology). Diffusion coefficients in solids range from 10^{-12} to 10^{-5} cm²/s (the latter being values for interstitial solutes at elevated temperatures) and therefore the times required to adjust the solid composition for this minimum thickness range from 10^5 to 10^{12} s. Even the shortest of these times is rather long for practical processing.

Even in the liquid state diffusivities are only of the order of 10^{-5} cm²/s, implying a processing time on the order of 10^5 s for a liquid 2 cm across (or 10^9 s for liquid of 2 m minimum dimension) if we had to rely on diffusion to adjust the composition. Fortunately, solutes can be transported in liquids by convection in addition to diffusion. The time to adjust a composition by convection is, very roughly, L/V, where V is a representative value for the speed of convection of the liquid. Speeds of 1 to 10 cm/s can be achieved without much difficulty, and therefore convection can be employed in the adjustment of even quite large volumes of liquid in reasonable times. In some instances this important motion of the liquid might be simply natural convection brought about by temperature or concentration gradients; more usually, deliberate steps are taken to stir the liquid, for example by injecting an inert gas into the liquid or by using electromagnetic forces.

The reader may have noted an exception to the generalization above (that refining be done prior to solidification); that exception is the electrorefining of copper (and a limited number of other metals, such as nickel), which is carried out starting with *solid* anodes. Of course, such electrorefining, described in Chapter 9, is accompanied by the dissolution of the metal of the anode and its re-formation as a cathode so that solid-state diffusion need not be relied upon.

Finally, we point out that any refining or alloying of metals produced by the hydrometallurgical routes described in Chapter 9 usually would first entail melting of the metal.

10.1.2 Degassing of Molten Metals

Let us first examine the removal of gases dissolved in metals. Molten metals produced in contact with air typically contain significant quantities of dissolved nitrogen, oxygen, and hydrogen (from water vapor in the air). It is usually necessary to remove these gases prior to solidification, because their solubility in the solid metal is much less than in the liquid. Consequently, on solidifying a metal with a high concentration of dissolved gas, voids are formed in the solid from evolved gas bubbles, and these would degrade the mechanical properties of a casting or require additional processing to eliminate them. Diatomic gases dissociate on solution in metals, for example,

$$H_2(g) = 2H(\text{in solution}) \tag{10.1}$$

for which we can write an equilibrium equation

$$K_a = \frac{a_H^2(\text{soln})}{p_{H_2}} \tag{10.2}$$

If the activity coefficient is independent of concentration, equation (10.2) can be rewritten

$$\text{concentration of hydrogen} \approx p_{H_2}^{1/2} \tag{10.3}$$

which is Sievert's law. This predicts that the amount of a gas dissolved in a metal can be reduced merely by reducing the partial pressure of that gas in the gas with which the metal is in contact (e.g., by contacting the metal with a vacuum). In practice, degassing

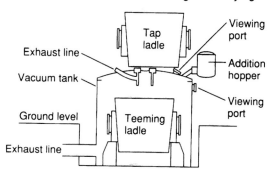

FIGURE 10.1 Spray degassing operation for removing gas dissolved in steel. (Courtesy Association of Iron and Steel Engineers, *The Making, Shaping, and Treating of Steel,* 9th ed.)

is not as simple as this. A hydrostatic calculation shows that the pressure at the bottom of molten steel 1.5 m deep with its top surface in vacuum is approximately 1 atm. The equilibrium pressure calculated from equation (10.2) would typically be well below atmospheric pressure, and therefore removal of gas as bubbles would take place only from (or near) the top surface.

One solution to this problem is to spray the molten metal into a vacuum chamber as shown in Fig. 10.1. As the metal droplets fall through the vacuum into a ladle placed to collect them, they are exposed directly to vacuum and degassing is promoted by a high surface-to-volume ratio of the droplet. Alternatively (as depicted in Fig. 10.2), an inert gas such as argon can be bubbled through the molten metal. The *total* pressure in bubbles at the bottom of the metal pool might be quite high, but the presence of the argon results in the partial pressure of the impure gas being below that necessary to remove the impurity. Other technologies have been developed for gas removal, but their

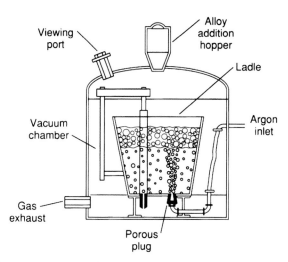

FIGURE 10.2 Argon degassing. (Courtesy Association of Iron and Steel Engineers, *The Making, Shaping, and Treating of Steel,* 9th ed.)

principles are no different from the ones discussed here. Note that, as shown in Fig. 10.1 and 10.2, the degassing apparatus may be a convenient point for adding deoxidizing or alloying agents. Also note that the processing vessels in both figures are ladles, the original purpose of which was merely the transportation of molten metal from one unit to the next. Nowadays a great deal of molten metal processing, such as desulfurization or deoxidation (see Section 10. 1.3) as well as degassing, takes place in these ladles. It has resulted in this technology, that is, the final adjustments of composition and impurity removal, becoming known as *ladle metallurgy*.

10.1.3 Removal of Dissolved Impurities by Reaction

Let us examine the last of the three impurities categorized in Section 10.1.1, those impurities such as carbon, sulfur, or phosphorus in steel that are present in the metal in solution in elemental form and which must be removed by chemical, rather than physical, steps. The reader will recall that in the case of steel, much removal of the impurities has already taken place in the steelmaking operation, where they were removed by oxidation. The subject of discussion here is the production of special grades of steel with even lower levels of these impurities. Another example of this type of impurity is iron in aluminum (typically, from small amounts of iron present in alumina and carbon anodes fed to the Hall cell, and from steel contamination of recycled aluminum scrap).

Impurities of this type can effectively be removed if a reagent can be found with which the impurities are more reactive than the host metal. For example, we might further reduce the quantity of carbon in a steel by additional contact with oxygen or an oxidizing agent such as Fe_2O_3. The procedure is ineffective if the host metal is more reactive than the impurity. For example, there is no hope of oxidizing iron impurities in aluminum because the aluminum has a much greater affinity for oxygen than does iron, as is evident from a glance at the Ellingham diagram (Fig. 2.12). Another example is provided by copper impurities in steel. This impurity results largely from copper wiring in recycled automobile bodies and similar scrap, and cannot be removed by oxidation because copper has much less affinity for oxygen than iron.

The ease of removal of impurities by chemical routes also depends on the level of impurity in the metal. For example, reduction of the sulfur level in a steel, say from 1.0% to 0.99%, is readily achieved during the steelmaking operation (which typically brings the sulfur level down to a few hundredths of a percent); reduction from 0.012% to 0.002% (i.e., the same quantity of sulfur but at lower concentration) is much more difficult because the driving forces for mass transport and chemical reaction are now much less.

Oxygen in steel presents a special case. This impurity can be expected from the fact that the steelmaking operation entails blowing oxygen onto or through the metal. One way of removing the oxygen is by reaction with carbon also present in the metal:

$$C + O = CO(g) \tag{10.4}$$

forming carbon monoxide that leaves as a gas bubble. Alternatively, *deoxidizing agents* such as silicon or aluminum can be added to the steel to produce oxides, for example,

$$2Al + 3O = Al_2O_3 \tag{10.5}$$

The resulting steel is referred to as *killed steel* because the steel now has little tendency

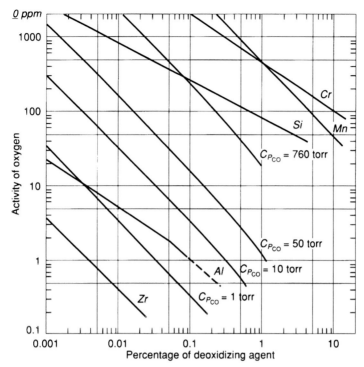

FIGURE 10.3 Equilibrium deoxidation plots for various steel deoxidizers at 1600°C. (From C. Jorgenson and I. Thorngren, *Thermodynamic Tables for Process Metallurgists,* Almqvist & Wiksell, Stockholm, 1969, p. 21.)

to bubble as gas is released during solidification. The impurity oxygen is now mostly present as oxide particles which can be physically separated from the molten metal. We can write an equilibrium for reaction (10.5):

$$K_a = \frac{a_{Al_2O_3}}{a_{Al}^2 a_0^3} \tag{10.6}$$

or

$$\log_{10} a_0 = \frac{-\log_{10} K_a}{3} - \frac{2}{3}\log_{10} a_{Al} \tag{10.7}$$

taking the Al_2O_3 to be a pure separate phase. A plot of this result, and analogous ones for other deoxidizing agents, appear in Fig. 10.3. Note that the horizontal axis is the amount of aluminum (or other deoxidizer) *in solution* at equilibrium. The amount of aluminum actually added would be greater than this by an amount dependent on the quantity of Al_2O_3 formed, which would itself depend on the initial oxygen content. Figure 10.3 also contains lines representing the equilibria for reaction (10.4) at four different CO pressures. Clearly, the removal of oxygen as CO is effective if the CO pressure can be kept low; such removal has the advantage that no subsequent removal of oxide particles is required. The techniques described in Section 10.1.2 on degassing (exposure to vacuum or bubbling argon through the melt) will therefore effectively remove oxygen and carbon.

Removal of sulfur from steel melts is best achieved under reducing conditions (when the carbon content is high) and in the presence of basic (high-CaO) slags. The reaction is

$$CaO \text{ (in slag)} + S \text{ (in metal)} + C \text{ (in metal)} = CaS \text{ (in slag)} + CO(g) \quad (10.8)$$

The requirement of reducing conditions means that sulfur removal is frequently carried out *prior* to steelmaking, that is, on the molten iron from the blast furnace, in which the carbon level is high. Other reducing agents, such as magnesium, injected into the steel with lime are used to bring sulfur levels in the steel down to low values by reactions such as

$$CaO + Mg + S(\text{in steel}) = CaS + MgO(\text{in slag}) \quad (10.9)$$

Magnesium granules and powdered lime would be suspended in a gas stream (such as nitrogen or natural gas) and injected beneath the surface of the molten metal through a refractory lance. By this and similar techniques (such as the use of calcium carbide rather than magnesium as a reductant), the sulfur content of the hot metal can be reduced to approximately 0.01%. Following steelmaking, a further reduction in sulfur can be achieved (to levels in the range of 10 ppm) by contacting the steel with fresh CaO-containing slags so that residual carbon reacts with sulfur according to equation (10.8).

In contrast to the reducing conditions required for the removal of sulfur from steels, their dephosphorization is carried out under oxidizing conditions, although basic slags are used here also. The reaction is

$$3CaO(\text{in slag}) + 2P(\text{in metal}) + 5O(\text{in metal}) = Ca_3(PO_4)_2 \text{ (in slag)} \quad (10.10)$$

and this happens to a considerable extent in steelmaking. If further removal of sulfur is required, the ladle of steel is treated with a fresh batch of slag containing oxidizing agents such as Fe_2O_3. This is done *prior* to degassing, when the oxygen level in the metal is high. In this way phosphorus levels down to 10 ppm or even lower are achievable.

10.1.4 Removal of Particles from Liquid Metals

As we have seen, deoxidation of steels by addition of deoxidizing agents results in the formation of particles of oxides. Some of the deoxidizing agents can also react with sulfur in the steel to form sulfide particles. Other oxide particles might arise from breakage or erosion of refractory bricks used to line ladles, from oxides such as lime particles injected during ladle metallurgical operations, and so on, and from oxide coatings naturally present on deoxidizing agents or alloying agents as they are added to the melt. Finally, droplets of molten slag, entrained into the metal, may be present. It is desirable to separate these *nonmetallic inclusions* from the melt prior to solidification; they usually would have a detrimental effect on the mechanical properties of the solidified steel. Given sufficient time, the particles, being less dense than the metal, will float to the surface, where they can be skimmed off or be trapped in a layer of molten slag. The particles are usually sufficiently small that Stokes' law (which we met already in Chapter 4) is obeyed:

$$V_t = \frac{g\Delta\rho d^2}{\Phi 18\mu} \qquad (10.11)$$

where

V_t = terminal velocity of the particle
g = acceleration due to gravity
$\Delta\rho$ = density difference between particle and liquid
d = particle diameter
μ = liquid viscosity
Φ = shape factor of the particle

Using representative particle sizes (1 to 20 μm) and the physical properties of steel (density = 7 g/cm³, viscosity = 0.07 g/cm · s) and of oxide (density = 2 to 4 g/cm³) yields rise velocities that are too low for all but the largest and least dense of these oxides to float out of a representative pool of metal of, say, 1 m depth in practical times. In practice it is found that by agitating the melt (e.g., by injecting argon) inclusions *can* be removed in a few minutes. The explanation is probably that the agitation brings about both inclusion coalescence (increasing their rise velocity) and causes the convective transport of inclusions to the surface and walls of the melt.

PROBLEM

A centrifuge is suggested for cleaning a liquid metal containing inclusions that will float. The centrifuge is a simple cylindrical drum of radius R that can be rotated about its long axis. The centrifuge is operated batchwise with a volume (V) of metal poured into it. The centrifuge is then sealed, spun for a fixed time (at a speed sufficient to generate a layer of metal of uniform thickness, δ, on the walls of the drum that is stationary with respect to these walls), and then an arm positioned within the drum sweeps the inclusions off the inside surface. For fixed V, drum height L, and angular velocity ω show that the time to free the melt of inclusions is inversely proportional to R^2.

Solution

Unit mass in a gravitational field suffers a force g, while unit mass in a centrifuge suffers a centrifugal force $\omega^2 r$, where r *is* the distance from the axis. We can therefore replace g in equation (10.11) by $\omega^2 r$ to get the inclusion rise velocity (toward the centrifuge axis), that is,

$$-\frac{dr}{dt} = \frac{\omega^2 r \Delta\rho d^2}{\phi 18\mu}$$

or for convenience we will put

$$-\frac{dr}{dt} = kr$$

with obvious meaning for the constant k. The last inclusions to leave the melt will be ones that start their journey from the outermost portion of the melt (i.e., at $r = R$). Henceforth we will concern ourselves only with these inclusions. Integrating the equation above and putting $r = R$ at $t = 0$, we get

$$\ln R - \ln r = kt$$

Let t_c be time to remove these last inclusions, that is, for them to reach $r = R—\delta$. Hence

$$-\ln\frac{R-\delta}{R} = kt_c$$

Using the approximation $\ln(1 + x) \approx x$,

$$\frac{\delta}{R} = kt_c$$

But $V = 2\pi R \delta L$ (if $\delta \ll R$); therefore

$$t_c = \frac{\delta}{kR} = \frac{V}{2\pi kLR^2}$$

that is, for fixed V and L the time to centrifuge the inclusions is inversely proportional to R^2.

Of course, filtration is another possible way of removing inclusions and is widely practiced in the aluminum industry. One example of this technology is the SNIF (Spinning Nozzle Inert Gas Flotation) unit, where part of the removal of oxide particles is by argon bubbles dispersed into the metal by revolving nozzles. The oxide particles attach to the argon bubbles and are brought to the surface. Aluminum and other metals have a high surface energy, and therefore the rejection of an oxide particle into a gas bubble is favored. The argon bubbles also have a degassing function, in that they aid the removal of dissolved hydrogen. Metal flows out of the unit from an underflow and through a porous alumina filter that serves to remove most of the remaining oxide particles.

10.1.5 Argon-oxygen Decarburization

One operation that is best presented in the present context is the *argon-oxygen decarburization* (AOD) unit used in the manufacture of stainless steel. Stainless steels (alloys of iron, chromium, and nickel) can readily be produced by mixing the constituent metals in the right proportions; an open arc furnace is a convenient device for doing this. However, a very suitable source of the chromium is ferrochromium made by reduction of chromite (an oxide ore of iron and chromium) using carbon, as described in Chapter 8. The product of this reduction is ferrochromium containing a large amount of carbon. It is possible to remove this carbon in a separate operation by reaction with chrome oxide or chromite, but this conversion of high-carbon ferrochromium to low-carbon ferrochromium raises the cost significantly. A second possibility is to use the high-carbon ferrochromium in the open arc furnace, where the ferrochromium is melted with nickel and steel scrap, and then attempt to oxidize out the carbon in a subsequent step, for example by injecting oxygen. Unfortunately, this results in a loss of the valuable chromium content because the undesired reaction

$$2Cr + O_2 = 2CrO(\text{in slag}) \tag{10.12}$$

occurs along with the desired reaction

$$2C + O_2 = 2CO \tag{10.13}$$

If, however, the oxygen is injected together with argon, the equilibrium activities of oxygen, carbon monoxide, chromium, and carbon are shifted so as to promote reaction (10.13) and discourage reaction (10.12). Consequently, high-carbon ferrochrome can be used as feed to the open arc furnace and in a separate operation, the AOD unit, excess carbon removed from the iron chromium-nickel alloy, yielding stainless steel. The interested reader will find a problem aimed at demonstrating the efficiency of the AOD unit at the end of this chapter. The AOD unit is rather similar to the Q - BOP in design, being a pear-shaped, refractory lined, vessel with a steel shell and a mechanism for tilting so that the charge from the arc furnace can be poured in and the finished stainless steel poured out. The argon-oxygen mixture is blown in through the bottom of the vessel. Processing of a typical batch of 200 tons would require approximately 90 minutes. The AOD process is used for the production of stainless steels low in carbon (L, ELC, and K grades).

10.2 Solidification of Metals

On first examination, solidification may appear to be quite a simple operation; all that is necessary is to allow the molten metal to cool and it will solidify. In fact, several quite complex phenomena occur during solidification, and these must be understood and manipulated in order to produce solidified metal with the desired properties at a high rate. In what follows we will be concerned with the bulk solidification of metals, that is, the production of large solid pieces of fairly regular shape (cylinders, rectangular parallelepipeds, etc.) that are subsequently fashioned into articles of commerce (sheet steel for making automobiles, for example) by metal processing operations. Production of castings (e.g., automobile engine blocks) is beyond the scope of this book, although much of what is presented here is relevant to that industry. Rapid solidification is treated in Chapter 16.

10.2.1 Simple Example of Solidification: Segregation

Consider the *unidirectional* solidification of an alloy depicted in Fig. 10.4. solidification is proceeding in one direction (the x direction) starting from a chilled surface (at $x = 0$). The other surfaces bounding the metal are insulated. We shall assume that the metal contains a small amount of just one solute; the relevant part (solute-poor part) of the phase diagram is sketched in Fig. 10.5. Consider what happens when the original molten metal has a composition and temperature given by point A. As this metal is cooled, its temperature eventually reaches point B at the coolest part of the metal (i.e., at $x = 0$). At

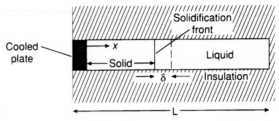

FIGURE 10.4 Unidirectional solidification of a liquid containing a solute.

this point (the liquidus line) solidification begins with the solid formed having a composition given by point C (on the solidus line) if equilibrium is maintained at the solidification front. Because this solid formed contains less solute than the liquid from which it formed, solute is "rejected" from the solid into the liquid, thereby increasing the solute content of the latter. If equilibrium is maintained at the solidification front, the point representing the composition and temperature in the liquid will move along the liquidus line (toward B') while the point representing the solid formed at the interface moves along the solidus (towards C'). We see therefore that there is a tendency for the solid formed during the final stages of solidification to be richer in solute than that formed earlier: this is referred to as *segregation*. Generally, segregation in the solidified metal is undesirable because the properties of the solidified metal will depend on the solute content and there may be other difficulties associated with segregation, such as the difficulty of accurately determining the solute content.

The degree to which segregation occurs depends on the phase diagram (gap between liquidus and solidus curves) and on mass transport of solute. That mass transport is by convection and diffusion in the liquid and by diffusion in the solid. Perhaps we can minimize segregation if mass transport is rapid. The solute rejected into the liquid will then find its way into the entire remaining liquid volume and will change its concentration to the minimum extent; concentration gradients in the solidified metal would be smoothed out by diffusion as solidification proceeds. Let us first show that the latter is unlikely. For bulk solidification of metals, L would have a minimum value of a few centimeters. The thickness of the solidified metal, X, except during the initial moments of solidification, would be a centimeter or more. In Section 10.1.1 we saw that transport

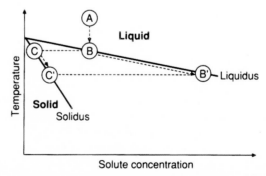

FIGURE 10.5 Phase diagram and paths taken by composition of liquid and solid phases, assuming equilibrium.

of a solute through only a centimeter or so of solid requires extremely long times. Because casting of this thickness of metal would be completed in a few minutes to an hour or so at most, we see that solid-state diffusion will not eliminate large-scale segregation *(macrosegregation)* during solidification. However, as we shall see below, segregation also occurs on a microscale, and this segregation is affected by solid-state diffusion.

In the liquid phase, natural convection due to temperature gradients is a likely mechanism of mass transport. Let us consider a simplified description of reality where this convection is sufficient to keep the liquid well mixed; we will call the solute concentration in the liquid C_1. If the liquidus and solidus lines in Fig. 10.5 are straight lines, we can write

$$K = C_s^* / C_1^* \qquad (10.14)$$

where K is a constant known as a partition ratio, C_s, is the concentration of the solute in the solid, and the asterisks indicate the equilibrium values at the interface. Because the liquid is well mixed $C_1 = C_1^*$. Let us suppose that in its original liquid state, the metal occupies a volume of unit cross-sectional area and length L; and consider unit time during which the solidification front advanced by an amount dX/dt. The volume of liquid solidified is therefore dX/dt and the quantity of solute removed from the liquid by this solidification is $C_s^* dX/dt$. Using our material balance equation with a control volume consisting of the liquid, we get

output of solute into solid + accumulation of solute in liquid = 0

Because the volume of liquid is $(L - X)$, it contains a mass $C_1 (L - X)$ of solute and the accumulation term becomes $d[C_1 (L—X)]/dt$. We can then write

$$-C_s^* \frac{dX}{dt} = \frac{d}{dt}\Big[C_1(L - X)\Big] = (L - X)\frac{dC_1}{dt} - C_1 \frac{dX}{dt} \qquad (10.15)$$

or invoking (10. 14) and rearranging. we obtain

$$\frac{dX}{L - X} = \frac{dC_1}{(1 - K)C_1} \qquad (10.16)$$

This can be integrated and the initial condition $C_1=C_o$ at $X=0$ used to obtain

$$C_1 = C_o\left(\frac{L - X}{L}\right)^{K-1} \qquad (10.17)$$

whence

$$C_s^* = KC_o\left(\frac{L - X}{L}\right)^{K-1} \qquad (10.18)$$

Note that *(L—X)/L is the fraction of liquid remaining.*

These last two equations predict that for $K < 1$ (the case sketched in Fig. 10.5), the

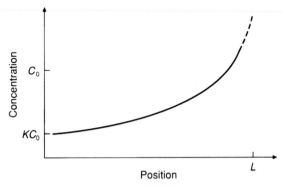

FIGURE 10.6 Segregation of solute resulting from unidirectional solidification when solute rejected into remaining liquid is well mixed in that liquid.

concentrations become infinite at the final stage of solidification. Of course, this is unrealistic; at higher solute concentrations the phase diagram will deviate from the simple one of Fig. 10.5, (e.g., a eutectic composition will be found). Nonetheless, equations (10.17) and (10.18) serve to illustrate that very marked deviations from the initial composition can be expected during the final stages of solidification. Figure 10.6 is a sketch of the solute distribution expected from equation (10.18). We see that maintaining a well-mixed liquid pool is *not* an effective way of minimizing segregation as we had at first supposed.

Now consider another case in which the solute is not rejected into the whole liquid volume but rather into a small volume near the solidification front, which we shall assume is well mixed with concentration C_i. For example, consider Fig. 10.4, where a small volume of width δ is bounded by the dashed line. During the initial stages of solidification there will be unsteady-state behavior (the initial transient) as the concentration in this volume builds up. This is followed by a quasi-steady state, where the inflow of solute through the surface shown by the dashed line (from liquid of

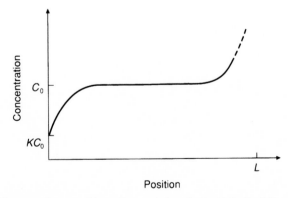

FIGURE 10.7 Segregation resulting when solute rejected into liquid is confined to small volume at solidification front.

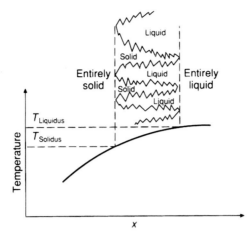

FIGURE 10.8 Dendrites in the partially solidified region (mushy zone) between solid and liquid.

concentration C_o) is matched by the output through the solidification front. that is,

$$C_s^* = C_o \qquad (10.19)$$

The concentration in the liquid in the small volume near the solidification front is C_o/K. Finally, the surface represented by the dashed line reaches the position $x = L$ and we have a final transient. The resulting solute distribution appears as in Fig. 10.7. Clearly, we have avoided macrosegregation over most of the solidified metal. In practice, it is difficult to confine the solute to a small well-mixed region near the interface. However, the description provided here also pertains to the practical case where solute rejected from the interface diffuses out into a nearly stagnant layer of fluid (a boundary layer) adjacent to the solidification front. Most of the diffusing solute atoms diffuse only a short distance in advance of the advancing solidification front and therefore do not penetrate through the boundary layer. Furthermore, in some casting technologies it is possible to separate the liquid near the solidification front from the rest of the liquid volume (e.g., in the continuous casting described below), and the concept applies to remelting operations treated in this chapter as well as the zone refining of Chapter 16.

10.2.2 Other Solidification Phenomena

In the simple case treated in Section 10.2.1 we assumed that the interface between the solid and liquid was sharply defined and planar. In reality, this is seldom the case; as shown in Fig. 10.8, the region between solid and liquid is usually an extended one across which the temperature drops from the liquidus to the solidus temperature. This *mushy zone is* partially liquid and partially solid. The solid in the mushy zone is frequently *dendritic* in structure, that is, consisting of intergrowing treelike entities rooted in the solid and known as dendrites. Even though we may start by solidifying the metal on a planar surface, an instability occurs that brings about faster growth of some interface regions. A qualitative description of how this happens follows.

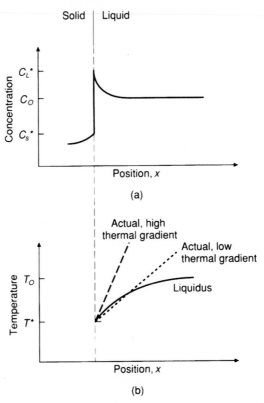

FIGURE 10.9 Constitutional supercooling and the stability of a solidification front.

Consider solidification of an alloy with the phase diagram of Fig. 10.5; solidification is proceeding with a planar surface. Furthermore. the liquid is treated as quiescent, so that solute atoms rejected at the solidification front are transported away from that interface by diffusion, resulting in the concentration profile of Fig. 10.9(a). Now examine the curve marked liquidus in Fig. 10.9(b). These are the liquidus temperatures corresponding to the composition curve of Fig. 10.9(a); the liquidus temperature is higher at a point farther from the interface because the concentration is *lower* there. Two straight lines in Fig. 10.9(b) indicate actual thermal gradients at the solidification front that might be obtained (e.g., by varying the cooling applied to the metal or the superheat, i.e., temperature above the liquidus temperature when the process is started). The dashed line (high temperature gradient) represents a stable situation; a protrusion growing out from the planar surface will have, at its tip, the local temperature given by the dashed line (i.e., the tip will be *above* the liquidus temperature at that point). It will therefore melt back, or at least not advance farther, until the rest of the solidification front catches up. On the other hand, if the temperature gradient is the one given by the dotted line, an instability occurs. The tip of a protrusion will have a temperature *below* the local liquidus temperature, and further solid growth will occur, at least until the protrusion is sufficiently long that its tip temperature crosses the liquidus line. This phenomenon is known as *constitutional supercooling,* the word "constitutional" indicating the key role played

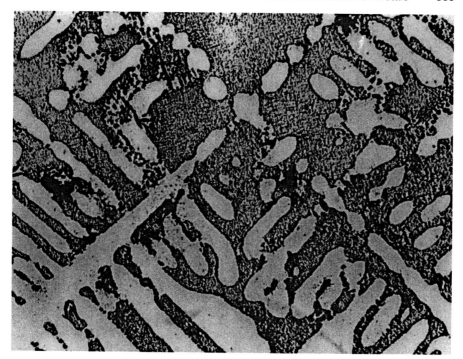

FIGURE 10.10 Dendritic structure formed by solidifying an aluminum-silicon alloy.

by composition variation. It is responsible for dendrites and other solidification front structures, although other phenomena play a role in shaping those structures.

Metals are such good conductors that it is difficult to maintain steep temperature gradients, particularly if the volume of metal is large. Consequently, dendrites occur readily in alloy solidification. The liquid trapped between growing dendrite arms becomes enriched in solute rejected from the dendrites, and segregation on a microscopic scale *(microsegregation)* then arises. Frequently, it is possible to homogenize the alloy during or subsequent to solidification by solid state diffusion because the length scale of this segregation is sufficiently small. Frequently, for example when the solidified alloy is two-phase, it is possible to see the dendrites in polished sections of the solid. Figure 10.10 is a micrograph of such a microstructure.

The formation of porosity by gas evolution during solidification was mentioned earlier. Gas evolution is undesirable if the released gas is trapped within dendrite structures. Other phenomena relevant to solidification are shrinkage and cracking. Metals (with the exception of bismuth) become more dense on solidifying (i.e., they shrink). For example, in the case of aluminum alloys, this shrinkage is over 6%. solidification proceeds inward from the outer surface of a metal pool, and if the whole outer surface is allowed to solidify, shrinkage accompanying the solidification of the remaining liquid can only be accommodated by the formation of voids or the deformation of the outer surface. We see, therefore, how shrinkage can result in stresses that may be sufficient to fissure the solidified metal (which has little mechanical strength close to its solidus temperature). Additional stresses arise from temperature gradients.

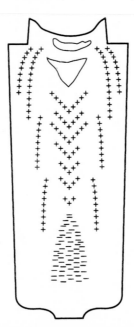

FIGURE 10.11 Macrosegregation in a killed steel ingot; + denotes region of positive segregation, - denotes region of negative segregation. [From T. E Bower and M. C. Flemings, *Trans. AIME 239*, 1620 (1967). Permission of the Iron and Steel Society.]

10.2.3 Casting Technology: Ingot Casting

The traditional way of casting metals has been simply to pour the molten metal into a mold and allow it to solidify by losing heat to the mold and its environment. Molds are typically made of metal (e.g., cast iron) and are sufficiently conductive and massive that the inside mold surface stays below the melting point of the mold metal on contacting it with the metal to be cast. In the case of steel this ingot casting uses molds with inside dimensions of, say, 1 to 2 m × 1 to 2 m which are 2 to 4 m high. Molten steel is "teemed" from a ladle into a series of ingot molds resting on cars that can be moved along a railroad track by a shunting engine. Solidification would be complete after 2 to 10 hours, although the molds are usually stripped from the ingot after only partial solidification because the ingots are then already self-supporting. Then the ingots are placed in furnaces known as *soaking pits* which serve both as holding furnaces (between casting and the hot-rolling operations described later in this chapter) and for homogenizing the temperature distribution in the ingot.

Figure 10.11 is a sketch of an ingot in section showing regions of (positive) macrosegregation toward the center and top of the ingot, which is the last part to solidify. Toward the bottom center is a region low in solutes (negative macrosegregation). One explanation for the latter region is that it is a part into which fragments of broken dendrites, formed early in the solidification(and therefore low in solute), fall and accumulate. A third feature of the ingot is the shrinkage void at the top. This section of the ingot would be cut off prior to rolling. The voids are minimized and formed at the

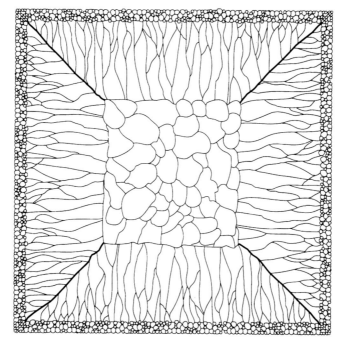

FIGURE 10.12 Sketch of ingot structure showing chill zone (outside), columnar zone, and equiaxed zone (in center). (From G. Derge, ed., *Basic Open Hearth Steelmaking*, 3rd ed., AIME, New York, 1964. Permission of the Iron and Steel Society.)

top by *hot topping*. A refractory collar is placed at the top of the ingot mold. Relatively poor heat conduction into the refractory means that the liquid steel within this collar is the last to solidify. This liquid steel can then "feed" molten metal into the ingot below it, preventing a shrinkage cavity in the ingot.

Figure 10.12 is another view of an ingot in section, this time seen from above. The figure depicts the grain structure seen (although somewhat less clearly) in many ingots. The outer chill zone results from rapid solidification as molten steel is poured against a cool massive mold. Thereafter the solidification front takes on a dendritic structure and these dendrites grow, predominantly along their long direction (perpendicular to the mold wall) to produce the columnar grains occupying the major part of the cross-sectional area seen in Fig. 10.12. Finally, an equiaxed zone appears near the center. It is formed during the final stages of solidification when cooling rates and thermal gradients are low. Under these circumstances the final solidification tends to occur by the growth (and eventual interlocking) of grains suspended in the liquid. Nuclei for these grains may be fragments of dendrites broken off during earlier stages of the solidification. The equiaxed structure is desirable in the subsequent rolling operations because less working is required to produce the desired fine grain structure and because excessive amounts of columnar structure tend to lead to the occurrence of cracks during rolling. Sometimes (e.g., in the casting of aluminum) nucleating agents are added to promote nucleation of grains ahead of the solidification front, resulting in an equiaxed structure.

Alternatively, vigorous agitation (such as produced by electromagnetic stirring) can be used to break up the dendrites, forming a larger number of nuclei throughout the bulk of the liquid and resulting in a more equiaxed structure.

Apart from the imperfections just described, ingot casting suffers from another major drawback. Very considerable expenditures of money, time, labor, and equipment are required in the handling of large numbers of ingots. They must be cast, have their molds stripped, be held in soaking pits, be rolled down to size, and be shuttled between these various operations using overhead cranes and other mechanical equipment. Ingot casting is still used in older steel plants but is being rapidly supplanted by continuous casting (see below), except for casting ingots that are to be forged into large steel shapes such as shafts for large turbines.

Ingot casting of aluminum, copper, and other metals is still widely practiced, although the sizes of these ingots are very much smaller than those of typical steel ingots. Such articles as copper wirebars (used in wire production) or aluminum pigs are cast in iron molds merely by pouring the metal into the mold, skimming any oxide (dross) from the surface, and letting the metal solidify. The solidified metal is then tipped from the mold. Frequently such operations are mechanized so that the molds revolve continuously on a wheel, undergoing the various casting steps.

10.2.4 Continuous Casting of Steel

Figure 10.13 is a schematic diagram of the continuous casting of steel. Molten metal is teemed from a ladle into a *tundish,* the primary function of which is to serve as a holding tank and provide a steady flow of steel to the mold(s). The mold is open at the bottom except for solidified metal, and rollers continuously pull and guide the solidified metal from the mold, through water sprays and a radiant cooling section, to a cutoff machine where the steel is cut into lengths. The mold, usually of copper with a cooling jacket through which water is pumped, is mechanically oscillated up and down to minimize difficulties of the partially solidified steel sticking in the mold. Starting the caster entails bringing a dummy bar of solid steel up into the mold. Steel is then poured onto this bar as it is withdrawn by the rollers.

Note that at its center the steel is still liquid when it exits the mold. It is therefore essential to avoid breaking the solidified shell at this point because the resulting breakout of molten steel would be dangerous for operators and equipment as well as ruining the cast. In recent years a great deal of work has been done to characterize the heat transfer from steel to mold that is a major factor determining shell thickness.

Continuous casters usually produce billets (roughly square in section and in the size range 50 by 50 mm to 300 by 300 mm) or slabs (rectangular in section and in the size range 50 by 250 mm to 300 mm by 2 m), although casters producing steel of much thinner section ("thin slab casters" or "strip casters") have been developed in recent years. It is seen, then, that continuous casters can produce large tonnages of steel in small sections, thereby eliminating some of the early stages of rolling used in ingot processing. Furthermore, the extensive handling of the ingot processing route is eliminated, with considerable savings. Additionally, the reader will note that the volume of liquid in contact with the solidification front is only the relatively small volume in the mold; that in the tundish and ladle is isolated from solute rejected at the solidification front, diffusion or convection up the liquid metal flowing rapidly down from tundish to

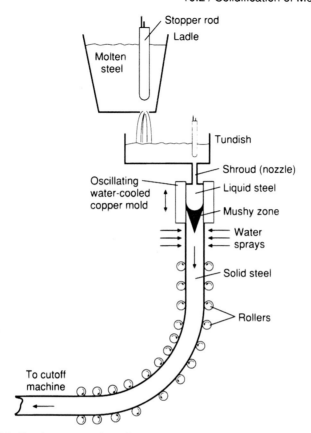

FIGURE 10.13 Continuous casting of steel.

ladle being impossible. In our treatment of macrosegregation in Section 10.2.1 we saw how this use of a small liquid volume in contact with the solidification front will result in a more uniform composition in the solidified metal. Finally, there is little danger of shrinkage cavities forming because a liquid pool is always present to "feed" the metal into shrinkage volume, and by special techniques (such as electromagnetic stirring) an equiaxed structure can be promoted over a columnar structure. However, the solidification is relatively fast with steep thermal gradients, resulting in high thermal stresses and dangers of crack formation. Proper design of the machine so as to minimize thermal gradients in regions where the steel is weak can alleviate this problem.

10.2.5 DC Casting and Electromagnetic Casting of Aluminum

Direct chill (DC) casting of aluminum, sketched in Fig. 10.14, bears a superficial resemblance to the continuous casting of steel. Principal differences are that tundishes are not used in DC casting (the aluminum flowing directly to the caster from a holding furnace through a trough), the mold is not oscillated, and the casting is more nearly a batch operation than a continuous one. That is, the products of the operation are large ingots approximately rectangular in section (say, 2 m by 1 m) and perhaps 5 m tall that

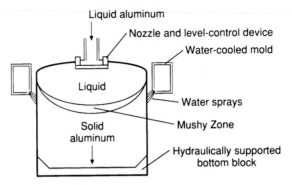

FIGURE 10.14 Direct chill (DC) casting of aluminum.

are cast one at a time by the machine. A cast is started by placing the steel bottom block into the mold and then beginning the flow of aluminum as the bottom block is withdrawn.

Electromagnetic casting of aluminum is a development which can be thought of as eliminating the mold of the DC caster. Instead, the molten metal is supported, at the side, by electromagnetic forces. These forces are produced by alternating current (a few kiloamperes at a few kilohertz) flowing in an inductor, as shown for two common designs in Fig. 10.15. An alternating magnetic field is produced which induces alternating currents in the metal. The interaction between the magnetic field and these currents generates the forces supporting the liquid metal. The principal advantage is the elimination of surface imperfections caused by the mold.

10.2.6 Remelting Technology: VAR and ESR

In many instances it is necessary to melt and cast metals prior to selling them to a customer. For example, in the manufacture of titanium a titanium "sponge" is produced by the reduction of titanium tetrachloride by sodium or magnesium. These lumps of porous, friable titanium are unsuitable for manufacturing parts for the chemical and aerospace industries (the major end users of this metal). Consequently, the titanium sponge is compacted into a rod which is then melted to form a nonporous ingot that can be further processed to parts (e.g., by machining). This is usually done by *vacuum arc*

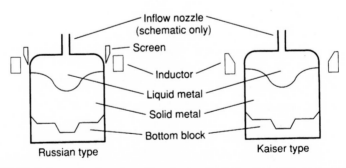

FIGURE 10.15 Schematic diagram of two types of electromagnetic caster used commercially for aluminum.

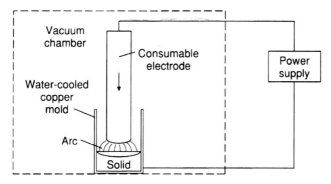

FIGURE 10.16 Sketch of vacuum arc remelting.

remelting (VAR), illustrated schematically in Fig. 10.16. The titanium sponge is a *consumable electrode* in a circuit where an arc is struck between this electrode and a shallow liquid pool of titanium. Heat generated in the arc causes melting of the titanium, which drips into the liquid pool. The latter is contained by a water-cooled mold, typically of copper, and continuously solidifies so that a solid ingot of titanium is built up in the mold. The consumable electrode is moved downward as the operation proceeds. Because the mechanical properties of titanium are severely degraded by traces of oxygen or hydrogen (which can be formed from water vapor), it is necessary to carry out the operation in a vacuum chamber. This has the advantage of reducing impurity levels in the finished ingot as volatile impurities (such as chlorides left over from sponge production) are vaporized and any dissolved gases evolve from the falling metal droplets.

The technology is not restricted to titanium; even some special grades of steel are refined in this way. Note that the fraction of metal that is molten at any instant is only a small part of the total; following our discussion of macrosegregation in Section 10.2.1, this implies that macrosegregation can be avoided provided that the consumable electrode is of nearly uniform composition.

Electroslag remelting (ESR) is an alternative technology to VAR for producing special steels with low levels of inclusions and other impurities. The equipment is sketched in Fig. 10.17 and bears a superficial resemblance to VAR in that a consumable electrode of the metal to be melted is used. Heating is provided by electric current and the solidified

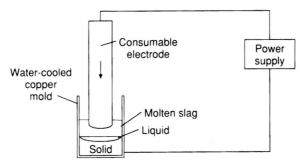

FIGURE 10.17 Sketch of electroslag remelting.

ingot is formed in a water-cooled copper mold. The principal differences between ESR and VAR are that the former does not employ a vacuum system but instead, protects the metal from gas oxidation by means of a molten *slag* (consisting largely of calcium fluoride with smaller quantities of CaO and Al_2O_3). It is the passage of electricity through this resistive slag, rather than an electric arc, that provides the heat necessary to melt the electrode. In one variant of this technology the mold moves upward as solidification proceeds so that tall ingots can be produced. The molten metal falls as droplets through the molten slag, providing (along with the horizontal slag-metal interface) a large surface area for mass transport. Consequently, some refining of the metal can be achieved in ESR, notably removal of nonmetallic inclusions and desulfurization [by reaction (10.8)].

At the present time, VAR and ESR units producing ingots of up to a few hundred tons are in service. A recent interesting development is *central zone ESR*. In this technology the central core of an ingot, cast by conventional methods as described in Section 10.2.3, is punched out (while the ingot is still hot and readily worked), eliminating the macrosegregated zones depicted in Fig. 10.11. The core of the ingot is then filled by using it as the "mold" for electroslag remelting of a consumable electrode of the same composition as the ingot.

10.3 Finishing Operations

In terms of the production of bulk metals, this book has now almost run its course. We have seen how chemical reactions can be carried out to convert compounds (ores or concentrates) into metals, how these metals are refined, and how they are solidified. We conclude this chapter with a short treatment of how metals are finished, that is, how large ingots, which are typically not articles of commerce, are shaped into sheets, rods, and other shapes that can be sold to manufacturing industries such as the automobile industry. These operations are really the province of the metal processor and physical metallurgist. Furthermore, the operations and phenomena involved are numerous and are sufficient for a book in themselves. Consequently, the treatment given here is brief, and for a fuller exposition, the interested reader is referred to the books listed at the end of the chapter. We have restricted attention to the finishing of steel because the processing of other major metals (copper and aluminum, for example) differs in detail rather than concept.

10.3.1 Plastic Deformation and Hot Working

The important physical phenomenon in finishing operations is *plastic deformation*. If a solid object is subjected to a stress (i.e., a force distribution that tends to deform the object rather than accelerate it), the response of the body at low stresses is usually elastic, as sketched in Fig. 10.18 for a simple tensile experiment. In this elastic region the object will resume its original dimensions if the stress is removed. At higher stresses, metals exhibit *plastic deformation,* shown in Fig. 10.18 to the right of the *yield point*. The precise shape of the curve depends on the metal (including its microstructure) and the way the stress is applied; the important point is that if the yield stress is exceeded, the object will *not* resume its original dimensions when the external stress is removed. Consequently, by applying high enough stresses, we can permanently shape metals to suit our needs.

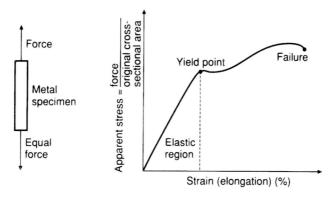

FIGURE 10.18 Mechanical behavior of metals in tension.

The stress needed to reach the yield point can be greatly reduced by raising the temperature of the metal. The force that it is necessary for the shaping machinery to apply would thereby be decreased. Furthermore, because the work done in plastic deformation is proportional to the area under the stress-strain curve (or more precisely, the area of a hysteresis loop traced out on the stress-strain plot during the deformation cycle), raising the temperature of the metal will result in the shaping machinery doing less work. Finally, it is clearly advantageous for the shaping to occur without the metal of the shaping machinery itself being plastically deformed. This is readily achieved if the metal to be worked is kept hot and the machinery is kept cold.

It is worthwhile stressing that the ability to deform the metal plastically is one of the principal differences between metal processing and ceramic processing. At ordinary temperatures, most ceramics fracture in tension below their yield points, and therefore it is impossible to produce ceramic shapes by plastic deformation. Of course, this does not preclude deforming ceramic powder mixes *before* they have been sintered (e.g., the extrusion of clays to produce sewer pipes, or the shaping of powder compacts in dies, as discussed in Chapters 11 and 12).

Finally, it is noted that plastic deformation, particularly near room temperature *(cold working),* has a profound impact on the microstructure of the metal (e.g., its grain size and the number of dislocations). These changes are topics in physical metallurgy and are not treated here.

10.3.2 Forging

Figure 10.19 is a simplified block diagram of steel finishing operations from ingot casting onward. Moving from left to right in this diagram, the section (i.e., the minimum dimension) of the steel is reduced. However, for some applications, steel parts of a large section are needed. Examples are shafts for electric generators and ships, rolls for steel rolling mills, and crankshafts for large stationary engines. These large pieces of steels, perhaps several hundred tons in weight, are produced from cast ingots by *forging.* The term derives from the blacksmith's forge, where red-hot lumps of metal are pounded into shape between a hammer and an anvil. The difference is primarily one of scale; in the forging press the hot ingot is shaped between *die blocks* which are repeatedly driven

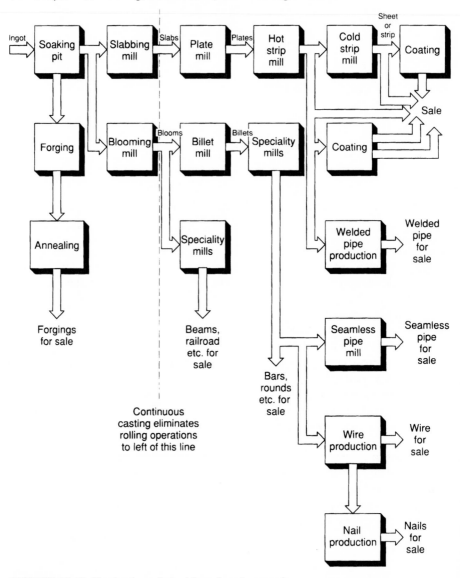

FIGURE 10.19 Production of steel from ingot onwards.

toward each other hydraulically. Mechanical manipulators allow the repositioning of the steel between each stroke of the press. By using die blocks of various shapes (e.g., the *swage dies* shown in Fig. 10.20) and appropriate rotation/translation of the steel, quite complicated shapes can be produced. Hollow shapes can be produced by forging between a horizontal "expanding bar" (which is inserted through the steel and, despite its name, does not itself expand) and an upper die.

The ingot is transferred to the forging press from the soaking pit, where its temperature is homogenized. Typically, the forging would be subjected to heat treatment following

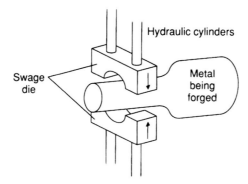

FIGURE 10.20 Schematic diagram of press producing a forging of round section. Dies shown open. Manipulators supporting and positioning metal and the frame of the forging press not shown.

the forging operations in order to eliminate stresses resulting from the forging and to achieve the desired microstructure.

10.3.3 Rolled Products

A large number of steel products are sold by steel companies in the form of long pieces of constant cross section. Examples are steel beams used in construction and sheet steel used in the production of automobile bodies. All of these are produced in rolling mills of various designs. Each of the units called a mill in Fig. 10.19 is a rolling mill. The blooming and slabbing mills shown in this figure are *primary mills;* their task is to reduce hot steel ingots to slabs (50 to 250 mm thick by 1 to 2 m wide) or *blooms* (roughly square in cross section and 150 to 500 mm thick). Figure 10.21 is a schematic diagram of a primary mill which is rolling hot steel that is being passed between the

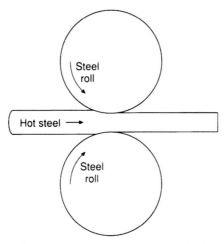

FIGURE 10.21 Schematic diagram of a primary rolling mill. Tables for supporting the steel drives for the rolls, frame (housing), and manipulators not shown.

rolls from left to right. Powerful electric motors drive the rolls, which are positioned by screw mechanisms so that a small reduction of the steel thickness occurs by plastic deformation as the steel passes between the rolls. The mill shown (partly) in Fig. 10.21 is a *single-stand, two-high mill,* indicating that it has a single set of two rolls. To achieve further reduction in the steel section after one pass through the rolls, it is necessary to either pull the steel back over the top of the upper roll or to reverse the direction of rotation of the rolls. The latter would be the more common practice, in which case we have a *reversing mill.* In the case of blooming mills the steel is turned through 90° about a horizontal axis on each pass by hydraulically driven *manipulators* so that the cross section remains approximately square. The manipulators, tables equipped with electrically driven rollers on either side of the stand to support the steel, and the housing are not shown in Fig. 10.21.

Slabs eventually become products that are wide but thin in cross section. This is achieved by additional rolling mills similar in principle to that shown in Fig. 10.21. The major difference is that these secondary mills usually have multiple stands and are not reversing; that is, the reduction in section is achieved by passes through several stands rather than by several passes through one stand. As the section of the steel is reduced, its speed through the rolling mills and its length greatly increase. The *plate mill* would produce steel that is perhaps 5 to 20 mm thick and 1 m wide. A significant fraction of such steel would be sold to customers, while the rest passes on to a hot strip mill, which would, by further rolling, produce strip that is about 2 mm thick. Again, a fraction of this mill's product may be sold, perhaps after coating as described below, while a fraction goes to a final mill, where it is cold rolled to thicknesses ranging down to 0.3 mm (cold-rolled sheet or strip). Tubular products (e.g., pipe) can readily be made from steel strip by machines that bend it along its length into a cylinder and weld along the join.

Blooms, on the other hand, are used to produce items whose cross sections are more nearly isometric (e.g., railroad rail). Again the usual first step is further reduction in section by hot rolling to yield *billets* that would be mostly in the range of 50 to 150 mm across. Billet mills are frequently three high, as shown (looking in the direction of travel of the steel) in Fig. 10.22. Elevating tables on either side of the stand permit the steel to be fed toward the reader between the bottom and middle rolls and away from the reader between the middle and top rolls. In this way it is unnecessary to reverse the direction of rotation of the rolls, saving the time and energy required in their deceleration and acceleration. Note that the rolls of the billet mill are not flat but contoured, with the steel passing through smaller gaps on each pass. Blooming mill rolls usually are contoured in a similar way, with slabbing mill rolls being practically flat.

Specialty mills produce shapes that are not of rectangular or square cross section. It should be clear that if, for example, the shapes of the gaps between the rolls in the mill of Fig. 10.22 were circular rather than nearly square, the rolled steel produced by the mill would be circular in section. This is exactly how a specialty mill for producing rounds works. Figure 10.23 indicates the shape of the steel (and hence of the roll contours) for a mill producing I-beams. Specialty mills produce a host of shapes from billets or blooms, such as channels, railroad rail, angle, triangular, round, hexagonal, and other shapes. By cutting a pattern into the rolls, it is possible to produce lengths that have (repeated) irregularities in their cross sections, such as concrete reinforcing bars. Mills producing the smallest sections would cold work the steel.

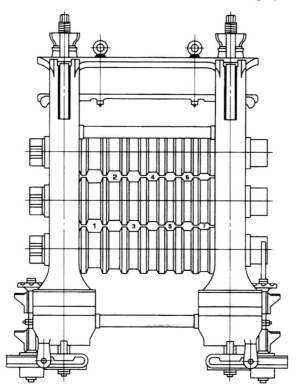

FIGURE 10.22 Rolls of a three-high billet mill seen in the direction the steel travels. (Courtesy Association of Iron and Steel Engineers, *The Making, Shaping, and Treating of Steel,* 9th ed.)

A fraction of the rounds from the specialty mills is used for the production of wire. Rounds 5 to 7 mm in diameter are pulled through a series of dies (tungsten carbide-lined holes which are slightly smaller in diameter than the approaching steel) to achieve the necessary diameter. A fraction of this wire is made into nails.

Another use for rounds, particularly of larger diameters, made in specialty mills is the production of seamless pipe. This is a hot-working operation that starts with the piercing of the steel on a mandrel positioned between two rollers that are slightly inclined to each other and the axis of the mandrel, as indicated in Fig. 10.24, where the direction of travel of the steel is from right to left. The round having been pierced, other hot-rolling operations using mandrels are employed to adjust the wall thickness and diameter and to ensure roundness. Seamless pipe produced in this way would range in diameter from, say, 100 mm to 1 m. By cold drawing through dies equipped with mandrels to shape the inside, seamless tube of smaller sections can be produced, including rectangular and irregular cross sections, although hot extrusion of billets is an alternative technique. In the latter the hot billet is first pierced by a mandrel and then forced through a die by a ram.

Referring again to Fig. 10.19, two things should be noted. During hot working the steel cools rapidly. This is particularly true when the section of the steel has been greatly reduced, because then the surface area per unit mass of steel is high and the steel is

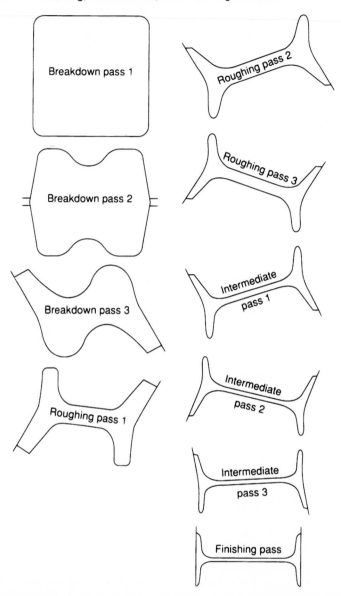

FIGURE 10.23 Shape of cross section of steel on each pass through a specialty mill rolling I-beams. (Courtesy Association of Iron and Steel Engineers, *The Making, Shaping, and Treating of Steel,* 9th ed.)

moved rapidly (increasing the rate of heat transfer to the environment). Consequently, the steel must be reheated, usually more than once, in its progression from left to right in Fig. 10.19. Finally, the vertical dashed line in this figure indicates those operations eliminated by continuous casting. These are the soaking pits and primary mills; an advantage of continuous casting is apparent.

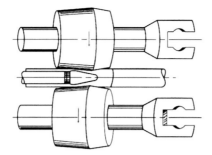

FIGURE 10.24 Piercer producing seamless steel tube. (Courtesy Association of Iron and Steel Engineers, *The Making, Shaping, and Treating of* Steel, 9th ed.)

10.3.4 Application of Finishes
(Tinning, Galvanizing, and Anodizing)

Because of its ready tendency to rust, most steel is sold with some kind of protective coating. The protection may simply be intended as temporary, to last until the steel is processed further (e.g., in ship construction), or it may be intended as a longer-lasting coating that will be protective following further fabrication (e.g., galvanized steel used in manufacturing garbage cans). Only the application of tin and zinc, perhaps the most important coatings, is discussed here. Paint is also an important coating for steel, but the technology needs no elaboration in this book.

Tin-coated steel strip, widely used in the food packing industry, is usually produced by the electrodeposition of tin from acid electrolytes. Figure 10.25 is a sketch of a plating unit that is at the heart of the tinning process. Coiled steel from a strip mill is welded to the end of the previous strip and uncoiled. It then runs continuously over

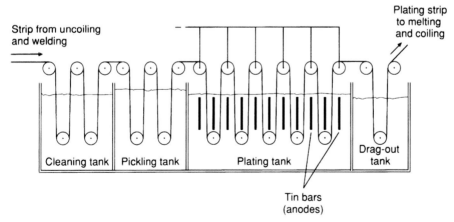

FIGURE 10.25 Schematic diagram of the plating section of an electrotinning line. Electrical connections (anodic) to tin bars not shown, nor are rinsing units (sprays and brushes) between tanks. (Courtesy Association of Iron and Steel Engineers, *The Making, Shaping and Treating of Steel,* 9th ed.)

rollers and through a series of tanks containing various solutions, with water sprays (and sometimes brushes) being used to minimize transfer of solution from one tank to the next. The first tank is intended to clean mainly organic matter (e.g., oil from rolling mills) from the strip and is usually strongly alkaline detergent solution. After rinsing, the strip then enters a "pickling" tank, which serves to remove iron oxide from the strip. Hot sulfuric acid is used for this purpose, although in some lines electrolysis is used to clean the steel. The steel can be made either anodic, in which case electrodissolution of a small amount of steel below the oxide layer serves to free the oxide that falls from the steel, or cathodic, in which case hydrogen evolved at the surface serves the same purpose. Electrolytic pickling is typically carried out using acid solutions, such as hydrochloric acid at low temperatures.

Following another rinse the steel enters the plating tank, where it passes over a series of rollers and through an acidic electrolyte. The rollers are connected to the negative side of a dc power supply of several tens of thousands of amperes, with the positive side connected to tin bars dangling into the electrolyte as shown in Fig. 10.25. The tin bars are therefore anodic and react to produce stannous ions in solution:

$$Sn = Sn^{2+} + 2\varepsilon^- \tag{10.20}$$

The contact between the steel and the rollers serves to complete the connection between the steel and the power supply, resulting in the steel being cathodic and the occurrence of the reaction

$$Sn^{2+} + 2\varepsilon^- = Sn \tag{10.21}$$

at its surface, producing a tin coating. Current densities of a few thousand amperes per square meter and strip speeds of 10 m/s would be typical of electrotinning lines using acid electrolytes.

The strip then enters a *drag-out tank,* which is intended to minimize electrolyte loss due to the solution being carried out of the plating tank by the moving strip. The simplest form of drag-out tank is a container filled with water. Periodically or continuously the water (now turned into a dilute solution by dragged-out electrolyte from the plating tank) would be concentrated by evaporation and returned to the plating tank. In this way, loss of valuable tin by "drag out" can be avoided. More complicated drag-out tanks would involve flow of solution countercurrent to the strip.

Tin coatings formed by electrodeposition lack the bright luster required for most applications, and therefore the next step in the operation is to melt the coating. This is achieved by passing alternating current through the strip (either by induction heating or by contact with rolls connected to an ac power supply) or in a gas-fired radiant heater. Not only does the tin melt to form a bright finish, but it alloys with the steel substrate, improving the adherence of the coating. The strip is finally sheared into lengths, which are then coiled ready for sale.

Zinc-coated steel *(galvanized steel) is* widely used in the construction industry (e.g., for ducting or as a roofing material) and its use in the automobile industry is growing. The traditional way of producing this material is simple. Steel strip is passed through cleaning and pickling operations similar to those used for electrotinning and then through a pot of molten zinc *(spelter)* usually containing other alloying agents. Zinc adheres to the strip as it leaves the pot, and the thickness of the film of molten zinc is controlled by jets of air blown across the surface of the strip. The strip cools as it passes along a conveyor, solidifying the zinc, and is then cut into lengths and coiled for sale.

Electrogalvanizing, wherein the zinc is deposited from aqueous solutions, is also an important technology. The advantages of electrogalvanizing are the precision with which the thickness of the zinc layer can be controlled, and an adherence to the substrate steel that is usually superior to that of hot-dipped coatings. From our discussion of the thermodynamics and kinetics of electrodeposition of zinc in Section 9.4.4 it is clear that the development of this technology was not a simple task; it is much easier to produce hydrogen on electrolyzing an aqueous electrolyte containing zinc than it is to deposit zinc. The current densities achievable in electrowinning cells without excessive hydrogen evolution are 300 to 500 A/m$^{2.}$ This would be too low a current density to give an acceptable zinc deposit thickness in a reasonable residence time in an electrogalvanizing process. Consequently, an important aspect of electrogalvanizing is maintaining a very high relative velocity between the electrolyte and the strip. This promotes mass transfer of zinc ions to the surface of the strip and therefore much higher current densities than in electrowinning. The high relative velocities are achieved by jetting the electrolyte onto the steel surface. It has been found possible to deposit alloys (e.g., zinc-manganese) with desired corrosion-resistance properties in electrogalvanizing lines. Two different types of electrochemical cell are in use for electrogalvanizing. In one the anodes are zinc slabs, connected to the positive side of the power supply and undergoing electrodissolution (analogously to the tin bars used in electrotinning). In the other, the anodes are inert lead alloy anodes and the zinc enters the solution by chemical dissolution of zinc in acid in a separate unit. In other respects the electrogalvanizing lines resemble electrotinning lines, in that they contain units for uncoiling and cleaning the strip prior to plating, for handling electrolyte drag-out, and for coiling the finished trip.

Finally, mention is made of another electrolytic technology: *anodizing,* which is used for finishing metals, notably aluminum. In this technology the aluminum is made anodic in an electrolyte containing metal ions that readily oxidize to colored oxides. A thin layer of aluminum oxide interpenetrated with the colored oxide is produced that is decorative as well as corrosion resistant.

10.4 Concluding Remarks

It is hoped that for metals, this chapter has served as the final link between the earth, which is the original source of our materials, and everyday life. The layperson has little knowledge of ores, concentrates, mattes, solutions of metal salts, and molten metals. Consequently, it is not unusual if these subjects, discussed in earlier chapters, are remote from the reader's daily experience. In the present chapter, however, we have learned how finished metals are produced in forms suitable for final shaping into end uses. The reader is encouraged to glance around and realize how metallic articles within reach or view (keys, coins, beverage cans, paper clips, automobile bodies, electric heaters, door knobs, watch straps, etc.) may easily be fashioned from the products discussed in this chapter.

LIST OF SYMBOLS

a_i	activity of species i
C_l, C_o	concentration of solute in liquid, concentration of solute initially
C_s	concentration in solid
g	acceleration due to gravity
K	partition ratio
K_a	equilibrium constant
L	length or thickness of metal
p_i	partial pressure
t	time
v_t	terminal velocity
X	position of solidification front
ϕ	shape factor
μ	liquid viscosity
$\Delta\rho$	density difference

Superscript

*	at solid-liquid interface

PROBLEMS

10.1 **(a)** The oxygen content of 100 tons of molten steel at 1600°C is to be reduced from 50 ppm to 10 ppm by the addition of aluminum. From thermodynamic calculations, what is the minimum quantity of aluminum required?

(b) The quantity of aluminum you have calculated in part (a) is used in plant trials and found to be insufficient to yield steel of 10 ppm oxygen. The plant manager storms into your office and accuses you of incompetence. What response can you make to defend yourself against her accusation?

10.2 Hydrogen is to be removed from a steel melt by blowing argon through it.

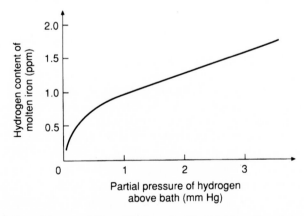

(Courtesy Association of Iron and Steel Engineers, *The Making, Shaping, and Treating of Steel*, 9th ed.)

How many kilograms of argon are required per kilogram of steel if the hydrogen content is to be reduced from 1.5 ppm to 1.0 ppm? Base your calculation on the figure shown here and on the assumptions that the melt is well mixed and in equilibrium with the gas mixture leaving the melt (which is at atmospheric pressure).

10.3 Calculate the rise velocity of (a) a 5 μm oxide inclusion particle of density 4 g/cm³, and (b) a gas bubble of 1 mm diameter, in molten steel of density 7 g/cm³ and viscosity 0.07 g/cm s, assuming that both are spherical and obey Stokes' law.

10.4 Copper from the converters described in Chapter 8 contains considerable amounts of dissolved gases, particularly oxygen. Nowadays the copper is deoxidized prior to casting by injecting natural gas, but even well into the twentieth century a cruder technique was used. This consisted of stirring the copper with "green poles" (unseasoned timber). Discuss the phenomena occurring during green poling.

10.5 Assume that in the AOD the chromium content of the stainless steel melt is determined by the equilibrium

$$Cr_2O_3 + 3C\left(\text{in solution in steel}\right) = 3CO(g) + 2Cr\left(\text{in solution in steel}\right)$$

Calculate and plot the weight percent chromium in the steel as a function of carbon monoxide partial pressures from 0.01 to 1 atm. Prepare curves for carbon contents of 0.5, 1, and 4% at 1873 K. Assume that activities are equal to mole fractions for carbon and chromium in solution in steel. Free energies of formation (from pure elements) are

$$Cr_2O_3 = -652 \text{ kJ/mol} \qquad CO = -276 \text{kJ/mol}$$

The free energies of solution at the dilutions in question are

$$C = -57 \text{ kJ/mol} \qquad Cr = -68 \text{ kJ/mol}$$

10.6 Explain why ice cubes are seldom transparent. Your answer should explain why they sometimes are.

10.7 A typical steel plant might produce 1 million tons per year of steel. If this production were all cast in ingot molds, provide an order-of-magnitude estimate of the number of molds, personnel, and floor area required for such an operation.

10.8 Certain steel parts (e.g., automobile camshafts) can be produced by either casting or forging. Discuss the relative costs of the two operations and compare the expected quality of the products.

10.9 Discuss the factors affecting the thickness of the steel shell exiting the mold of a continuous caster.

10.10 The electromagnetic casting of aluminum described in this chapter is a commercial success. There has been much interest in applying electromagnetic casting to steel. Discuss the difficulties that will be faced in such casting, compared to the casting of aluminum.

10.11 Calculate the maximum weight of tin deposited per square meter of steel strip in a tinning operation where 4000 A/m² of current passes into the steel as it moves at 10 m/s through 30 m of electrolyte. Assume that the tin ion that is discharged to form the coating is Sn^{2+}.

REFERENCES AND FURTHER READINGS

FLEMINGS, M. C., *Solidification Processing,* McGraw-Hill, New York, 1974.

The Making, Shaping and Treating of Steel, 10th ed., Association of Iron and Steel Engineers, Pittsburgh, PA, 1984.

MINKOFF, I., *Solidification and Cast Structure,* Wiley, New York, 1988.

HOSFORD, W. F., and R. M. CADDELL, *Metal Forming: Mechanics and Metallurgy,,* Prentice-Hall, Englewood Cliffs, NJ 1983.

KURZ, W. and D. J. FISHER, *Fundamentals of Solidification*, Trans Tech Publications, Aedermannsdorf, Switzerland, 1992.

CHAPTER 11

Production of Powders

11.1 Introduction

In Chapter 4 we described the physical aspects of powders and, from the examples that were shown, it was clear that a wide variety of shapes and agglomeration states was possible. Much of the powder's physical characteristics are determined by the way it is prepared. The preparation method will then have profound consequences for the fabricability of the ceramic or metal part from powder compacts. Often the variations in powder properties from batch to batch, prepared by the same method in the same factory, are sufficient to cause unacceptable variations in product properties. This susceptibility of fine powders to seemingly minor, and often unrecorded or undetected processing variations, is a major difficulty in the production of reliable and reproducible ceramic products.

A wide variety of techniques have been used for powder fabrication. These can be broadly classified as mechanical or as physicochemical methods. The modern trend in the production of powders for high-performance ceramics is toward chemical methods. Although often more expensive than mechanical techniques, these methods generally allow for better control of particle size distribution, degree of agglomeration, and purity.

We will discuss some of the most common powder fabrication methods. The majority of these are relevant to the production of brittle materials, such as ceramics. A few methods for the production of metal powders are discussed at the end of the chapter.

11.2 Mechanical Methods for Powder Production

In Chapter 1 a brief description was provided of how ores are mined and subjected to physical (mineral) processing to produce concentrates which are chemically processed to produce metals. These same methods are widely used to produce ceramic powders, particularly powders for such applications as the production of porcelains, whitewares (household wash basins etc.), refractory bricks, and so on. Extensive descriptions of these materials and their production methods can be found in the book by Norton (1974). Jaw crushers, gyratory crushers, ball mills, and so on, turn large boulders into particles in the size range 0.1 mm to 25 μm or so, at the rate of hundreds or even thousands of tons per hour. We will not repeat our discussion of such devices but assume that we have a stock of particles with a diameter not in excess of 0.1 mm. The most common way of producing fine ceramic powders from such a feed is further grinding in a ball mill. As described in Chapter 1, this is a drum rotated about a horizontal axis into which are placed the material to be ground and large, dense balls or rods. The latter are known as *grinding media*. As these balls tumble over each other, the particles between balls are shattered as the balls impact each other. This type of comminution clearly works best for brittle materials. However, as the reader is probably aware, metals that are normally ductile are embrittled by cold working, hence many metal powders can also be produced by grinding. Milling of ceramics can be wet or dry and batch or continuous. Batch processing would be common when relatively small quantities are to be produced or long grinding times are required.

The milling medium, consisting of balls (or rods in the case of a rod mill), should be of as high a density as possible. This provides for the most effective collisions. The materials choice for the milling medium is, however, limited by cost, wear resistance and the introduction of impurities into the powder from the medium. Zirconia balls (density 5.4 g/cm³), ranging from 1 to 20 mm are commercially available but are quite expensive. Steel balls (7.7 g/cm³) are often used when the contamination caused by their wear can be removed by leaching the ground powder with acid. Alumina balls (3.6 g/cm³) are relatively cheap, but they usually contain a fair amount of silica (10 to 20 vol %), which may provide undesirable contamination. The silica in these alumina balls is added because it makes their fabrication easier. Porcelain balls (2.3 g/cm³) are quite cheap, but wear easily. Tungsten carbide balls (14.8 g/cm³) are also used. To have the least contamination, we should have the milling medium of the same composition as the powder, with the mill lining made of some material that does not wear or is easily removed, such as polyurethane or hard rubber. All of these are commercially available.

The productivity of a rotating ball mill depends on many factors. The first is mill speed (rotations per minute). Ideally, the grinding medium should be carried to the top of the rotating mill and then dropped. This leads to a critical mill speed when gravitational and centrifugal forces just balance (see Problem 11.1). This speed (in revolutions per unit time) is given by $(g/R)^{1/2} /2\pi$, where g *is* the gravitational acceleration and R the mill radius. Rotating ball mills are typically operated at 60 to 70% of the critical speed. The prediction of the efficiency of milling and of the rate of grinding has been the subject of innumerable investigations. By and large, however, the process does not lend itself well to a fundamental analysis, and we have to be satisfied with empirical correlations. In one such relationship the rate of grinding, which is defined by the amount of surface area generated per unit mass of powder per unit time of grinding, is expressed as

$$\text{rate} = A\rho R^{1/2} \frac{d}{r} \tag{11.1}$$

where

A = property of the material being ground and of the mill construction
ρ = density of the milling medium
R = mill radius
r = radius of the milling medium
d = powder particle diameter

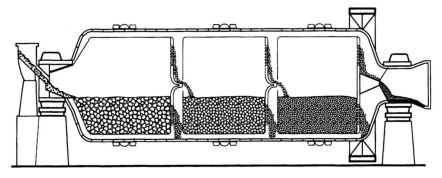

FIGURE 11.1 Three-compartment tube mill. (Reprinted with permission from J. K. Beddow, *Particulate Science and Technology,* Copyright 1980 Chemical Publishing Co., Inc., New York.)

We see from this equation that the rate is higher for smaller grinding balls. Of course, there is a lower limit on the size of the grinding medium. The tumbling balls must have sufficient kinetic energy to break the largest particles in the feed. This leads to the practice of staged grinding, in which the powder, as it is ground finer, is moved to different compartments in the same mill or to other mills that contain smaller milling media. An example of a mill with compartments is shown in Fig. 11.1. Wear of the milling medium can be substantial, especially when liquids have been added to the mill charge. It is not unusual to have 5 to 10 wt % of the milling medium worn away after 24 hours of milling in a ball mill, leading to substantial contamination of the powder.

In a typical rotary ball milling operation we would fill the mill approximately half full with the milling medium. This provides the maximum number of balls in the mill consistent with the need for the balls to fall a significant distance within the mill before impacting another ball. Approximately 25% of the mill volume can then be occupied by the powder charge before the falling balls start to drop onto a loose bed of powder rather than a nearly rigid mass of balls covered by a thin layer of particles. For wet grinding, slurries of approximately 30% solids are fed to the mill and approximately 40% of the mill volume would be occupied by such a slurry.

A glance at the empirical equation (11.1) shows that the rate of grinding drops as the particle size diminishes, so that a practical grinding limit might be reached after a long period of grinding (say, one or two days).

The efficiency of ball milling is notoriously low. At most, 1 or 2% of the energy input goes into the creation of new surface area. The rest is lost to friction and heating. Improving the efficiency of grinding operations continues to be a subject of investigation, especially in the mineral processing industry. The grinding efficiency can be strongly affected by additives to the powder. The most common grinding aid is water, to which various small amounts of detergentlike substances are added. The advantages of wet grinding are significant: a reduction in power consumption (up to 25% less for achieving the same particle size as that achieved with dry grinding), the possibility of achieving smaller particles at the practical grinding limit, and better mixing for multiphase powders. The disadvantages are that the powder will have to be dried, and that mill wear is high compared to dry grinding, leading to increased contamination.

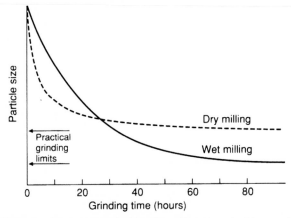

FIGURE 11.2 Particle size versus grinding time. (Reprinted with permission from J. K. Beddow, *Particulate Science and Technology* Copyright 1980 Chemical Publishing Co., Inc., New York.)

For dry grinding, the particle size reduction typically proceeds faster initially than for wet grinding, but particle sizes remain larger in the grinding limit than for wet grinding. A schematic comparison of the particle size evolution for wet and dry ball milling is shown in Fig. 11.2. Small amounts of detergents, as little as 0.01 vol % can sometimes increase the mill efficiency dramatically [e.g., a factor of 2 in surface area (Onoda, 1976)]. An example is shown in Fig. 11.3. Caking of powder onto the balls and the mill wall reduces the rate of grinding. The improvement resulting from addition of surfactants can be ascribed to the elimination of such caking (i.e., the particles are kept in suspen-

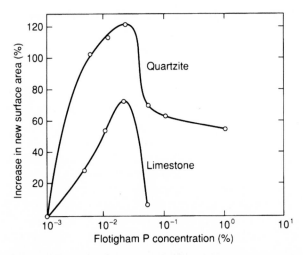

FIGURE 11.3 Effect of Flotigam P, an organic dispersing agent, on grinding of quartzite and limestone in a rod mill. (From R Somasundaran, "Theories of Grinding" in Ceramic *Processing Before Firing, G. Y.* Onoda and L. L. Hench, eds., Wiley, New York, 1978, p. 115.)

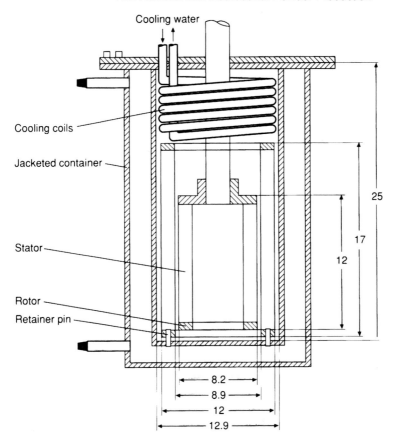

FIGURE 11.4 Internal details of a 5-inch attrition mill. Dimensions in cms. (Reprinted by permission of J. L. Pentecost.)

sion). The drop in grinding rate at high surfactant concentration is caused by foaming; this carries the powder to the top of the mill, where it is no longer hit by the balls.

There are a host of other types of mechanical methods by which powder can be comminuted; more than we want to describe here. An important one is the *vibro-mill,* which is a drum filled with balls and powder much like the rotary ball mill, but which is vigorously agitated by a machine that works somewhat like a paint shaker. The collisions between the balls are much more violent and the grinding rates can be 10 times faster than in the rotary mill. It is then possible to use polymeric balls effectively as a milling medium. This gives carbonaceous contamination that can be burned out in subsequent heating of the powder in an oxidizing environment, keeping final contamination low. Another common mill that is used in the laboratory as well as in industry is the attrition mill. In this mill, the grinding medium, usually very small balls, is stirred very vigorously by steel paddles or bars. Attrition milling rapidly gives a very fine particle size, but contamination from the agitator arms can be troublesome. An attrition mill is shown in Fig. 11.4.

11.3 Production of Powders from Solutions

The reader will probably have produced solids from solutions in the chemical laboratory. There are two general ways in which this is done, evaporation of the solvent, and precipitation by adding a second reagent solution (or gas or solid) that reacts with the first. For example, we can precipitate magnesium hydroxide from a solution of magnesium chloride by adding sodium hydroxide solution. Let us start by considering the phenomena that occur in these operations. In both cases the concentration of the solid to be precipitated is increased to, or above, the saturation value. If the solution is free of dirt particles and the walls of the container are clean and smooth, it is possible for the concentration to exceed the equilibrium saturation by quite a large amount, and we then have a supersaturated solution. Typically, the introduction of only a few particles into this solution would bring about the nucleation of crystals, and these crystals would grow rapidly, diminishing the supersaturation and therefore the tendency for new nuclei to form. The nuclei formed could then continue to grow as evaporation is continued or more reagent is added. In many cases imperfections in the walls of the container can provide nucleation sites. It is seen then that the important phenomena are nucleation and growth of crystals, and we have already treated these topics in Chapter 6.

It should be clear that if our objective is to produce small crystals of a uniform size, we shall need to control the nucleation and growth of the crystals. For example, it would be necessary to avoid the haphazard nucleation on dirt particles that results from working with solutions that are not clean. Ideally, we would like all the nuclei to form at the same time and subsequently grow at the same rate; the resulting particles will then all be the same size.

A particular difficulty becomes evident if we attempt to prepare particles by precipitation. Let us first consider a batch reactor. At the point in our reactor where the second solution is added there will be danger of locally exceeding the concentration necessary for nucleation. If the nuclei produced are carried into the rest of solution and do not dissolve before the bulk of solution reaches saturation, they will grow at that time, preventing or delaying nucleation throughout the rest of the solution. The consequence is likely to be larger particles than desired, with a broad size distribution that depends on precisely how the second solution is added and the liquid agitated. As a general principle, then, we should like to avoid concentration variations with position in carrying out this kind of precipitation. Temperature variations can have similar adverse effect, and it is necessary to avoid these.

In the preceding paragraph we envisioned a *batch* reactor where precipitation is being carried out by adding a reagent to an agitated solution. If the objective is uniform fine particles by control of nucleation and growth, what is important is that each "packet" of fluid (and particles therein) experience the same environment of composition and temperature as a function of time. The easiest way to achieve this in a batch reactor is to ensure that the reactor is well mixed. Now let us consider *continuous* reactors. The reader will recall from Chapter 7 the two extremes of such reactors: perfectly mixed and plug flow units. If our reactor approximates a perfectly mixed unit, we shall have difficulties because different packets of fluid will have different residence times and therefore will experience different environments over time, *even if mixing is excellent.* We see, then, that for continuous reactors, mixing should be minimized—the exact opposite of what is best for a batch reactor. In the plug flow unit reactor mixing is minimized and we should expect a real continuous precipitation reactor to give us more uniform particles, the more closely it approaches this ideal.

Let us illustrate these concepts by an example from the work of Jean et al. (1987). These investigators studied two kinds of continuous reactor: packed beds and static mixers. We have met the former before in this book. The latter consist of pipes through which the solution or suspension is pumped; the pipes contain helixes designed to promote mixing. It is clear that in both cases plug flow is only approximated, because some mixing is involved. This mixing is more significant at low flow rates than at high flow rates; for example, at low flow rates there is time for significant axial diffusion of solution components during the long residence time in the reactor. We should expect, therefore, that as the flow rate increases, the particle size distribution at the reactor outlet would become narrower, and this is what Jean et al. (1987) observed, as illustrated in Fig. 11.5.

If we use evaporation rather than precipitation, a similar difficulty can arise if, during evaporation, different packets of liquid suffer different histories of concentration and temperature. Let us examine some of the ways in which evaporation can be carried out to produce practical powders.

11.3.1 Direct Evaporation

A simple example of this method is the production of table salt, where water is allowed to evaporate and the salt eventually crystallizes out. A problem may arise in the crystallization of mixed salts. If the salts are of very different concentrations or solubilities, precipitation will occur in stages, and segregation of the salts results. Consider, for example, the evaporation of a dilute solution of sodium bicarbonate and sodium chloride. Sodium bicarbonate will crystallize first, and if the solution is unstirred, will form a layer at the bottom of the solution that is subsequently overlain by sodium chloride when that eventually crystallizes. This macrosegregation of the different compounds then creates undesirable heterogeneity in the mixture, and this has to be removed by extensive mixing or regrinding. Stirring the drying solutions can help in the homogenization of the resultant powders. However, even here segregation would occur—merely on a finer scale; crystals of one salt, perhaps as much as a few millimeters across, can be interspersed with similarly large crystals of the other salt.

We are therefore interested in the size of the crystals produced and would usually like the crystals to be as small as possible so that the segregation is on a micro or nano scale. The reader will recognize that small crystal sizes will result if nucleation is rapid and growth slow. In this way a multitude of small crystals are formed rather than a few large ones. Both nucleation and growth rates increase with increasing supersaturation, but the effect on the former is usually larger. The faster the evaporation rate, the faster the nucleation relative to the growth and the finer the resultant powder.

11.3.2 Spray Drying

A solution of the salts is sprayed into a hot chamber and the resultant powder is collected with a cyclone separator. Examples of spray nozzles that produce a fine mist, or aerosol, are shown in Fig. 11.6a and b. The droplet diameter decreases with increasing

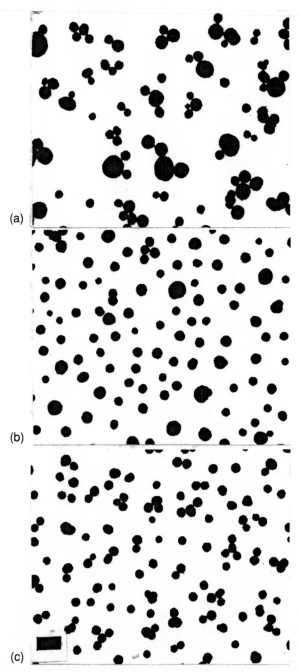

FIGURE 11.5a TEM micrographs of typical TiO$_2$ powers produced with a 98-cm static mixer reactor at flow rates of (a) 5 cm³/min, (b) 70 cm³/min, and (c) 160 cm³/min (bar = 1mm). [From J. H. Jean, D. M. Goy, and T. A. Ring, Continuous Production of Unagglomerated TiO$_2$ Particles, *Ceram. Bull.* 66(10)1518 (1987). Reprinted by permission of the American Ceramic Society, Columbus, OH.]

gas velocity and with increasing gas/liquid ratio. Droplets with a size from above 100 μm to well below 10 μm can be produced. Ultrasonic atomization devices are an alternative. A schematic of a spray drying operation is shown in Fig. 11.6c. Filtered air is preheated by passing it through a heat exchanger, where heat is transferred from combustion gases. The solution is sprayed into the hot air in the drying chamber and the particles produced, suspended in the gas stream leaving the drying chamber, pass to a baghouse, where they are trapped on bag filters. The hot gases pass through the filters and are vented through a fan. Periodically, the bag filters are discharged (by shaking or reversing the air flow through them) and the particles that fall from them are conveyed by a sidestream of filtered air to a final collector, where they can be discharged into drums.

If conditions such as temperature, concentration in the feed solution, chamber design, and gas flow rate are well controlled, powder consisting of spherical agglomerates can be obtained. The primary particle size in these agglomerates can be very fine, 0.1 μm or even less. This is another example of fine particles being generated by rapidly bringing the solution to a state of supersaturation. Such powders are used in ferrites (magnetic materials). It should be clear that the only segregation possible in these powders, except for any introduced subsequently, is microsegregation on the scale of the solid agglomerate formed by the evaporation of a single droplet. If the solution is sufficiently dilute, that length scale would be much smaller than the size of the droplet when it is formed at the nozzle. This is an important principle; we can avoid segregation, or at least minimize the length scale over which it occurs, by ensuring that the volume (e.g., droplet) from which the solid forms is as small as possible. If the length scale of the segregation is small, there is hope of eliminating the segregation in subsequent processing. For example, the time required for diffusion to eliminate the segregation is proportional to the square of the length scale, and very small-scale segregation can usually be eliminated in this way. We have discussed this topic in Chapter 6.

If conditions are not properly controlled, fluid may become trapped inside the spherical agglomerates, leading to undesirable defects, such as large voids within the agglomerates that cause problems in subsequent sintering. The atmosphere in the hot chamber

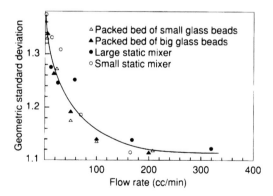

FIGURE 11.5b Geometric standard deviation versus flow rates utilized [From J. H. Jean, D. M. Goy, and T. A. Ring, Continuous Production of Unagglomerated TiO$_2$ Particles, *Ceram. Bull.* 66(10) 1518 (1987). Reprinted by permission of the American Ceramic Society, Columbus, OH.]

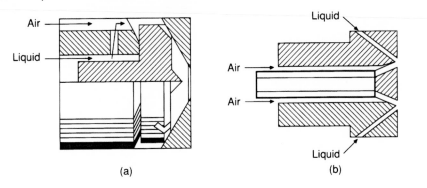

(a)   (b)

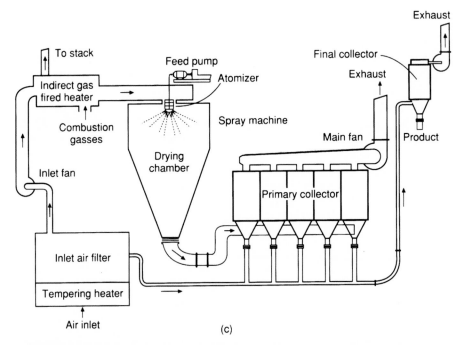

(c)

FIGURE 11.6 Two-fluid atomizers: (a) internal-mixing type; (b) external-mixing type; (c) schematic of a production spray dryer. (Reprinted with permission from *Encyclopedia of Chemical Technology Vol.* 21, 3rd ed., Wiley, New York, 1979.)

may be made reactive, for example oxidizing, so that oxide particles are formed rather than salts. In that case one speaks of spray roasting or evaporative decomposition. For example, both zinc and nickel oxide can readily be produced by spray roasting acetate solutions into an oxidizing environment. Nor are we limited to aqueous solutions; oxides have been produced by spray roasting alcohol solutions of alkoxides into oxidizing environments. Finally, it is possible to spray-dry slurries rather than solutions. In that event the apparatus serves to remove the liquid phase in a way that minimizes the opportunity for agglomeration of the suspended particles. Figure 11.7 illustrates an

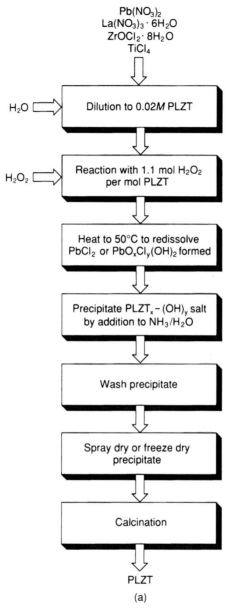

FIGURE 11.7a Flow diagram for the preparation of PLZT powders. (Reprinted with permission from R. Schwartz, D. Eichhart, and D. Payne, *Better Ceramics Through Chemistry: II, Mater. Res. Soc.* 73. Copyright 1986 Materials Research Society, Pittsburgh, PA.)

application of this kind where spray drying is used in the production of PLZT (lead-lanthanum-zirconium-titanate, an important electroactive ceramic). Some powder resulting from a spray drying operation is shown in Fig. 11.7.

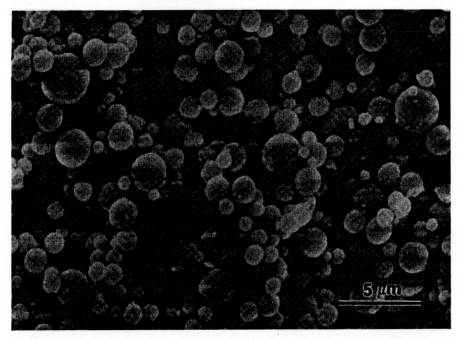

FIGURE 11.7b SEM photomicrograph of spray-dried lead zirconate titanate. (Reprinted with permission from R. Schwartz, D. Eichhart, and D. Payne, Better Ceramics Through Chemistry: II, *Mater. Res. Soc.* 73. Copyright 1986 Materials Research Society, Pittsburgh, PA.)

1 1.3.3 Freeze-Drying

Freeze-drying is widely used for the processing of a variety of foods, such as powdered milk and coffee, and has also been applied successfully to the preparation of ceramic powders. First, a solution of the salts of interest is prepared, and this is sprayed by an atomizer, as an aerosol, into a very cold liquid mixture, (e.g., hexane and dry ice, or liquid nitrogen). The aerosol droplets are thus rapidly frozen before they have an opportunity to coalesce, are removed from the hexane, and are put into a cooled vacuum chamber before they have a chance to melt. Under the action of the vacuum, the carrier fluid (solvent) sublimes, leaving powder that may have as much as 60 m²/g of surface area.

It is worthwhile comparing the phenomena occurring during freeze-drying with those occurring during evaporation of a solution. The crystals are formed during the freezing step in freeze-drying. Because the solution is cooled extremely rapidly and therefore rapidly becomes supersaturated, the particle size is extremely small. Sublimation of the solvent is then carried out at a low temperature, eliminating any chance of redissolution or sintering of the particles. The evaporation of a solution, even in the spray dryer discussed above, would be a comparatively slower approach to supersaturation. Hence freeze-drying yields much finer particles with a large surface area per unit mass. Again, segregation is limited to a maximum length scale, in this case the dimension of the droplet when it freezes. From these remarks it should be clear that it is important to

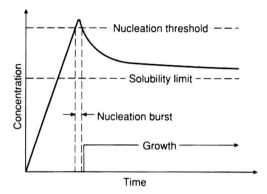

FIGURE 11.8 Variation of concentration in a solution where the concentration is first increased to the point of nucleation (e.g., by evaporation) and then declines as a precipitate grows.

avoid melting of the solution during the drying phase of freeze-drying and that control of heat transfer to the particles is an important aspect of equipment design.

11.3.4 Precipitation

It is possible to control precipitation reactions to such a degree that the concentration of the solute exceeds that for nucleation for only a brief period. This is achieved by bringing the solution into supersaturation, either by changing the temperature, the salt concentration, the pH, or by exploiting the slow release of some hydrolysis products in a water solution. The last-named technique is particularly beneficial because the same supersaturation can be established throughout the whole of the solution, provided that temperature gradients can be avoided. An example would be the thermal decomposition of urea in water,

$$(NH_2)_2CO + 2H_2O = (NH_4)_2CO_3 \qquad (11.2)$$

which results in an increase in pH and the generation of carbonate ions, either or both of which may bring about precipitation, depending on the cation in solution. At sufficient supersaturation, exceeding the concentration threshold for homogeneous nucleation, a large number of nuclei are formed suddenly (see Fig. 11.8). The formation of the nuclei drops the solution concentration to a level below that at which nucleation occurs, but enough excess solute remains to permit further growth of the existing nuclei. If the solution is kept uniform, subsequent growth of all particles proceeds at the same rate, giving powders of exceptionally uniform size distribution.

To control the initial formation of the nuclei precisely it is necessary to work under ultraclean conditions so that the necessary high degree of supersaturation is obtained before the particles are nucleated. Typically, the solutions are first passed through a very fine filter to remove all specks of dirt, and reaction vessels are cleaned meticulously.

A variety of particle sizes and shapes can be produced, depending on the reaction conditions. Moreover, the particles can be agglomerates of much finer primary particles. Development of a complete understanding of the factors and processes that de-

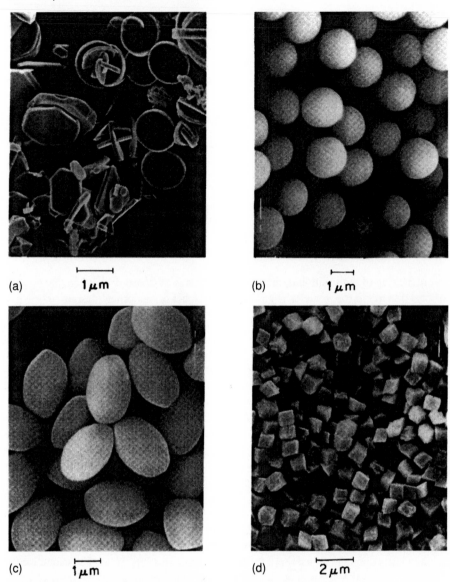

FIGURE 11.9 (a) Hematite (a-Fe$_2$O$_3$) particles obtained by heating at 250°C for 1 hour a solution 0.04 M in Fe(ClO$_4$)$_3$, 0.2 M triethanolamine (TEA) 1.2 M in NaOH, and 0.5 M in H$_2$O$_2$. (b) Cadmium sulfide (CdS) particles obtained by growing a "seed" sol prepared by aging a solution of 0.0012 M in Cd(NO$_3$)$_2$, 0.24 M in HNO$_3$, and 0.005 M in thioacetamide (TAA) at 26°C for 14 hours. To these "seeds" 12.5 cm^3 of 0.05 M TAA was added and kept at 26°C for 1.5 hours. (c) Iron (111) oxide particles obtained by heating at 100°C for 7 days a 50% ethanol/water solution 0.019 M in FeCl$_3$, 0.0012 M in HCl, and 5 x 10^{-5} M in NaH$_2$PO$_2$. (d) Cadmium carbonate particles obtained by mixing 5 cm^3 of a solution 2 x 10^{-3} M in CdC$_2$ with 5 cm^3 of a solution 10 M in urea at room temperature. (From E. Matijevic in Ultrastructure *Processing of* Ceramics, *Glasses and Composites,* L. Hench and D. Ulrich, eds., 1984, Wiley, New York.)

termine the sizes and shapes in the precipitation reactions remains as a challenge. Striking examples of uniform particles obtained by decomposition of some salt solutions are shown in Fig. 11.9.

Precipitation of mixed oxides is possible. For example, in the fabrication of some ferrites (magnetic ceramic materials used for recording heads and magnetic memory cores of which nickel ferrite is an example), one starts with a mixed solution of iron and nickel sulfates in water. The solution is kept at 70 to 80°C and then brought to a pH of around 11 with ammonium hydroxide. A very fine mixed hydroxide precipitates that, after rinsing to remove residual sulfate, can be dried to a powder with a particle size between 0.05 and 1 µm.

A particularly interesting case is that of stabilized and partially stabilized zirconia. The tetragonal phase of pure zirconia is stable above 1000°C, while the monoclinic phase is stable below that temperature. Consequently, pure zirconia is not a practical ceramic; on cooling a zirconia part, sintered at higher temperature, through 1000°C, the stresses accompanying the volume change of the transformation cause the material to fall apart. However, the tetragonal phase can be stabilized down to room temperature by adding small amounts of other oxides (e.g., yttria) in solid solution, and zirconia is then a practical material. A second possibility is to add very small quantities of yttria so that the zirconia is "partially stabilized." In this state the metastable zirconia will undergo transformation to monoclinic if it is sufficiently disturbed. Consequently, transformation takes place ahead of the tip of any crack growing into the ceramic part, and the stress causing the crack to propagate is partly opposed. In this way, zirconia is toughened [made less sensitive to the presence of microcracks or other flaws (Evans, 1984)]. It is clear that to be effective in either stabilized or partially stabilized zirconia, the yttria must be uniformly dispersed. A zirconia part containing yttria-free regions would be almost as useless as a pure zirconia part. Furthermore, the uniformity must extend down to a microscale; grains of the ceramic that contain insufficient yttria could crack on cooling, introducing microcracks into the structure.

The usual method of preparing stabilized and partially stabilized zirconia is by precipitation from aqueous solution using ammonia to adjust the pH. An alternative route is the removal of solvent water from mixed aqueous solutions of yttrium and zirconium sulfates by dispersing the solution into a large volume of ethanol (O'Toole and Card, 1987). The fine solid particles of mixed sulfate can then be calcined to the oxide. Finally, these ceramic powders can be prepared from mixed alkoxides by spray pyrolysis (Ishitawa, 1986) or hydrolysis (Fegley et al., 1985).

11.3.5 Sol-Gel Reactions

When ultrafine colloidal dispersions, consisting of particles 1 to 100 nm in diameter (a *sol),* lose fluid, they can turn into a highly viscous mass that responds elastically to low applied stresses. Such a mass is called a *gel.* When chemical methods are used to turn solutions of metal compounds, typically organometallics, into gels, we are dealing with the sol-gel process. We will see that in many cases the gelled material does not consist of isolated, identifiable colloidal particles, but rather of a three-dimensional continuous network of oxygen-metal bonds, much like a polymer (Yoldas, 1982). Highly reactive and pure ceramic powders can be prepared from such gels. With care, a big monolithic piece of gel may also be turned into a corresponding monolithic piece of dense ceramic. Aluminum silicate whiskers are prepared commercially by a sol-gel process.

In nearly all sol-gel processes one starts with solutions of metal alkoxy compounds. Alkoxides result from the reaction of metals with the OH group in alcohols, with the liberation of hydrogen:

$$nROH(\text{alcohol}) + Me(\text{metal}) \rightarrow (RO)_n Me(\text{alkoxide}) + (n/2)\,H_2(g) \qquad (11.3)$$

Often catalysts are necessary to make the reactions rapid. For example, reaction of aluminum metal with an excess of isopropyl alcohol will proceed around 80°C in the presence of a small amount of $HgCl_2$ that is needed, presumably, to break down the protective oxide layer that forms on the aluminum. A number of alkoxides from a very wide range of metals are commercially available in high purity. To make metal oxide powders from these organometallic precursors, we would start with a solution of the metal alkoxide in alcohol. A good choice is the same alcohol as that used for the alkoxide formation. To this solution water is added, either pure or diluted with some more alcohol. Two reactions then occur, which may be written, taking an alkoxide of a trivalent metal as an example, as

1. Hydrolysis

$$(RO)\text{-}Me\text{-}OR + HOH \rightarrow (RO)_2\text{-}Me\text{-}OH + ROH(\text{alcohol}) \qquad (11.4)$$

2. Condensation

$$(RO)_2\text{-}Me\text{-}OH + RO\text{-}Me\text{-}(RO)_2 \rightarrow (RO)_2\text{-}Me\text{-}O\text{-}Me\text{-}(OR)_2 + ROH \qquad (11.5)$$

The relative rates of the hydrolysis and of the condensation reactions can be influenced significantly by the presence of an acid (e.g., HCl) or a base (e.g., NH_4OH). Small quantities of acid will increase the rate of the hydrolysis reaction, while bases increase the rate of the condensation reaction (Zarzycki, 1984). Presumably this is because the reactions are actually more complex than depicted in equations (11.4) and (11.5), perhaps involving steps where hydroxyl ions and hydrogen ions are taken up and then released. The differences in the relative rates of the hydrolysis and condensation reactions can be used to affect the microstructure of the material that eventually results from the sol-gel fabrication process.

The remaining alkoxy groups -OR of the condensation product in equation (11.5) can hydrolyze again and go through another condensation reaction. If, for example, a trivalent metal atom has all three of its alkoxy groups undergo condensation, it will be linked by oxygen bridges to three other metal atoms. In this way, cross-linked, three-dimensional networks of metal-oxygen bonds can be formed.

What we end up with will depend on the relative rates of the hydrolysis and condensation reactions, the concentration of compounds in the solutions, the type of alkoxide, the reaction temperature, the relative amount of water added, and the pH of the solutions. For example, titanium alkoxide solutions will be readily converted to a colloidal dispersion of amorphous titania particles by the rapid addition of excess water. In this way all three metal-alkoxy bonds are hydrolyzed for most alkoxide molecules before there is much linking of the alkoxide molecules. However, highly polymerized titania gels have been obtained by using less water and by slowing condensation so that the partly hydrolyzed alkoxide molecules have a chance to diffuse to the growing polymers. This slowing is achieved by using high-molecular-weight alkoxy groups and acid catalysts. The time required for the gelation to develop fully can vary very strongly, from a few seconds to several days, depending on the temperature and the ratio of the water to alkoxide. Less water increases the gelation time dramatically.

Most metal alkoxides, dissolved in anhydrous alcohol, will react completely to form the condensation product when water is added, so that virtually no unreacted hydroxyl groups remain. The silicon alkoxides are notable exceptions, and clear solutions can be maintained for long times in which both the hydrolysis product and unreacted silicon alkoxide can coexist in significant concentrations. If to such a solution another metal alkoxide is added that reacts readily with hydrolysis products, we could prepare interesting complexes of mixed oxides in solution. This process is called solution complexation. Upon further addition of water, the complexed solution could be pushed into complete condensation polymerization. This can give powders or monolithic gels of extraordinary uniformity in chemical composition (i.e., the scale of segregation can be shifted down to the atomic level).

As an example, one can consider the formation of mullite by a sol-gel process (Mazdiyasni and Brown, 1972). In this process, silicon tetraethoxide in anhydrous ethanol is partially hydrolyzed by the addition of water. An aluminum alkoxide is added to produce solution complexation. Complexation can be a slow step and the clear solution needs to be stirred from a few hours to a few days. Then water is added in excess, leading to the polymerization condensation, forming a monolithic gel. This gel can then be dried and calcined to give mullite. At low temperatures the resultant mullite is amorphous. Upon heating to 1100 or 1200°C, the amorphous mullite transforms to crystalline mullite.

Practically, one works with a significant excess of anhydrous alcohol in the alkoxide solutions, so that when the gel forms, it contains only about 5 vol % of oxide. After drying at 100°C for times up to several days, a very porous material results that has about 10 to 15% of the theoretical density of mullite. If cracking of the gel is to be avoided, much slower drying at lower temperatures may be required.

After drying a monolithic gel, a material is obtained that has continuous, ultra-fine porosity. Depending on the details of the preparation steps, the porosity may be nearly monosized with a pore diameter of a few hundred nanometers or less. If this porous material is produced by drying in a common air drying chamber, such as an oven, then it is known as a *xerogel*. An alternative method of drying is to use a supercritical fluid. This is fluid above its critical point (where the distinction between a gas and a liquid disappears). Such fluids have unusual properties, such as high miscibility with common solvents, so that the alcohol of the gel can readily be removed by contacting it a few times with the supercritical fluid. By releasing the pressure, that supercritical fluid leaves as a gas, leaving the dried porous gel, which in this case is known as an *aerogel*. The principle of supercritical drying is illustrated in Fig. 11.10, which provides critical temperatures and pressures of some common solvents. Supercritical drying using carbon dioxide is especially interesting, because of its low critical temperature and pressure and its ability to act as a solvent for a wide variety of liquids, including water and alcohols.

An important difficulty in the fabrication of ceramic bodies directly from an appro-

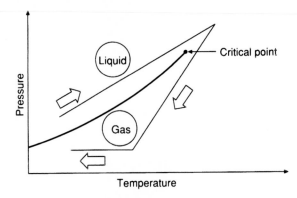

Solvent	Critical Temperature (°C)	Critical Pressure (atm)
Water	374.1	218.3
CO$_2$	31.0	72.9
Ethanol	243.0	63.0
Methanol	240.0	78.5
Ammonia	132.5	112.5
Acetone	235.5	47.0
Freon (CC1$_2$F$_2$)	111.5	39.6

FIGURE 11.10 Critical temperatures and pressures of some common solvents.

priately shaped monolithic gel is the management of shrinkage. These problems are most severe in the drying stage and in early densification, when the material is mechanically weak and large shrinkages are involved. Conventional drying of gels in air sometimes takes as much as five weeks, but the addition of certain chemical drying agents can speed up this process considerably. These *drying control chemical additives* (DCCAs) are compounds such as formamide or oxalic acid that facilitate the removal of water while avoiding cracking of the monolithic gel. The mechanism by which these additives work is presently uncertain. Possibilities include their effects on the surface tension or viscosity of the alcohol-water mixture permeating the gel, or more likely the "propping open" of the pores by the residues formed when solutions containing these additives are evaporated from the pores. After the water removal stage, which may have been shortened from a few days to a few hours, an additional heating step must be included to remove the remaining organic before final densification can proceed. If this organic burn-out step is not carried out, we may end up with a strongly discolored ceramic after densification, containing undesirable porosity.

Subsequent calcining of the xerogels or aerogels at 500°C leads to further shrinkage to densities around 30% of the theoretical full density of the ceramic. Final densification can usually be carried out significantly below the temperature at which conventionally prepared powder compacts are sintered because of the enormous surface area which provides a great driving force for sintering (Rahaman et al., 1988).

11.3.6 Other Solution-based Methods

Other methods have been exploited for the production of at least small quantities of powders from aqueous solutions; for example, passing the solution through a flame sufficiently high temperature is reached to evaporate the solvent rapidly and react (e.g., oxidize or thermally decompose) the solid formed.

In Chapter 9 we encountered solvent extraction where metal ions can be loaded into organic solvents containing suitable reagents (such as the LIX group of reagents). On contacting with water at elevated temperature in an autoclave, these organic phases undergo *hydrolytic stripping,* whereby an oxide or hydroxide is precipitated and the organic phase is regenerated. It is even possible to produce mixed oxides in this way (Doyle-Garner and Monhemius, 1985).

A difficulty in using solution methods for the production of multicomponent powders may arise when the component precursors convert to the useful precipitate under very different reaction conditions. For example, barium-alkoxides hydrolyze much more readily than yttrium or copper alkoxides, making the formation of uniform colloidal particles difficult. An interesting solution to such a problem has been devised, in which two different stable solutions, containing the reactants, are brought into contact with each other as high-velocity, impinging streams. The conditions are chosen so that upon collision, the solutions react instantaneously, thereby not allowing time for reactions to occur sequentially. To assist the reaction further, at the same time limiting the product particle size, the jets are collided inside a hollow, ultrasonically agitated horn. Figure 11.11 shows the method schematically. Highly uniform oxide particles, 0.1 mm or less in size, of multicomponent ZnO varistor composition and of the superconducting ceramic $YBa_2Cu_3O_{7-x}$ have been prepared this way (Voight and Bunker, 1988). The method is particularly well suited for continuous powder production.

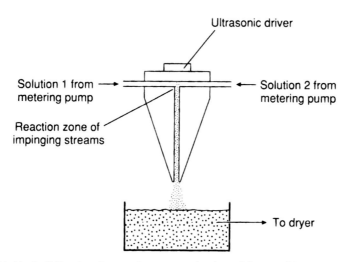

FIGURE 11.11 Colliding jets for continuous production of fine, multicomponent powders (after J. Voight, 1988).

11.4 Production of Powders by Solid-Solid and Gas-Solid Reactions

A number of powders can be prepared by direct reaction in the solid state or by gas reactions. Examples of these are the fabrication of silicon carbide and silicon nitride.

11.4.1 Silicon Carbide Production

Silicon carbide has many uses. It finds application, for example, as an abrasive, as a refractory furnace liner in some pyrometallurgical operations, as electrical heating elements, and as a high-temperature structural ceramic. It can be prepared by the direct reaction of silica sand (quartz) and carbon (coke). This method, known as the Acheson process, is carried out on a large scale (see the *Metals Handbook* for a detailed description). In its crudest form, the sand and coke mixture is piled around a long graphite cylinder in a trough that may be 20 m long, 3 m wide, and 2 or 3 m deep, holding over 100 tons of material at a time. Electricity (as much as 5 MW) is passed through the central graphite core. The overall reaction is

$$SiO_2 + 3C \rightarrow SiC + 2CO \qquad (11.6)$$

An examination of the Ellingham diagram of Chapter 2 shows that this reaction should proceed at temperatures of somewhat over 1500°C, but core temperatures of well over 2000°C are frequently employed. Significant side reactions occur, such as

$$SiO_2 + C \rightarrow SiO + CO \qquad (11.7)$$

SiO is a gas at the temperatures involved. This gas escapes from the reacting mass and oxidizes to SiO_2 on contact with air. In addition, SiO_2 and carbon form by the reverse of equation (11.7) as the gas mixture produced by this reaction cools on reaching the cooler outer regions of the reacting mass. The Acheson process is therefore one that produces a great deal of fine dust.

Sawdust is often added to the mixture to provide the pores through which the CO escapes. The reacted mass around the central graphite core is then removed by chipping and raking and the SiC washed with reagents such as hydrofluoric acid, which removes residual silica. The Acheson process produces approximately half a million tons of product per year worldwide. One drawback of the process is that it has a poor energy efficiency. Furthermore, for demanding applications such as high-temperature structural ceramics, better powders are needed and modifications have been made to suit small-scale operation. In one example, SiC was prepared from a silica gel, derived from silicon ethoxide in a sugar solution, by reacting the calcined mixture in a graphite crucible at 1600 to 1700°C. Oxidation of the powder mixture in air at about 600°C removes residual carbon, and washing with hydrofluoric acid removes unreacted silica. The resultant silicon carbide powder has a primary particle size below 1 μm, but the powder is agglomerated.

A more refined approach to the Acheson process is the reaction of silica vapor with carbon black. Silica is kept in a crucible below a bed of carbon black that is sitting on a grid. At about 1600°C the silica will start to evaporate and react with the carbon bed to produce silicon carbide powder. The qualities of the resulting silicon carbide powder depend significantly on the morphology of the carbon black particles. Silicon powders have also been produced by thermal decomposition of such vapors as CH_3SiCl_3 in hydrogen. Reactions carried out in a plasma, such as that between CH_4 and $SiCl_4$, are also possible routes to silicon carbide powders.

11.4.2 Silicon Nitride Production

Most silicon nitride is produced by the reaction of nitrogen and silicon in the temperature range 1100 to 1400°C:

$$3Si + 2N_2 = Si_3N_4 \tag{11.8}$$

Frequently, hydrogen is mixed with the nitrogen to minimize oxidation due to any oxygen introduced into the system, and ammonia is occasionally used as a substitute for this gas mixture. The silicon is in the form of a packed bed of metallurgical-grade silicon (produced in a submerged arc furnace as described in Chapter 8). Silicon nitride occurs as α, β or amorphous Si_3N_4, and the β form is stable at practical temperatures. The α or amorphous forms sinter more readily and reaction conditions are selected to minimize the amount of β silicon nitride produced.

Purer silicon nitride is produced by the gas-phase reaction of ammonia with silicon tetrachloride. The overall reaction is

$$3SiCl_4 + 16NH_3 = Si_3N_4 + 12NH_4Cl \tag{11.9}$$

but consists first of the formation of silicon diimide $Si(NH)_2$ and its thermal decomposition.

Silicon nitride can be prepared by the reaction of silane and ammonia in the gas phase. Ammonia has to be used because nitrogen is too unreactive. Direct combustion of these gases can, for example, be done in a burner that is somewhat similar to a regular acetylene-oxygen torch. One ends up with large amounts of hydrogen and voluminous quantities of smoke, consisting of very fine particles of silicon nitride. These have to be passed through a good cyclone collector or even an electrostatic precipitator to be captured. A variant is the reaction of silane with ammonia under irradiation of a CO_2 laser. The overall reaction can be written as

$$\underset{\text{(silane gas)}}{3\,SiH_4} + \underset{\text{(ammonia gas)}}{4NH_3} \rightarrow \underset{\text{(silicon nitride)}}{Si_3N_4} + 12H_2 \tag{11.10}$$

The laser beam does not merely provide the heat necessary to bring the reactants to the reaction temperature; rather, the incident radiation is chosen so that its frequency matches the stretching frequency of the bonds in one or more of the reactants. That bond is thereby stimulated directly to undergo reaction, producing a smoke of silicon nitride particles with a mean diameter of around 50 nm.

Some of these gas reactions tend to produce very low density, stringy agglomerates that have to be broken down mechanically. The laser-stimulated gas reactions have the advantage that reactions can be well controlled by manipulating such parameters as wavelength and intensity of the laser beam, together with concentration of gaseous reactants. The method is presently used only for specialty powders.

11.4.3 Production of Other Ceramic Powders at High Temperature

Many other ceramic powders can be produced by high-temperature gas-phase or gas-solid reactions. Examples are boron nitride production by reaction of boron trichloride with ammonia:

$$BCl_3 + NH_3 = 3HCl + BN \tag{11.11}$$

the production of titanium diboride by reaction of titanium tetrachloride with boron trichloride and hydrogen:

$$TiCl_4 + 2BCl_3 + SH_2 = TiB_2 + 10HCl \tag{11.12}$$

and the production of tungsten carbide by reaction of tungsten hexachloride with methane and hydrogen:

$$WCl_6 + H_2 + CH_4 = WC + 6HCl \tag{11.13}$$

Fine alumina powders have been prepared commercially by the thermal decomposition of ammonium alum:

$$(NH_4)_2SO_4\, Al_2(SO_4)_3 = 2NH_3 + Al_2O_3 + 4SO_2 + H_2O \tag{11.14}$$

11.5 Production of Metal Powder

The methods using powdered metals for manufacturing of various objects are often referred to as *P/M technology*. In their oldest form they were practiced as early as 3000 B.C. in Egypt, India, and Africa for making tools from sponge iron. Much of the early commercial efforts, in the beginning of the nineteenth century, were for making platinum objects. Modern applications include the so-called cemented carbides, such as tungsten carbide sintered with 5 to 10% cobalt powder for machine tool bits, porous metal bearings, tungsten rocket nozzles, and a variety of mechanical parts from steel powder for automobile engines, as well as high-performance jet engine parts forged from superalloy powders. The use of P/M methods rather than casting permits the fabrication of parts with much finer and more uniform microstructures. This is because cast parts are prone to the macrosegregation discussed in Chapter 10, and such segregation can be avoided in P/M processing. Care must be taken, however, to avoid an alternative form of segregation where, for example, the pouring of dissimilar powders into a die results in the denser particles segregating to the bottom.

Two major types of metal powder production are in use: atomization of molten metal, and chemical methods, including gaseous reduction of oxides and decomposition of carbonyl complexes. A variety of methods used for the spraying of the liquid metal are depicted schematically in Fig. 11.12. When oxidation of the sprayed metal is a problem, the spray can be produced in vacuum or under inert gas with quenching into an oil bath. An example of metal powders fabricated by atomization of molten metal is shown in Fig. 11.13. High-purity powders with a low oxygen content can be produced this way, with particle sizes in the range 1 μm to a few tens of micrometers.

Gaseous reduction of metal oxides has been used to produce powders of iron, cobalt, and nickel. The reducing gases are either hydrogen or carbon monoxide. Water and carbon dioxide, respectively, can be used to fix the oxygen partial pressures of the gases; the oxygen partial pressure is found in many cases to affect the structure of the resulting metal

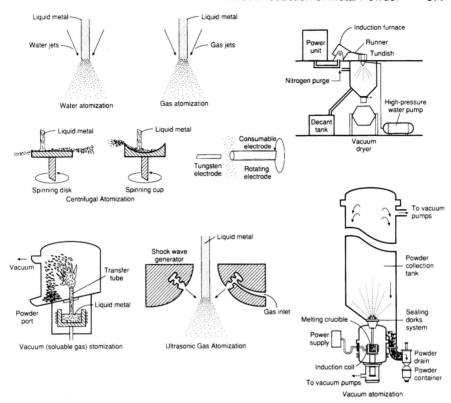

FIGURE 11.12 Various schematics for production of metal powders by atomization. (Reprinted with permission from *Metals Handbook, Vol. 7,* 9th ed., *Powder Metallurgy,* American Society for Metals, Metals Park, OH, 1984.)

powder. Much of the background pertaining to the gaseous reduction reaction was discussed in Chapter 3. In general, the lower the reaction temperature, the finer the resultant powder. This was realized quite early [in the powder production technology]. Osann observed this as early as 1830, when he first produced copper powder by reduction of $CuCO_3$.

Metal powders, particularly nickel and iron, can be produced by thermal decomposition of carbonyls (see the *Metals Handbook).* These highly toxic gases are produced by reaction of the metal with carbon monoxide at temperatures of a few tens of degrees and at elevated carbon monoxide pressures. The carbonyls can then be decomposed at temperatures of a few hundred degrees, regenerating the carbon monoxide. The kinetics of the thermal decomposition are worth discussing in the context of the concepts laid out in Chapter 3. Fig. 11.14 shows a sketch of a heated surface in contact with a gas containing a metal carbonyl. If the surface temperature is not excessive, the decomposition reaction takes place on the substrate surface itself, coating it with the metal. In this case the reaction is controlled by the chemical steps at the surface, and the concentration gradient across the boundary layer is small. At higher surface temperatures the kinetics of the surface reactions are accelerated, concentration of the carbonyl at the surface is low, and mass transfer becomes rate controlling. At yet higher temperatures

FIGURE 11.13 SEM of a vacuum-atomized nickel-based powder (-100 mesh). (Reprinted with permission from *Metals Handbook, Vol. 7,* 9th ed., *Powder Metallurgy,* American Society for Metals, Metals Park, OH, 1984.)

heat transfer into the gas from the surface becomes significant and the carbonyl molecules decompose before reaching the surface. Under these circumstances the metal is formed around nuclei produced away from the surface (i.e., a powder results). Manipulation of the substrate temperature therefore enables the chemistry of carbonyl decomposition to be exploited to produce coatings or powders.

Metal powders can also be generated electrochemically; for example, electrolysis of aqueous solutions of metal salts, using a cathode to which the metal will not adhere, typically results in metal powders (see the Metals Handbook).

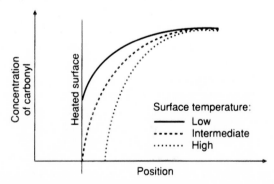

FIGURE 11.14 Concentration profile of nickel carbonyl in a gas phase in contact with a heated surface on which decomposition takes place to produce nickel and carbon monoxide.

11.6 Production of Other Important Ceramic Powders

We have already encountered (in Chapter 7) the Bayer process, which is used to produce alumina from bauxite. Although much of that alumina is used in the production of aluminum (in the Hall cell described in Chapter 9), a significant fraction remains as oxide and is used in the manufacture of high-grade refractories, spark plug insulation, crucibles, cutting tool bits, and other products. In the Bayer process the alumina is selectively leached by aqueous alkaline solution at high pH, leaving behind most of the impurities (silicon and iron compounds) of the bauxite. After separating the solution from residual solids (Fig. 7.1), it is diluted with water and seeded with crystals to precipitate hydrated alumina, which is calcined in a kiln to form alumina. A similar technology is used to obtain zirconium dioxide (zirconia) from impure zirconia containing sands or deposits of zircon ($ZrSi_2O_4$).

Production of magnesium oxide from seawater and calcined dolomite was mentioned in Chapter 9. After most of the sodium chloride has been removed from the seawater by evaporation, calcined dolomite (CaO•MgO) is added to raise the pH and yield magnesium hydroxide, which can be calcined to the oxide.

11.7 Concluding Remarks

In this chapter we have examined several techniques for the production of powders for use in ceramic processing or powder metallurgy. The reader will recognize that the techniques described show considerable variation in the quality of the materials produced and the cost of production, the two usually going hand in hand. In many instances the costs of the reagents vary enormously, so that, for example, the cost of the silane used in reaction (11.10) would be far higher than the cost of metallurgical silicon used in the production of silicon nitride by nitriding silicon. Energy costs would also be significant, because many of the techniques we have discussed require the use of high temperatures. Finally, capital costs may be high for some of the more advanced techniques, such as laser processing; the rate of production would typically be small relative to the cost of the equipment, making the capital cost per unit of production high. In some cases the high cost of production is justified by the superior quality, (e.g., narrow size distribution and freedom from impurities) of the powders produced.

LIST OF SYMBOLS

d diameter
g gravitational acceleration constant
R radius
ρ density

PROBLEMS

11.1 Determine the critical rotation speed for a ball mill at which the centrifugal forces just balance gravity.

11.2 The centrifugal force acting on a body of mass m rotating about a center at a distance r with an angular velocity ω, is $m\omega^2 r$.

    (a) Consider a ball mill containing just one ball. If θ is the angle that a line from the ball to the center of the mill makes with a vertical line drawn down from the center, derive an equation giving the value of θ (call it θ_c) at which the ball will leave the wall, assuming it to be smooth.

    (b) Write and run a computer program to calculate the point of impact of the ball in part (a) with the mill. From this, plot the kinetic energy of the ball on impact as a function of mill speed (expressed as a percentage of the critical rotational speed at which the centrifugal force just balances gravity at the top of the mill).

    (c) Now consider a real ball mill where many layers of balls are present. Would the value of θ_c for a ball in contact with the mill be changed by the presence of the other balls? Explain your answer.

11.3 A 0.05 M solution of magnesium sulfate is spray-roasted, using a nozzle that produces 50-mm-diameter droplets, to produce MgO particles. If the agglomerates produced have a porosity of 50%, what diameter agglomerate would be expected from the roaster? Nonporous MgO has a density of 3.6×10^3 kg/m^3 .

11.4 In this chapter we have emphasized the significance of producing fine, unsized ceramic particles. We encountered an interest in producing unsized particles much earlier in this book, in Chapter 1. In that chapter, techniques for separating mineral particles were briefly discussed and it will be recalled that many of these techniques work better if the particles are of similar size. Consequently, it is usual in mineral processing to encounter units that produce particles coupled to size classification devices, as shown.

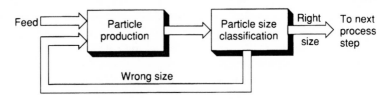

Conceivably, size classification could be coupled to any of the techniques for powder production described in the present chapter. Discuss how this might be attempted and the difficulties you would expect to encounter.

11.5 You wish to make some titania (TiO_2) powder by the sol-gel method, starting with a 10 vol % solution of titanium ethoxide. Assuming that the reaction is stoichiometric, how much water must you add? Describe the process in detail.

11.6 You wish to make a disk of dense mullite 5 in. across and 1 in. thick, using the sol-gel process, which yields a gel monolith. Calculate the volume of the liquid vessels required for the synthesis. Assume the following: you start with a 40 vol % solution of the alkoxide in butanol, you add about 100 mol of water per mole of alkoxide, and the oxide content of the gel generated in the first synthesis step is about 10%.

11.7   You are considering setting up a unit to produce monosized titania powders by the sol-gel method in a continuous process. Consulting the available literature as necessary, discuss the issues involved in the design of such a production unit.

11.8   A ball mill, using Al_2O_3 balls, reduces particle diameter from 100 mm to 5 μm. The following data are available:

> Ball mill diameter 3 ft
> Ball mill length 4 ft
> Ball mill load:
> Al_2O_3 balls: 50% full
>   Powder charge: 30% of ZnO powder
>   ZnO powder density: 40% of theoretical (5.6 g/cm³)
>   ZnO surface energy: 1000 erg/cm²

Assuming that all the energy input for milling goes into lifting the Al_2O_3 balls to the point where they drop down, estimate the mill efficiency in terms of surface area energy increase per unit of energy input. State whatever additional assumptions you may have to make.

11.9   Consider a drying slab of monolithic gel of thickness 2L. The retained solvent is water. Upon drying, stresses will develop in this slab due to water concentration gradients that result from emptying the larger pores first. Assume that these stresses follow a parabolic distribution ($\sigma = ax^2 + b$), where x is the distance from the center of the slab, $x = 0$, to the surface, $x = L$. Note that the slab is in a state of self-stress. Assume further that the stress at the outside surface is determined by the capillary tension produced by the water menisci in the finer pores, which have a radius of 100 angstroms, and that the gel is fully saturated with water in the center of the slab. Water has a surface tension of 0.073 N/m. What is the magnitude of the compressive stress in the center of the slab? What is the magnitude of the tensile stress at the surface of the slab? At what value of x/L is the stress equal to zero? [*Note:* This is a worst case. In practice, the water concentration gradient will decrease with slower drying. See also Zarzycki (1984).]

REFERENCES AND FURTHER READINGS

BEDDOW, J. K., *Particulate Science and Technology,* Chemical Publishing Co., New York, 1980.

BRINKER, C., D. CLARK, and D. ULRICH, eds., *Better Ceramics Through Chemistry,* 11, Material Research Society, 1986.

CHIN, G., ed., *Advances in Powder Metallurgy,* American Society for Metals, Metals Park, OH, 1982.

DOYLE-GARNER, F. M., and A. J. MONHEMIUS, *Metall. Trans. 16B,* 671 (1985).

EVANS, A. G., *Ceramic Firing Systems,* Noyes Publications, Park Ridge, NJ 1984.

FAMILY, F., and D. LANDAU, eds., *Kinetics of Aggregation and Chelation,* North Holland, Amsterdam, 1984.

FEGLEY, B., Jr., P. WHITE, and H. V. BOWEN, *Am. Ceram. Soc. Bull. 64,* 1115 (1985).

HENCH, L., and D. ULRICH, eds., *Science of Ceramic Chemical Processing,* Wiley, New York, 1986.

ISHIZAWA, H., et al., *Am. Ceram. Soc. Bull. 65,* 1399 (1986).

JEAN, B., R. GOY, and T. RING, *Am. Ceram. Soc. Bull. 66,* 1517 (1987).

JOHNSON, D. W. Jr., Nonconventional Powder Preparation Techniques, *Am Ceram. Soc. Bull* 60, 221 (1981).

JONES, W., ed., *Fundamental Principles of Powder Metallurgy,* Edward Arnold, London, 1976.

MAZDIYASNI, K., and L. M. BROWN, *J. Am. Ceram. Soc. 55,* 548 (1972).

MCCOLM, I. J., and N. J. CLARK, *Forming, Shaping and Working of High-Performance Ceramics,* Chapman & Hall, London, 1988.

Metals Handbook, ASM American Society for Metals, Metals Park, OH.

NORTON, F., *Elements of Ceramics,* 2nd ed., Addison-Wesley, Reading, MA, 1974.

ONODA, G. and L. HENCH, eds., *Ceramic Processing Before Firing,* Wiley, New York, 1978.

O TOOLE, M. P. and R. J. CARD, *Am. Ceram. Soc. Bull. 66,* 1486 (1987).

RAHAMAN, M., L. DEJONGHE, and P. TEWARI, *J. Am. Ceram. Soc.* 71, C-338 (1988).

VOIGHT, J., and B. BUNKER, unpublished work at Sandia National Laboratory, 1988.

YOLDAS, B., *Non-Cryst. Solids 51,* 105 (1982).

ZARZYCKI, J., in *Ultrastructure Processing of Ceramics, Glasses and Composites,* L. Hench and D. Ulrich, eds., Wiley, New York, 1984.

Powder Compaction

12.1 Introduction

In Chapter 5 we mentioned the importance of the arrangements of the particles in a powder compact with respect to the perfection of the microstructure that results after sintering. Colloidal methods proved to be very fruitful in manipulating the structure of the starting powder compact, the green compact, especially for submicrometer powders. In many industrial applications, where mass production is desired, colloidal compaction methods have not yet been incorporated or can be adopted only with sacrifice of higher fabrication costs. To date, many articles prepared from powders are still made by various mechanical consolidation methods, such as die pressing, isostatic compaction, and extrusion. We describe these methods in a little more detail in this chapter. As before, the structure of the consolidated green powder compact will have a very strong effect on the properties of the final sintered product, and the factors that affect the homogeneity of the green compact need to be examined.

We start out by considering how simply shaped particles can be arranged in compacts, and then point to some practical factors that interfere in obtaining the idealized particle packings. Then we deal with the additional problems that are introduced when compacting powders in dies. Finally, we describe various casting methods as well as injection molding.

12.2 Packing of Particles

A first approach to understanding how the structure of the green state compact comes about is to look at how equally sized spheres can be packed. The simplest ways are the regular, crystal-like packings where ordered three-dimensional arrangements are achieved. In order of increasing packing density these are simple cubic, body-centered cubic, face-centered cubic, and hexagonal close-packed arrangements. Examination of the geometry of these packings will yield the packing density as well as the number of nearest neighbors—the coordination number—for each particle. The following results are obtained:

Packing geometry	Density (fully dense = 100)	Coordination number
Simple cubic	52	6
BCC	68	8
FCC	74	12
HCP	74	12

In powder compacts we cannot expect the particles to sit in such regular arrangements, even if they were all monosized, except under extraordinary circumstances. Small particle clusters might approach this crystal-like packing, but overall we should expect more random arrangements. Such random arrangements have also been studied for a variety of differently shaped particles: equally sized spheres, different sized spheres, and even mixtures of differently shaped particles such as rods and spheres. In most cases, computer experiments can give considerable insight as to what we should expect. In the most complex cases people are still resorting to experiments with model materials.

In the simplest case of perfectly spherical, monosized particles, we expect intuitively two different situations: If we pile the particles randomly into a box, we should get a lower packing density if we do not allow rearrangement than when we shake the system and let the particles settle in as favorable a position as possible with respect to overall density, while still being unable to produce a crystal-like packing arrangement. You might want to liken this difference to the difference between a liquid and a crystalline material; in fact, this similarity is more than casual. Two different random packings are then distinguished: a *loose random packing* and the settled *dense random packing*. Computer experiments and trials with ball bearings have yielded a loose random packing density of about 60% (with reference to a solid without any holes), and about 64% for the dense random packing. In practice, this would need to be compared with what is known as a *poured density* and a *tap density*. As the descriptive name indicates, in the latter case one pours powder in a graduated cylinder of some sort, then taps or vibrates the beaker to enable the particles to settle. After approximately 200 taps, the powder does not settle any further; this is the tap density. Some experiments on copper powder are summarized below and simultaneously show the strong effect of the particle shape on the resulting packing densities:

Powder Shape	Poured density	Tap density
Spherical	50.4	59.4
Irregular	25.8	35.2
Flakes	4.5	7.8

Source: Metals Handbook 9th ed., Vol. 7.

An important reason for the significant difference between the practically achieved tap densities and the calculated ones is that real powders cannot rearrange fully and that large voids, often tens to hundreds of times larger than those in the calculated

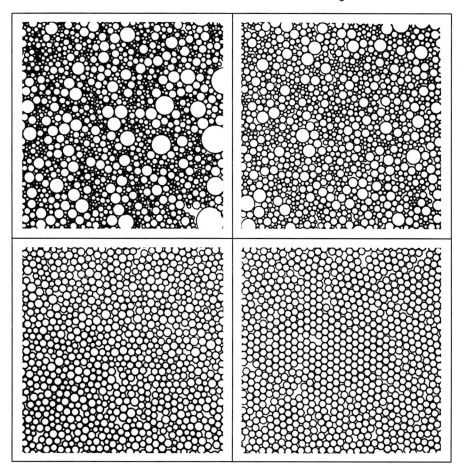

FIGURE 12.1 Four typical packings of multisized circles. [From R. Burk and R. Apte, A Packing Scheme for Real Size Distributions, *Am. Ceram. Soc. Bull.* 66(9), 1390 (1987). Reprinted by permission of the American Ceramic Society, Columbus, OH.]

structures, are not eliminated by any finite amount of tapping or shaking. Tap density is sometimes given by powder manufacturers, and high tap densities are generally seen as a characteristic of a high-quality powder.

Although we might have near-monosized powder, there will be a spread of particle sizes around the mean size. This spread significantly influences the packing density achieved. Figure 12.1 shows some two-dimensional computer experiment results for packing of non-monosized circles with a log-normal size distribution of standard deviation σ. For strictly monosized particles, crystal-like arrangements are achieved with high packing density. As soon as σ assumes some nonzero value, the packing density drops rapidly but then starts rising again as smaller particles can be fitted into the interstices of the larger ones. This is evident in Fig. 12.2. It would then seem that we would actually be better off by having wide particle size distributions, at least from the

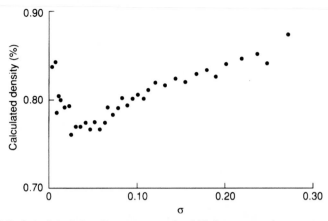

FIGURE 12.2 Calculated density versus peak width for one peak.

point of green packing density; and we might wonder why attempts should be made to make monosized particles. The packing density itself is, however, a misleading parameter for predicting subsequent behavior in densification and microstructural development during sintering. With increasing width of particle distribution the system becomes heterogeneous on a continually larger spatial scale. This is evident from a comparison of the pictures in Fig. 12.1. It is this increasing scale of heterogeneity that introduces ill effects in the subsequent sintering process, and heterogeneous final products are likely to result. To maintain the benefit of a higher green density with wide particle size distributions, it is necessary to keep the spatial heterogeneity down, at least to the same scale as the particles. Although this is not impossible in principle, it becomes increasingly difficult with decreasing mean particle size. Nevertheless, many manufacturers have not yet adopted the use of monosized particles and achieve acceptable product quality through intense mixing of the usual powders.

Interesting powder packing phenomena occur in simple bimodal mixtures of powders. Here we blend two fractions of particles with different sizes; each fraction has a very narrow size distribution. For such a binary mixture of powders, a maximum packing density (or, conversely, a minimum packing volume) is often observed at about 30 wt % of the fine fraction. Westman and Hugill attempted to explain this as follows: Starting from the coarse fraction, which forms an open skeletal framework, we could fit fine particles inside the interstices of this skeleton without increasing the external size of the compact. This procedure would lead to an extrapolated infinite density or zero specific volume, and produces the line *C-B* in Fig. 12.3. Starting from the fine fraction, insertion of the dense coarse powder particles would lead to an extrapolated 100% density, line *F-A* in Fig. 12.3. The lines intersect at about 30% fines if the specific volumes, the volume for a given weight of powder, of the pure coarse and the pure fine fractions do not differ significantly. This simple reasoning produces reasonable agreement with experimental results on binary mixtures. The question of spatial homogeneity remains a problem, although it is less severe than that of wide particle size distributions. One could carry this line of reasoning further and attempt to fill up the hierarchy of interstices in the powder compact with continually finer particles, thus

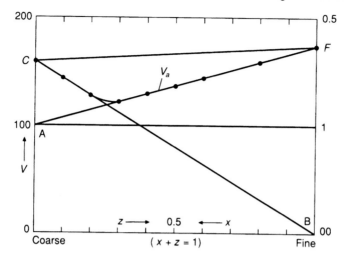

FIGURE 12.3 Diagrammatic representation of the Westman and Hugill hypothesis to account for the apparent density of binary mixtures of particles. Vertical axis is volume per unit mass with fully dense solid being 100. (From P N. Rhines, "Dynamic Particle Stacking" in Ceramic *Processing Before Firing*, G. Onoda and L. Hench, eds., Wiley, New York, 1978, p. 327)

using simple geometrical methods to generate a distribution of particles that would lead to very high green density. Practically, however, little is gained by this since particles do not locate where they ideally would need to be. The problem is analogous to that of the wide particle size distributions mentioned before.

A first requirement for compacts of non-monosized particles will then be to approach spatial homogeneity as much as possible by extensive mixing, before the compact is made. After this, the powder will be further compacted to increase its green density prior to sintering, and hopefully to eliminate, as much as possible, large voids that might have been produced accidentally. Large voids can be reduced significantly by the age-old process of kneading. This is a process that involves extensive shear of the powder mixture, and can be achieved by designing the forming procedure of the green compact so that it includes large-strain shear deformation. For example, a fat cylinder of powder mixed with some sticky binder (organics) could be re-pressed into a thin plate. The large voids that may have been present in the original compact are then squeezed shut, provided that the pores in the compact remain continuous so that gas is not trapped.

The production of short-fiber composites in which whiskers such as silicon carbide are dispersed in a ceramic matrix of regular powder, such as alumina, presents special packing problems. This may be inferred from the data on the comparison of the poured and tap densities shown above. Whiskers tend to stack to very low relative densities, with decreasing packing density at increasing ratios of their length L over their diameter D, or at increasing aspect ratio, L/D. This problem was considered by Milewski. Figure 12.4 shows some data on the relative tap density of whisker compacts as a function of aspect ratio. At large aspect ratio the whiskers have a strong tendency to tangle. They may also form bundles or loose clumps. In the final product we may then

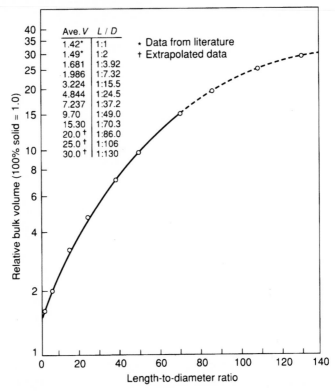

FIGURE 12.4 Packing curve for fibers with varied *L/D* ratios. [From J. Milewski, *Adv. Ceram. Mater.* 1 (36) 37 (1986). Reprinted by permission of the American Ceramic Society, Columbus, OH.]

find large pores spanned by whiskers, with little or no matrix powder. Conversely, we would also find regions that are quite dense but do not contain whiskers. More extensive powder mixing is necessary to improve the structure of such ceramics. Table 12.1 gives some packing densities of model material mixtures of fibers with a spherical powder. The factor R is the ratio of the diameter of the spheres divided by the diameter of the fibers. Best packing results are clearly obtained for low-aspect-ratio whiskers that are thin compared to the sphere diameters, at lower whisker contents.

In some cases it has been found that good results can be obtained for fabrication of whisker/powder ceramic composites if the matrix powder and the whiskers are ball-milled together for a while. Hard whiskers, such as SiC, can actually retain a useful anisotropy despite the milling, perhaps around 10 for the length/diameter ratio. In that case, simple ball milling can provide excellent mixing.

12.3 Powder Compaction

When an initial powder compact is further compacted by mechanical means, shear deformation will usually accompany the density change. In the simplest case, the pow-

TABLE 12.1 Experimental solid contents at 25%, 50%, and 75% fiber loading for fiber-sphere packing

Fiber L/D	Percent fibers	R value									
		0	0.11	0.45	0.94	1.95	3.71	6.96	14.30	17.40	∞
	25	68.5	68.5	65.4	61.7	61.0	64.5	70.0	74.6	76.4	82.0
3.91	50	76.4	74.6	67.2	61.7	60.2	64.1	67.5	72.5	74.5	75.7
	75	78.2	69.5	64.5	61.0	59.5	62.5	64.2	66.7	67.2	67.1
	25	68.5	68.5	64.5	61.0	58.5	59.9	64.5	73.5	74.6	80.6
7.31	50	76.4	71.4	67.5	58.8	55.5	56.6	58.8	65.4	67.1	67.1
	75	66.3	61.7	60.0	55.0	52.8	53.5	54.6	57.2	58.2	57.4
	25	68.5	66.7	63.7	59.9	54.6	50.3	50.5	54.1	57.5	65.0
15.52	50	61.7	55.6	51.8	50.7	45.5	42.0	42.4	44.3	44.3	48.1
	75	41.0	40.4	37.9	38.2	37.3	35.7	35.5	36.0	36.8	38.2
	25	68.5	66.5[a]	61.5[a]	55.5[a]	47.5	45.5	40.2	42.7	44.7	50.5
24.50	50	40.0	39.0	38.0	36.0	34.0	32.7	30.3	31.8	31.8	33.5
	75	26.4	26.3[a]	26.2[a]	25.8[a]	25.5[a]	25.2	24.3	25.0	25.6	26.2
	25	50.0	48.0[a]	45.0[a]	42.0[a]	39.4	37.7	33.8	33.1	39.2	41.3
37.10	50	25.7						22.6	22.6	22.6	25.6
	75										

[a] Estimated values (extrapolated data).

Source: J. Milewski. Efficient Use of Whiskers in the Reinforcement of Ceramics, Adv. *Ceram. Mater.* 1(1) 40 (1986). Reprinted by permission of the American Ceramic Society.

der is enclosed in some deformable capsule, like a thin rubber balloon, and immersed in a liquid that is put under high pressure. This process is isostatic compaction. Although the applied external stresses are hydrostatic, the powder compact cannot change volume in the same way as can a compressible material such as a gas. Particle rearrangement and fracture, collapse of pores and voids, and collapse and fracturing of agglomerates are all processes that can accompany the volume decrease of the powder compact. Despite the microscopic complexities of the compaction process, a relatively simple pressure-volume relationship is often observed for the compaction of brittle powders. When a powder is compressed for the first time, the pressure-volume relationship can be expressed by a normal or virgin consolidation relationship of the type

$$P = \alpha + \beta \ln\left(\frac{1}{1-\rho}\right) \tag{12.1}$$

where P is the applied hydrostatic pressure, ρ the relative density of the compact, and α

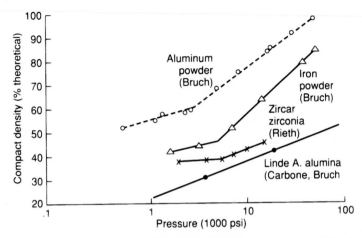

FIGURE 12.5 Variation of relative compact density with log pressure. (Reprinted with permission from J. Reed in *Treatise on Materials Science and Technology: Ceramic Fabrication Processes,* Academic Press, New York, 1976.)

and β are constants that depend on the initial density and on the nature of material. For plastically deforming materials such as metal powders, similar types of relationships can be found, where now the constants and α and β relate to the yield strength of the powder particles.

If the powder has previously been compressed, it will be considered over consolidated and respond elastically until the pressure of the previous compaction has been exceeded. Equation (12.1) results from fitting curves to experimental data. Many other relationships have been proposed in the literature, but equation (12.1) is just as good as any; further, it is relatively simple. In the compaction of ceramic materials, breaks in the log-linear relationship [equation (12.1)] can sometimes be observed at high pressures. This is usually thought to be associated with the fracture and collapse of looser agglomerates in the compact. Figure 12.5 shows volume-pressure relationships for some ceramic powders in which some breaks were observed.

12.4 Die Compaction

When powders are compacted in a die, a new difficulty enters: friction between the die wall and the powder. This friction is inherent in die compaction and places significant limitations on the size and shapes that can be fabricated. Die wall friction occurs because the stress applied by the die plunger is transmitted to the die walls. Thus a frictional stress opposes the applied one at the contact plane of the powder with the die. This will lead to stress gradients in the compact and hence to green density gradients, which can be quite troublesome. The severity of the friction between the powder and the wall depends to a large degree on how extensively the powder gets embedded in the wall or fits in the microscopic roughness of the wall. Wall friction is therefore decreased if the die wall is very hard; tool steel or even tungsten carbide is used and is

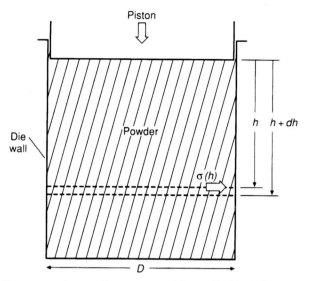

FIGURE 12.6 Geometry of a powder compacted in a cylindrical die.

surface-finished to a high degree of smoothness so that the microscopic roughness is significantly smaller than the powder particle size. Addition of some lubricants, such as polyvinyl alcohols or zinc stearate, can greatly reduce die wall friction and thus leads to improved spatial homogeneity of the density of the green compact. Die wall friction has been measured experimentally by several people, for example by measuring the force necessary to push a slab of compacted powder between two plates made out of the die material. Measurement of the force necessary to push this plate forward as a function of the known pressure on the side plates can then give the coefficient of friction. For low-friction conditions the slippage or failure occurs right at the die-powder interface. For conditions of high friction the failure occurs in the powder, adjacent to that interface; this situation is undesirable.

Detailed calculation of the stress and accompanying density gradients that are present in a powder during die compaction is difficult, but a qualitative idea can be obtained if we ignore the radial stress gradients. Let us consider the geometry shown in Fig. 12.6. If we consider an incremental slice of the cylindrical compact with a thickness dh at a depth, h the local axial stress, when taken to be uniform over the compact cross section, is $\sigma(h)$; the resulting radial stress acting against the die wall is $\sigma_r(h)$. If the coefficient of friction between the die wall and the powder is μ, an opposing axial force will be generated of magnitude

$$F = \mu\sigma_r(h) \tag{12.2}$$

per *unit area of wall*. The total opposing force, F_t along the periphery of the thin slab at h is then

$$F_t = F \pi D d h \tag{12.3}$$

The upward force at the periphery due to friction must equal the difference between the downward force exerted on the top of the slab by the powder above it and the downward force exerted by the slab on the powder below it.

$$F_t = \frac{\pi D^2}{4}\sigma(h) - \frac{\pi D^2}{4}\sigma(h + dh) = \frac{\pi D^2}{4}\left(-d\,\sigma\right) \qquad (12.4)$$

The wall stress $\sigma_r(h)$ will be proportional to the axial stress $\sigma(h)$

$$\sigma_r(h) = B\sigma(h) \qquad (12.5)$$

where the proportionality coefficient B is determined by the elastic properties, such as the effective Poisson ratio, of the powder compact. In detail, B would be a function of the powder density, for example, but we will leave that complication aside. A combination of equations (12.2)-(12.5) then gives upon integration

$$\ln\frac{\sigma(h)}{\sigma_0} = -\frac{4\mu Bh}{D} \qquad (12.6)$$

The proportionality coefficient B can vary from 1 for the purely hydrostatic stress in the slab (never realized in a powder in a die) to some low value. $B = 0.5$ would be a reasonable guess. The coefficient of friction can also vary strongly depending on the degree of lubrication. For a dry, unlubricated powder, μ might be as much as 1. For a lubricated powder it could be as low as 0.1. Taking the case of the lubricated powder with $\mu = 0.1$ and $B = 0.5$, we could now consider the maximum value of h for which the stress $\sigma(h)$ differs by no more than 10% from σ_0. Inserting those numbers in equation (12.6) would give you the maximum permissible ratio of the die depth to its diameter:

$$\left(\frac{h}{D}\right)_{max} \approx 0.5 \qquad (12.7)$$

In practice this is about the accepted limit for a single-action die in which the powder is compressed by one piston only against the stationary die bottom. For a double-action die in which top and bottom pistons move to compress the powder, the practical limit to h/D is usually taken to be around 1. Special care should be taken in filling the die evenly with loose powder before the compression cycle, since uneven filling requires more internal powder movement, which can lead to additional difficulties in obtaining reasonably homogeneous compacts. This is especially true when pressing thin wafers, where lateral motion of the powder becomes very difficult, resulting in fracture of the wafer due to elastic spring-back when the die pistons are removed upon unloading. This is because the thin, pressed plate expands as the die pistons are removed, but since the plate is still confined by the die wall, it will buckle and break. Removing the die while maintaining pressure on the pistons, and then removing the pressure on the pistons when the plate can expand without obstruction, can reduce the problem of fracturing significantly.

Gradients in green density after compaction can also lead to fracture of the compact when it is removed from the die. The cracks tend to run parallel to the piston surface in

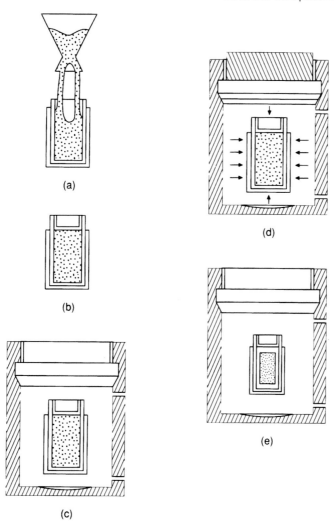

FIGURE 12.7 Wet-bag pressing principle: (a) fill up flexible bag tooling with material to be compacted; (b) close and seal bag tooling; (c) bag tooling in pressure medium and pressure vessel; (d) pressurize; (e) resulting compact, after decompression. (Reprinted with permission from G. Austin and D. McTaggert in *Treatise on Materials Science and Technology: Ceramic Fabrication Processes*, Academic Press, New York, 1976.)

the middle of the compact, and are inclined to be at an angle near the end of the compact. The latter effect is called *end-capping*. Sometimes the compact is seemingly flawless after compaction but actually contains these internal powder fractures, which then develop into severe cracklike flaws during sintering. Adding as little as 0.25 wt % of some lubricant to the powder and decreasing the compaction stress can provide dramatic improvements.

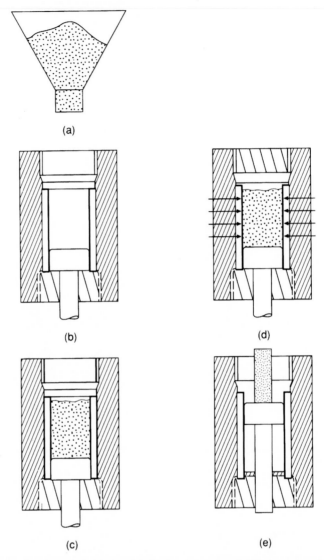

FIGURE 12.8 Dry-bag tool: (a) material to be compacted; (b) pressure chamber including dry bag; (c) pressure chamber filled; (d) pressurizing; (e) ejection of compact after decompression. (Reprinted with permission from G. Austin and D. McTaggert in *Treatise on Materials Science and Technology: Ceramic Fabrication Processes,* Academic Press, New York, 1976.)

12.5 Isostatic Compaction

In isostatic compaction the powder is enclosed in a flexible mold such as a thin rubber bag, and then compacted in a pressure vessel filled with oil. This is a batch process called *wet bag tooling,* which does not easily lend itself to automation. An alternative is *dry bag tooling,* in which the powder is enclosed in a thick rubber mold that is compressed by fluid or gas. The compressed part can then be removed with a plunger, making the process suitable for mass production. For example, spark plugs are made that way by compressing a porcelain mix around a metal core. The schematics of the two processes are depicted in Figs. 12.7 and 12.8.

12.6 Casting Methods

Powder casting methods have been in use for a long time, well before the related colloidal consolidation received attention. In fact, the colloidal consolidation (discussed briefly in Chapter 5) might be described simply as a refined casting technique.

12.6.1 Slip Casting

Casting Suspensions: Slips

For the casting methods we would start out with a suspension of powder in a liquid. This suspension is called a *slip.* It is a smooth liquid mixture (hence its name) with a consistency like heavy cream. Often the suspension is stabilized by a variety of polymeric additives. The intent is to produce a slip of relatively low viscosity with the highest concentration of solids. The additives can strongly affect the viscosity of a slip; unfortunately, there is presently no universal additive that serves all ceramic powder suspensions equally well. One example is a commercial motor oil additive developed by the Chevron Company, Oloa 1200, that seems to give good results even in very low concentrations (below 0.2 wt %). Many others have been used, including some containing inorganic ions. The latter include sodium silicates or phosphates, which may not be desirable for high-performance ceramics that should avoid contamination, although they are effective slip stabilizers. Soaplike surfactants with pH adjustment can be very effective in allowing very high solids contents in the slip. In some cases it has been possible to prepare usable slips that contain as much as 55 to 60 vol % solids. One problem with high solids contents is that the slip may start to exhibit undesirable rheological properties, such as a yield point and an increasing viscosity with increasing shear rate (dilatancy; see Chapter 5). A good practice would be to characterize the rheological properties of the suspensions with a viscometer or similar device to assure reproducibility. The more sophisticated rheometers unfortunately tend to be costly, but simple ones that measure the torque needed to stir a suspension at a fixed rate with a paddle can already be quite useful in quality control.

Slip Casting Process

In the slip casting process, the slip is poured onto a porous mold, traditionally made out of the microporous plaster of Paris (hydrated calcium sulfate). The liquid soaks into the mold, due to the capillary action of the pores, and a deposit or cast forms. The rate of the deposit formation can be derived by simple considerations.

Looking at Fig. 12.9 at a unit area of cross section, we can call Q the total amount of liquid removed per unit time from the suspension. This will move the suspension down by dH, depositing the solids contained in an equal volume of suspension onto the po-

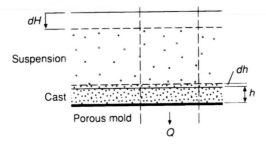

FIGURE 12.9 Geometry of formation of a cast from a suspension.

rous mold. We can consider the cast to contain all the same capillaries, lined up parallel, instead of the actual pore structure. We further consider any cast to be some photographic enlargement of this idealized structure, assuming that the radius of the capillaries is proportional to the radius of the particles in the cast, and that the number of capillaries per unit area is fixed by the particle radius. Then the total flow of liquid seeping through unit area of the cast per unit time, Q, will be the amount of liquid coming through one capillary per unit time, q, times the number of capillaries per unit area, N:

$$Q = Nq \tag{12.8}$$

Where

$$N = \frac{p\alpha}{R^2} \tag{12.9}$$

p is the porosity of the cast, α is a geometrical parameter that is nearly equal to $1/\pi$, and R is the diameter of the capillary.

According to Poiseuille's law for liquid flow through capillary of radius R, we have

$$q = \frac{\pi \Delta P R^4}{8\eta h} \tag{12.10}$$

where q is the flow rate, ΔP is the pressure difference over the capillary, η is the viscosity of the fluid, and h is the length of the capillary. The porosity p of equation (12.9) can then be rewritten as

$$p = \frac{NR^2}{\alpha} \tag{12.11}$$

since the cross-sectional porosity is equal to the volumetric porosity. Thus, combining equations (12.8)-(12.11), we get

$$Q = \frac{\lambda p \Delta P R^2}{\eta h} \tag{12.12}$$

where λ is a proportionality constant that contains α as well.

If in a suspension of volume V we have V_S solids and V_L liquid, then the solids fraction, f, in the suspension is

$$f = \frac{V_S}{V} \tag{12.13}$$

and

$$-Q = \frac{1-f}{f} \frac{dV_S}{dt} \tag{12.14}$$

Thus since we are considering a unit cross-sectional area of the cast making $dV_S = -(1-p)dh$, from equations (12.12) and (12.14) we get

$$h = \left[\frac{2\lambda}{h} \frac{p}{1-p} \frac{f}{1-f} \Delta P \right]^{1/2} R t^{1/2} \tag{12.15}$$

While more refined treatments can be constructed to find expressions for the casting rate, the major features of the process are contained in equation (12.15). The cast thickness, h, depends parabolically on time, and the rate of the cast formation decreases with decreasing porosity and particle radius if we assume that the resistance to liquid flow in the mold is negligible and that the radius of the capillary pore network, R, is proportional to the particle size. The latter assumptions are not usually too far off. Equation (12.15) also indicates that increased casting rates can be achieved if gas pressures were applied above the cast, since this increases ΔP. This method is used and is called *pressure casting.*

One problem in slip casting can be that the fluid is actually not coming through the cast uniformly; instead, large channels develop through which the fluid flows preferentially. This is more likely to occur for very dense casts. Thus the slip casting process itself may introduce green-state heterogeneity despite our efforts to make the slip uniform or to keep the particles monosized. Pressing the slip mechanically when it is still wet, while allowing the fluid to drain, can help in closing large pores and channels that developed accidentally.

Tape Casting

In many applications (e.g., substrates for electronic packaging) thin plates of ceramics are needed. A common way to produce these is by tape casting. In this method, the slip is spread onto a removable support such as a paper or a plastic tape by a carefully controlled blade called a *doctor blade.* The principle of the method is shown in Fig. 12.10. The ceramic can be very thin, perhaps less than 10 μm. The method is especially suited for mass production of thin, flat shapes which can be stamped out of the tape. To achieve successful production by this technique, much attention needs to be given to proper formulation of polymeric additives. These are necessary as binders and as materials that allow some flexibility of the cast tape without causing fracture. For long tapes, for example, it may be practical to wind them on a spool and unroll them again later.

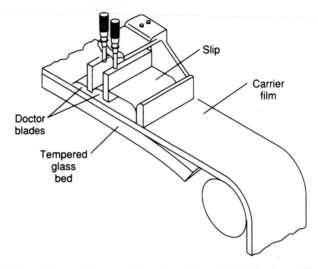

FIGURE 12.10 Dual doctor-blade configuration. [From R. B. Runk and M. J. Andrejco, *Am. Ceram. Soc. Bull.* 54, 199-200 (1973). Permission of American Ceramic Society.]

12.7 Extrusion Methods

An attractive way of mass-producing ceramics from doughlike powder/additive starting mixtures is to use the extrusion methods that are common in the polymer industry. In this method, the powder mix is injected into a die. Complex parts can be formed, depending on die design. Control of the rheological properties of the starting mix is again crucial, and most of the successful formulations are company secrets. In one approach, the viscosity maximum of the powder/organic mix was seen as the optimum viscosity for injection molding, striking the best compromise between highest formability at lowest additive content. Two examples are shown in Fig. 12.11. Here, oleic acid, a polyunsaturated fat that looks like an oil, was used as the additive. The dry powder becomes rapidly sticky as the oil is added, increasing the torque necessary to stir the mix at a fixed rate. Too much oil will decrease the viscosity, hence the maximum. The maximum has been called the *critical ceramic powder volume* (CCPV). The maximum viscosity depends on the nature of the powder, as one would expect.

Polymer additives could produce a thermosetting or thermoplastic mix. Substantial amounts of polymer are usually necessary, which need to be burned out later. One of the major difficulties in the process is, in fact, the binder removal. To avoid bloating or introduction of other deleterious flaws in the binder removal process, heating rates and temperatures need to be highly controlled. Often, a minimum of several hours is needed to burn out the binder without causing damage to the green compact. The injection process itself can also introduce a host of undesirable defects. Most of these are a direct or indirect consequence of die wall friction. The high shear deformation during injection can lead to shear failure in the powder mix, producing defects that cannot be removed during the sintering process. Also, mold filling is important and must be de-

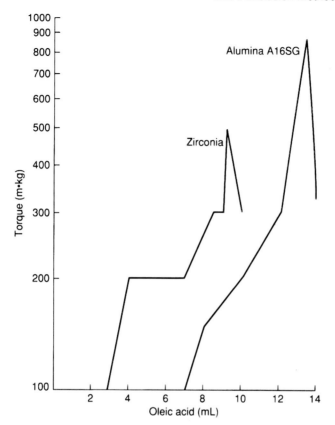

FIGURE 12.11 Changes in torque with incremental additions of oleic acid. The maximum peak for each powder is described as the CPVC point. (From C. J. Markhoff, B. G. Mutsuddy, and J. W. Lennon, "A Method for Determining Critical Ceramic Powder Volume Concentration in the Plastic Forming of Ceramic Mixes" in *Advances in Ceramics*, Vol. 9, J. Mangels, ed., 1984, p. 248. Reprinted with permission of the American Ceramic Society, Columbus, OH.)

signed carefully so that empty corners will not be left nor internal lamellar flaws produced where the flowing powder mixes join. The high-volume production potential of the techniques makes it one to which considerable development efforts are currently devoted.

A schematic of an injection molding machine and some of the parameters that affect its functioning are shown in Fig. 12.12.

LIST OF SYMBOLS

B	proportionality constant
F	force
f	solids fraction

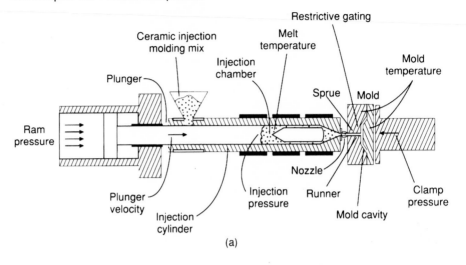

(a)

Fundamental Variable	Machine Variable
Material temperature	Melt temperature Injection pressure Plunger velocity Mold geometry Mold temperature
Flow rate (shear rate)	Plunger velocity Mold geometry Injection pressure
Cavity pressure	Injection pressure Melt temperature Plunger velocity Mold geometry
Cooling rate	Melt temperature Mold temperature

(b)

FIGURE 12.12 (a) Schematic of a plunger injection-molding machine identifying the principal machine variables. (b) Relation of fundamental injection-molding variables to plunger machine variables. (From J. Mangels and W. Trela, "Ceramic Components by Injection Molding" in *Advances in Ceramics,* Vol. 9, J. Mangels, ed., 1984, p. 227. Reprinted with permission of the American Ceramic Society, Columbus, OH.)

N	number of capillary pores per unit cross-sectional area
P	applied hydrostatic pressure
ΔP	pressure differential over the length of the capillary
p	porosity
Q	liquid flow rate
q	liquid flow rate per capillary capillary

R	radius
V	volume
α, β	constants
λ	proportionality constant
μ	coefficient of friction
η	viscosity
σ_r	stress normal to the die wall

PROBLEMS

12.1 Develop a relationship between the solid contents of a ceramic powder-binder mix and its linear shrinkage when sintering to a specific endpoint density.

12.2 Derive the radius relationship for three generations of spherical particles that would fit precisely in the interstices of a closed packed hexagonal stacking spheres and the interstices that they in turn create.

12.3 Assume that the radial pressure distribution follows a parabolic relationship for powders compacted in a die. Rederive equation (12.7) with this assumption.

REFERENCES AND FURTHER READINGS

CUMBERLAND, D. J. and R. J. CRAWFORD, *The Packing of Particles,* Elsevier, Amsterdam, 1987

MANGELS, J. A., ed., *Forming of Ceramics,* Advances in Ceramics, Vol. 9, American Ceramics Society, Columbus, OH, 1984.

ONODA, G., and L. HENCH, eds., *Ceramic Processes Before Firing,* Wiley, New York, 1978.

WANG, F. Y., ed., *Ceramic Fabrication Processes,* Academic Press, New York, 1976.

WESTMAN, A. E. R. and H. R. HUGILL, *J. Amer. Ceram. Soc.,* Vol. 13, 767 (1930)

13

Sintering of Powder Compacts

13.1 Introduction

When we have compacted a powder to some shape, turning it into a useful part is, in principle, quite simple. All that should be necessary is to put the compact in a furnace for awhile, between 0.5 and 0.8 of its melting temperature, and hold it there for a few hours. Since elimination of internal surfaces means lowering the free energy of the system, diffusion should redistribute the atoms inside the powder compact, slowly trying to fill up the voids (pores). You could expect a final product that has some pores in it and in which the grain size has increased, in some cases significantly, compared to the initial particle size. Typically, if we start with a powder particle size of 1 μm or less, we might have a final grain size of 10 to sometimes 100 μm. Since there was no massive melting in the powder mix (we allow for some powder mixtures to have some liquid present during sintering) the compact retains its form, and the final shape is a shrunken version of the initial compact shape. For large parts we will need to guard against sagging during sintering and judicious support may have to be used.

The essentials of the sintering process are simple. The difficulty comes when we start to be demanding about the quality of the product. For example, if we require high mechanical strength, then flaws, such as pores or large grains with cracks in them, must not be allowed to develop or remain. The search for improved properties which fully exploit the potential advantages that some ceramics or metals offer is the reason we must look more closely at the process of sintering. We have to understand how the process works on a detailed, even atomic scale to identify the important parameters and to find how we can manipulate them. To achieve this, we want to identify what causes the powder compact to densify, the *driving force,* and how matter is transported, the *mechanism.* Our approach is to derive an expression for the densification rate of the powder compact, taking all the process variables properly into account. If we do this, we should at the same time include a description of the microstructural evolution of the sintering compact. This is the more important point: A prediction of the sintering rate is in itself not particularly useful, because if a powder compact does not sinter fast enough, we could simply increase the processing temperature. More useful is to have a predictive model of the evolution of the microstructure as a function of the density that has been reached. A question we would ask is, for example: What is the grain size distribution, the pore size distribution, and the pore spatial distribution as a function of density? This looks like quite a difficult question, and it is. A complete, quantitative universal model describing the entire sintering process and its accompanying microstructural evolution is not very likely to be developed, since each material may have idiosyncrasies that cannot be incorporated. It is particularly in the latter that practitioners in industry have much valuable experience; however, for the present we shall be concerned with some fundamental aspects of sintering that are common to all densifying powder compacts. A clear understanding of these fundamentals provides a framework that allows time saving in process optimization or in developing improved processing techniques, even for the more complex systems. The question of densification is considered in this chapter; the microstructure evolution that accompanies sintering is discussed in Chapter 14.

13.2 The Sintering Process

When the powder compact is sintered, its density, equal to its total externally measured volume divided into its mass, increases. Often, the density of a practical compact, expressed as a fraction of the maximum achievable compact density (i.e., when all pores are eliminated), increases linearly with the logarithm of time below densities of about 90 to 95% of the theoretical maximum. Compacts of coarse powders densify more slowly than do compacts of fine powders. A point may be reached where the density practically does not increase with further heating. This limiting density is most often related to the way the particles were arranged in the starting powder compact, and it may be possible that a poorly structured fine-powder compact reaches a lower limiting density than that of a better packed coarse-powder compact. This is depicted schematically in Fig. 13.1.

Initially, the powder compact consists of identifiable particles with void space (i.e., porosity) in between. At the onset of sintering, this porosity has the form of a continuous three-dimensional network of considerable complexity. Upon heating to the sintering temperature, the particles will smoothen somewhat and start to fuse together, a sequence

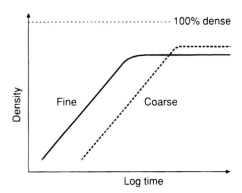

FIGURE 13.1 Schematic of sintered density of a powder compact as a function of time; a power-law dependence on time is observed for a large part of the densification.

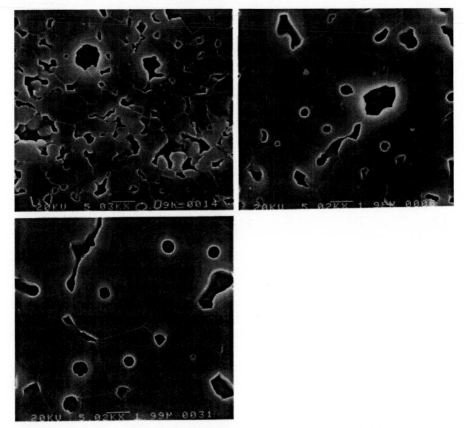

FIGURE 13.2 Microstructure evolution during sintering of an MgO powder compact.

of events shown in Fig. 13.2. Liquid does not have to be involved in this, although it may; even in the solid state, matter will be transported by solid-state diffusion to accomplish this fusion. After awhile, we will have a three-dimensionally connected network of solid matter that does not look much like the initial particle configuration. Instead, we can describe the compact now as a continuous assemblage of grains, intertwined with a continuous pore network. Still later, the continuous network of pores starts to break up into individual pores that continue to shrink. Pores are progressively eliminated, leading to larger and larger pore spacings as sintering proceeds. Ideally, all pores would eventually be removed, but in reality that rarely happens. All the while, the grains continue to grow. In many cases, when the pores become isolated, growing grains may engulf them so that these pores become separated from grain boundaries. This is quite undesirable, because these pores are nearly impossible to remove by continued heating. The essential aspect of the process described here is that internal free surface is eliminated, while at the same time grain growth and coarsening of the pore structure occurs. An example of a desirable microstructure of a sintered ceramic is shown in Fig. 13.3(a); that of an undesirable one, in which pores have been trapped inside grains, is shown in Fig. 13.3(b).

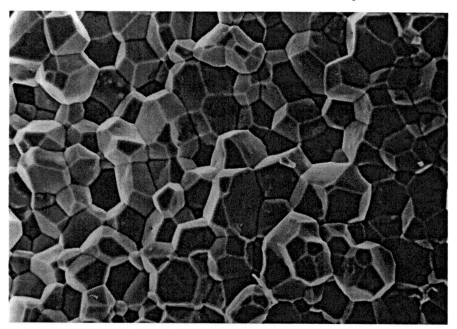

FIGURE 13.3a Microstructure of ZnO, sintered to nearly 100% density, showing a uniform grain size.

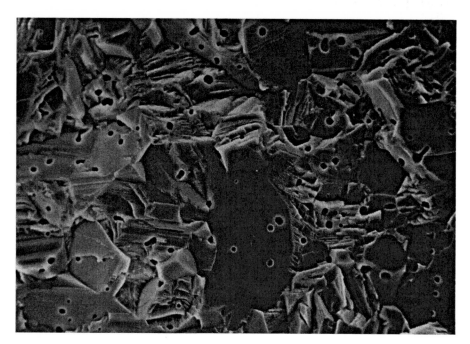

FIGURE 13.3b MgO sintered to nearly 100% density, showing a large spread in grain size, as well as pores trapped within grains.

13.3 Sintering Mechanisms

To describe the way in which matter is transported, we use Fig. 13.4. This figure shows a two-dimensional cross section of some arbitrary pores adjacent to a grain boundary. Looking at the various paths that atoms can take, we realize that a continuous network of grains can only shrink (i.e., the compact shrinks) when matter is removed from between the grains and deposited on the pore surface.[1] For glass powders, which cannot contain grain boundaries, densification usually occurs through viscous flow, involving a deformation of the particles.

The broad distinction that we need to make is between *densifying* and *nondensifying mechanisms.* These have been tabulated in Fig. 13.4. The nondensifying mechanisms are those that merely move matter around on the surfaces of the pores, or move matter from one grain to the next, without causing the grain centers to move closer together as required for shrinkage of the compact. Evaporation/condensation and surface diffusion can be troublesome nondensifying mechanisms, because they can compete with the densifying mechanisms. One should try to choose sintering conditions such that the nondensifying mechanisms are not very active, since they dissipate free energy, otherwise available to drive densifying mechanisms, and produce undesirably coarse microstructures. Grain boundary diffusion and volume diffusion, shown in Fig. 13.4, are the most important densifying mechanisms in sintering.

13.4 Driving Forces for Sintering

A major process in a sintering compact is the elimination of the internal free surface and hence of the pores. In this way, the free energy of the compact can decrease. The rate of matter movement should then be described by a solution to Fick's diffusion equation (Chapter 6), where the gradient of the chemical potential, the *driving force, is* determined by how much free energy the system releases for an incremental increase in density brought about by the movement of atoms to the pore surfaces.

Our approach will be to develop the concept of the *sintering stress,* which is the *equivalent externally applied stress* that would have the same effect on the densification rate as the action of grain boundaries and pores. Once we have succeeded in defining the sintering stress as an equivalent externally applied stress, combinations of the sintering stress with applied mechanical stresses should be easier to handle. The formulation of the sintering process in terms of mechanical stresses rather than chemical potential differences will have some advantages because mechanical stress effects will arise in the densification of nonhomogeneous systems and will be easier to visualize.

13.4.1 Relation of Surface Curvature to the Sintering Stress

A first step in the analysis is to determine the effects of surface curvature on the difference

---

[1] Removal of matter from the interior of crystalline grains is, in principle, also possible, but this is an unlikely mechanism: It would involve dislocation movement and this requires more stress than is generated by the shrinking pores.

Grain boundary

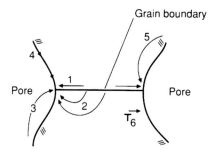

Mechanism	Densifying	Non-densifying
1. Grain boundary diffusion	×	
2. Volume diffusion	×	
3. Evaporation/condensation		×
4. Surface diffusion		×
5. Coarsening by volume diffusion		×
6. Dislocation (T) creep	×	

FIGURE 13.4 Schematic representation of the various atom transport paths (mechanisms) for two sintering particles.

between the chemical potentials of atoms in a flat surface and those in a curved surface, and then to express this difference as a mechanical stress.

If we consider a semi-infinite reference solid, as in Fig. 13.5, and transfer an atom from its flat surface to a curved surface, such as the surface of a solid sphere, the radius of this spherical particle grows, and consequently its surface area increases. The work required to transfer some atoms from the flat surface to the surface of the sphere is the surface energy γ_s times the corresponding surface area change. The volume change of the spherical particle, dV, for the transfer of dn atoms is equal to the atomic volume, Ω, times dn: Ωdn. Thus

$$dV = 4\pi R^2\, dR = \Omega dn \qquad (13.1)$$

The work per atom transferred is, by definition, the chemical potential change, $\Delta\mu$, due to the accompanying change in surface area:

$$\Delta\mu = 2\gamma_s \frac{d\left(\text{surface area}\right)}{dn} = \gamma_s \cdot 8\pi R\, dR \frac{\Omega}{dV} \qquad (13.2)$$

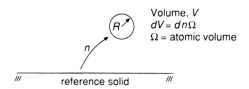

Volume, V
$dV = dn\Omega$
Ω = atomic volume

reference solid

FIGURE 13.5 Transport of n atoms from the flat surface of a semi-infinite reference solid to the curved surface of a solid sphere.

Note here that removing atoms from the infinite, flat, reference surface does not result in any change in surface energy at that surface, because replacing the missing surface atom with one of the atoms from the bulk does not involve any shape change and, hence, any energy expenditure for this hypothetical, semi-finite material. This artifice provides a handy way of defining a reference surface for thermodynamic problems of this sort, and we can then set the reference surface at some convenient invariant chemical potential (e.g., zero).

Combining equations (13.1) and (13.2) and eliminating dn immediately gives the chemical potential of the atom in a curved surface with respect to the flat reference surface as

$$\Delta\mu = \frac{2\gamma_s\Omega}{R} \tag{13.3}$$

This is the famous Kelvin equation. It can be generalized[2] for any curved surface that is described by its two principal radii of curvature, R_1 and R_2, as

$$\Delta\mu = \gamma_s\Omega\kappa \tag{13.4}$$

where the curvature κ is

$$\kappa = \frac{1}{R_1} + \frac{1}{R_2} \tag{13.5}$$

Notice that as we have defined it here, the curvature of a convex surface is positive. A concave surface would have a negative curvature, implying, correctly, that the chemical potential of atoms in such a surface would be less than that of atoms in a flat surface. You will notice that we arrived at equation (13.2) by letting the system do a minuscule amount of work, during which it is prescribed that dV, the incremental volume change, is described correctly by equation (13.1). This can only be true, of course, if the solid is taken to be a structureless piece of "jelly," a continuum from which we took dn "atoms" away in the form of dn small pieces with a volume Ω. Equation (13.2) will thus describe the average change in the energy per atom. If we did not take this approach, we would start generating a "fine structure" in the dependence of the free energy of the sphere on the number of atoms in it. This phenomenon becomes important for ultrafine particles consisting of only a few atoms. We will ignore this problem by assuming that, on the average, all detailed configurations of atoms in the surface are always present so that we do not have to deal with local minima or maxima in chemical potential.

We now attempt a connection between chemical potential and mechanical stress. If such a stress existed, we could imagine it as acting as a hydrostatic stress, σ, externally applied to the surface of the sphere. If dn atoms were transferred from the reference surface to the sphere's surface, the volume change, dV, of the sphere would lead to work, equal to $\sigma\,dV$, performed by this stress. The relation of this stress to the chemical

[2] A derivation of this generalization can be found, for example, in A. W. Adamson, *Physical Chemistry of Surfaces*, Wiley, New York, 1976, p. 5.

potential (per atom) is then established by equating the mechanical work with the thermodynamic work:

$$\sigma \, dV = \Delta\mu \, dn = \Delta\mu \, \frac{dV}{\Omega} \qquad (13.6)$$

or

$$\sigma = \frac{\Delta\mu}{\Omega} \qquad (13.7)$$

The physical reality of such a stress can easily be visualized if we think of a soap bubble on an open-ended pipe: The bubble will shrink because the higher pressure of the air inside the bubble. brought about by the surface tension of the spherical soap film that wants to shrink in size, will push the air out. If you knew the surface tension of this film, γ_s, you could, using equations (13.3) and (13.7), calculate precisely what the pressure of an inert gas inside a free soap bubble of a given size must be to prevent shrinking.

13.4.2 Equal and Unequal Pores in Solids and the Sintering Stress

Equal Pores

If we represent a porous solid as one in which the pores are all the same, evenly spaced and spherical with a radius R. and we ignored changes that occurred in grain boundary area as the solid densified, equation (13.7) would describe the driving force (i.e.. the equivalent externally applied stress, $\sigma_{equivalent}$) for sintering of a porous solid appropriately described by the geometry that led to equation (13.3). Combination of equations (13.3) and (13.7) gives readily

$$\sigma_{equivalent} = \frac{2\gamma_s}{R} \qquad (13.8)$$

This would be a good expression for the driving force for densification, or sintering stress, of a porous glass, which contains no grain boundaries, in which monomodal (single-sized) pores were unifonnly distributed.

Unequal Pores

We now ask what the driving force (i.e., sintering stress) for densification of this glass would be if all the pores were not the same size, although the pore centers would still be located on a homogeneous lattice, such as, for example, a simple cubic arrangement. To find the answer we could use the same procedure as the one that led to equation (13.3). Suppose that we have N pores per unit volume in our glass-like substance, and we give each individual pore the label i, ranging from 1 to N. If the pores form a simple cubic array in space, we could imagine each pore at the center of a little cube of material, such that the entire porous solid would be made up of an assemblage of all the same

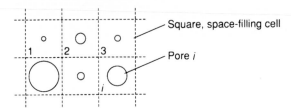

FIGURE 13.6 Pores of different sizes distributed over a regular lattice.

cubes with different size pores at their center. A two-dimensional analog is shown in Fig. 13.6. Each of these space-filling cells would then have to shrink by the same amount, to maintain strain compatibility. Thus each pore, regardless of its size, must shrink by the same amount, dV, in the geometry that we are considering here. The volume change of a pore of radius R_i is then

$$dV_i = 4\pi(R_i)^2 \, dR_i = dV \tag{13.9}$$

while the corresponding surface area change of pore i is

$$da_i = 8\pi R_i \, dR_i \tag{13.10}$$

For the entire volume V of the porous solid, we then have a total incremental work performed, dW, equal to

$$dW = \gamma_s \, dA = \gamma_s \sum_i^N da_i \tag{13.11}$$

Equations (13.9) and (13.10) show again that

$$da_i = \frac{2dV_i}{R_i} \tag{13.12}$$

so that

$$dW = 2\gamma_s \, dV \sum_i^N \frac{1}{R_i} \tag{13.13}$$

or, since in the present geometry we must have

$$\Delta V = N \, dV \tag{13.14}$$

where ΔV is the total volume change for our initial unit volume, it follows from equations (13.13) and (13.14) that

$$dW = 2\gamma_s \Delta V \frac{1}{N} \sum_i^N \frac{1}{R_i} \tag{13.15}$$

If ΔV is now put equal to the atomic volume Ω, then dW corresponds to the average free-energy change per atom, $\Delta\mu$.[3] Therefore, we end up with an expression for the average chemical potential of the atoms for a spatially ordered collection of unequal pores inside a solid that looks quite similar to that for equal pores, equation (13.3):

$$\Delta\mu = 2\gamma_s \Omega \left[\frac{1}{N} \sum_i^N \frac{1}{R_i} \right] \qquad (13.16)$$

except that now we have to use the harmonic mean of the pore sizes. Note that the quantity in brackets in equation (13.16) is half the mean curvature in our pore system and we are therefore back to equation (13.4) with κ the mean curvature.

Irregular Pore Distributions

We have stressed the regularity of the spatial distribution of the pores in our discussions so far. You might ask what the consequences would be if the pores were not distributed regularly. This is a more complex question and requires specific information on the relationship between the pore spacing and the pore sizes. In principle, the procedure that led to equation (13.16) can be used again, but the result will not look as simple, except for some special pore size-pore spacing distributions. Eventually, $\Delta\mu$ would be a function of a lot of microstructural factors. In one simple case, where all the pores are the same size, you could easily show that the distribution does not matter. For this, you again consider the porous solid to be made up of space filling cells—no longer cubes but having a shape that is determined by the relative position of the pores with respect to one another—each containing one pore. The volumetric strain of each cell i, dV_i/V_i, again has to be the same or else there would be strain mismatch, and we would be storing elastic energy in the system. In the end, you find that the size distribution of V_i does not matter (Problem 13.5).

In all that we have discussed so far, we have prescribed the solution of the system to occur only along *one specifically selected* geometrical path; that is, the volume and surface changes of every pore are determined by equations such as (13.9) and (13.10) based on simple geometric arguments about how the total volume change of the solid is to be distributed among the pores. In fact, when we consider the case of pore size distributions, the system is not in full equilibrium. Equilibrium would have been reached when matter had been redistributed such that all the pores were finally eliminated and we had only a spherical solid left. It is, however, legitimate to consider what energy changes occur when we proceed along one specified structural path. We then talk about *constrained equilibrium*. In sintering, the microstructural evolution is often limited in this way, and driving forces for densification must be considered together with the possible geometrical path of the system. As an analogy, the force that you experience skiing down a hill, accelerations aside, depends on the local slope (i.e., the incremental free-energy change for an incremental change in position) and the various frictions, not on the difference in height between the top and the bottom of the hill.

---

[3] Note that we ignore here the minuscule change in free energy that we really should have included when we changed the *external* size of the body by this transfer process; however, since the external dimension of the body is so much larger than that of the pores inside it, we can safely ignore this contribution.

A complication that we have not considered so far is the possible stress interaction between pores. This would be equivalent to the forces that two skiers would exert on each other if one were an expert and the other a novice and they were tied together so that they came down the hill at some average speed. In sintering solids, mismatch stresses will be generated inside the solid if one region tries to densify faster than another, and these mutual stresses will affect how fast the local sintering can proceed. We examine this question later.

Another complication that we have ignored is that in real materials the surface energy is not isotropic. If the anisotropies are not too extreme, the overall energy of the system will not vary by much when the pores or particles are not exactly in their geometry of lowest energy, and it is legitimate to use some approximate average surface energy and not worry about the specific equilibrium shape that a particle would need to assume if its surface energy were anisotropic.

13.4.3 Crystalline Materials: Grain Boundary Effects

Effect of Grain Boundaries on Equilibrium Pore Shapes

For a crystalline material the grain boundaries play a role that must be included when determining the magnitude of the driving force for densification. A first step in assessing the effects of the presence of grain boundaries in polycrystalline, porous powder compacts is to consider the hypothetical case of a pore connected to three grain boundaries in an infinite solid, as depicted in Fig. 13.7. Equilibrium will impose two important requirements on the geometry: (2) the chemical potential of the atoms in the pore surface must be the same everywhere, and (2) no net force is to be present at the junction of the grain boundaries and the pore surface.

The first requirement, for an istrotropic solid, is equivalent to saying that the curvature of the pore surface is the same everywhere; therefore, the pore surface must consist of circular arcs in two-dimensional models. This immediately simplifies the kind of pore shapes we should consider and avoids turning two-dimensional problems[4] into a tedious exercise in variational calculus.

The second requirement deserves a little more attention. It will lead to a specific angle of intersection at the surface-grain boundary junction between the surface and the grain boundary. For the moment we limit ourselves strictly to the very junction and see how the surface and the grain boundary geometry must be, so that no net force results at the junction. The situation can be represented with vectors, as shown in Fig. 13.8, in which the magnitude of the vectors is proportional to the surface tension or to the grain boundary tension, and the direction of the vectors is tangential to the pore surface, or is in the grain boundary plane. The geometry at the junction with the requirement of no net force immediately leads to

$$2\gamma_s \cos\frac{\psi}{2} = \gamma_b \qquad (13.17)$$

---

[4] In three dimensions we can have the same overall curvature by a combination of positive and negative curvatures, and the three-dimensional surface is therefore much more complex than a combination of spherical surface segments.

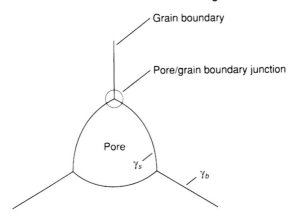

FIGURE 13.7 Geometry of a pore surrounded by three grains in two dimensions.

where γ_s is the surface energy, γ_b the grain boundary energy, and Ψ the dihedral angle.

Now let us examine the free-energy changes that occur when sintering takes place and the pore shrinks. There will be a decrease in free energy resulting from a decrease in the pore surface but an *increase* due to an increase in the grain boundary area. Whether the pore shrinks or not is dependent on the former (decrease) being greater than the latter (increase). For example, in a problem at the end of the chapter you are asked to show that for a flat-sided pore the decrease in energy of the pore surface exactly matches the increase in grain boundary energy. Consequently, the sintering stress resulting from flat-sided pores is zero. It is only a little more difficult to show (by sketching or by geometry) that if the sides of the pores are concave to the pore, the reduction in surface energy is greater than the increase in grain boundary energy and the pores will therefore tend to shrink. Conversely, convex pore walls will cause the pore to tend to grow.

The next step in this two-dimensional geometrical development of the thermo-dynamics of sintering is to recognize that for a given dihedral angle the curvature of the pore walls is determined by the number of grain boundaries intersecting the pore. For example, you may remember from elementary geometry that the angles of an N-sided regular polygon are equal to $[N(180) - 360]/N$. Consequently, if the pore walls are to be flat, in our two-dimensional approach, there must be exactly $360/(180-\Psi)$ grains surrounding the pore. If there are more grains surrounding the pore, the walls will be

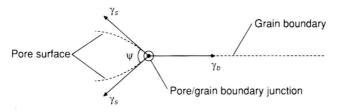

FIGURE 13.8 Balance of surface and interface forces at the point where a grain boundary intersects a pore. y is the dihedral angle, g_s is the surface tension, and g_b is the grain boundary tension.

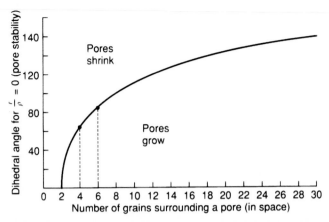

FIGURE 13.9 Conditions for pore stability as a function of pore coordination number. (Reprinted with permission from W. D. Kingery, H. Bowen, and D. Uhlmann, *Introduction to Ceramics*, Wiley, New York, 1976.)

convex; if fewer grains, the pore walls will be concave. You can see this by redrawing Fig. 13.7, keeping the dihedral angle the same (at approximately 120°) but with 10 grains surrounding the pore.

These geometric considerations can be extended to three dimensions (i.e., polyhedral pores) and Fig. 13.9 gives the results of geometric calculations for three dimensions. If we know the dihedral angle and the number of grains surrounding a pore (which we call the *pore coordination number),* we can use Fig. 13.9 to tell us the curvature of the pore walls and hence the tendency of the pore to shrink or grow. If the point corresponding to the dihedral angle and pore coordination number falls exactly on the curve, the pore walls are flat and consequently the sintering stress is zero. If the point is above the curve, the walls are concave and the pore tends to shrink. We see that for a given dihedral angle we should have as few grains as possible surrounding a pore if that pore is to shrink during sintering. Conversely, for a given pore coordination number, we would need to have a high dihedral angle if we expect to remove pores during sintering.

From this discussion it follows that materials that have a high grain boundary energy compared to their surface energy may be difficult to densify, especially if the pore size is large compared to the grain size, because then many grain boundaries intersect the pore. One practical aspect that emerges here is then that in making powder compacts for sintering, we should always strive to keep the pore size small compared to the grain size for the green compact. This usually means trying to achieve a high green density and a uniform microstructure of the powder compact. Thus densification could proceed only after the pore coordination number has become aufficiently small, and the free energy of the system can decrease again if atoms move from the boundary to the pore surface. This can be accomplished when the grains grow larger with respect to the pore size (i.e., when grain growth occurs). If we had a system with fine grains and large pores, chances are that we would not have any densification unless some grain growth were permitted first. However, we realize that this does *not* imply that we should seek a coarse grain size to achieve densification. Normally, when we increase the particle size in a powder compact the pores increase in size as well, and we do not come out ahead. The coordination number argument therefore is most relevant for large pores that might have been introduced inadvertently in the compact.

Sintering Stress and the Stress Intensification Factor

To determine the exact relation between the sintering stress and the structure of the porous solid, we could follow the same procedure as we did for a glass (a solid without grain boundaries), but there is one twist: The atoms are now removed from specific locations in the material, the grain boundaries, when the pores are filled up. In addition, the diffusion problem associated with describing sintering will have to be formulated in terms of atom transport along the grain boundaries to the pore surfaces. We will need some way of relating the sintering stress, that is, the equivalent externally applied stress to the mean stress on the grain boundaries themselves. The factor relating an externally applied stress to the resulting stress on a grain boundary in a porous solid is of a geometric nature. For a hydrostatically applied external stress the factor is the ratio of the grain boundary area when it is assumed that the boundaries would continue straight through the pores, divided by the actual grain boundary area (in which the pores make holes). This geometrical factor has been called the *stress intensification factor* [5], ϕ. It will come up each time we want to relate the mean stress on the grain boundary to a stress that we can measure externally on the sintering compact. The stress intensification factor, ϕ, thus allows us to go freely from the grain boundaries to the exterior surface of the sample. For an applied hydrostatic stress σ_a, the mean grain boundary stress σ_m is thus

$$\sigma_m = \sigma_a \phi \tag{13.18}$$

If, for example, we would call the mean stress on the grain boundary resulting from the action of the surface tension and the grain boundaries, $\phi\Sigma$, the sintering stress would be equal to Σ.

If all the pores are nicely spherical and randomly distributed through the porous solid, ϕ is easily found. Indeed, the area fraction porosity for a random planar cut through such a solid is exactly equal to the volume fraction porosity. Then ϕ = total cross-sectional area/load-bearing cross-sectional area would simply be equal to 1/(1-porosity). This would be the correct expression for, say, a glass that contains isolated nearly spherical pores. When the pores are no longer spherical, and as the grain boundary intersection distorts the pores, that simple relationship can no longer be true. In general, the relationship would be expected to be complex. However, some computer calculations have been done (Beere, 1975) on the detailed equilibrium shape that a uniform pore network should assume in a porous polycrystalline solid. The grain boundaries will intersect the pores and distort their shape in order to satisfy the balance of forces exerted at the intersection point, as we discussed above. An example of a lowest-energy shape assumed by a continuous, uniform pore network in a polycrystalline solid is shown in

---

[5] This is a somewhat unfortunate term since a nearly similar one but with different meaning is in use with fracture mechanics (the stress intensity factor).

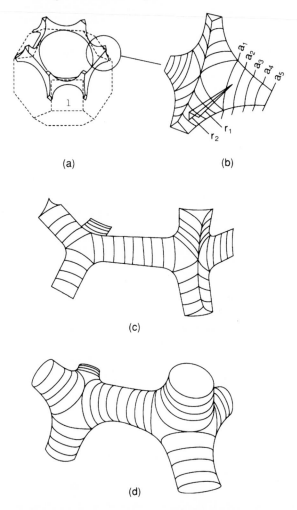

FIGURE 13.10 (a) Idealized system of porosity extending round the edges of a tetrakaidecahedron grain. (b) Unit of porosity situated on a corner of the tetrakaidecahedron. (c) Predicted shape of two adjacent corner units when the dihedral angle is 30° and the volume fraction of pores is 2%. (d) Predicted shape of two adjacent corner units when the dihedral angle is 180° and volume fraction of porosity is 10%. (Reprinted with permission from W. Beere, *Acta. Metall.* 23, 131 (1975).]

Fig. 13.10. This is the result of an extensive computer calculation that minimized the energy associated with the surface area as its shape was varied. The results from the computer calculations can be fitted to some analytical expressions. The equation (Vieira and Brook, 1984)

$$\phi = \exp(aP) \tag{13.19}$$

describes the calculated values rather well over a relatively wide density range from about 0.4 to 0.95 of theoretical. P is the fractional porosity and a is a parameter that depends on the dihedral angle only, in the geometry adopted by Beere. An interesting aspect of equation (13.19) is that ϕ does not depend on the scale of the system. Experimental verification of equation (13.19) for ϕ was recently obtained (Rahaman et al., 1986).

One observation that we should make at this point is that a in equation (13.19) must also depend on other geometric factors in addition to the dihedral angle. For example, if a pore has many grains surrounding it, such as for an isolated large pore in a fine-grained matrix, it would assume a more and more spherical shape, so that the geometry for which equation (13.19) was calculated is not satisfied. In fact, in that case we would need to have $\phi \rightarrow 1/(1—\text{porosity})$, as we found for the simple case of spherical pores in glass. We must then conclude that a will also depend on the pore coordination number, the number of grains around a pore. An expression similar to equation (13.19) but including a range of pore coordination numbers has not yet been developed.

Sintering Stress and Polycrystalline Powder Compact Structure

We can again consider the sintering stress Σ to arise as the result of the geometric changes inside the porous solid as it changes volume. The energy changes accompanying the geometry changes that result from densification for the porous polycrystal are

$$\Sigma dV = \gamma_s \, ds + \gamma_b \, db \tag{13.20}$$

where s and b are the total pore surface and the grain boundary area in the compact, and γ_s and γ_b are the respective surface and grain boundary energies. The problem now is to construct some representative geometry of the porous sample and to figure out what the right-hand side of equation (13.20) should be like. To do this, we should transfer some incremental number of atoms from the grain boundary into the pore and see what the incremental geometrical changes are for our model. Numerous studies have been done that may be regarded as being specialized cases of the right-hand side of equation (13.20). The simplicity of the expression hides all the complexities, however, and it is probably fair to say that, so far, not a single model has been devised that properly describes this part of the equation over a wide range of density changes of the powder compact. Some models deal, for example, with the growth of the neck between two contacting spheres, as shown in Fig. 13.11. While useful in some respects, such models obviously ignore the complications of interactions in multiparticle powder compacts. Their geometry is also considerably removed from the complex pore structure that is present during any stage of densification. Nevertheless, they were useful in the early

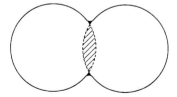

FIGURE 13.11 Neck growth between spheres with mutual approach of centers. Half of the material in the shaded area must be put into the neck area that is black.

developments of the theories of densification that focused on the kinetics of sintering.

The concept of sintering stress Σ and its relation to the mean grain boundary stress $\phi\Sigma$ now puts us in a position to address densification from a more microscopic point of view: the solution of Fick's diffusion equation in geometries appropriate to the sintering mechanisms.

13.5 Densification by Grain Boundary Diffusion

The detailed modeling of densification by diffusional processes was first addressed by Kuczynski (1949) and has been examined many times. Let us first consider the transport of atoms out of the grain boundaries by diffusion that is strictly limited to the grain boundary itself. This is, in fact, not an uncommon case. We refer here to the two-dimensional geometry shown in Fig. 13.4.

To get the dimensions in agreement with a three-dimensional geometry, we imagine that the two-dimensional geometry extends to a large distance perpendicular to the paper. We will also assume that the grain boundary remains flat and of constant thickness δ_b. Although not known accurately, δ_b is usually taken to be 1 nm, a value in fair agreement with what can be observed, for example, with a high-resolution transmission electron microscope.

We shall now carry out a mass balance on the segment of grain boundary between x and $x + \Delta x$ as shown in Fig. 13.12. Let us consider unit dimension perpendicular to the paper. Atoms will be diffusing into the left-hand face of this segment and out of the right-hand face. Atoms will also be entering from the grains abutting the upper and lower faces of the segment. If the number of atoms per unit volume of grain boundary is not changing, we have steady state and

$$\begin{array}{c}\text{input from}\\\text{left face}\end{array} + \begin{array}{c}\text{input from}\\\text{upper face}\end{array} + \begin{array}{c}\text{input from}\\\text{lower face}\end{array} = \begin{array}{c}\text{output from}\\\text{right face}\end{array}$$

or

$$\delta_b \cdot J\Big|_x + \frac{\Delta x}{\Omega}\left(\frac{-dY}{dt}\right) = \delta_b \cdot J\Big|_{x+\Delta x} \tag{13.21}$$

where J is in atoms/(length^{2}time) and Y is the separation between grain centers.[6]

Gathering the flux terms on the right, dividing through by Δx, and letting $\Delta x \to 0$ yields

$$\frac{1}{\Omega}\frac{dY}{dt} = -\delta_b \frac{dJ}{dx} \tag{13.22}$$

We will employ Fick's first law, written with the chemical potential gradient as the

[6] The quasi-steady-state condition requires that the grain boundary stay flat. The same amount of matter per unit area and per unit time therefore must enter the grain boundary everywhere. Removal of unequal amounts is opposed by the mechanical stress that such uneven removal would generate.

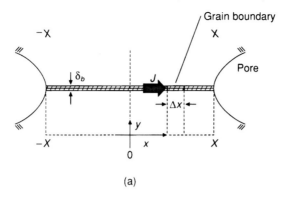

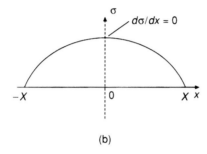

FIGURE 13.12 (a) Geometry of the material transport through a grain boundary to the pore surface. d_b = grain boundary thickness. (b) Schematic of the stress distribution on a grain boundary during sintering.

driving force (see Section 6.2.6).

$$J = -\frac{D_b}{\Omega kT}\frac{d\mu}{dx} \qquad (13.23)$$

where D_b is the grain boundary diffusion coefficient. In Section 13.4.1 we related the local chemical potential of atoms in a surface to a local stress, σ, acting perpendicular to that surface. An identical approach can be taken to yield a relationship between the chemical potential of atoms in a grain boundary and the stress on the boundary. When differentiated this becomes

$$\frac{1}{\Omega}\frac{d\mu}{dx} = \frac{d\sigma}{dx} \qquad (13.24)$$

Substituting (13.23) and (13.24) in (13.22) yields

$$\frac{d^2\sigma}{dx^2} = \frac{kT}{\Omega\delta_b D_n}\frac{dY}{dt} \qquad (13.25)$$

All the terms on the right of (13.25) are independent of x, so we can readily integrate twice to give

$$\sigma = \left(\frac{kT}{\Omega \delta_b D_b} \frac{dY}{dt} \right) \frac{x^2}{2} + C_1 x + C_2 \tag{13.26}$$

where C_1 and C_2 are integration constants. Symmetry around $x = 0$ requires that we cannot have odd powers of x in the expression for σ, so C_1 must be zero. The stress distribution that equation (13.26) describes is parabolic and can be sketched as shown in Fig. 13.12.

To solve the problem completely, we need to consider the mechanical force balance at the grain boundary. To do this we need to use a somewhat artificial reasoning. Recall that upon defining the sintering stress, we represented the system of pores and grain boundaries that we are considering here as one in which the grain boundary and surface energies were artificially set equal to zero and an equivalent external stress was applied equal to i. In this equivalent representation we must take the hoop stress acting in the surface of a pore as equal to zero, since its effect is already included in Σ. Thus $\sigma = 0$ at $x = -X$ and at $x = X$.[7]

Substituting either one of these conditions in (13.26) yields

$$C_2 = -\frac{\alpha X^2}{2} \tag{13.27}$$

where

$$\alpha = \frac{kT}{\Omega \delta_b D_b} \frac{dY}{dt} \tag{13.28}$$

(a negative number since Y is diminishing with time).

The mean stress on the grain boundary, $\phi \Sigma$, can be obtained by integrating (3.26) from $x = 0$ to $x = X$ and then dividing by X:

$$\frac{1}{X} \int_0^X \sigma(x) \, dx = \phi \Sigma \tag{13.29}$$

Carrying out the integration and substituting for C_2 using (13.27) gives

$$\phi \Sigma = -\frac{\alpha X^2}{3} \tag{13.30}$$

Writing out α in full from equation (13.28) and rearranging, we obtain

$$\frac{dY}{dt} = -\frac{3 \Omega \delta_b D_b \Sigma \phi}{kTX^2} \tag{13.31}$$

---

[7] Another approach, derived from considering the formation of a weld or neck between two perfect spheres is to find the chemical potential of the atoms at the exit point of the grain boundary from geometrical considerations (Exner, 1979). We will not adopt this "two-sphere" model here since the detailed neck geometry is not known in sintering bodies.

If we now put the center-to-center distance, Y, of the two grains equal to the grain size G, and further consider the solid to be made up of a string of grains in series, the uniaxial strain rate of the sintering solid, $\dot{\varepsilon}_\rho$, could be written as

$$\dot{\varepsilon}_\rho = \frac{dY / dt}{G} = -\left(\frac{3D_b\delta_b\Omega}{G\,kTX^2}\right)\Sigma\phi \tag{13.32}$$

One difficulty that remains here is that the pore spacing X is not readily accessible experimentally and would need to be obtained from examination of laboriously prepared metallographic sections of the porous compact. However, G can be related to X through the geometrical parameter, which appeared in equation (13.18), and we will write[8]

$$\phi = \frac{G^2}{X^2} \tag{13.33}$$

At present, this merely displaces the problem to finding ϕ as a function of the degree of densification, but later we will see that this can, in fact, be done. We should then be able to write the uniaxial strain rate due to densification in the following form:

$$\dot{\varepsilon}_\rho = -\left(\frac{K_b}{G^3}\right)\Sigma\phi^2 \tag{13.34}$$

where Σ is the sintering stress for densification, an equivalent externally applied stress, as in equation (13.20). K_b is the temperature-dependent kinetic factor associated with grain boundary diffusion. Finally, we could write this equation in an even simpler form:

$$\dot{\varepsilon}_\rho = \frac{-\Sigma}{\eta_\rho} \tag{13.35}$$

where η has the dimensions of a viscosity and may be called the *densification viscosity*. It contains the kinetic parameter D_b, which is the grain boundary diffusion coefficient; the instantaneous grain size G; and the stress intensification factor ϕ. It can be shown, without too much difficulty, that in a cylindrical geometry, where we consider the circular-disk grain boundary between two spherical grains, the dependence of the densification strain has the same form as that obtained in equation (13.34).

13.6 Densification by Volume Diffusion

Volume diffusion of atoms from the grain boundary to the pore surface is more difficult to treat rigorously. The usual approach is a simplification in which one replaces D_b, the grain boundary diffusion coefficient, by the volume diffusion coefficient, D_v and replaces the grain boundary thickness, δ_b, by half the pore separation, X. This will lead to a

---

[8] Strictly speaking, in the two-dimensional geometry considered so far, $f = G/X$, but (13.33) is appropriate for an isotropic solid in three dimensions.

different dependence of the densification strain rate on the instantaneous grain size

$$\dot{\varepsilon}_p = -\left(\frac{K_v}{G^2}\right)\Sigma\phi^m \tag{13.36}$$

for a volume diffusion mechanism. Here K_v, the kinetic factor, is that for volume diffusion K_v. A detailed derivation would show that $m = 3/2$ in this case.

13.7 Creep of Porous Solids and the Shape Factor Φ

Creep is the slow deformation of a solid under the action of an applied stress. When creep deformation of a solid is controlled by diffusion rather than by plastic deformation, the creep rate conforms to an expression of the form

$$\text{creep deformation rate} = \frac{\text{applied stress}}{\text{creep viscosity}}$$

Equation (13.35), describing the linear contraction rate of the densifying powder compact, looks a lot like a simple creep equation. The similarity is not accidental, and we should examine the connection between creep of a porous powder compact, under the action of applied stresses, and densification, under the action of the sintering stress. Earlier we came to the conclusion that the sintering stress, Σ, acts like an applied compressive hydrostatic stress of the same magnitude.[9]

For uniaxial compression of a porous compact by an externally applied stress, σ_a, where deformation is controlled by diffusional processes, the uniaxial creep rate, $\dot{\varepsilon}_c$, can be written as

$$\dot{\varepsilon}_c = -\frac{\sigma_a}{\eta_c} \tag{13.37}$$

where η_c is the creep viscosity. We could now compare the creep rates and the densification rates of identical samples as a function of the density of the powder compact to see if experiments indeed show some relationship. This has been done for a number of different materials and the results are shown in Fig. 13.13. At a fixed uniaxially applied stress, the ratio of the densification strain rates over the creep strain rates are remarkably constant for a wide range of density.

---

[9] To be completely accurate, we have to realize that whereas this equivalent externally applied hydrostatic stress would produce an instantaneous densification rate that is equal to the effect of the pores and grain boundaries, it may well take the compact along a somewhat different microstructural path. The reason is that the sintering stress is an average of all the local effects of the pores and grain boundaries, and that there might be some dependence of the evolution of the densifying compact on exactly how these local effects are distributed. Thus, applying an equivalent externally applied stress may not produce exactly the same stress distribution at the grain level, and may lead to some different microstructural paths. We do not expect these differences to be significant. The details of this point are still a matter of research at present. For now we assume complete equivalence.

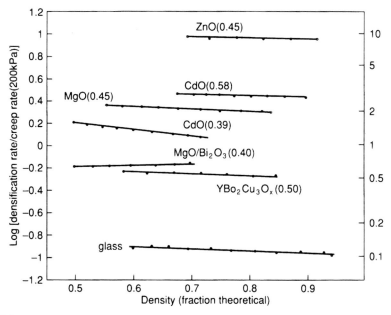

FIGURE 13.13 Sintering rate over constant-stress creep rate for various ceramics and for glass. The applied stress for creep was 0.2 MPa. Numbers in parentheses are initial densities.

If we look at the densification rate equation with an externally applied hydrostatic stress σ_h, we should be able to write

$$\dot{\varepsilon}_\rho = -\frac{\Sigma + \sigma_h}{\eta_\rho} \qquad (13.38)$$

The hydrostatic stress component produced by an applied uniaxial stress σ_a follows from mechanics:

$$\sigma_h = \frac{\sigma_a}{3} \qquad (13.39)$$

Thus for a low applied uniaxial stress we would find the densification rates practically unaffected, and the effect of the presence of such a stress during densification would simply be to superimpose a constant volume creep deformation on top of the densification. This has also been verified experimentally. At least for applied uniaxial stresses that are much smaller than the sintering stress, we do not have to take into account the effect of the shear stresses on the densification.

The effects of the uniaxial applied stresses on densification have been analyzed further and tested experimentally. An experimental comparison of equations (13.37) and (13.38), giving the ratio of the creep rate over the densification rate, then allows for a determination of the ratio of the creep viscosity over the densification viscosity. A detailed theoretical analysis (Scherer, 1987) has shown that

$$\frac{\eta_c}{\eta_\rho} \leq 2 \tag{13.40}$$

From all the information that we have available to date on the relationship between the creep rate and the densification rate of porous compacts it is apparent that a very close relationship exists between low-stress creep viscosities, as defined by equation (13.37) and the densification viscosity as defined by equation (13.35). In a way, this is not too surprising, since a low uniaxially applied stress would only serve to bias the transport processes involved in densification.

To make the connection more formal, we could look back to equation (13.29) and observe that extra matter would be transported from the grain boundary due to the applied stress σ_a, which exerts a mean effective stress σ_{eff} on the grain boundary that is equal to

$$\sigma_{eff} = \sigma_a \phi \tag{13.41}$$

since ϕ is the factor that relates the load-bearing cross section at the grain boundary to the external cross section that disregards the existence of pores. The strain rate that results from the applied stress alone (i.e., the creep rate) now follows from the new force balance where the surface tensions, γ_s and γ_b, have been excluded, or put to zero. Thus equation (13.29) will now read

$$\frac{1}{X}\int_0^X \sigma(x)\ dx = \sigma_a \phi \tag{13.42}$$

Since we are dealing with simple superposition of effects, we can continue the derivation similar to the one that gave equation (13.31) and including

$$\sigma = 0 \text{ at } x = \pm X \tag{13.43}$$

Evaluation of C_2 in (13.26), and completing the derivation, finally gives for the creep strain rate,

$$\dot{\varepsilon}_c = \frac{dY/dt}{G} = -\left(\frac{3\Omega D_b \delta_b}{kTGX^2}\right)\sigma_a \phi \tag{13.44}$$

Thus we see that we have arrived at a creep rate expression that is similar in all respects to that for the densification rate, equation (13.32), except that Σ has been replaced by σ_a. For the particular grain boundary diffusion case that we have considered here, where the effective stress follows clearly from the geometry, η_c and η_ρ are identical.

In a general geometry of an actual powder compact the effective stress no longer follows so clearly from the applied one, but we still expect the same relationship between the densification viscosity and the creep viscosity, apart from a numerical constant. This constant would need to be found from experiment, and that was exactly what was discussed when we talked about equation (13.40). The only reason that we cannot be more precise in that relationship is that, to date, measurements have not yielded sufficient data.

From equations (13.44) and (13.33) we should now be able to write for creep of a porous solid described well by Fig. 13.12, in the case of grain boundary diffusion as the dominant transport mechanism,

$$\dot{\epsilon}_c = -\left(\frac{K_c}{G^3}\right)\sigma_a\phi^2 \tag{13.45}$$

If we substitute the relationship between stress intensification factor and porosity given by equation (13.19), we get

$$-G^3\dot{\epsilon}_c = K_c\sigma_a\exp(2aP) \tag{13.46}$$

Dividing throughout by the initial grain size (G_o) cubed, taking logarithms of both sides, and using

$$P = 1 - \text{relative density} \tag{13.47}$$

we obtain

$$\ln\frac{-G^3\dot{\epsilon}_c}{G_o^3} = \ln\frac{K_c\sigma_a}{G_o^3} + 2a - 2a(\text{relative density}) \tag{13.48}$$

That is, a plot of the left-hand side versus the relative density should yield a straight line with a slope (-2a) independent of the applied stress and the current grain size. Figure 13.14 is such a plot (with the minus sign of the ordinate assumed) and the data fall on straight lines as expected. The parameter a can be obtained from the slope and substituted in equation (13.19) to yield ϕ as a function of the porosity.

An alternative way of treating creep and densification data is to use equations (13.35) and (13.37). Measurement of the creep rate $\dot{\epsilon}_c$ at applied stress σ_a gives η_c, the creep viscosity. If we make the approximation that the creep viscosity and the densification viscosity are the same, we have the densification viscosity. Finally, multiplication of this viscosity by the densification rate yields the densification stress Σ.

An example of sintering stresses thus determined is shown in Fig. 13.15. The sintering stress appears to be relatively constant for a large part of the densification process, on the order of 1 or 2 MPa for CdO powder, and decreases with decreasing initial density at constant particle size. Σ should increase with decreasing particle size at constant initial compact density, but this has not yet been verified by experiments such as those discussed here.

13.8 Interrelation of Various Transport Mechanisms: Sintering Maps

So far we have considered only the two principal mechanisms by which matter gets transported away from the grain boundary or from the interior of the grain to the pore surface; these were the densifying mechanisms. The nondensifying mechanisms will interfere with the rate at which densification can proceed by coarsening the microstructure of the system at constant density. When the microstructure coarsens,

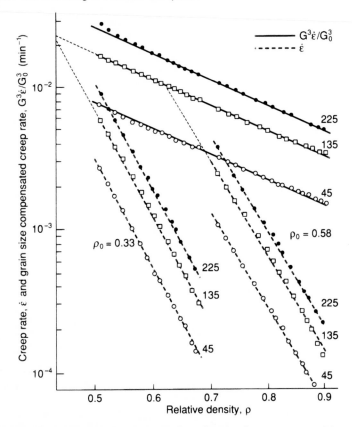

FIGURE 13.14 Dependence of creep rate, $\dot{\varepsilon}$, and grain-size-compensated creep rate, $G^3\dot{\varepsilon}$ / G_0^3 on relative density for initial densities 0.39 and 0.58 at stresses shown (in kPa). Data for CdO.

large pores and large grains grow at the expense of small ones, and the compact's structure evolves to one in which the pores are farther apart and their radii are increased. Examination of the expression describing the rate for the densifying mechanisms, equations (13.34) and (13.36), indicate that such coarsening would slow down the densification process rather rapidly. The increase in the pore separation would have a particularly strong influence because the densification rates are proportional to the inverse second or third power of this spacing, depending on whether the densification is by volume or grain boundary diffusion. If, for example, transport of matter between a large pore and a small pore could occur, the large pore could grow at the expense of the small one, because the chemical potential of the atoms in the surface of the small pore is lower (more negative) than that of the large pore [equation (13.3)]. This is a form of coarsening. Coarsening could be especially fast if the pores were connected by an open channel, as they are during much of the sintering process, and the vapor pressure of the material is high or the surface diffusion is very rapid.

In a first, simplified approach to coarsening by an evaporation/condensation mechanism, whereby atoms evaporate from a surface of high curvature to settle on one

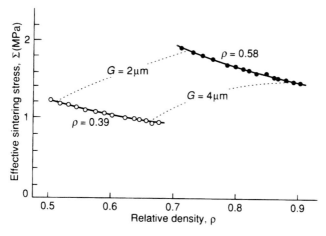

FIGURE 13.15 Sintering stress versus relative density for compacts of CdO of two differing initial densities r.

of low (more negative) curvature, we would expect that the coarsening rate would be proportional to the inverse of the pore distance, because we are dealing with simple vapor transport over that distance. We would further expect that the difference in pressure between the small pores and the large pores would also depend on the overall *scale* of the system. The latter dependence could be obtained from the relationship between vapor pressure and the difference in the chemical potential of the atoms in a curved surface compared to a flat surface, $\Delta\mu$, equation (13.4):

$$\Delta\mu = \gamma_s \Omega\kappa = kT \ln\frac{p_R}{p_o} \tag{13.49}$$

where p_R is the equilibrium vapor pressure over the curved surface and p_o is that over the flat surface. The relationship for the equilibrium pressures over two differently curved surfaces, p_1 and p_2, having a curvature difference $\Delta\kappa$, would then be

$$\ln\frac{p_1}{p_2} = \frac{(\Delta\kappa)\gamma_s \Omega}{kT} \tag{13.50}$$

We should expect that for pore distributions that relate to one another by a simple factor of scale, $\Delta\kappa$ should be proportional to 1/scale of the system. Thus for such pore distributions, $\Delta\kappa$ is proportional to $1/X$, where X is the mean interpore distance. Many pore distributions in real compacts relate approximately through such a scale factor.

Because the difference in pressure, Δp, for the usual range of pore sizes is small and $\ln(1 + x) \cong x$, when x is small compared to 1, we can rewrite equation (13.50) as

$$\ln\frac{p_1}{p_2} = \ln\frac{p_2 + \Delta p}{p_2} \cong \frac{\Delta p}{p_2} = \frac{(\Delta\kappa)\gamma_s \Omega}{kT} \tag{13.51}$$

or

$$\Delta p \sim \Delta \kappa \sim \frac{1}{X} \qquad (13.52)$$

The coarsening rate due to evaporation/condensation should then be proportional to $(1/X)\Delta p$ and thus to $1/X^2$. Therefore, for densification controlled by grain boundary diffusion, for which $\dot{\varepsilon}_\rho$ is proportional to $1/X^3$, we should expect to be able to decrease undesirable coarsening due to the evaporation/condensation, by decreasing the particle size. In other words, both coarsening and densification will proceed more rapidly at smaller particle size, but the increase in the rate of the latter will be much greater than the increase in the rate of the former. Furthermore, transport through the vapor increases rapidly in significance once vapor pressure becomes appreciable, so that lowering the sintering temperature will also decrease the coarsening rate due to evaporation/condensation with respect to the densification rate. We should, however, be aware that other coarsening processes may occur that become *more* important with decreasing temperature, so that dropping the densification temperature may not be a sure cure for systems that seem to coarsen excessively. We discuss coarsening and microstructure development more fully in Chapter 15.

An argument similar to the one above can be constructed for the case of coarsening by surface diffusion, and again, a smaller particle size favors the densifying mechanisms. However, the temperature change now has a different effect: surface diffusion usually has a lower activation energy than either volume diffusion or grain boundary diffusion, and as a result, lowering the temperature can increase the relative importance of coarsening by surface diffusion.

Since powder compacts always coarsen during densification, it is possible for the relative importance of the coarsening to change as sintering progresses. If, for example, evaporation/condensation is active, we might have some rapid densification initially, but sintering could slow down considerably when coarsening has increased the grain size and therefore the relative importance of the evaporation/condensation mechanism. It is becoming evident here that a complete detailed treatment of interactions of various transport mechanisms during sintering will be a very difficult task. Research in this area is continuing.

The problem of interaction of transport mechanisms has usually been considered only for the very earliest stages of densification, where the sintering process supposedly could be described by the growth rate of the contact area, the "neck," between two perfectly spherical particles. For this highly restrictive geometry, the contributions of the various mechanisms depicted in Fig. 13.4 to the growth of diameter of the neck between the spheres. One can make various assumptions about the particle size and all the transport coefficients, and then construct a map in which the relative neck size is plotted versus the homologous temperature ($T/T_{\text{melting point}}$). The fields in which a particular mechanism dominates the neck development have as boundaries the lines where two adjacent mechanisms contribute equally to the neck growth rate. An example of a sintering map constructed along these general principles is shown in Fig. 13.16. Such maps may suggest temperature-density-time paths that could avoid conditions where nondensifying mechanisms are active. The practical utility of such maps is severely limited, however, since transport coefficients are almost never known with sufficient

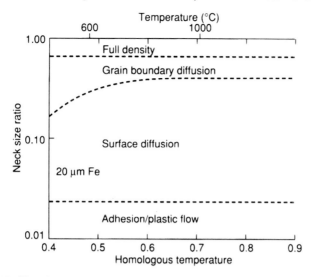

FIGURE 13.16 Sintering diagram for 20-μm iron powder, showing the dominant sintering mechanism for various neck size ratios and sintering temperatures. (From R. German, *Powder Metallurgy Science*, 1984, reprinted with permission of Metal Powder Industries Federation, Princeton, NJ.)

accuracy for specific ceramic systems, and powder compact densification is often poorly described by a two-particle model. Sintering maps are generally not available for practical powders; they would need to be determined for each system, which would involve an enormous amount of work. However, they do clearly represent the conceptual relationships between the various mechanisms.

13.9 Densification of Heterogeneous Powder Compacts: Differential Densification

In practical powder compacts, complete homogeneity is never achieved. For example, even if we have strictly monosized particles, local packing variations are invariably present. If the powders are not monosized, local packing density variations as well as clusters of differing particle sizes will be unavoidable. In a way, homogeneity depends on the scale at which we examine the powder compact. The coarser the scale of observation, the more uniform the compact will appear. It is not a simple matter to put this statement into a quantitatively useful form. It is even more difficult to link such a statement quantitatively to the densification behavior of a powder compact, and interesting research challenges remain in this area. In some relatively simple cases, however, the effects of heterogeneities on densification of a powder compact can be contemplated. We will consider a few of these cases.

A first case is when we have a powder compact that consists of simple mixtures of particles of two different sizes. As a guess, we might propose that such mixtures would densify in a way that is described by a linear combination of the independent densification of the two separate size fractions. This "rule of mixtures" behavior has actually been observed for some bimodal mixtures of alumina powders (Smith and Onoda, 1984). An interesting observation in that case was that the coarse powder fraction was relatively inactive, and most of the densification and grain growth took place in the fine powder fraction. As a result, upon sintering, some of these bimodal powder mixtures tended toward homogeneity rather than away from it. An example is shown in Fig. 13.17.

A second case is when undeformable, dense particles are distributed in a homogeneously densifying matrix. This corresponds, for example, to the compact structure of a composite material made up of a powder matrix with dispersed, inert particles that were added to modify the properties of the final product. The inert particles are typically a lot larger than the matrix particles and present an obstacle to densification. The problem arises when the matrix tries to shrink around an inclusion that itself cannot shrink. As a result, a backstress will be generated in the matrix that opposes the sintering stress, thus slowing down the overall sintering process. This process can be considered as one of *differential densification,* meaning that some regions inside the powder compact densify at a different rate than do other regions. The regions will interact because they exert stresses on each other.

An experimental example of differential densification is shown in Fig. 13.18 for an irregular two-dimensional raft of copper spheres (Exner, 1979). In a detailed statistical analysis of the evolution of two-dimensional rafts like this (Weiser and DeJonghe, 1986), it became apparent that the rearrangement of the particles came about because some regions densify faster than others. The regions that densify faster exert tensile stresses on neighboring slower regions, and these stresses combine with the local sintering stress as in equation (13.38). If these differential densification stresses are larger and opposed to the local sintering stress, pores at that location will grow rather than shrink despite the fact that, overall, the powder compact is shrinking. Therefore, differential densification can have a significant effect on the microstructural evolution and the densification of a powder compact. The more homogeneous the compact structure, the less differential densification and its undesirable consequences.

Recent studies on the densification of particulate composites indicate that the backstresses that arise as a result of differential densification depend on the ratio of the creep viscosity over the densification viscosity, η_c/η_ρ, of the inclusion-free matrix. In one study (De Jonghe and Rahaman, 1988) it was found that the densification rate $\dot{\varepsilon}_{m\rho}$ of the matrix of a particulate composite could be expressed as

$$\dot{\varepsilon}_{mp} = \frac{\Sigma}{\eta_{mp}} \cdot \frac{1}{1 + 4\left(\eta_c / \eta_\rho\right) f_i / (1 - f_i)} \tag{13.53}$$

where f_i is the volume fraction of the inert, dispersed phase.

Thus, for a particulate composite, the matrix densification rate would be proportional to that of the matrix without any dispersed phase, multiplied by a function that depends on the ratio of the creep rate over the densification rate and decreases with increasing volume fraction of the inclusion f_i. We could assess the importance of this function by inserting the maximum ratio of η_c / η_ρ that has been found or can be expected on the basis of some viscoelastic analyses (Scherer, 1987; De Jonghe and Rahaman,

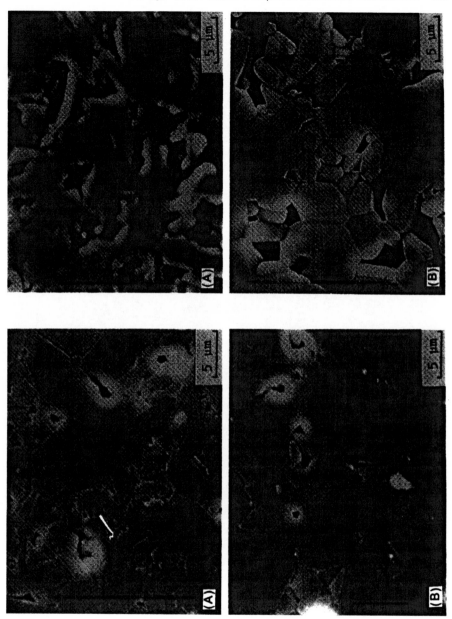

FIGURE 13.17 Scanning electron micrograph of sintered alumina compacts that consisted of 0.8 weight fraction of fine alumina (lower pair) and 0.5 weight fraction of fine alumina (upper pair) after (A) 60 min and (B) 360 min at 1600°C. (From P. Smith and G. L. Messing, Sintering of Bimodally Distributed Alumina Powders, *J. Am. Ceram. Soc. 67*, (4), 241 (1984). Reprinted with permission from American Ceramic Society, Columbus, OH.)

0.5 mm

Presintered	16 h	64 h	256 h
$x/a = 0.15$	$x/a = 0.36$	$x/a = 0.44$	$x/a = 0.58$

(a)

0.5 mm

Presintered	2 h	5 h	25 h
$x/a = 0.34$	$x/a = 0.40$	$x/a = 0.46$	$x/a = 0.60$

(b)

FIGURE 13.18 Evolution of the structure of two planar arrays of copper spheres during sintering. Two different starting densities, expressed as x/a, are shown. $x =$ neck diameter between the particles of diameter a. (From H. Exner in *Reviews on Powder Metallurgy and Physical Ceramics*. Permission of Freund Publishing House Ltd., Tel Aviv.)

1988) at some volume fraction of inclusions, f_i. This value would be around 2. Experimentally, the presence of inclusions can decrease the densification rather significantly, as is shown in Fig. 13.19 for SiC inclusion in ZnO. An interesting observation in these data is that the densification rate of the composite compacts tends to zero at inclusion volume fractions that are about 0.5 or even less. The disagreement with equation (13.53) stems from the failure to recognize that the undeformable inclusions can form a continuous, three-dimensional skeleton, preventing the compact from densifying. At inclusion levels approaching 25 to 30 vol % of the green compact, this undeformable skeleton formation begins to dominate the densification behavior (Lange, 1987). However, the effect is strongly dependent on how uniformly the second phase is dispersed within the matrix, and is especially strong for nonspherical inclusions such as whiskers. In practice, then, achieving a highly uniform dispersion of the second-phase inclusions becomes the most difficult but also the most rewarding goal for processing of particulate composites.

In the densification of such composites, equation (13.53) clearly shows that whatever is beneficial to the densification of the pure matrix will also be beneficial to the densification of the composite. This is a useful point since, in general, much more is known about the densification behavior of the pure matrix than about that of the composite.

A further observation, which examination of equation (13.53) would lead to, is that the densification rate of the composite would be independent of the particle size of the dispersed phase. This is *not* in agreement with most experimental findings; decreasing the particle size of the dispersed phase generally decreases the densification rate of the compact (Weiser and De Jonghe, 1988; Tuan et al., 1989). In fact, when these particles are much smaller than the grain size, they tend to be preferentially located at the grain boundaries and can arrest sintering (and coarsening) completely. This effect can be exploited for the fabrication of highly porous materials that have to be exposed to high temperatures (such as some catalysts, catalyst supports, or electrodes in high-temperature fuel cells), where densification and coarsening would be deleterious.

13.10 Liquid-Phase Sintering

So far we have treated powder compacts in which everything is solid during densification. An important parameter that determines the densification rate is the grain boundary diffusion coefficient multiplied by the grain boundary thickness. If we could manage to replace that grain boundary with a film of liquid many times thicker than the grain boundary thickness, δ_b, we might greatly increase the densification rate because, in general, diffusion rates in liquids are a lot higher than in solids. This can be achieved in liquid-phase sintering.

In a typical liquid-phase sintering process, some liquid, say 10 to 30 vol %, is present during densification. The liquid is supposed to penetrate between the particles to produce the intergranular liquid film we just mentioned. Despite the fact that liquid is present

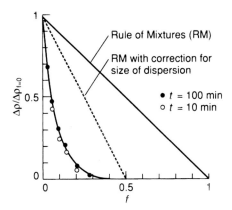

FIGURE 13.19 Normalized change in density versus volume fraction of SiC compared to the prediction of the rule of mixtures for a composite consisting of a ZnO matrix with dispersed SiC particles.

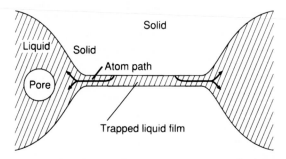

FIGURE 13.20 Schematic of the densification mechanism during liquid-phase sintering where the liquid leaves a thin grain boundary film between grains.

between all the grains, the powder compact usually does not collapse under its own weight unless, of course, a lot of liquid is present. The capillary effects of the liquid still succeed in holding the compact together, although its creep viscosity can be reduced significantly.

In general, one can distinguish four stages in the evolution of liquid-phase sintering (Kingery, 1959).

1. As the temperature of the compact is raised, liquid will form at some point and penetrate between the powder particles, possibly breaking up agglomerates.
2. The broken-up agglomerates and the particles coated with liquid rearrange extensively under the action of the surface tension of the liquid. The process is somewhat similar in appearance to an extreme form of differential densification. During this stage there can be extensive liquid redistribution.
3. Solid dissolves at the wetted contact points between the particles and precipitates elsewhere. This stage is known as the solution/precipitation stage. It leads to particle/particle approach and hence to further densification.
4. Final stage: Isolated pores are partly filled with liquid. Densification continues until the maximum possible density is reached, when all pores are eliminated.

In general, stages 1 and 2 proceed comparatively rapidly, perhaps in less than a few minutes. In detail, these stages are complex, since they involve the redistribution of liquid and solids. Depending on the amount of liquid, a substantial amount of densification can take place in these stages. It would, in fact, be possible to reach the final stage immediately if there is enough liquid to fill all the interparticle porosity that would remain after the particles had rearranged to form a continuous network. If, for example, the solids could rearrange to form a solid skeleton that, by itself, would have 30 vol % porosity, then obviously 30 vol % liquid would be required to reach full density after stage 2. For less liquid, as is often the case, stage 3 takes most of the time and thus merits special consideration.

The essentials of liquid-phase sintering are shown in Fig. 13.20. We consider the liquid initially trapped between two spherical particles. A liquid bridge is formed, and the particles will move together under the action of the surface tension of the liquid, which has created a meniscus with a curvature κ_l. The excess pressure, P_l, in the liquid

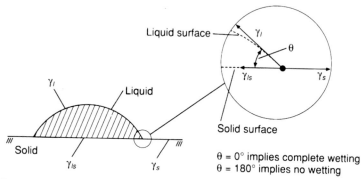

FIGURE 13.21 Geometry of a stationary drop of liquid on a solid surface. q is the contact or wetting angle.

is readily found from the Kelvin equation, equation (13.3) or (13.4), which together with equation (13.7) give

$$P_l = -\gamma_{ls} K_l \qquad (13.54)$$

where γ_{ls} is the solid-liquid interfacial tension. The force with which the two particles are drawn together will then depend on this pressure, on the surface tension of the liquid, and on the liquid-solid contact area.

To have the geometry properly defined we will have to know the *contact angle, θ,* between the liquid and the solid. The meaning of this angle is shown in Fig. 13.21. It follows from a balance of the surface tensions and the interface tensions of the gas-solid-liquid triple junction. In Fig. 13.21 we see a drop of liquid sitting on top of a flat solid surface with a surface tension γ_s. Such an immobile drop is called a *sessile drop*. The interfacial surface tension between the liquid and the solid is γ_{ls}. Since we have equilibrium, the chemical potential of the atoms in the free surface of the liquid must be the same, which in turn must mean that the curvature of the free liquid is constant. Thus the liquid must form a spherical cap (gravity effects are ignored here; they can distort the droplet—very large drops would flatten to form a pancake). Treating the surface tensions as vector forces that must be in mechanical equilibrium, and assuming that the vertical force component at the triple junction is properly compensated by an insignificant bending of the solid, we can readily arrive at Young's equation

$$\cos\theta = \frac{\gamma_s - \gamma_{ls}}{\gamma_l} \qquad (13.55)$$

where θ is the wetting angle. Equilibrium *requires* that the configuration at the liquid-solid-gas triple junction satisfy the wetting angle. Figuring out what the curvature of the liquid is in Fig. 13.21 is then an exercise in analytical geometry.

Once the particles are in close proximity, pressure gradients start to develop in the thin liquid film that is trapped. The reason for this is twofold: first, to make a liquid flow through a small opening requires a pressure (capillary streaming pressure or Poiseuille pressure); and second, the possibility of dissolving the atoms of the solid during grain center-to-center approach will lead to a stress gradient, in much the same way as for solid-state sintering. These stress gradients will determine the gradient of the chemical potential for the solid atoms at the liquid-solid interface as solid atoms are removed from in between the nearly touching grains and transported toward the free meniscus of the liquid. From there, the atoms could continue their journey and redistribute themselves over the free pore surface. In the analysis of liquid-phase sintering it is usually assumed that an insignificant amount of free energy is consumed in the latter step, so that the transport of matter through the intergranular liquid film can be regarded as rate controlling. All the while, barring any short-range repulsive effects of the approaching interfaces, the liquid film should get progressively thinner with time. Thus liquid-phase sintering involves competition between the rate at which the liquid film gets squeezed out from between the solid particles and the rate at which the solid dissolves at the contact points and is transported away. The resistance to fluid flow through a capillary increases rapidly as the capillary gets narrower, and for film thicknesses on the order of 1 to 5 nm, fluid flow is usually much slower than the solution-precipitation process. One often finds evidence of such films if the sintered materials are examined with high-resolution transmission electron microscopy. Another reason for a finite liquid fihn remaining between the grain boundaries is that it may be energetically unfavorable to remove it. The kinetics of the grain boundary film thinning and the concurrent mass transport has been analyzed in some detail, and the interested reader is referred to the relevant literature (Marion et al., 1987).

It is important to stress the significance of the wetting angle in liquid-phase sintering. Going back to equation (13.54), it is not difficult to construct a geome try in which, when ~ is large, we could end up with the liquid actually *opposing* densification when the meniscus reaches a positive curvature. In such cases the densification of the liquid-phase sintering system would cease at an intermediate density. This phenomenon has been observed for tungsten particles that were sintered in the presence of some liquid copper (Gessinger et al., 1973). The wetting angle is an important factor and should be zero to avoid this problem entirely. Another requirement for liquid-phase sintering is that the solid dissolve to some extent into the liquid. Obviously, if the solid is insoluble in the liquid, no matter can be transported through the liquid film and we do not achieve any benefit from liquid-phase sintering.

13.11 Concluding Remarks

In this chapter we have discussed some of the fundamental aspects of densification of powder compacts. The essential aspects of the process are simple: the elimination of the void space due to matter transport driven by a reduction in surface energy. A more detailed look at the process of mass transport and the factors that influence it has been considered here. Broadly, we needed to distinguish between densifying and nondensifying mass transport. We used the concept of a sintering stress to define the driving force for sintering, and experiments now indicate that this driving force, although system specific, is nearly constant, over most of the densification process for most systems. From this, we proceeded to analyze the sintering process in terms of solutions to the diffusional mass transport problem. In some cases we were able to illuminate the effects that structural heterogeneities have on the densification process. We also discussed briefly some aspects of liquid-phase sintering. To complete the picture we will need to look more closely at the microstructural evolution during densification; this matter is discussed in Chapter 14.

LIST OF SYMBOLS

a_i	surface area of pore i
a	parameter depending on dihedral angle
D_b	grain boundary diffusion coefficient
f_i	volume fraction of inert dispersed phase
G	grain size (grain diameter)
J	atom flux
K_b	kinetic factor for grain boundary diffusion
K_c	kinetic factor for creep
K_v	kinetic factor for volume diffusion
K	Boltzmann constant
N	number of pores per unit volume
P	porosity
P_l	excess pressure in liquid
P_o	equilibrium vapor pressure over flat surface
P_R	equilibrium vapor pressure over curved surface of radius R
R	radius
t	time
V	volume
X	radius of contact area between two sintering particles
Y	center-to-center distance between two sintering particles
δ_b	grain boundary thickness
$\dot{\varepsilon}_c$	creep strain rate
$\dot{\varepsilon}_\rho$	densification strain rate
γ_b	grain boundary energy per unit area
γ_{ls}	solid-liquid interface energy per unit area
γ_s	surface energy per unit area
η_c	creep viscosity
η_ρ	densification viscosity
κ	surface curvature
μ	chemical potential per atom
Ω	atomic volume
ϕ	stress intensification factor
Ψ	dihedral angle
θ	contact angle
Σ	sintering stress (equivalent externally applied stress)
σ	stress
σ_a	externally uniaxial applied stress
σ_h	externally applied hydrostatic stress

PROBLEMS

13.1 Assuming that a powder has a surface energy of 1000 erg/cm^2, calculate the maximum amount of energy that is available for densification of particles of 0.5 μm diameter, compacted to a green density of 0.55 of theoretical, and assuming that there is no grain growth. The grain boundary energy is 550 erg/cm^2.

13.2 An inert gas is trapped within a liquid bubble. The surface energy of the liquid is 300 erg/cm^2. Calculate the pressure of the inert gas inside the bubble.

13.3 Calculate the difference in the equilibrium vapor pressure for two particles, one with a 0.1-μm radius and the other with a 10-μm radius. Describe what would happen if these two particles were to be suspended in the same closed box.

13.4 Calculate the mean curvature for an ellipse as a function of eccentricity.

13.5 A porous solid contains equally sized pores. Show that the spatial distribution of the pores does not affect the sintering stress.

13.6 Consider a pore with a constant volume. Assuming that the pore takes on the shape of an ellipsoid of revolution, calculate the surface energy of the pore as a function of the ratio of the small axis over the large axis.

13.7 Consider a pore in a two-dimensional space, surrounded by three grains. Demonstrate by a virtual work argument that if we consider the local geometry of this pore, the condition for pore stability (no tendency to grow or to shrink) implies that the pore sides will be flat. Note here that when the dihedral angle exactly matches the geometry, it must be 60°. Note that the pore coordination number and the dihedral angles have to match just right to achieve this condition of pore stability, as you will note immediately if you try to find a stable pore configuration for a dihedral angle of, say, 75°.

13.8 Show that the general form of the sintering rate as shown in equation (13.35) is retained if grain boundary diffusion is considered for a boundary that has the form of a circular disk rather than a line. You will need to consider diffusion in cylindrical coordinates (see Chapter 6) and argue that there is no singularity at the center to find the form of equation (13.35).

13.9 Derive equation (13.36).

REFERENCES AND FURTHER READINGS

ALEXANDER, B., and R. BALLUFFI, *Acta Metall. 5,* 666 (1957).

BEERE, W., *Acta Metall. 23,* 139 (1975).

COBLE, R., J. *Appl. Phys. 32,* 787 (1961).

DE JONGHE, L., and M. RAHAMAN, *Acta Metall. 36,* 223 (1988).

EXNER, H., *Powder Metall. Phys. Ceram. 1,* 1 (1979).

GESSINGER, G., H. FISCHMEISTER, and H. LUKAS, *Acta. Metall. 21,* 715 (1973).

JOHNSON, D., and I. CUTLER, J. *Am. Ceram. Soc. 46,* 541 (1963).

KINGERY, W., J. *Appl. Phys. 30,* 301 (1959).

KUCZYNSKI, G., *Phys. Rev. 75,* 344 (1949).

LANGE, F., J., *Mater. Res. 2,* 59 (1987).

MARION, J., C. HSUEH, and A. EVANS, J. *Am. Ceram. Soc. 70,* 766 (1987).

MESSINGER G., and G. ONODA, J. *Am. Ceram. Soc. 67,* 328 (1984).

RAHAMAN, M., L. DE JONGHE, and R. BROOK, J. *Am. Ceram. Soc. 69,* 53 (1986).

SCHERER, G., J., *Am. Ceram. Soc. 70,* 719 (1987).

SMITH, J., and G. ONODA, J. *Am. Ceram. Soc. 67,* 238 (1984).

TUAN, W., E. GILBART, and R. BROOK, J. *Mater. Sci. 24,* 1062 (1989).

VIEIRA, J., and R. BROOK, J. *Am. Ceram. Soc. 67,* 450 (1984).

WEISER, M., and L. DE JONGHE, J. *Am. Ceram. Soc. 69,* 822 (1986).

WEISER, M., and L. DE JONGHE, J. *Am. Ceram. Soc. 71,* C125 (1988).

14

Microstructure Development During Sintering

14.1 Introduction

The engineering properties of materials are strongly affected by their microstructure. The microstructural features of importance are many: grain size and grain shape, pore size and location, distribution of second phases (such as precipitates), and others. For polycrystalline ceramics, for example, it is frequently found that the fracture strength is proportional to the inverse square root of the grain size. Metals show a similar relationship for the stress at which they yield. As another example, metal alloys can be strengthened by the presence of dispersed second-phase particles. A well-known case is that of aluminum, a very soft material in its pure form, which can be hardened by the incorporation of a few percent of copper. The added copper, after appropriate heat treatment, forms very small Al-Cu precipitates that make plastic deformation of the alloy more dif-ficult than that of the pure aluminum. In another example, the high-temperature creep resistance of sintered silicon nitride depends strongly on the nature and amount of a thin glassy layer, perhaps only 2 nm wide, that is often present at the grain boundaries.

For ceramics the microstructural features are particularly dependent on the quality of the starting powder and the way this is compacted. Property-limiting flaws are often accidentally introduced in the initial forming stages of the ceramic body. These include the incorporation of large voids, foreign objects such as ball mill debris, dust, hairs, or even insects, and atmospheric contaminants such as adsorbed water or carbon dioxide. Much can be gained by working in clean conditions and with powders for which the particle size distribution, the shape, and the packing have been carefully controlled. Assuming that adequate environmental conditions and proper compaction procedures are in effect (and unfortunately this is less often the case than one would expect) further microstructural manipulation has to be done during the sintering process itself. The aim is to achieve the desired microstructure for the purpose at hand, which usually means as high a density, as small a grain size, and as uniform a microstructure as possible. Large grains could, for example, be detrimental because thermal expansion anisotropy would lead to very significant local stresses around those grains, which can induce local fracture from which catastrophic failure can begin. Large pores also can be origins of failure, especially if they are located at the surface of the sample. Such imperfections are the cause of wide variations in the properties of the sintered bodies.

In many metallurgical materials, such as steels, the desired microstructure can be obtained by a combination of mechanical deformation and heat treatment. For example, steel can be made tougher if it is hammered extensively or heat treated; this has been known for centuries. Some ancient samurai swords reveal a microstructure that indicates a sophisticated knowledge of the effects of heat treatment and mechanical deformation. For ceramic materials, the choice of processes that can alter the microstructure after sintering is far more limited: the brittleness of ceramics prevents their mechanical deformation at ordinary temperatures; strengthening by the formation of second phases with precipitation reactions is, more often than not, unsuccessful because, in contrast to metals, the internal strain that these precipitates produce around them as they form cannot be as easily accommodated, often leading to crack initiation. For most ceramics, the control of microstructure must therefore be incorporated in the sintering process. This complicates processing considerably. Further, technical ceramics are often not single phases, and chemical reactions may occur together with sintering. For example, alumina can be toughened by the incorporation of SiC whiskers. However, at the elevated temperatures needed to densify alumina-silicon carbide composites, reaction between the silicon carbide and some oxygen that may be present leads to the formation of silica, which, in turn, reacts with the alumina to produce glassy aluminosilicates. As a result, the silicon carbide whiskers disappear. Processing methods must then be adopted that minimize such interreaction (in the alumina-silicon carbide case a reducing atmosphere is necessary). Clearly, the availability and use of an accurate phase diagram for the processing of reacting systems is a primary requirement. Unfortunately, nonequilibrium phases are often formed in reacting systems, so that even accurate equilibrium phase diagrams may only be partially useful. Evidently, reactive sintering, although technologically important, is so complex that a detailed predictive analysis is still a long way off. Instead of attempting to deal with all aspects of microstructure formation during the processing of ceramics from powders, we will limit ourselves to considering the microstructural development for one relatively simple case: the densification of a single-phase ceramic. Even in this case there is still much technically relevant research that remains to be done.

In considering the development of the microstructure during the densification of a single-phase ceramic, the important issues that need to be addressed concern the grain size, the pore size, the pore location, and the interplay between coarsening of the microstructure and densification.

First we will treat grain growth in dense and porous materials. We follow this with an examination of the coarsening of the pore network in a sintering powder compact. The interaction between isolated pores and moving grain boundaries is then analyzed, since this determines the limits of densification.

14.2 Grain Growth

Grain growth occurs in both dense and porous polycrystals at sufficiently high temperatures. The elimination of grain boundary area will lower the material's free energy, and this provides a driving force for grain enlargement. First we should look at grain growth in dense polycrystals, since these are free of the complications that an evolving pore network adds.

14.2.1 Grain Growth in Dense Polycrystals

Grain growth in dense polycrystals requires the change of grain shapes by movement of grain boundaries together with a decrease in the number of grains per unit volume. The details of the movement of these grain boundaries is difficult to characterize fully since they depend on the local geometry of the growing and disappearing grains. A few general observations can, however, be enumerated:

1. Large grains will grow at the expense of small ones.
2. The process of grain growth is limited by diffusional motion of the atoms across the grain boundaries and thus is temperature-activated.
3. The rate of grain growth decreases with increasing grain size.
4. Often, the coarsened microstructure looks much the same as the one at an earlier time, except for a magnification factor; this property of the evolution of a grain size distribution is referred to as *self-similarity* and is characteristic of *normal grain growth.*
5. In some cases self-similarity is not preserved and a few large grains develop rapidly. This usually undesired evolution is called *abnormal grain growth.*
6. The narrower the grain size distribution, the slower the grain growth rate.

We consider grain growth first in a very simplistic manner, as shown in Fig. 14.1, ignoring the geometrical complications of satisfying the equilibrium dihedral angles at grain boundary junctions. We take this quasi-two-dimensional model to be extended into the paper for a length l. The cylindrical central grain, of radius R, is surrounded by n nearest neighbors. An increase dR in the grain radius R will involve a decrease of $nldR$ in area of the boundaries between the grains that are the neighbors of the growing grain. The area of the growing grain's boundary is $2\pi Rl$, and its incremental change is then $2\pi l\, dR$. The driving force for growth of the central grain may be interpreted as a

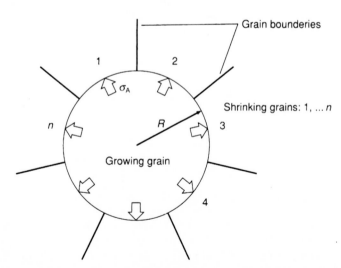

FIGURE 14.1 Idealized, two-dimensional geometry for a single circular grain surrounded by grains.

pressure, σ_{A}, acting on the boundary of that grain. The magnitude of this pressure can be found from a virtual work argument, equating the work performed by that pressure, $-\sigma_{A}dV$, to the energy changes associated with the changing grain boundary areas:

$$\sigma_{A} \cdot l\,2\pi R\,dR = -\gamma_{b}2\pi l\,dR + \gamma_{b}nl\,dR \qquad (14.1)$$

where $dV = 2\pi rRdR$, so that the driving force per unit area of grain boundary, σ_{A}, becomes

$$\sigma_{A} = \frac{-\gamma_{b}(2\pi - n)}{2\pi R} \qquad (14.2)$$

where γ_{b} is the energy per unit area of grain boundary. Note that σ_{A} has the dimensions of stress or pressure, as we should expect from the Kelvin relation [equation (13.3)]. The rate of grain boundary advance, or dR/dt, will then be proportional to this driving force. If we assume that the proportionality is a linear one, we can define a *grain boundary mobility*, M_{b}, as the ratio of the grain boundary velocity to the driving force. The grain boundary mobility is a material-specific parameter. It relates to the microscopic process of atoms jumping to their new positions when a grain boundary sweeps by.

For the rate of grain growth, one can then write

$$\frac{dR}{dt} = \frac{\alpha M_{b}\lambda_{b}(n - 2\pi)}{R} \qquad (14.3)$$

where α is a constant for a particular geometrical configuration; for the present simplistic example α is equal to $1/2\pi$. We should expect different values for more realistic geometries. α will, for example, depend on th e value of the dihedral angles between the grain boundaries; as you will note, our simple example assumes that the dihedral angle is 180°. For a self-similar microstructure n *is* invariant, and if we assume that equation (14.3) describes the average growing grain, we immediately obtain a parabolic grain growth equation by integrating equation (14.3) with respect to time t:

$$R^2 = R_{o}^2 + 2\alpha M_{b}\lambda_{b}(n - 2\pi)t \qquad (14.4)$$

where R_{o} *is* the initial grain size. Despite its simplicity, this model contains most of the elements of many more complex models that have been proposed in the literature; these all arrive eventually at an expression that is similar to equation (14.4). A rather complete review of this subject has been given by Atkinson (1988). Relevant parameters are as follows:

M_{b}: a grain boundary mobility that is dependent on the temperature in a manner characteristic of a temperature-activated process, i.e., $M_{b} = M_{o}\exp(-E_{b}/kT)$, where M_{o} is a constant and E_{b} is the grain growth activation energy.

$\alpha(n-2\pi)$: a function that expresses the local geometry and is independent of the average grain size for self-similar distributions. The higher the grain coordination number of a particular grain, the higher its driving force for growth. If the grain coordination number is below a critical number (in our quasi-two-dimensional case this number is 2π), that grain would shrink rather than grow. If we had a grain arrangement in which all the dihedral angles were satisfied and matched the grain coordination number (i.e., flat grain boundary segments at equilibrium), we could have no grain growth. In two dimensions this is the case for a dihedral angle of 120° when all the grains are identical hexagons. In three dimensions this condition cannot be satisfied entirely for space-filling polyhedra with isotropic grain boundary energies, but grain growth would still be comparatively slow in compacts consisting of nearly-identical grains.

γ_b: the grain boundary energy to which the grain growth rate is proportional.

In practice, the grain growth rates in dense systems do not behave as well as described by equation (14.4). Often the mobility depends in a more complex way on temperature because of the presence of impurities that may affect the grain boundary motion. Impurities that are adsorbed on the boundary and themselves move with difficulty will impede boundary motion as long as they remain with the boundary. If, however, the boundary experiences a large driving force for movement, it may break away from its slow-moving impurities. The boundary will largely ignore the randomly distributed impurities it then encounters. This is because for randomly positioned impurities there will be, on the average, as many impurities pushing a grain boundary segment away as there are holding that boundary segment back. In the case of such a grain boundary, the grain boundary mobility would suddenly go up. Thus M_b may also be some complicated non-single-valued function of the driving force (which is proportional to $1/R$ here). As a consequence, a practical form of equation (14.4) is often written as

$$R^n = R_o^n + B\exp\left(\frac{-E_b}{kT}\right)\left(t - t_o\right) \tag{14.5}$$

where n has been found to be between 2 and 3 and may depend on temperature; R_o is the grain radius at time t_o.

In some cases, specific boundaries may have an abnormally high mobility. This may occur, for example, as a result of a local impurity accumulation at the advancing boundary, forming a liquid film in which atoms can be transported very rapidly compared to solid-state diffusion, or because some special boundary orientation is particularly mobile. Such boundaries can then sweep through the material with abnormal rapidity. As a result, one may develop a microstructure in which few very large grains are embedded in a fine-grained matrix in which the usual grain growth proceeds at a comparatively low rate. This process of growth of a few isolated grains is called abnormal grain growth; it becomes a problem at increasing temperatures when liquid phases are likely to form. An example of a microstructure in which abnormal grain growth has occurred is shown in Fig. 14.2. You can contrast this microstructure to that to of Fig. 13.3(a), which is representative of normal grain growth.

14.2.2 Grain Growth in Porous Compacts

When looking at the evolving microstructure of a sintering ceramic powder compact as shown in Fig. 13.2, it is evident that there is considerable evolution in the microstructure: grains as well as pores increase in size while decreasing in number as densifica-

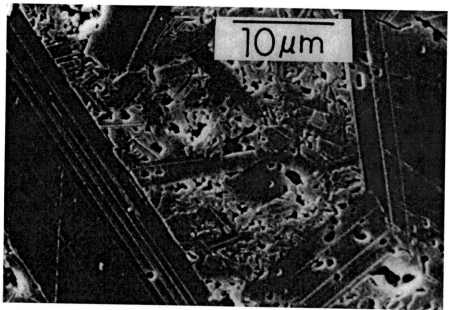

FIGURE 14.2 Abnormal grain growth in a sodium β-alumina sintered ceramic.

tion proceeds. Grain growth and pore shape change therefore interact in sintering powder compacts, and the microstructural evolution of the system can be expected to be considerably more complex than for the dense polycrystals. The overall process of pore and grain size increase is generally referred to as *coarsening*.

We should examine how the grain size and the pore size can be manipulated - hopefully, independently. An example where such independence was achieved is shown in Fig. 14.3 for nickel-zinc ferrite (Okazaki, 1976). For these ferrites, the desired microstructures were obtained by controlling the time-temperature schedule during the sintering or by applying pressure (hot pressing; see Chapter 15). In general, such a degree of control is difficult to achieve. We now consider grain growth in porous compacts and will look at the coarsening in the pore network.

Grain growth will occur when the system can reduce its energy by eliminating a grain boundary area. If we look first at a grain boundary between two particles [Fig. 14.4(a)], we note that this grain boundary will have difficulty moving. Its displacement, as shown in this figure, will actually increase its area: this corresponds to going uphill energetically, which cannot happen spontaneously. For the grain boundary to sweep into a neighboring grain, extra energy must be liberated somehow, and one way in which this can occur, described by Edelson and Glaeser (1988), is shown in Fig. 14.4(b) and (c). In this case, elimination of grain boundaries by the advancing boundary releases enough energy to make the overall process favorable. As shown in Fig. 14.4c, the advancing grain boundary might even sweep by some intra- or interagglomerate pores and cause them to be trapped inside a grain, an undesirable feature. This process is less likely to occur early in the sintering, because continuous-pore channels pin boundaries very effectively.

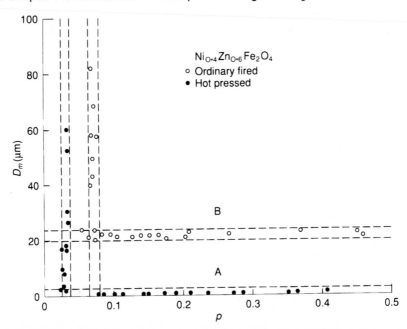

FIGURE 14.3 Examples of specimens with controlled porosity p and with controlled grain size D_m. (From K. Okazaki and H. Igarashi, "Importance of Microstructure in Electronic Ceramics" in *Ceramic Microstructures '76*, R. Fulrath and J. Pask, eds., Westview Press, Boulder, CO, 1977, p. 566.)

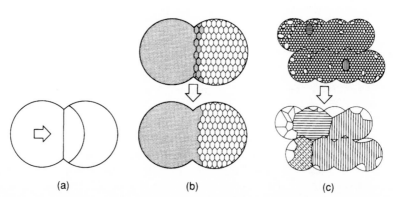

FIGURE 14.4 Growth of intra-agglomerate grains through polycrystalline necks. (a) Grain growth increases total grain-boundary areas. The dihedral angle constraint at the free surface creates a boundary curvature that opposes grain growth. (b) Growth of a large grain proceeds across a polycrystalline neck by incremental consumption of smaller grains. (c) Exaggerated intra-agglomerate crystallite growth results in the development of large grains in the early stage of sintering. The resulting grain structure may contain some large grains that grow abnormally during interagglomerate sintering. [From L. Edelson and A. Glaeser, "Role of Particle Substructure in the Sintering of Monosized Titania," *J. Am. Ceram. Soc.* 71 (4), 227 (1988), reprinted with permission from the American Ceramic Society, Columbus, OH.]

As the densification proceeds, pores become isolated and small pores may then be engulfed, as a result of grain growth, in polycrystalline regions where only those small pores remain. This mechanism of grain growth is most likely to occur when large differences in grain size are present, as is also implied in Fig. 14.4(b).

We should, in general, expect grain growth to occur whenever local densification has produced dense regions that can support such grain growth. The rate of grain growth will depend on the driving force (which is the amount of grain boundary eliminated for an incremental boundary movement), on other energy changes brought about by the movement of the boundary (pore surface area changes), and on the ease with which the boundary can move (its mobility—similar to the grain growth in dense polycrystals). If we look again at Fig. 14.4(b), it would be possible, for this specific geometry, to calculate how fine the grain size in the right-hand grain must be to compensate for the increase in the area of the sweeping boundary when it leaves its original position (Problem 14.1), given the grain boundary energy and the surface energy. The process described here is similar to recrystallization, which usually refers to the change of the grain size in pore-free fine-grained materials by the development and growth of isolated large grains.

Another possible way in which a boundary can sweep into neighboring grains is when these grains are small compared to the growing one and surface diffusion assists in rounding the cluster. This process for grain growth in porous compacts was the one considered originally by Greskovich and Lay (1972); it is shown schematically in Fig. 14.5. Whether or not the grain boundary can sweep into the neighboring grain depends again on whether or not the geometry permits it; for an incremental movement of the

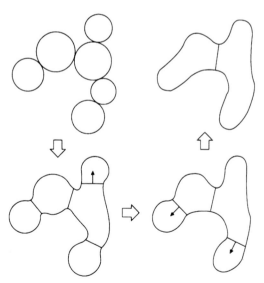

FIGURE 14.5 Diagram illustrating grain growth in cluster of particles by surface diffusion. Arrows on grain boundaries indicate direction of boundary movement. [From C. Greskovitch and K. Lay, "Grain Growth in Very Porous Al$_2$O$_3$ Compacts," J. Am. Ceram. *Soc.,* 55 (3), 145 (1972) reprinted with permission from the American Ceramic Society, Columbus, OH.]

boundary, the total energy must go down. A problem at the end of this chapter lets you work through a two-dimensional example relating to this type of grain growth.

Numerous experiments have shown that under isothermal conditions, the time dependence of grain growth in porous compacts can be expressed similar to the one for grain growth in dense materials:

$$G^n = G_o^n + K(t - t_o) \qquad (14.6)$$

where G is the grain size after time $t-t_o$, and G_o is the grain size at time t_o. K is the coarsening rate constant, which depends on temperature in much the same manner as a diffusion coefficient. Similar to equation (14.4), K can be expressed as

$$K = K_o \exp\left(\frac{-E_c}{kT}\right) \qquad (14.7)$$

where K_o is a constant and E_c is the coarsening activation energy.

The exponent n in equation (14.6) has been found to range between 2 and 5 for different materials, with 3 being most common. Note that the last case is an interesting relationship. If we take such a grain growth rate relationship, assume that $G_o << G(t)$, and use it in the equation describing densification by a grain boundary diffusion mechanism, equation (13.34), we would get the often observed *densification power law*, which puts the density increase [in equation (13.34) expressed as $\Delta\varepsilon$ corresponding to the strain that accompanies the change in density $\Delta\rho$] proportional to a power of time.

$$\Delta\rho \approx t^m \qquad (14.8)$$

where $\Delta\rho = \rho - \rho_o$ with ρ the instantaneous density and ρ_o the initial density of the powder compact. This is, at the very least, an interesting coincidence.

14.3 Pore Evolution During Densification

An example of a pore network in a partly sintered ceramic is shown in Fig. 14.6. This picture is a stereo pair of an epoxy replica of the pore network, obtained by pushing the epoxy into the open pores at an applied pressure of about 20000 psi (Chu, 1989). You can view it by using stereo viewers commonly found in materials characterization labs. After the epoxy was cured, the ceramic was dissolved, leaving a casting of the connected pore network. Pores that are already closed off can, of course, not be revealed by this method, but the stereo picture gives a clear idea of the complexity of the pore network in the intermediate density regime of sintering. The porosity is initially all connected, but as densification proceeds, more and more isolated pores—closed porosity—are produced. Qualitatively, the change from open to closed porosity may be sketched as in Fig. 14.7 (Coleman and Beere, 1975). The more homogeneous the microstructure of the compact, the later and more sudden the switch to closed porosity will be. In highly homogeneous compacts, a very large fraction of the total porosity may be maintained as an open pore network even when the compact has sintered to a density of as much as 90% of theoretical. Sintering to full density is facilitated by such

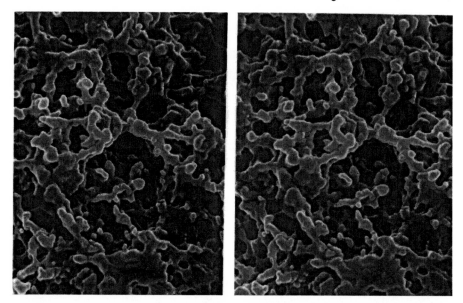

FIGURE 14.6 Stereographic pair of a replica of the pore space in a partially densified ceramic compact. The material was ZnO sintered to a density of 73% of theoretical. The pores in this material were filled with epoxy under a pressure of 20,000 psi and, after curing, the ZnO was dissolved. Stereo viewers are needed to obtain the three-dimensional effect.

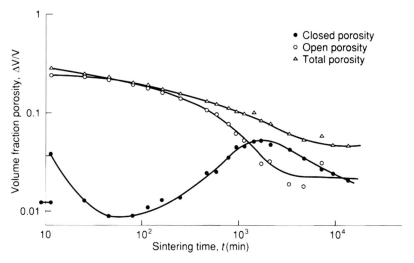

FIGURE 14.7 Sintering curve of a compact of UO_2 annealed at 1400°C. The initial drop in closed porosity is due to rapid densification inside high-density agglomerates. [Reprinted with permission from S. Coleman and W. Beere, *Philos.* Mag. 34, 1403 (1975).]

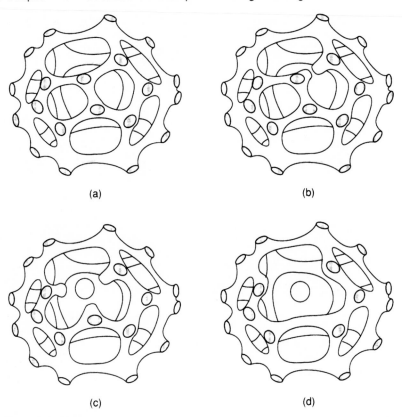

(a)   (b)

(c)   (d)

FIGURE 14.8 Sequence of pore channel collapse in a porous sintering solid. The sequential collapse can be envisioned as the coarsening process leaving a larger pore network as well as isolated pores. (Reprinted with permission from P. Rhines and R. Dehoff, in *Mater. Sci. Rev. 16, Sintering and Heterogeneous Catalysis*, Plenum Press, New York, 1983.)

homogeneity, but for some applications high density-open porosity ceramics might well provide a useful material.

The evolution of a pore network during densification, as shown in Fig. 14.6 is difficult to quantify. A first approach, and a common one, is to determine the average pore diameter as it appears in a polished cross section of the compact. Far less frequently have efforts been made to do a complete stereological characterization of the evolution of the pore network. This characterization requires serial sectioning and polishing as well as advanced image analysis and quantitative stereology methods. Work by Rhines and DeHoff (1983) has characterized the change in the pore network as a topological decay process in which a pore network, as shown in Fig. 14.8, changes by the collapse of pore channels and the reforming of a new network of lower connectivity, as sketched in Fig. 14.8(b). Contemplation of the stereo picture, Fig. 14.6, shows some microstructural features that probably resulted from such a process.

A major feature of the pore changes during the intermediate stage of densification can be seen in Fig. 13.2: The pores get fewer and larger. Such an evolution is universally observed in sintering powder compacts.

Without dealing with all the details of the geometrical changes that the real pore structure reveals, it remains useful to define a relationship that describes coarsening in a way that can be related to the sintering rate equations that were developed in Chapter 13. Qualitatively, coarsening of the pore structure can result from (1) densification itself and (2) from nondensifying processes, such as evaporation/condensation or surface diffusion. The coarsening relationship must be expected to have the same form as the grain growth equation, equation (14.6). Thus the average distance between the pores is taken to be equal to the grain size C; we could propose to write, under isothermal conditions,

$$G^n = G_o^n + \left[A\exp\left(\frac{-E_d}{kT}\right) + B\exp\left(\frac{-E_s}{kT}\right) \right](t - t_o) \tag{14.9}$$

where A and B are system-specific constants; E_d is the activation energy of the diffusional transport of atoms, which is also responsible for densification, and E_s is the activation energy characteristic of the nondensifying transport processes that contribute to coarsening. Equation (14.3) implies, for example, that coarsening can proceed even when there are only negligible density changes occurring in the samples (i.e., when A is near zero).

Since sintering is often carried out in nonisothermal conditions, we could rewrite equation (14.9) in a more general form from which the instantaneous grain size may be obtained for any time and temperature history:

$$G^n(T,t) = G_o^n + A\int_0^t \exp\left(\frac{-E_d}{kT}\right) dt + B\int_0^t \exp\left(\frac{-E_s}{kT}\right) dt \tag{14.10}$$

You will note that equation (14.9) is recovered from equation (14.10) under isothermal conditions (T = constant). The value of $G(T,t)$ can then be obtained by numerical integration once a temperature-time relationship has been established. One of the most useful, yet simplest relationships would be a constant heating rate, for which

$$T(t) = \beta t + T_o \tag{14.11}$$

and

$$dt = \frac{dT}{\beta} \tag{14.12}$$

Substitution of equation (14.12) into (14.10) then shows that for a constant heating rate, the mean grain size G will be inversely proportional to the heating rate when $G \gg G_o$. Note also that for a given compact, for which A, B, E_d and E_s are fixed, the right-hand side of equation (14.10) will be only a function of temperature. The corresponding densification rate follows from inserting equation (14.10) in the expression for the densification rate [equation (13.34) in Chapter 13], which would then read

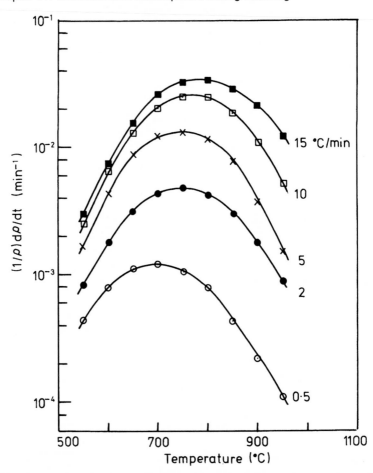

FIGURE 14.9a Densification strain rate versus temperature in a constant-heating rate experiment for a ZnO powder compact. The heating rates were between 0.5 and 15°C per minute.

$$\dot{\varepsilon}_\rho(T,\,t) = -K_o\left[\frac{\exp(-E_d\,/\,kT)}{G(T,t)^n}\right]\,\Sigma\phi^2 \qquad (14.13)$$

where $G(T,t)$ *is* given by equation (14.10), and where Σ is the sintering stress discussed in Chapter 13. An example of some densification data collected during constant-heating-rate sintering of ZnO is shown in Fig. 14.9, together with a numerical evaluation based on equations (14.10) and (14.13).

The actual density at time t and temperature T follows after some algebra (Chu, 1990) from integration of equation (14.13) for the known T-t relationship. In a constant-heating-rate experiment, when G has grown significantly larger than G_o, the densification rate then becomes proportional to the heating rate, β, as follows from equations (14.10), (14.12), and (14.13), and to a function of temperature while the

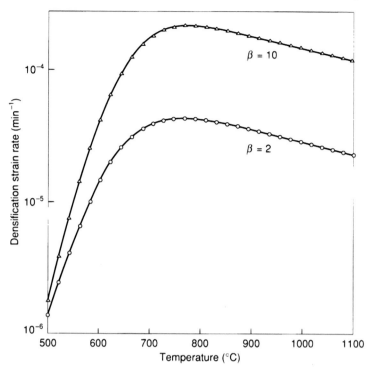

FIGURE 14.9b Calculated densification strain rate as a function of temperature for two different heating rates, versus temperature. b is heating rate in °C/min.

density becomes only a function of temperature and does not depend on the heating rate. This interesting result is borne out by experiment, as shown for ZnO in Fig. 14. 10.

The activation energy for the nondensifying processes, E_s, is usually significantly lower (a factor of 2 or more) than E_d, and the coarsening rate can therefore be manipulated relative to the densification rate by changing the time-temperature schedule. This permits an important means of controlling the microstructure of a sintered compact. If a coarse microstructure is desired, sintering would have to be carried out at lower temperatures. For obtaining a fine microstructure, one would need to heat up as rapidly as possible to a high temperature. The control of the microstructure by adjusting the sintering temperature such that the densification rate is changed relative to the heating rate is sometimes called *ratio-controlled sintering*. Combinations of low temperature coarsening up to some intermediate sintered density, followed by a high-temperature completion of the densification process, can also be used effectively in manipulating the microstructure of sintered ceramics (De Jonghe et al., 1988, Chu, 1990).

14.4 Pore Movement and Pore Breakaway

When the pores become isolated, as is the case in the later stages of densification, a

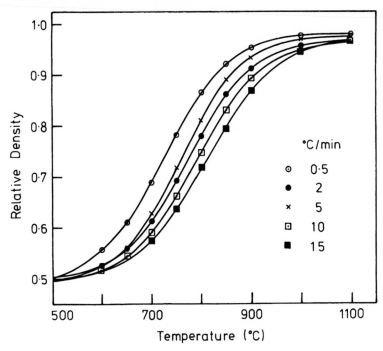

FIGURE 14.10 Density increments versus temperature at different heating rates for ZnO. The incremental sintered density is not strongly dependent on the heating rate.

different phenomenon comes into play. On the one hand, grain growth will continue to be hindered by the pores; on the other hand, the isolated and small pores are less effective in impeding grain boundary movement than are the large continuous pores. The interaction between the grain boundaries and the pores now determines whether a growing grain will sweep by a pore or not. If the pore is to remain attached, it will either have to stop the advancing grain boundary or it will have to move with it. We first consider the case where a grain boundary, subject to some force trying to move it, is held up by a shape-invariant and immobile spherical pore. This situation was first considered by Zener (cited by Smith, 1964) in a simple model.

The force acting on the grain boundary will be reflected in its curvature, as shown in Fig. 14.11(a). If the radius of curvature is R, then, according to the Kelvin equation, the stress exerted on the boundary should be

$$\sigma = 2\frac{\gamma_b}{R} \tag{14.14}$$

where γ_b is the grain boundary energy. To get the mechanical force balance, indicative of equilibrium, we consider a single spherical pore located in a grain boundary segment as shown in Fig. 14.11(b). The pore will leave a hole where it intersects the grain boundary, and it is the change in the position of the boundary that will change the size of this hole and thus produce a change in energy of the system.

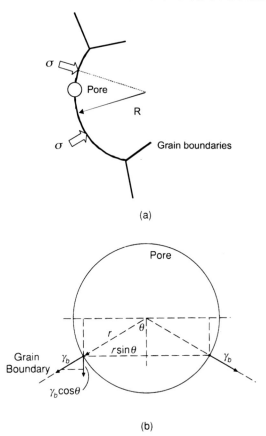

FIGURE 14.11 (a) Geometry of a pore on a moving grain boundary. The curvature of the boundary indicates the presence of an equivalent stress, $\sigma = 2\gamma_b/R$, where g_b is the grain boundary energy and R is the spherical radius of curvature of the boundary. For a cylindrically curved boundary $\sigma = \gamma_b/R$. (b) Detailed geometry of the pore/grain boundary intersection.

Contemplation of Fig. 14.11(b) will reveal that the force F, exerted by the pore of radius r, on the grain boundary is

$$F = \gamma_b \cos \theta (2\pi r \sin \theta) \qquad (14.15)$$

where r is the pore radius and θ is the angle of interception shown in Fig. 14.11(b). F will be maximum when $\cos \theta \cdot \sin \theta \sim 1/2$, so that

$$F_{max} = \pi r \gamma_b \qquad (14.16)$$

The number of pores, N, necessary to stop the advancing boundary, which is subjected to a driving force equal to γ_b/R per unit area of boundary, follows from the mechanical equilibrium when

$$NF_{max} = 2\frac{\gamma_b}{R} \tag{14.17}$$

where N is that number of pores in a unit area of grain boundary. Combining equations (14.16) and (14.17) gives

$$N = \frac{2}{\pi r R} \tag{14.18}$$

In this condition, the pores are taken to be all on the grain boundaries, so that the overall porosity in a unit volume can be found from the pore size, the grain size, and N. The total porosity p is

$$p = N_{total}\frac{4\pi r^3}{3} \tag{14.19}$$

provided that all the pores are the same size, with N_{total} the total number of pores on all the grain boundaries contained in a unit volume. When all the grains are taken to be of the same size, the total grain boundary area per unit volume is A/C, with A a geometrical factor (e.g., $\lambda = 3$ if all the grains are equal cubes). Thus, from equation (14.18), N_{total} will be

$$N_{total} = \frac{\lambda}{G}\frac{2}{\pi r R} \tag{14.20}$$

From (14.19) and (14.20) we can then obtain a critical porosity, p_{crit} below which pore breakaway is likely to occur. Equation (14.19) expresses the condition for which the number of pores of size r in the boundary is just sufficient to stop the boundary. If the porosity were less, the boundary would sweep by the pores. Thus we actually have $p_{crit} = p$ in equation (14.19), or

$$p_{crit} = \frac{8\lambda r^2}{3RG} \tag{14.21}$$

With all the simplifying assumptions (especially the details of the geometry) that led to this relation, we can only take this expression as approximate. Nevertheless. it is useful in giving some qualitative trends. The critical porosity should obviously be as low as possible if we do not want pores trapped inside grains. This would correspond to maintaining small pores compared to the grain size (i.e., more pores at some fixed porosities) while trying to maintain R large (i.e., the driving force for grain growth is kept low). The latter corresponds to having the grain size distribution as narrow as possible. Indeed, the radius of curvature, R, of the boundary is not directly predictable from the average grain size, G. Instead, R depends on the *local difference* in the grain size. When a large grain is surrounded by fine grains, R will be small and the driving force for local grain growth will be high.

Thus R would depend on the inverse of width of the grain size distribution as well as the average grain size. The grain size distribution in the final stages of sintering depends strongly on the narrowness of the grain size distribution in the initial powder

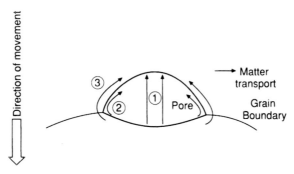

FIGURE 14.12 Possible matter transport paths for a pore moving with a grain boundary. 1, Evaporation/condensation; 2, surface transport; 3, bulk transport.

compact and on the steps that were taken to suppress grain growth, especially abnormal grain growth. All this is again related to the homogeneity and the particle size of the initial powder compact and underscores the need to make efforts toward achieving a high degree of perfection in the fabrication of powder compact.

Keeping the grain size small is clearly desirable if one wants to achieve high sintered densities, as equation (14.21) shows. One way to obstruct grain growth is to introduce impurities, called *sintering aids,* in the ceramic that are adsorbed on the grain boundaries and reduce their mobility. Indeed, if we could succeed in selecting an impurity that is strongly bound to the grain boundary but moves with difficulty, grain growth could be reduced significantly. If at the same time this impurity did not lower the sintering rate too much, a beneficial effect would result. Notable successes have been the introduction of MgO (Coble, 1961) and Y_2O_3 (Rhodes, 1981) for the sintering of alumina and the introduction of some alumina for the sintering of zinc oxide. Impurity levels of only a few hundred parts per million can often produce rather dramatic effects. Unfortunately, the choice of such impurities is largely empirical at present.

When pores are isolated and small, they may well be dragged along by a moving grain boundary. The reason is that a pore attached to a moving grain boundary will change shape, as shown in Fig. 14.12. The leading surface of the pore becomes less strongly curved than the trailing surface, creating a chemical potential difference that can drive matter transport. This transport can occur through the vapor phase, over the surface, or through the bulk. The relative mobility of the pores decreases rapidly with increasing size. For matter transport over the internal surface of the pore, the mobility is proportional to r^4, where r is the pore radius; for matter transport through the lattice or through the vapor phase, the mobility would be proportional to r^3 (Brook, 1976). Clearly, pores will have a better chance of being removed during sintering if the grain boundaries are slowed down by the use of impurities, second-phase particles, or by the proper choice of sintering temperature.

Pores moved along by grain boundaries may meet and flow together, forming a larger pore. This process is thought to contribute to pore coarsening in the late stages of densification. A schematic of this process (Kingery and Francois, 1965) is shown in Fig. 14.13.

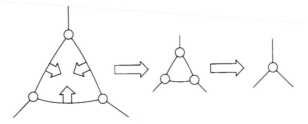

FIGURE 14.13 Pore coalescence (After Kingery and Francois.)

14.5 Concluding Remarks

In this chapter we have dealt with some aspects of microstructural development, particularly from the point of grain growth and pore evolution. Simple models assisted in providing qualitative predictive models, which often is all we really need in practice.

The simple models also provided some indications as to how to proceed when controlling the microstructural evolution during sintering. The importance of the homogeneity of the initial structure of the powder compact was again evident. Consideration of an improved expression for coarsening led to a more useful equation describing densification for nonisothermal conditions.

LIST OF SYMBOLS

A, B	constants
E_b	grain growth activation energy
E_d	activation energy for densifying processes
E_s	activation energy characteristic for nondensifying processes
F	force
G	grain size
l	length
M_b	grain boundary mobility
n	number of grain nearest neighbors
p	porosity
R	grain radius
T	temperature
t	time
V	grain volume
α	geometrical constant
β	heating rate
γ_b	grain boundary energy
λ	geometric parameter
σ_A	stress acting on grain boundary

PROBLEMS

14.1 Consider a grain surrounded by eight grains, as in Fig. 14.1, but with a dihedral angle of 120°. Calculate the driving force for grain growth of that grain,
as a stress per unit area, when all the grain boundary energies are equal.

14.2 The grain sizes of a dense ceramic were determined to be as follows:

Temperature (°C)	Time (min)	Grain size (μm)
950	120	15
1250	45	22

The original grain size at time = 0 was 8 μm. Predict what the grain size will
be after 30 minutes at 1450°C.

14.3 Consider a grain boundary between two spherical particles of equal radius.
Assume that one grain is polycrystalline and the other single crystalline. Determine the relative grain size of the polycrystalline particle that would allow
grain growth of the single-crystalline particle.

14.4 Develop an expression for the critical porosity assuming that pores (of radius
r) are randomly distributed and randomly intersected by grain boundaries,
when the grain boundary curvature is put equal to the grain radius.

14.5 Assume that grain growth in a dense, polycrystalline ceramic is related to the
width of the grain size distribution σ, by

$$G^2 = G_o^2 + M_b \sigma_s t$$

G = average grain size at time t
G_o = initial grain size at time 0
M_b = mobility of the grain boundaries
σ_s = width of the steady-state, self-similar grain size distribution

Assume further that $\sigma_s = \sigma_o [1 - A \exp(-Bt)]$, where A and B are materials-specific
constants. Show that a large, abnormal grain with a size G_L embedded in the
fine-grained matrix of grain size G will grow faster than G only when $\sigma < \sigma_s$,
but that G_L will grow slower than G for $\sigma \geq \sigma_s$. [Hint: Read this article first: G.
Thompson, H. Frost, and F. Spaepen, *Acta Metall. 35*, 887 (1987).]

REFERENCES AND FURTHER READINGS

ATKINSON. H. V., *Acta Metall.* 36, 363 (1988).

BROOK, R. J., "Controlled Grain Growth," in *Ceramic Fabrication Processes,* (Treatise on Materials Science and Technology, Vol. 9), F. Wang, ed., Academic Press,
New York, 1976.

CHU, M. -Y., Ph.D. thesis, University of California at Berkeley, 1989.

COBLE, R., *J. Appl. Phys.* 32(5),793 (1961).

COLEMAN, Sand W. BEERE, *Philos. Mag.* 31,1403 (1975).

DE JONGHE, L., M. RAHAMAN, and M. F. LIN, in *Ceramic Microstructures '86,* J.

Park and A. Evans, eds., Plenum Press, New York, 1988, p.447.

EDELSON, L., and A. GLAESER, *J. Am. Ceram. Soc. 71*, 225 (1988).

GRESKOVICH, C., and K. LAY, *J. Am. Ceram. Soc. 55,* 142 (1972).

KINGERY, W., and B. FRANCOIS, *J. Am. Ceram. Soc* . 48,546 (1965).

RHINES, F., and R. DEHOFF, *Mater. Sci. Res.* 16,49 (1983).

RHODES, W., *J. Am. Ceram. Soc* . 64(1),19 (1981).

SMITH, C. S., *Metall. Rev.* 9,1 (1964).

CHAPTER 15

Densification Technology

15.1 Introduction

During sintering of ceramic or metal powder compacts, the useful properties of the materials are developed and the part takes on its final shape. After sintering, only finishing processes, such as surface grinding or a follow-up heat treatment to remove residual stresses, may remain. Perhaps the most significant factor determining the rate of progress in densification technology has been the attainment of increasingly higher furnace temperatures, and the improved control of the furnace environment. Some structural ceramics, such as silicon carbide, require temperatures over 2000°C to get sufficient densification; other ceramics, such as zinc oxide varistors, require exceptionally close control of the sintering temperature to achieve optimum properties; in yet others, such as the lead zirconate titanate family of electrooptical and ferroelectric ceramics, the furnace atmosphere must be controlled to avoid evaporation of volatile constituents.

Two broad classes of densification technology methods can be distinguished: those that simply expose the samples to heat (i.e., pressureless or free sintering) and those that apply pressure during sintering (i.e., pressure-assisted sintering). In this chapter we describe some of the technology and methods that are used in the practice of free and pressure-assisted sintering.

Sintering processes generally involve several steps: binder removal to get rid of the organic materials that were added for the powder compaction, calcination to convert powders to the right compound, and densification.

15.1.1 Binder Removal

A first step is the removal of the binders and other forming aids, (e.g., die lubricants) that may have been used in the compaction of the ceramic or metal powder. The binder removal is sometimes also referred to as *thermal debinding*. The initial heat-up is typically to a few hundred degrees Celsius, and in this processing step the organic material contained in the compact will melt, possibly boil, evaporate, or thermally decompose to form a carbonaceous residue. Rapid heat-up must be avoided, especially if the organic content of the compact is high, since this can lead to considerable pressure buildup inside the compact, which can cause it to fracture catastrophically or even to explode.

Usually, heat-up rates between 1 and 10°C per minute are sufficient to avoid these problems. The possibility of forming carbonaceous residues increases if organic binders are removed under reducing conditions.

Residual carbon is usually undesirable since, at higher temperatures, it may react with the powder to form unwanted phases (e.g., oxycarbides in oxide ceramics). In some instances, however, retaining carbon is necessary. An example of this is the sintering of silicon carbide, in which the sintering aids are small amounts of boron and carbon. The carbon is usually introduced by dissolving some pitchlike material in toluene, drying, and heating in nonoxidizing conditions to convert the pitch to carbon. This procedure assures excellent distribution of the carbon and promotes uniform densification of the SiC to nearly theoretical densities (Greskovich and Prochazka, 1987).

The total time involved for complete binder removal may be from 1 or 2 hours to several days. In general, a high powder compact density and a high degree of filling of the pores with binder, large parts, as well as lower temperatures, will lengthen the time required for complete binder removal (German, 1987).

15.1.2 Calcination

A second step in the sintering process is calcination. This is heating of the powder compact, after binder removal, at a temperature well below the temperature at which rapid sintering starts, but sufficiently high to have some chemical reactions occur. These chemical reactions might, for example, be the decomposition of a ceramic precursor, such as the conversion of an alkoxide or a carbonate to its corresponding oxide, or the interreaction of various component oxides to form the desired ceramic phases. Clearly, the powder structure and morphology may be substantially altered during the calcination treatment when new phases develop. In general, one should expect that lower calcination/conversion temperatures decrease the particle size of the resulting reaction product. This is an effect that is quite commonly observed in phase transformations or solid-state reactions, and is thought to result from the tendency of a reaction to proceed at a maximum rate. The maximum rate would then follow from a trade-off between shorter diffusion distances (i.e., a finer transformation microstructure) and the decrease in the thermodynamic driving force as caused by the accompanying decrease in interface and surface areas. Although the effect is quite general and has been known for many years, a definitive explanation is still lacking.

The time required for the calcination/homogenization treatment will depend very strongly on the homogeneity of the starting powder mixture; the more intimate and homogeneous the mix, the shorter the calcination time required. This may be easily deduced when it is remembered (Chapter 6) that the mean diffusion distance after a time t increases with the square root of t. To achieve a higher degree of homogeneity, calcination treatments are sometimes repeated, with ball milling of the reaction products between each treatment. Densification of the compact during calcination should in this case be avoided by appropriate choice of the calcination temperature; otherwise, remilling becomes too difficult. Such treatments have been employed, for example, to prepare more homogeneous powder compacts for the fabrication of yttrium-barium-copper oxide ceramic superconductors from mixtures of oxides and carbonates (Dou et al., 1988).

15.2 Free Sintering or Pressureless Sintering

Sintering without the application of pressure is known as free sintering or pressureless sintering. A wide variety of furnaces are available commercially for the attainment of the sintering temperature in free sintering, accommodating small research samples to parts several feet in diameter (PMEA, 1977). The most common types are electrical resistance furnaces. Temperatures up to 1200°C are readily achieved with metal alloy furnace windings, tolerating both oxidizing and reducing conditions. The next temperature plateau is around 1450°C when SiC heating elements are used. $MoSi_2$ elements can deliver furnace temperatures up to about 1550°C in air for extended periods, and make brief excursions to perhaps 1650 or 1700°C. As a rule, furnace element lifetimes go down drastically with increasing furnace temperatures. Lanthanum chromite elements may also be used in air to deliver steady temperatures of around 1750 to 1800°C. Beyond these temperatures various refractory metal heating elements, such as tantalum, molybdenum, or tungsten, can be used in reducing or inert environments, up to about 2000°C. Care must be exercised with these metallic elements to avoid contacts with carbon-containing environments, since the refractory metals tend to form carbides, destroying the integrity of the elements. Above 1800°C furnace insulation is achieved effectively by the use of a series of closely spaced sheet metal radiation shields, since heat transfer is dominated by radiation at high temperatures. Graphite furnace elements, together with graphite fiber insulation, can also be used at temperatures up to 2750°C. Oxygen contamination of the furnace environment needs to be avoided in this case, as it can provide a mechanism for deposition of significant quantities of soot in the colder furnace parts as a result of reaction with the heating elements. Still higher temperatures may be reached with imaging furnaces, where mirrors focus the radiation of a heat source (such as an arc discharge or even the sun) onto a small sample.

A different principle of heating ceramics or metals is by microwave heating. When the microwave radiation is tuned to the absorption maximum of the ceramic (usually in the gigahertz range), effective and rapid heating may be achieved. In the past, the precise control of temperature tended to be somewhat more difficult in the microwave heating processes than for simple resistance furnaces. When using thermocouple temperature monitoring and control, one problem is to suppress interference of the microwave radiation with thermocouple functioning. Infrared optical thermometers have, however, been perfected that provide reliable measurement and control. Effective heating by microwave radiation can also be achieved in the megahertz range if the ceramic is first preheated to become somewhat conducting. Metals are sufficiently conducting that preheating is not necessary. The art in such microwave sintering is to find ways to supply the microwave energy such that uniform heating occurs.

For large furnaces, as employed in the fabrication of ceramic parts such as sinks and other whiteware, gas heating is commonly used with good economy. A significant factor in determining the economy of the sintering process resides in the thermal mass of the sintering furnace itself. Efficient design will minimize the thermal mass of these furnaces. In addition, flue gases may be recirculated to minimize heat losses to the environment.

The most important process variables in free sintering are the temperature-time history and atmosphere control. The process is one in which a part to be sintered is intro-

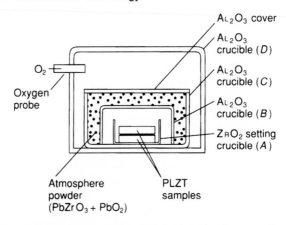

FIGURE 15.1 Atmosphere sintering approach for lead-lanthanum-zirconium titanate (PLZT). [From G. S. Snow, *J. Am. Ceram. Soc.* 56, 479, (1973). Permission of the American Ceramic Society, Columbus, OH.]

duced in a furnace, either manually or on a conveyer tray. If it is important to avoid contamination of the part as may result from reaction with supports, the part can be placed on coarse powder of the same composition. If volatilization of some components may occur during sintering, further protection may be necessary by completely enclosing the part in a crucible that contains coarse powder of the right composition. An example of a method for maintaining the proper atmosphere during the sintering of a ceramic for an electronic application, a lead zirconate titanate, is shown in Fig. 15.1. Lead oxide is volatile as well as poisonous, so it must be contained during the sintering. At the same time, the powder mixture surrounding the sample, a mix of lead zirconate and lead oxide, maintains the proper stoichiometry. Another example, shown in Fig. 15.2, shows schematically how sodium β-alumina tubes are protected during sintering. This ceramic exhibits very high ionic conductivity at temperatures even below 300°C, and is used as a solid electrolyte in batteries that work with molten sodium and sodium polysulfides as electrodes. At the temperatures needed to sinter these solid electrolytes, about 1500°C (Sudworth, 1986), sodium oxide would evaporate rapidly, thereby destroying the electrolyte properties. Encapsulation in alumina tubes is effective in laboratory practice, although the alumina capsule slowly reacts with the powder and can only be used a limited number of times before it falls apart. In industry, encapsulation in MgO or in platinum tubes has therefore been used (Sudworth, 1986). Although the initial cost is higher, especially for platinum, the longer capsule life in this application can actually lead to a cost saving in mass production compared to the less expensive alumina capsules.

A suitable temperature controller will take the furnace through the appropriate heat cycle. This cycle may have several stages: for example, an initial heat-up and low-temperature "soak" either for chemically homogenizing the part or to develop a desirable starting microstructure, a rapid temperature rise to the "sintering temperature", a slow cool to a post-treatment temperature, and finally cooling to room temperature. More complex heat treatments may, of course, be necessary. Also, heat-up

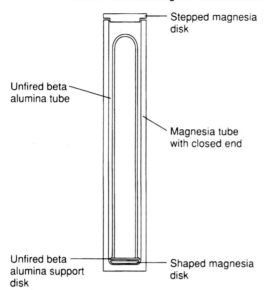

Stepped magnesia disk

Unfired beta alumina tube

Magnesia tube with closed end

Unfired beta alumina support disk

Shaped magnesia disk

FIGURE 15.2 Magnesia system for firing 33 mm by 360 mm [3-alumina tubes. (Reprinted with permission from J. H. Duncan, R. S. Gordon, R. W Powers, and R. J. Bones in *The Sodium Sulfur* Battery ed. by J. L. Sudworth and A. R. Tilley, Chapman & Hall, London, 1980, p. 107.)

rates may have to be controlled carefully to develop the required properties. When very fast heat-up rates are called for, the part is usually introduced into the furnace—which is kept at the sintering temperature—by a pushrod. In some cases it may be possible to design the furnace and its heat zone such that the part can move through at a constant speed and come out finished at the other end. This allows for a continuous production scheme. Such furnaces are known as stoker furnaces. It is important to limit rapid temperature changes for parts that are thick. since temperature gradients will then be present from the surface to the interior of the part. An exercise at the end of this chapter lets you calculate what the magnitude of such gradients can be. In the heat-up stage, too rapid heating may cause the formation of a dense skin on the part while the inside has not yet densified. Then, when the inside starts to densify and shrink, sufficient stress may develop to crack the part. A similar problem arises on cooling. where high tensile stresses may develop in the colder surface of the part when it is cooled too rapidly.

Atmosphere control can present special problems when the decomposition pressure of the material is high. As an example, Fig. 15.3 shows the nitrogen pressure necessary to avoid decomposition of either the silicon nitride matrix or reaction of the silicon carbide phase in a silicon nitride-silicon carbide composite (Nickel et al., 1988). In this case, the nitrogen pressure would have to be more than 100 atm to be able to sinter this composite above 2200°C. Keeping within the indicated stability range shown in this calculated stability diagram might even make it possible to prepare silicon carbide matrix composites containing silicon nitride whiskers, instead of the other way around.

Low oxygen pressures may have to be maintained over some materials during sin-

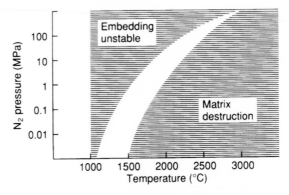

FIGURE 15.3 Result of a thermodynamical calculation for the manufacture of silicon nitride composite with included silicon carbide whiskers. The clear area represents the nitrogen (N_2) partial pressures as a function of temperature at which whiskers as well as the silicon nitride matrix would stay stable during the manufacture. Knowledge of stability conditions is an absolutely necessary precondition for the manufacture of high-performance ceramics. [*Powder Metallurgy International, Vol.* 20, No. 4, 26-28 (1988), Verlag Schmid GmbH, Freiburg, West Germany. Permission of the Max Planck Institute.]

tering, and this is often best accomplished by flowing gas mixtures such as hydrogen-water or carbon monoxide-carbon dioxide through the furnace. We met the Ellingham diagram in Chapter 2, and the diagram for a number of oxides was shown in Fig. 2.12. From this diagram the proper gas mixture can be found. If, for example, Fe_3O_4, magnetite, is held at 1000°C, this diagram will show that the hydrogen-water vapor composition should not be above a molar ratio of 1:10 if reduction to FeO is to be avoided (see Fig. 15.4). It may be necessary in the fabrication of some oxides, such as ferrites, to adjust the oxygen pressures at several points during sintering, to avoid unwanted reduction or oxidation reactions. For example, CO/CO_2 must not be exhausted into the workplace. Gas mixtures containing hydrogen must also be handled with care, although mixtures containing less than 4% hydrogen in inert gases are not explosive and can be handled safely, while small amounts may be vented without problems.

15.3 Pressure-Assisted Sintering

In many cases, pressureless sintering may not lead to sufficient densification. The difficulty may be that as a high sintered density is approached, grain boundaries start to sweep past the remaining pores, or that densification is so slow that too much coarsening occurs at the maximum permissible sintering temperature. This maximum temperature may, for example, be imposed by the thermal decomposition of the material. In other cases, as in particulate composites, densification may be obstructed by the presence of dispersed second phases or other intentional inclusions. When the sintering stress of the powder compact, usually a few megapascals or less in magnitude, is insufficient to bring about rapid densification, it is necessary to apply mechanical

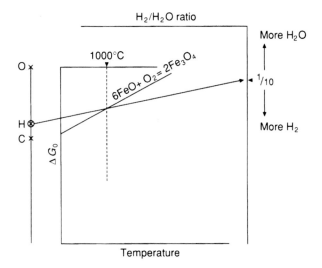

FIGURE 15.4 H_2/H_2O ratio from Ellingham diagram. To get PO_2 start at 0; to get CO/ CO_2, start at C.

pressure during sintering to achieve the required densities. In this pressure-assisted sintering (PAS), mechanical pressure has been applied at the sintering temperature in various ways:

1. Hot pressing: the powder is compacted in a die.
2. Hot forging: a powder compact is pressed between two rams.
3. Hot isostatic pressing (HIP): the powder compact is subject to high gas pressure inside a pressure vessel.
4. Hot extrusion: the powder is pressed through a die to form dense rods or tubes.
5. Hot swaying: the powder, encapsulated in a metal tube, is hammered down to a smaller-diameter, metal-cladded rod by a set of moving die segments.

Depending on the method, the applied pressures range from 10 to maybe 300 MPa (about 1500 to about 45,000 psi). In some experimental setups, much higher pressures have been used with special hot presses.

15.3.1 Hot Pressing

Hot pressing is the simplest form of PAS. A schematic is shown in Fig. 15.5. Typically, the powder is pressed into a die, much like cold die pressing, and heated while pressure is applied. The most common die material is graphite, which is both cheap and quite easy to machine. For common graphites the maximum applied pressure is about 35 MPa (5 kpsi), to avoid breaking of the die. Special high-density graphites with a fine grain size may be able to support applied stresses of perhaps three to four times this value (Waldron and Daniell, 1978). Graphite oxidizes rather slowly and so may be used in air for short periods below about 1200°C. Above this temperature a protective atmosphere of nitrogen or argon is necessary. Since graphite provides a strongly reducing atmosphere, it may be necessary to protect the powder from direct contact with the

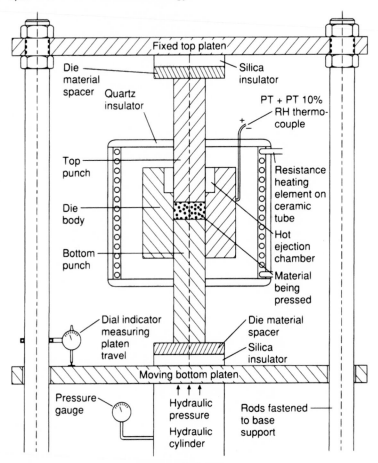

FIGURE 15.5 Schematic design of unit for pressure sintering (hot pressing) of ceramic materials. (Reprinted with permission from T. Vasilos and R. M. Springgs, "Pressure Sintering of Ceramics" in *Progress in Ceramic Science*, Vol. 4, J. Burke, ed., Pergamon Press, New York, 1966, p. 97.)

die. Thin foils of suitable metals such as Mo may be used for this purpose. One other problem may also be that the hot pressed piece sticks very strongly to the die wall and cannot be removed without damaging the die severely. In that case, graphite foil can be used to line the die and prevent die damage. Spraying the die wall with a material such as boron nitride can also be effective in preventing sticking and die damage.

The maximum tensile stress, σ_m, that a die can support is developed on the inside wall of the die and can be calculated for a hydrostatic stress condition (Timoshenko and Goodier, 1970):

$$\sigma_m = \frac{p_a\left(r_i^2 + r_e^2\right)}{r_e^2 - r_i^2} \tag{15.1}$$

This equation gives the maximum applied stress p_a on the rams, assuming that the resulting stress inside the powder compact will be hydrostatic. This condition is actually not satisfied in practice, and the die wall pressure will be less than the applied pressure some distance from the ram faces. r_i and r_e are the internal and the external dimensions of the die cylinder. As may be deduced from equation (15.1), the maximum stress decreases slowly with increasing die wall thicknesses exceeding the internal cavity diameter. A good rule of thumb is thus to make the die wall thickness equal to the diameter of the die bore and allow some safety factor in hot pressing, where the upper pressing limit is set by the tensile fracture strength of the die material. If the die is to fit inside a furnace, this rule of thumb indicates that the heating space in the furnace should have a diameter that is at least three times the diameter of the piece that is to be hot pressed. A few common die materials properties are listed in Table 15.1.

The same restrictions with regard to die wall friction and resulting stress gradients

TABLE 15.1 Nominal properties of die and ram materials at noted temperatures

Material	Condition	Thermal expansion (°C-1)	Modules of elasticity (mpsi)	Tensile fracture strength (kpsi)	Compressive strength (kpsi)	Creep rupture strength (kpsi)
Al$_2$O$_3$ (RT) $\rho > 99\%$	Al$_2$O$_3$ > 99%	9.0	52	30–50	>300	a
SiC (RT) $\rho > 99\%$		3.9	~65	~40	>200	a
Mo(TZM) (1000°C)	Stress relieved	4.9	34	80	a	28
Udimet 700 (1000°C)	Solution treated and aged	4.9	20.6	40	a	~10
Rene 41 (1000°C)		16.5	20	50 (925°C)	a	10
422M (700°C)		16.8		40	a	9
Graphite (conv.)	RT, $\rho = 1.7$	10.4	1–2	~4[b]	~10[b]	a
Graphite[c] (fine grained)	RT, $\rho = 1.86$	2–4		10[b]	21[b]	a
Graphite[d] (composite)	RT, $\rho = 1.34$	1.5–2.5	~3	15-20[b]	a	a

[a] Not normal failure mode.

[b] Normally increases with increasing temperature.

[c] POCO Graphite, Inc., P.O. Box 2121, Decatur, TX 76234.

[d] Carborundum Co., P.O. Box 577, Niagara Falls, NY 14302.

Source: Reprinted with permission from M. H. Leipold in *Treatise on Materials Science and Technology*, Vol. 9, *Ceramic Fabrication Processes,* Academic Press, New York, 1976.

apply for hot pressing as for cold die compaction, and property variations may arise when making thick pieces. Hot pressing is thus well suited for the fabrication of plates. It is possible to press several plates in one pressing, by using thin separators between them, like a stack of pancakes. It is obviously difficult to control the atmosphere inside a die, and therefore some materials may not be hot pressed successfully when significant atmosphere changes would be required during the heating/cooling cycle.

Although the applied pressure is typically larger than the sintering stress, complete densification may still not be achieved in the standard hot pressing (graphite die, 5000 psi applied stress) and small amounts of additives are mixed with the powder to speed up the densification, just as in free sintering. It is, of course, important that the additives not deteriorate the material's performance. One well known example is that of adding about 1 wt % of LiF to MgO. The LiF probably forms a liquid phase at the grain boundaries of the MgO, which enhances densification rates significantly. The LiF must be removed after densification, and this is achieved by heating the entire sample for some time in vacuum at about 1200°C, causing the LiF to evaporate away (Rice, 1967). At this low temperature, grain growth in the dense MgO is slow, and the grain structure obtained after hot pressing remains virtually unmodified. In other cases, additives would lead to enhanced densification, but the properties of the ceramic are not maintained. A typical example is the use of a silicate glass in the densification of alumina. The silicate glass at the grain boundaries enhances densification but at the same time reduces the resistance of the ceramic to slow, high-temperature deformation (creep).

One possible problem in pressure-assisted sintering is that gases can get trapped in closing pores. If these gases are insoluble in the ceramic or metal and cannot diffuse away, substantial pressure may build up in the pores during further densification. The ceramic may start to swell or even break when used at elevated temperatures. A good antidote for this problem is to do the hot pressing in an evacuated chamber. Most commercial hot pressing machines have this capability.

At the relatively low applied stresses used in hot pressing (say, about 5 to 20 times the sintering stress), plastic deformation[1] is usually not a dominant creep mechanism for most ceramics, and hot pressing can then be regarded in much the same way as free sintering as discussed in earlier chapters, but with the applied stress simply added to the free-sintering stress.

The directionality of the applied stress during hot pressing may affect the microstructural development and cause preferential orientations of the grains and enhanced grain growth in specific directions. This is called *texture formation*. It may be detected by microscopy or x-ray diffraction, and usually makes the materials properties anisotropic. For materials that tend to develop plate-shaped grains, most often the grains will align with the plates perpendicular to the direction of the applied stress.

The shape of the parts that can be readily hot pressed is clearly very limited. The most suitable shapes are cylinders, although more complex parts have also been made by surrounding the part inside the cylindrical die cavity by coarse powder (e.g.. boron nitride) that does not densify well and does not react with the workpiece (Lange and Terwilliger, 1973).

---

[1] In plastic deformation the applied stress causes the material to flow by the movement of dislocations rather than by diffusional transport of atoms.

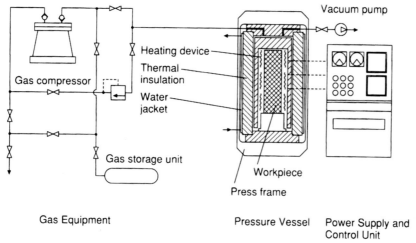

FIGURE 15.6 Schematic of HIP system. (Reprinted by permission of the publisher from T. Fujikawa in *Fine Ceramics*, S. Saito, ed., p. 50. Copyright 1985 by Elsevier Science Publishing Co., Inc., Amsterdam.)

15.3.2 Hot Forging

In hot forging a powder compact is placed between two rams and pressure is applied, shaping the compact into a disk. The applied pressures are not limited by a die wall and can be significantly higher than in hot pressing, so that plastic deformation in the compact can occur. The nature of the deformation is such that the forged part experiences significant shear, and as a result considerable texture formation and also recrystallization may occur. This is especially the case for metallic materials, which tend to deform plastically more readily than ceramics. Since plastic deformation occurs more readily under the higher applied forging stress, relatively complex shapes, such as curved ceramic domes, may be pressed. Hot forging is also used extensively in the fabrication of mechanical parts, such as gears, from metal alloys that are difficult to process otherwise. In fact, hot forging can be considerably more economical than machining or even casting, and may lead to parts with greater homogeneity and properties superior to those obtained with traditional forming methods.

15.3.3 Hot Isostatic Pressing

In hot isostatic pressing (HIPing) a powder compact is placed inside a furnace that is contained in a cold-wall pressure vessel. A cross-sectional sketch of a HIP furnace is shown in Fig. 15.6. The temperature and the gas pressure can be raised to as much as 2700°C and 140 to 200 MPa (20,000 to 30,000 psi). The powder sample is either encapsulated in an evacuated and sealed refractory metal can (such as molybdenum or tantalum), or presintered to closed porosity. Adsorbed gases on the powder surfaces can be a problem in the encapsulation method, since they cannot escape from the sealed van. Vacuum degassing of the powder in the capsule prior to sealing it is effective in removing the adsorbed gases.

Since the temperatures and pressures are high, plastic deformation plays a more important role in HIPing than in hot pressing. Yielding of the powder particles at their mutual contact points will be more prominent in the early stages of densification, when the density of the powder compact is low so that the contact stresses between the particles, produced by the applied stress, are high. It has been calculated (Helle et al., 1985) that plastic yielding of a porous material would continue until its relative density ρ_{pl}, reaches a value given by

$$\rho_{pl} = 1 - \exp\left(\frac{-3p_a}{2\sigma_y}\right) \tag{15.2}$$

where p_a is the applied pressure and σ_y the applied stress at which the fully dense material would yield plastically. Putting in some numbers readily shows that, for example, if p_a, is twice the plastic yield stress σ_y, a density of 95% of theoretical would be reached by plastic yielding alone. Since this yielding is not time-dependent process, in contrast to diffusional processes, such density would be reached soon after the pressure is applied. Yield strengths generally go down with increasing temperature and can become very low near the melting point of the materials. When the pressure is applied from the beginning of the heating cycle and densification by diffusional processes is ignored, the density reached by plastic HIPing would be determined by the temperature dependence of the yield stress of the material.

If transient temperature gradients are present, as is likely during the heat-up of thick parts, densification may proceed from the outside to the inside, as was the case for free sintering. In HIPing this can give rise to undesirable shape changes of the part after densification (Li et al., 1987). This happens especially when the densification rate or the yield stress depends strongly on temperature. In the latter case, many of the undesirable shape changes may be avoided by letting the sample temperature equilibrate before applying the full densification pressure. In some simpler HIP equipment this procedure is, however, not practical, as the HIP is pressurized cold to a few thousand psi and then the gas inlet valve is closed. Achieving the necessary high pressure at the sintering temperature relies on the heating up of the HIP furnace, so that pressure and temperature cannot be controlled independently. However, the problem of forming a dense outside layer due to temperature gradients in the sample can be reduced if the sample is pressureless presintered to as high a density as possible.

HIPing is inherently a batch process. and its economy in mass production depends significantly on the size of the pressure vessel cavity. HIPing is used extensively in the fabrication of cemented carbides[2] which are difficult to consolidate without forming residual pores. Residual pores are particularly deleterious in surfaces required to have high wear resistance.

Containerless HIPing is particularly useful since it eliminates the necessity of the costly encapsulation procedure. In this case the ceramic or metal part is first sintered to closed porosity and then the final densification is carried out in the HIP.

---

[2] Cemented carbides consist of a mixture of carbide particles, such as WC, held together with 5 to 25 vol % of a metal such as Co. and are widely used as cutting tools for lathes.

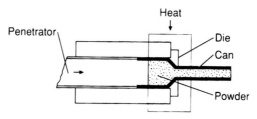

FIGURE 15.7 Powder extrusion is accomplished using canned powder and a penetrator ram. (Reprinted with permission from *Powder Metallurgy Science, by* R. M. German, Powder Metallurgy Science, Metal Powder Industries Federation, Princeton, NJ, 1984.)

15.3.4 Hot Extrusion

Hot extrusion forces a metal-clad powder compact through a heated die, as shown schematically in Fig. 15.7. Densification results from the reduction in the diameter of the workpiece as it is forced through the die. Reduction ratios (initial cross section divided by extruded cross section) of 10 or more are usually necessary to reach full density in extruded ceramics. Empirically. it is found that the force necessary to extrude the clad powder compact is proportional to the logarithm of the reduction ratio, as well as to the cross-sectional area of the feed rod (German, 1984). Significant texture formation can also develop in extruded wires or rods as a result of the anisotropic deformation during extrusion.

15.3.5 Hot Swaging

Hot swaying is related to hot extrusion, and also starts out with a metal-clad preform of the powder. This preform is heated up and then passed through a swaying machine. The swaying operation consists in hammering the preform with a split die. This noisy operation is usually carried out in several passes, with sets of progressively smaller die diameters. Intermediate heat-ups may be necessary to maintain sufficient plasticity in the compacted powder, as well as in the cladding, so that fracture does not occur. Long, metal-clad ceramic wires may be produced in this way.

15.4 Concluding Remarks

In this chapter we have briefly reviewed some of the most common methods used to achieve the desired properties of materials prepared from powder compacts. Thermomechanical processing is especially effective and allows greater flexibility in balancing density against microstructural coarsening. Where it may be very difficult or impossible to prepare translucent or transparent ceramics—obtained only at near-zero porosity—by pressureless sintering, PAS often will do the job. The only drawbacks of PAS versus pressureless sintering are that it is a more expensive manufacturing method and that the shape of the sintered part is limited in its complexity.

LIST OF SYMBOLS

p_a	pressure, applied hydrostatic pressure
r	radius
ρ_{pl}	relative density reached by plastic deformation
σ_m	maximum tensile stress supported by die wall
σ_y	yield stress of dense material

PROBLEMS

15.1 A mixture of powders A and B is to be converted to the compound AB by calcination. The interreaction may be described in a simplified manner as one in which a layer of AB forms on spherical B particles from a uniform supply of A. The initial particle size of B is 2 μm. Laboratory experiments on a flat-plate geometry showed that the layer of AB grew on a plate of B at a rate of 0.2 μm per hour at 700°C. after 30 minutes of reaction. following a parabolic rate relationship. At the same time, particle growth may occur during calcination. At 850°C the initial particle size had grown to 3.8 μm in 1 hour. It was observed that particle growth could also be described by parabolic kinetics for which the rate factor, k, had a temperature dependence characterizable by an activation energy of 2×10^5 J/mol. It is deemed necessary to have the particle growth less than 20% of the original size of B, so that a useful product is obtained while the simple reaction model remains applicable.

What would be a proper choice for a calcination temperature and time if the reaction layer formation rate could be characterized as a temperature activated process with an activation energy of 1.3×10^5 J/mol?

[Note: First convert the flat-plate reaction rate to a spherical geometry; consult Chapter 5 or J. Crank, *Mathematics of Diffusion,* Oxford University Press, London, 1956. Then figure out the temperature that keeps particle growth within the set limit. To get the right conditions you will have to find the acceptable value of $kt^{1/2}$ and find a self-consistent solution with the rate of the reaction layer formation. Also read W. D. Kingery, H. K. Bowen, and D. Uhlmann, *Introduction to Ceramics,* Wiley, New York, 1975, pp. 402-430.]

15.2 A company wants to produce sintered FeO from powder. The procedure calls for the powder to remain as FeO at all times above 400°C, never changing to Fe or to Fe_3O_4. Give the limits of the hydrogen/water ratio as a function of temperature for this process.

15.3 Somebody made a die for hot pressing from alumina, with a die wall thickness of 1 in. and a bore of 1.5 in., to be used up to 15,000 psi. He based his design on the data supplied by the manufacturer of the alumina. The manufacturer of the block of alumina from which this die was laboriously ground assured you that the fracture strength of his material is 45 kpsi. The die was used at an applied load of 22,000 lb and it broke. Who is at fault here: the alumina manufacturer, the person who designed the die, or the person who put so much pressure on it?

15.4 A powder is encapsulated in a metal can and hot isostatically pressed at an elevated temperature at an applied pressure of 30,000 psi. Very early on in the pressing, you already get a density of 92% of theoretical, while subsequent densification goes rather slowly. What can you tell about the plastic behavior of this powder?

15.5 The figures below give the temperature profiles in a slab of material (large compared to its thickness) at a uniform initial temperature T_i that is plunged into a fluid at temperature T_f (assumed constant). The parameters appearing in these figures are the Biot number, $B_i \equiv hL/k$, and the Fourier number, $F_o \equiv \alpha t / L^2$, where h is the heat transfer coefficient between the slab and the fluid (assumed constant), L the half thickness of the slab, k the thermal conductivity of the slab, t is time, and α is the thermal diffusivity (thermal conductivity divided by the product of density and heat capacity). Estimate the maximum temperature gradient in a slab of ceramic 50mm thick at initial temperature 25°C 50 seconds after it is plunged into an environment at 800°C if the heat transfer coefficient is 1060 J/(m²•s•K), the thermal conductivity of the ceramic is 26.5 J/(m• s•K), its density is 3.2 x 10³ kg/m³, and its heat capacity is 1656 J/(kg•K).

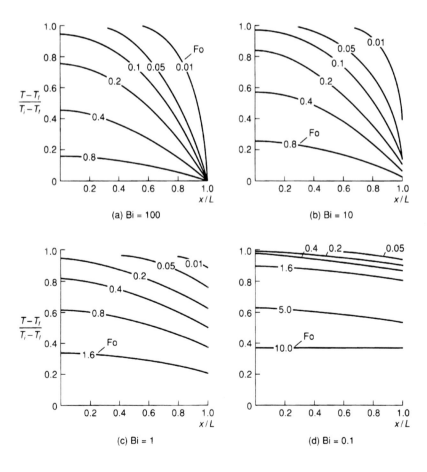

(a) Bi = 100

(b) Bi = 10

(c) Bi = 1

(d) Bi = 0.1

REFERENCES AND FURTHER READINGS

DOU, S., H. LTU, A. BOURDILLON, J. ZHOU, N. TAN, X. SUN, and C. SORRELL, *J. Am. Ceram. Soc.* 71, C239 (1988).

GERMAN, R. M., *Int. J. Powder Metall. 23*, 237 (1987).

GERMAN, R. M., *Powder Metallurgy Science,* Metals Powder Industries Federation, Princeton, NJ, 1984.

GRESKOVICH, C., and S. PROCHAZKA, *Mater. Sci. Res. 21,* 601 (1987).

HELLE, A. S., K. EASTERLING, and M. F. ASHBY, *Acta Metall. 33,* 2163 (1985).

LANGE, F., and G. TERWILLIGER, *J. Am. Ceram. Soc.* 52, 563 (1973).

LI, W.-B., M. F. ASHBY, and K. EASTERLING, *Acta Metall.* 35, 2831 (1987).

NICKEL, K., M. HOFFMANN, P. GREIL, and G. PETZOW, *Adv. Ceram. Mater .,* 557 (1988).

PMEA (Powder Metallurgy Equipment Association), *Powder Metallurgy Equipment Manual,* Metal Powder Industries Federation, Princeton, NJ, 1977.

RICE, R., in *High Temperature Materials and Technology,* I. Campbell and E. Sherwood, eds., Wiley, New York, 1967, p. 99.

SUDWORTH, J.L., and A.R. TILL EY "The Sodium Sulfur Battery" Chapman and Hall London, 1985.

TIMOSHENKO, S., and J. GOODIER, *Theory of Elasticity,* McGraw-Hill, New York, 1970.

WALDRON, M., and B. DANIELL, *Sintering,* Heyden & Son, London, 1978.

CHAPTER 16

Advanced Materials and Processes

16.1 Introduction

In the first seven chapters of this book, we attempted a succinct statement of the fundamentals of science and engineering thought necessary for an understanding of processes for producing materials. The remaining chapters up to the present one, have had as their focus the application of these fundamentals to the technology of what are frequently called "traditional" structural materials, such as steel, copper, alumina, and silicon carbide. These traditional materials are, in terms of economic measures such as tons produced or sales per year, the most important ones and likely to remain so well into the next century.

However, these economic measures do an injustice to an emerging class of materials that we, along with many others, have called *advanced materials.* Examples here are semiconductor-grade silicon and composite materials. The former is produced in relatively small quantities and its production employs few people. Nevertheless, without it our lives would be as different as they would be without aluminum. Furthermore, the pace of development of knowledge of these materials and the support available for research at least equals that of the traditional materials.

It is therefore appropriate that this book close with an examination of the technology used in producing advanced materials, including surface coatings. It is emphasized that in most instances these technologies do not differ in *their fundamentals* from technologies discussed in earlier chapters. In rapid solidification, for example, we are concerned with heat transport, fluid flow, and segregation, just as in conventional solidification. As a second example, in chemical vapor infiltration, treated in this chapter, the important phenomena (mass and heat transport, heterogeneous reaction kinetics, and morphological change) are just those encountered in the reduction of porous iron ore pellets. The contents of earlier chapters will therefore be of use in reading the present one, and the reader is urged to avoid isolating advanced materials from traditional ones.

16.2 Rapid Solidification

A treatment of the solidification of metals was provided in Chapter 10. In that treatment it was assumed that solidification was sufficiently slow that the solid phase which is formed is the thermodynamically stable one. In the solidification technology presented so far in this book, that assumption is valid (although the solid phase formed might not be the one that is stable down to room temperature, e.g. the case of carbon steels). This is because cooling rates obtained in conventional technology are rather slow, perhaps as low as 1 K/s, and this provides plenty of time for equilibrium to be achieved in the vicinity of the solidification front. The reader will recall that one unfortunate aspect of conventional solidification technology is that macrosegregation occurs; that is, there are compositional (and therefore property) variations over length scales that might be of the order of a meter or so.

Consider now what happens when a metal is rapidly solidified (to form a single phase alloy), for example by breaking it up into droplets of, say, 10 μm diameter, which are then solidified by cooling at rates that might be higher than 10^5 K/s. One advantage is immediately apparent; macrosegregation is impossible because the maximum length scale is 10 μm. Furthermore, compositional variations across the solidified particle may be largely eliminated, as can be seen from the simple calculation that follows.

Suppose that the droplet may be solidified by lowering its temperature by 100 K. At a cooling rate of 10^5 K/s the droplet would solidify in 1 ms. Compare this brief time with the time required to homogenize a liquid droplet by liquid-phase diffusion. The latter time is the square of the droplet diameter over the liquid phase diffusivity. Taking 10^{-5} cm²/s as a representative liquid-phase diffusivity, we see that the homogenization time is 100 ms (i.e., far longer than the solidification time). Of course, this calculation is a simplistic one, in that it ignores the possibility of convection in the liquid and the fact that the volume of liquid diminishes during the solidification. Nevertheless, it serves to suggest that during the solidification, solute rejected into the liquid at the solidification front will remain in the vicinity of the front and we shall get a solute distribution in the solid similar to that illustrated in Fig. 10.7. Therefore, over the length scale of the particle the composition will be homogeneous. The only remaining segregation possible is therefore that on a microscale (much smaller than the particle) which can readily be removed by annealing, as discussed in Section 10.2.2. This type of rapid solidification can therefore result in particles that are very uniform in composition and that can subsequently be compacted and sintered, just as ceramic powders are (see Chapters 12 to 15).

Furthermore, the rapid solidification has an influence on the size of the grains making up the particle. To understand this, we must examine the competition between nucleation and growth of a solid phase formed from a melt. If nucleation is much faster than growth, an enormous number of nuclei will form in the time taken for growing nuclei to impinge on each other and complete the solidification. The consequence is a fine-grained microstructure where the grain size is small. Nucleation rates are very dependent on the degree of supercooling, that is, to the extent to which the liquid has been cooled below its normal melting point. Provided that we are not too far below the normal melting point, the greater this extent, the greater the nucleation rate. On the other hand, growth, which is usually dependent on diffusion and heat transport, is only weakly dependent on the degree of supercooling. In rapid solidification we bring the liquid to a high degree of supercooling almost instantaneously, resulting in enormous nucleation rates and a fine-grained structure.

There is an interesting exception to the generalization above concerning the dependence of nucleation rate on supercooling. At very high degrees of supercooling the rate of nucleation again becomes small, and in some cases the supercooled fluid changes to a metastable solid state: the amorphous solid. An everyday example is window glass. Amorphous solids lack the order found in crystalline solids. Glass is amorphous at length scales greater than the atomic. This disordered state is metastable; it would, however, take extremely long times to get it to crystallize. It has been found that amorphous metals with unusual properties can be achieved by very rapid solidification.

Another interesting result achieved by rapid solidification is the production of metastable phases not achieved by normal solidification technology. Consider the solidification of an alloy that would normally yield a solid of two or more phases. Some such alloys on rapid solidification yield a *single* metastable phase because the nucleation and growth of the second phase has not been possible in the fleeting time of solidification. It should be stressed that materials produced from these rapidly solidified metal powders are not just laboratory curiosities but exhibit properties of great engineering significance. Typically, their mechanical properties are different from the normally solidified metal (e.g., they exhibit greater ductility), and frequently they exhibit improved electrical or magnetic properties or even corrosion resistance. For example, rapidly solidified alloys are now used as magnetic cores in high-frequency transformers, because of their very low coercive force (e.g., they reverse magnetization very easily).

16.2.1 Technology

One method of producing these fine droplets and rapidly solidifying them is *centrifugal atomization,* illustrated in Fig. 16.1. A jet of molten metal impinges on the center of a rapidly spinning disk. The jet breaks up into droplets that are flung off the disk at high speed into a cold stream of an inert gas such as argon.

Alternatively, more conventional atomization methods can produce fine droplets and chill them. One method is gas atomization, in which inert gas and molten metal are passed through a concentric nozzle similar in principle to that used on a paint spraying

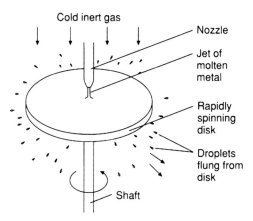

FIGURE 16.1 Production of the droplets of molten metal (which then solidify to powder) by centrifugal atomization.

machine. To achieve the necessary small droplet size and high cooling rates, high gas pressures and flow rates are necessary. Water has also been used for atomization of metals in this type of device.

Another type of atomization apparatus is one where a jet of metal enters a rapidly flowing stream of water. One convenient way of doing this is by use of a drum rotating rapidly about its axis so that centrifugal force holds a layer of water against the inside of the drum. A jet of molten metal is then directed into this layer by a stationary nozzle positioned inside the drum. The jet breaks up into small droplets as it enters the water, and these droplets are rapidly cooled.

Of course, it would be desirable to solidify metal rapidly into wire or strip. In fact, such wire or strip is now commercially available, for example, in the form of wire for metallic superconductors. In the case of strip the technology is seen to have an additional advantage as follows. In Chapter 10 we saw how continuous casting is an improvement over ingot casting because the former avoids many of the initial size reduction steps (rolling mills) on the way to finished product in the form of steel a few millimeters in thickness. Rapid solidification of strip, sometimes known as *direct casting* or *near net shape casting*, is the obvious extension of this concept. The objective is to eliminate *all* the rolling mills (or nearly all) by casting metal a few millimeters thick. One method under development is the vertical twin roll process shown in Fig. 16.2. In this technology the molten steel is solidified between two rapidly rotating water-cooled rollers spaced a few millimeters apart. Steel strip a few hundred millimeters in width cast by this method is commercially available at the time of writing.

16.3 Production of Silicon

Silicon is perhaps the most important material of the semiconductor industry. Semiconductor-grade silicon wafer sales worldwide totaled $13.1 billion in 2000. This is a relatively small amount compared to some of the other materials in this book, but the amount belies the significance of this material. Without high-purity silicon the semiconductor industry would not have advanced to its present state and we should probably not be enjoying the benefits of personal computers, microprocessor-controlled automobile ignition systems, and so on.

Of course, elemental silicon of much lower purity has been produced for the metallurgical industries in large quantities for many years. This *metallurgical grade silicon* is obtained by reduction of silica with carbon in a submerged arc furnace, as described in Chapter 8, and is the starting point for the production of semiconductor-grade material that we now describe.

Metallurgical-grade silicon typically contains impurities at the level of 10^2 to 10^3 parts per million by weight (ppm). Most notable among these impurities are aluminum, iron. and other metallic impurities that are present in the ore that is the source of this material. Specifications on semiconductor-grade silicon typically require the removal of these impurities to below 10^{-3} ppm, that is, a reduction in the levels of impurities by five or six orders of magnitude. That this can be done is the very considerable achievement of the Siemens process, whereby volatile chlorides of silicon are formed and then purified by distillation. The dominant volatile chloride is trichlorosilane ($HSiCl_3$), formed

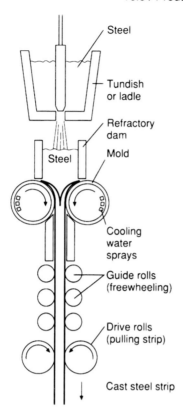

FIGURE 16.2 Vertical twin roll caster for casting steel strip.

by reacting silicon with HCl gas:

$$Si + 3HCl = HSiCl_3 + H_2 \qquad (16.1)$$

The reaction is then reversed to yield high-purity silicon.

Figure 16.3 illustrates the equilibria for the silicon chlorine-hydrogen system at atmospheric pressure and with a large excess of hydrogen present. At equilibrium the amount of silicon in the gas phase (as various chlorides) approaches the amount of chlorine if the temperature is sufficiently low, while at high temperatures there is little silicon in the gas phase. Reaction (16.1) can therefore be encouraged to go to the right by lowering the temperature (and by removing the H_2). The major silicon-containing species in the gas phase at these lower temperatures is then trichlorosilane. After purification of the trichlorosilane, reaction (16.1) can then be reversed by raising the temperature.

The technology is illustrated in Fig. 16.4. A batch of metallurgical-grade silicon is reacted with dry hydrogen chloride in a fluidized bed reactor at approximately 300°C. The gases leaving the reactor consist of the chlorides of silicon, hydrogen, and unreacted HCl, as well as the chlorides of impurities. After filtering to remove silicon particles, these gases pass into a condenser at approximately -30°C, where $HSiCl_3$ and other chlorides of silicon condense to form a liquid. Hydrogen and HCl pass through the

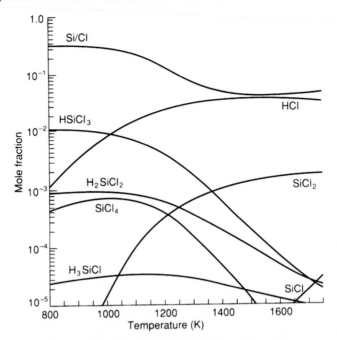

FIGURE 16.3 Equilibrium composition for a gas phase at 1 atm in contact with silicon. The chlorine/hydrogen ratio is 0.02. [Reprinted with permission from R. F. Lever, The Equilibrium Behavior of the Silicon-Hydrogen-Chlorine System, *IBM J. Res. Dev.* 8(4), 460-465 (1964). Copyright 1964 by International Business Machines Corporation.]

condenser into an absorber, where HCl is trapped by sodium hydroxide solution, the hydrogen passing on to be used later in the process. Impurity chlorides are insoluble solids at -30°C and form layers on the inside of the condenser. These layers are removed periodically when the condenser is shut down.

The chlorides of silicon are then separated by distillation, yielding a byproduct silicon tetrachloride stream and an overhead stream that is predominantly trichlorosilane. The latter stream is subjected to final purification by contacting it with sorbents such as activated carbon and alumina that trap residual impurities (notably the chlorides of boron and phosphorus). It is then mixed with hydrogen and passed over a silicon rod that is held at approximately 1000°C by passing electric current through it. The silicon rods, which have an initial diameter of about 5 mm, grow as silicon is deposited on them by the reverse of reaction (16.1). Gases exiting the reactor consist of HCl and unreacted hydrogen and silicon chlorides. A condenser serves to liquefy the last named, and these chlorides can then be recycled to purification. The gases pass on to the scrubber.

The polycrystalline silicon rods produced in this way are then melted and grown into single-crystal *ingots,* usually in the Czochralski apparatus shown in Fig. 16.5.

A single "seed" crystal oriented in the desired direction is mounted on a rotating shaft and contacted with the melt. The rotating crystal is then slowly withdrawn while

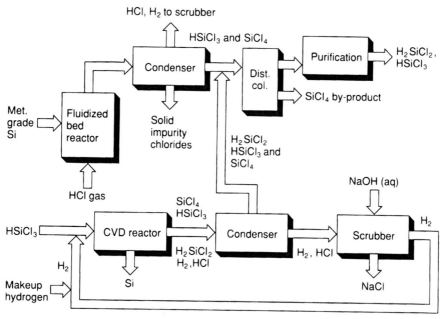

FIGURE 16.4 Siemens process for the production of semiconductor-grade silicon.

precise control of temperatures within the apparatus is maintained so that freezing occurs on the seed crystal and a long cylindrical single crystal is produced. The crystal is then sawn into wafers, which are the raw materials for the semiconductor industry.

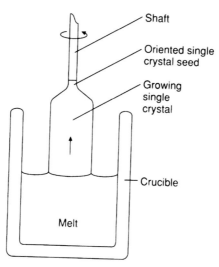

FIGURE 16.5 Czochralski process for growth of single crystal.

PROBLEM

What is the effect of total pressure on the composition of the gas mixture of Fig. 16.3 at 800 K?

Solution

An exact solution to this problem would entail the equilibria of all the species appearing in Fig. 16.3. resulting in a large number of equations that would have to be solved by computer. We shall content ourselves with a simpler solution where only the species appearing in equation (16.1) are considered. We can invoke equation (2.85) from Chapter 2. which tells us the effect of pressure on the equilibrium constant. K,

$$-\frac{1}{RT}\frac{\partial \Delta \tilde{G}_R}{\partial P} = \frac{\partial \ln K}{\partial P} = -\frac{\Delta \tilde{V}_R}{RT}$$

where $\Delta \tilde{V}_R$ is the volume increase accompanying reaction.

If we ignore the molar volume of silicon in comparison to the volumes of the gases involved. we obtain

$$\Delta \tilde{V}_R = \tilde{V}_{H_2} + \tilde{V}_{TCS} - 3\tilde{V}_{HCl}$$

where we have abbreviated trichlorosilane as TCS. If all gases are ideal,

$$\Delta \tilde{V}_R = -\frac{RT}{P} \quad \text{and} \quad \frac{\partial \ln K}{\partial dP} = \frac{1}{P}$$

Integrating, we have

$$\ln K = \ln P + \text{const.}$$

Let K when the total pressure is atmospheric be K_o. (As an exercise it is suggested that the reader show that the numerical value of K_o from the data of Fig. 16.3 is approximately 10^7.) Hence $\ln K_o = \text{const.}$ and

$$K = K_o P = \frac{P_{H_2} P_{TCS}}{P_{HCl}^3} \tag{a}$$

By our simplifying assumption the total pressure is

$$P_{H_2} + P_{TCS} + P_{HCl} = P \tag{b}$$

and we are told in Fig. 16.3 that the chlorine/hydrogen ratio (presumably atomic) is 0.02: therefore,

$$\frac{P_{HCl} + 3P_{TCS}}{P_{HCl} + P_{TCS} + 2P_{H_2}} = 0.02 \tag{c}$$

We now have three equations, (a), (b), and (c). that can be solved for the three unknown partial pressures P_{HCl}, P_{TCS}, and P_{H2} in terms of the total pressure P. However. to avoid some tedious algebra we shall obtain an approximate solution which recognizes that most of the gas present is hydrogen. This will be clear from Fig. 16.3; the mole fractions of the gases present have to add to 1 and, at 800 K, the next most abundant gas to hydrogen (TCS) has a mole fraction of 0.01. Therefore, most of the gas is hydrogen and we will put the hydrogen partial pressure equal to P, the total pressures, in equations (a) and (c). Hence

$$\frac{P_{TCS}}{P_{HCl}^3} = K_o \qquad \text{(d)}$$

and [from (c)]

$$0.98P_{HCl} + 2.98P_{TCS} = 0.04P \qquad \text{(e)}$$

Substituting (d) in (e) yields

$$2.98K_oP_{HCl}^3 + 0.98P_{HCl} - 0.04P = 0 \qquad \text{(f)}$$

Equation (f) is a cubic equation that has the real solution (from any mathematics handbook)

$$P_{HCl} = A + B \qquad \text{(g)}$$

with

$$A = \left(\frac{0.04P}{2 \times 2.98K_o} + \sqrt{Q}\right)^{1/3} \qquad B = \left(\frac{0.04P}{2 \times 2.98K_o} - \sqrt{Q}\right)^{1/3}$$

and

$$Q = \left(\frac{0.98}{3 \times 2.98K_o}\right)^3 + \left(\frac{0.04P}{2 \times 2.98K_o}\right)^2$$

The reader will be able to show that, provided that P is greater than about 0.1 atm, the second term in the equation for Q is more than three orders of magnitude greater than the first. Ignoring the first term gives

$$A = \left(\frac{0.04P}{2.98K_o}\right)^{1/3} \quad \text{and} \quad B = \text{zero}$$

Hence

$$P_{HCl} = \left(\frac{0.04P}{2.98K_o}\right)^{1/3}$$

and [from (d)]

$$P_{TCS} = \frac{0.04P}{2.98}$$

The reader will be able to show that these equations yield mole fractions matching those of Fig. 16.3 at 1 atm pressure. Furthermore, it is easy to show that at 0.1 atm total pressure the mole fraction of trichlorosilane is unchanged while that of HCl rises to 5×10^{-3} at 800 K (i.e., the silicon/chlorine ratio in the gas phase drops).

16.4 Zone Refining

In Chapter 10 the reader was given a short treatment of segregation as it occurs in solidification. In Section 10.2.1 we saw how solids of nearly uniform composition can be obtained by restricting the liquid volume into which a solute is rejected at the solidification front. Figure 10.7 depicts the solute concentration profile developed within the solid by this restriction of the liquid volume. Zone refining, illustrated in Fig. 16.6, is a technique that exploits the development of solute profiles such as that of Fig. 10.7. A relatively narrow molten zone is formed at one end of a crucible containing the material to be refined; this is done by means of a suitably designed heater that heats only a small part of the crucible. The heater is then moved along the crucible in the direction shown (or the crucible moved in the opposite direction) so that solidification takes place to the left of the pool and melting on the right.

One "pass" of the heater results in the solute concentration profile depicted in Fig. 10.7 and the upper solid line in Fig. 16.7. Further passes of the heater from left to right continue to "sweep" the impurities toward the right-hand end of the solid. Finally, the end of the solid rich in impurities can be cut off and rejected. The result is a solid containing much lower levels of impurities than the original material but which is macrosegregated (i.e., there are variations of solute concentration along the length of the remaining solid). Homogenization can be achieved either by melting and casting all the material or by *zone leveling*. The latter technique is similar to zone refining except that the heater is passed backward and forward rather than only in one direction. The effect is to reduce the macrosegregation, as will be seen by the reader on completing a problem at the end of the chapter.

Zone refining has been used extensively for the production of semiconductor-grade materials. In many cases it is desirable to eliminate impurities formed by contact between the molten material and the crucible. One way of achieving this is by floating-zone refining. Here the rod of material to be refined is positioned vertically and the dimensions of the liquid pool made sufficiently small that surface tension serves to keep the pool in place between the two solid sections.

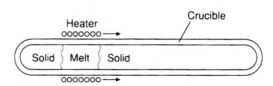

FIGURE 16.6 Zone refining.

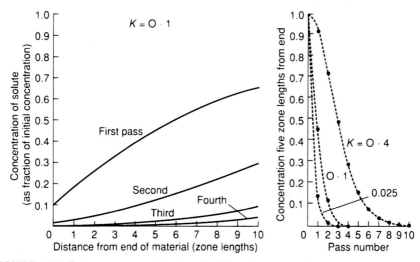

FIGURE 16.7 Solute concentration in solid following zone refining. Material has uniform initial concentration and is at least 11 zone lengths long (or longer for more than one pass).

16.5 Chemical Vapor Deposition

Chemical vapor deposition (CVD) is an established technology for the production of thin films, thick films, or even bulk materials. The technology has found its widest application in the semiconductor industry, but application of coatings to metal parts and optical components, as well as the production of ceramic materials, are other applications.

In CVD the desired material is formed by reaction between gaseous species. An example that we have already met in this chapter is the formation of silicon by reaction of hydrogen and trichlorosilane:

$$H_2 + HSiCl_3 = Si + 3HCl \tag{16.2}$$

The reaction is carried out at a heated substrate surface, and silicon is deposited on the surface. Other examples are shown in Table 16.1, and it is seen that materials ranging from metals (W) through semiconductors (GaAs) to dielectrics (SiO_2) can be deposited by this technique. Doping of materials can frequently be achieved during the deposition process; for example, phosphorus-doped GaAs can be produced by introducing phosphine into the reactor where GaAs films are being produced by reaction of arsine and trimethyl gallium. Deposition rates would usually lie in the range 1 to 100 μm/h.

From Chapter 3 the reader will recognize that reaction (16.2) and those of Table 16.1 are heterogeneous ones and are therefore influenced by the transport of molecules to the substrate. as well as by heat transport, and these transport phenomena become important in determining the uniformity of composition and thickness of the deposited film. Substrates are heated to make favorable both the thermodynamics (because many CVD reactions are endothermic) and the kinetics of the deposition reaction. Of course, the reactions may be more complicated than shown above. For example, in

Table 16.1 Some important CVD reactions

Reaction	Application
$WF_6 + 3H_2 = W + 6HF$	Metallization in semiconductor industry
$(CH_3)_3Ga + AsH_3 = GaAs + 3CH_4$	Gallium arsenide films for electronic applications
$SiH_4 + 2O_2 = SiO_2 + 2H_2O$	Dielectric films for semiconductor devices
$3SiH_4 + 4NH_3 = Si_3Ni_4 + 12H_2$	Dielectric films for semiconductor devices
$CH_3Cl_3Si = SiC + 3HCl$	Production of ceramic composites
$TiCl_4 + 2BH_3 = TiB_2 + 4HCl + H_2$	Production of bulk titanium diboride or composites
$NH_3 + BCl_3 = BN + 3HCl$	Production of pyrolytic BN parts
$3HSiCl_3 + 4NH_3 = Si_3N_4 = 9HCl + 3H_2$	Production of Si_3N_4 parts
$2(C_4H_9)_3Al = 2Al + 3H_2 + 6C_4H_8$	Aluminum deposition for protective films or composite production

silicon-chlorine-hydrogen systems, additional silicon-containing vapor species may occur besides trichlorosilane. The formation and decomposition of these additional species may be by homogeneous or heterogeneous reactions. For example, the reduction of tetrachlorosilane may proceed by two homogeneous reactions followed by a heterogeneous reaction:

$$SiCl_4 + H_2 = SiHCl_3 + HCl \tag{16.3}$$

$$SiHCl_3 = SiCl_2 + HCl \tag{16.4}$$

$$SiCl_2 + H_2 = Si + 2HCl \tag{16.5}$$

Nucleation of the main (deposition) reaction on the substrate (heterogeneous nucleation) is necessary if the film is to form, but gas-phase nucleation (homogeneous nucleation) should be avoided because it typically leads to particles that contaminate the film. Homogeneous nucleation is avoided by keeping the gas phase cool, although some heating of the gas by transfer from the heated substrate is inevitable. This homogeneous nucleation just above the substrate surface is minimized by using substrate temperatures that are no higher than necessary.

CVD has a number of process parameters that may be manipulated to achieve a deposit of desired properties. Among these parameters are gas-phase flow rate, substrate temperature, system pressure, nature of any carrier gases, and flow rates of reactive gases. In many cases it is possible to deposit films that are epitaxial with the substrate.

16.5.1 Technology

Figure 16.8 illustrates the reactors that have typically been used for chemical vapor deposition in the semiconductor industry. The reactors are semibatch in that the silicon wafers are processed as batches while the gases pass through continuously as a batch is processed.

The first three reactors illustrated are "cold wall" reactors, where the design attempts

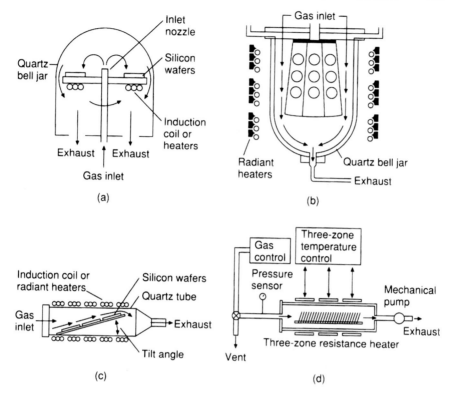

FIGURE 16.8 Typical reactors used in chemical vapor deposition: (a) pancake reactor; (b) barrel reactor; (c) horizontal reactor; (d) LPCVD reactor. [From D. W. Hess, K. F. Jensen, and T. J. Anderson, "Chemical Vapor Deposition: A Chemical Engineering Perspective", *Rev. Chem. Eng.* 3(2), 130 (1985). Permission of Freund Publishing House Ltd. London.]

to provide heat only to the substrate rather than the whole reactor. In this way the undesirable homogeneous nucleation can be minimized along with wasteful deposition on the reactor walls. This is achieved by local resistance heaters, by induction coils, or by radiant heaters. Temperature gradients existing in the reactor can cause natural convection, and in some cases this convective flow has been shown to cause nonuniform films. One way of correcting this problem (e.g., in the case of the barrel reactor) is to impose a rapid relative flow between gas and substrate by spinning the barrel.

The *low-pressure chemical vapor deposition* (LPCVD) apparatus shown in Fig. 16.8 exploits three advantages that are found at lower system pressure. As the pressure is lowered there is less tendency for natural convection to occur. Furthermore, diffusivities in the gas phase are approximately inversely proportional to the total pressure. Consequently, in LPCVD the deposition is more likely to be controlled by the chemistry of reaction than by transport in the gas phase, and therefore deposits will be more uniform (provided that substrate temperatures are uniform). Finally, some of the reagents used in CVD, particularly the organometallic compounds used in metal organic chemical vapor deposition (MOCVD), are not very volatile. In LPCVD the system pressure is typically well below the vapor pressure of the reagents used, and they can then be injected into the apparatus without a carrier gas. LPCVD reactors are typically "hot wall reactors," where the entire reactor is maintained at the reaction temperature.

16.5.2 Variants on CVD Technology

So far we have been treating reactors where the energy required to break the bonds of the reactant molecules is thermal energy derived from the heated substrate. An alternative way of supplying this energy is by generating a plasma within the gas phase. Some ways in which this is done are illustrated in Fig. 16.9. The substrate is positioned in a low-pressure system containing a plasma that is capacitively or inductively coupled to a radio-frequency power supply. The plasma consists of partly ionized gas, and the substrate is therefore showered by electrons and ions. The impact of these particles with gas molecules on or near the substrate surface provides the energy necessary for bond breaking. The primary advantages of the plasma-assisted chemical vapor deposition (PACVD) is that substrate temperatures can be kept much lower than in normal CVD, which means that films can be deposited on thermally sensitive substrates.

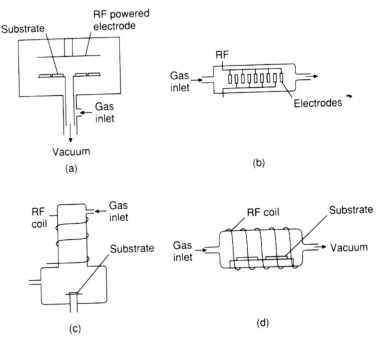

FIGURE 16.9 Basic configurations of plasma-assisted chemical vapor deposition reactors: (a) capacitively coupled, surface loaded; (b) capacitively coupled, volume loaded; (c) inductively coupled, substrates isolated from discharge; (d) inductively coupled, substrates within glow region. [From D. W. Hess, K. F. Jensen, and T. J. Anderson, "Chemical Vapor Deposition: A Chemical Engineering Perspective", *Rev. Chem. Eng.* 3(2), 130 (1985). Permission of Freund Publishing House Ltd. London.]

Alternatively, photon energy can serve to break chemical bonds during chemical vapor deposition. Of course, it is always possible to use photons to introduce thermal energy into the substrate, which then transfers it to adsorbed molecules or the adjacent gas phase. This is an everyday occurrence in CVD reactors with radiant heaters and has been used to deposit films locally using a laser to heat small regions of a substrate. However, it is also possible for photons to interact directly with the bonds of gas-phase or adsorbed molecules, supplying the energy necessary to break them. The technique has the advantage that by proper selection of the wavelength of the incident photons (from a laser, for example) the process can be made very specific for the desired reaction. Furthermore, as in PACVD, it becomes possible to carry out deposition onto thermally sensitive substrates.

16.5.3 Preparation of Bulk Materials and Chemical Vapor Infiltration

CVD can be used to manufacture parts of a material that can be formed by chemical vapor deposition. For example, boron nitride crucibles are made by the reaction shown in Table 16.1. The ammonia and boron trichloride are contacted with a heated substrate in the shape of the crucible. Wall thicknesses of several millimeters can readily be produced. Similarly, silicon nitride crucibles can be formed by reaction of trichlorosilane with ammonia.

A concept that has attracted much attention recently is the use of CVD in the production of composite materials. Such a material might consist of fibers (of carbon, for example) embedded in a matrix of metal (e.g., aluminum) or ceramic (e.g., silicon nitride). One approach is to coat the fiber by CVD in an apparatus where the heated fiber is pulled through the reactive atmosphere. The coated fiber is then fashioned into a composite, for example, by weaving or compaction followed by heat treatment to sinter the films.

An alternative method, known as *chemical vapor infiltration* (CVI), starts with the fibers preformed into the shape and dimensions of the finished composite part. The part is then placed in the reactive gases and held at temperature with the intention that chemical vapor deposition will fill the interstices in between fibers. To achieve this, the reacting molecules must first diffuse into the interior of the part and then react, with product gases diffusing out. Furthermore, the inward diffusion must take place against a net outward flow of gas because (see Table 16.1) typical deposition reactions generate more gas molecules than they consume. A further difficulty is that heat must be transported to the reaction site by conduction through the composite. This implies that the regions near the external surface are hotter than those deep inside. The reader who is familiar with Chapter 3 will recognize that because of this, and because gaseous reactant concentrations will be highest near the surface, there will be a tendency for most of the reaction to occur near the outside of the part with the production of a highly nonuniform part. The problem is exacerbated by the sealing off of the interior by the near-surface deposit as reaction proceeds. The difficulty is minimized by operating CVI at temperatures where the progress of reaction is controlled by the chemical step (i.e., at moderate temperatures), under which conditions infiltration rates are slow, affecting the economics of the technology. More promising avenues involve using forced

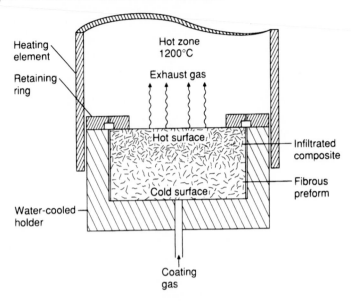

Heating element

Retaining ring

Hot zone 1200°C

Exhaust gas

Hot surface

Cold surface

Infiltrated composite

Fibrous preform

Water-cooled holder

Coating gas

FIGURE 16.10 Schematic diagram of chemical vapor infiltration process exploiting forced flow of the infiltrating gas. (From D. P. Stinton, "Ceramic Composites by Chemical Vapor Deposition", *Xth Intl. Conf. CVD*, PV 87-8, p. 1037, 1987. This paper was originally presented at the Fall 1987 Meeting of the Electrochemical Society, Inc. held in Honolulu, Hawaii.)

flow, rather than diffusion, to introduce reactants into the preform. One way of achieving this is to evacuate the system containing the compact being infiltrated and then rapidly allow the gaseous reactants to flow in. The cycle is repeated numerous times and this technique is known as pulse CVI. Alternatively, the preform can be mounted in a holder such as that shown in Fig. 16.10 and the gaseous reactants forced through the preform.

Chemical vapor infiltration is still under development at the time of writing. In production of many composites (glass fiber-reinforced polymers being an obvious example) it is unlikely to compete with simple mixing (or physical impregnation) or sol-gel techniques similar to those mentioned in Chapter 11.

16.5.4 Production of Optical Fiber

Optical fiber is now being widely used in both long-distance and local telecommunications. At the time of writing more than 8 million kilometers of optical fiber has been installed worldwide. A first step in the production of optical fiber bears some resemblance to chemical vapor deposition, and this is therefore an appropriate place to treat this topic. Optical fibers are long lengths (perhaps a few tens of kilometers long) of vitreous silica a few tens of micrometers in diameter. If the fibers are to transmit light with minimal loss, they must be free of defects such as bubbles and extraneous particles. Impurities such as copper, iron, and vanadium that 'color" the glass and thereby absorb light are eliminated by starting with the highly refined silicon chlorides encountered earlier in this chapter when the Siemens process for high-purity silicon was described. However, the fibers are doped with oxides of germanium and boron; these dopants are distributed nonuniformly across the fiber with the consequence that the outside of the fiber has a lower refractive index than the inside. With this variation of refractive index, light rays heading outward from the center are refracted back along the axis, resulting in minimal loss.

Figure 16.11 depicts the production of optical fiber starting with trichlorosilane or silicon tetrachloride. The technique illustrated is known as *outside vapor deposition (OVD)*. The silicon chlorides are vaporized and burned with oxygen in a burner. producing a flame that plays on a rotating mandrel or "target rod." A soot of silica particles forms on the mandrel, and by moving the flame back and forth in an axial direction a cylindrical porous *boule* of silica particles is built up. Growth rates in typical production machines are only of the order of 10 g/min. Dopants can be fed to the burner as their chlorides, and control of the flow enables the achievement of the desired refractive index profile. When the boule has been grown to the correct size for subsequent processing (typically, 1 m long by 100 mm diameter), the target rod is removed (exploiting differences in thermal expansion coefficient) and the boule is heated in an

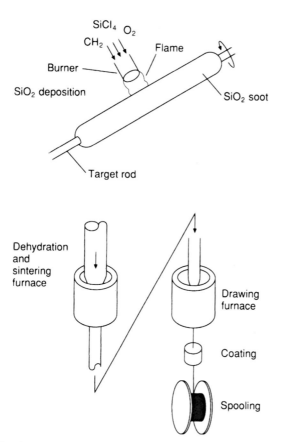

FIGURE 16.11 Production of optical fiber by outside vapor deposition.

atmosphere of helium and chlorine to remove adsorbed gases (particularly water). It is then sintered to full density at approximately 1400°C. The densified boule is then moved through a furnace, where the end of the boule is raised to 2200°C, and the thin optical fiber is drawn from this end. The fiber cools rapidly as it emerges from the furnace; it is given a protective coating of plastic and spooled.

There are alternative techniques exploiting the same chemistry, such as *modified chemical vapor deposition* (MCVD), where the silicon chlorides and oxygen are reacted on the inside of a quartz tube, or *vapor axial deposition* (VAD), where the SiO_2 soot is deposited from a burner playing on the end of a growing rotating cylindrical boule. Bulk forming techniques such as sol-gel processing, described in Chapter 11, are being examined as ways of producing porous silica preforms at higher rates than those achieved by oxidation of silicon chlorides.

16.6 Preparation of Composite Materials from Liquid Metals

In the preceding section we examined, very briefly, the production of composite materials by chemical vapor infiltration. Of course, there are other techniques for producing composites, some at a much more advanced state of development than CVI. A very simple approach is to mix the two (or more) phases that are to form the composite and then heat the mixture to the temperature where sintering occurs (perhaps with formation of a liquid phase or with densification aided by hot isostatic pressing). This simple approach works well for a few combinations of matrix material and reinforcement; the production of cobalt-tungsten carbide composites is an example of the successful application of this method on a commercial scale.

Another approach consists of intruding liquid metal into a ceramic preform (e.g., of woven silicon carbide fibers). This intrusion must be carried out under pressure or with the assistance of vacuum if the metal does not wet the ceramic, but there have been recent reports (Aghajanian et al., 1989) of methods in which pressureless infiltration of an aluminum alloy into ceramic preforms was achieved by control of process conditions. Specifically, the penetration was achieved by adding magnesium and operating under a nitrogen (or nitrogen-argon) atmosphere. Under these conditions the alloy wets the reinforcement and penetrates the preform without use of pressure or vacuum assistance.

Yet another straightforward technique is to disperse the reinforcing particles (or fibers) into the metal. If the metal is partially solidified to the point where it contains a large fraction of solidified metal grains but has not yet become rigid, the reinforcing particles can be effectively dispersed throughout the metal. Even if the reinforcing particles have a different density than the metal and are not wetted by it, the presence of the metal grains hinders the separation of the reinforcing particles prior to solidification. However, it is not possible to incorporate more than approximately 30 vol% of reinforcing phase in this way.

An elegant technique for the production of ceramic composites was developed by the Lanxide Corporation (Newkirk et al., 1987). The method is particularly interesting in the context of this book, exploiting many of the fundamental phenomena described in earlier chapters. The technology employs directed oxidation of metals to develop the

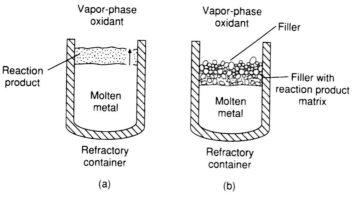

FIGURE 16.12 Schematic diagram of the formation of a matrix of oxide and unreacted metal by directed oxidation of molten metal (a). Oxidation in the presence of a filler is shown in (b). (Courtesy of LANXIDE Corp. Newark, DE.)

desired microstructures (which include both dense and porous materials, as well as composites). Figure 16.12 illustrates how this directed oxidation is carried out. In Fig. 16.12(a) a molten metal (e.g., an aluminum alloy) is being oxidized by a gaseous environment (e.g., air). If the temperature is maintained in the range 900 to 1350°C and the aluminum alloy contains a few percent magnesium and a group IVA element (Si, Ge, Sn, or Pb), the oxide film that forms is not protective. Instead, it has pores filled with molten metal and the metal "wicks" through the pores to the top surface of the film, where it is oxidized. The film of reaction product therefore thickens at the rate of up to a few centimeters per day until the aluminum alloy is consumed. The material produced in this way will henceforth be referred to as a *matrix;* it actually consists of two phases, the oxidation product, which is continuous and interconnected, and interspersed unreacted aluminum alloy. The ratio of phases is dependent on process parameters such as temperature.

Now consider what happens if a "filler" material is placed adjacent to the molten metal prior to the oxidation, as illustrated in Fig. 16.12(b). The filler material could be particles of a reinforcing material or a preform of fibers inert to the metal and oxidizing gas. Now oxidation proceeds within the interstices of the filler, and these interstices become filled with the matrix. Figure 16.13 is a micrograph of a cross woven silicon carbide fiber preform that has been filled with an alumina/aluminum matrix in this way. If a self-supporting preform (e.g., a porous slipcast ceramic) is used, this directed oxidation technology yields a composite part of the same size and shape as the preform. In other words, densification is achieved without dimensional changes.

The term *directed oxidation* should be taken to include all reactions where the metal gives up or shares its electrons so that composites with nitride or carbide matrices have been produced also. Among the composites made in this way are ones with matrices of Al_2O_3/Al, AlN/Al, ZrN/Zr, TiN/Ti, ZrC/Zr, and fillers of Al_2O_3, SiC, $BaTiO_3$, AlN, B_4C, TiB_2, ZrN, ZrB_2, and TiN.

Figure 16.14 illustrates a speculative explanation (Newkirk et al., 1986) for the unusual oxidation behavior of aluminum under the conditions employed in this di-

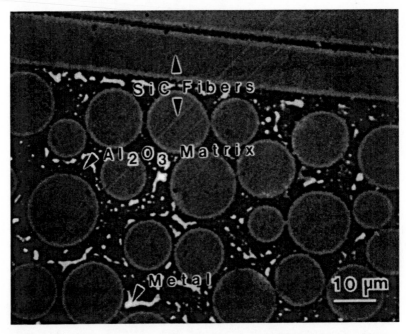

FIGURE 16.13 Micrograph of an Al_2O_3/Al matrix with a SiC fiber filler produced by directed oxidation. [From M. S. Newkirk et al., *Ceram. Eng. Sci. Proc. 8*(7-8) 879 (1987). Reprinted by permission of American Ceramic Society, Columbus, OH.]

rected metal oxidation (DIMOX™) technology. Two oxide grains are shown on the surface of a molten metal; the grain boundary energy is γ_B, while γ_{SL} and γ_{SV} denote the oxide-liquid and oxide-gas interfacial energies, respectively. If $\gamma_B > 2\gamma_{SL}$, the system can lower its free energy by the aluminum wetting the grain boundary (as shown in the lower figure), provided that $\gamma_{SV} > \gamma_{SL}$ (If $\gamma_{SV} < \gamma_{SL}$, the region between the grains will tend to fill with gas.) The aluminum that has penetrated to the surface of the film along the high-energy grain boundary regions will now oxidize, forming fresh Al_2O_3, and so on. The effect of the magnesium, group IVA element and temperature are therefore presumably on the grain boundary and interfacial energies, according to this hypothesis.

So far the directed metal oxidation technique has been described in terms of the metal being oxidized by a gas. It is possible for the oxidation reaction to be between the molten metal and a reactive preform. For example, Johnson and co-workers (1989) have produced platelet-reinforced ceramics of zirconium boride-zirconium carbide-zirconium in this way. Examples of parts produced in this way can be seen in Fig. 16.15.

The LANXIDE™ ceramic and metal matrix technologies described in the previous few paragraphs may be a relatively inexpensive way of producing dense composite parts in net shape or near net shape. They employ bulk processing techniques that have been traditional in foundries and ceramic processing for many years, rather than the expensive equipment and materials encountered elsewhere in the production of advanced materials.

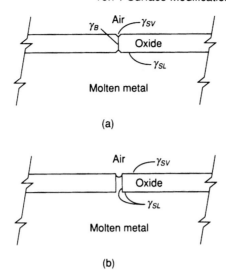

FIGURE 16.14 Schematic representation of conditions for nonprotective oxide formation due to grain boundary instability: (a) configuration with stable grain boundary; (b) configuration with unstable grain boundary ($g_{SL} < g_{SV}$ and $g_B > 2g_{SL}$). [Reprinted with permission from M. S. Newkirk et al., *J. Mater. Res.* *1*(1), 88 (1988).]

16.7 Surface Modification Processes

In many applications in the semiconductor industry as well as in wear protection of contacting surfaces, the essential properties are only developed after modification of a relatively thin surface layer, from a few monolayers to a few tens of micrometers thick. A wide range of surface modification techniques have been developed, apart from the CVD-type processes, to achieve the proper surface properties in a controlled manner. Here it is our intention to describe only a few of these.

We will deal with just four types of processes: liquid-phase epitaxy, plasma coating and ion plating processes, molecular beam epitaxy, and vapor-phase deposition of diamondlike coatings.

16.7.1 Liquid-phase Epitaxy

Liquid-phase epitaxy (LPE) is a method used in the semiconductor industry to deposit thin surface layers onto a substrate. In a typical case, epitaxial layers of GaAs or GaAlAs with various levels of impurity doping are formed by a combination of growth and deposition on a pure GaAs single-crystalline substrate wafer (Hsieh, 1980). Such surface-layer structures can be used as miniature lasers. The principle of the LPE method is relatively straightforward and is shown in Fig. 16.16. The substrate to be coated is positioned in a graphite holder such that it is level with the surface of the holder. The coating melt, a melt of GaAs containing the required doping elements and therefore having a melting point a little lower than that of the purer substrate GaAs, is held in the slider and is brought into contact with the substrate at well-controlled temperatures for

FIGURE 16.15 Examples of composite parts produced by the LANXIDE directed oxidation technology. (Photo courtesy of LANXIDE Corp. Newark, DE.)

some predetermined short interval of time as the slider is moved across the substrate holder. The furnace atmosphere is pure hydrogen, to prevent oxidation. The melt itself contains the doping elements above their solubility limit, and with control of temperature, this automatically gives precise control of the liquid composition, since the doping element concentration is then fixed by the liquid/impurity-precipitate equilibrium. With time, an epitaxial layer of doped GaAs develops on the substrate. The process can be repeated with some other melt, yielding devices that have a series of thin, epitaxial layers on their surfaces. The exact thickness of the growth layers is important and will depend on the time-temperature history, as well as on the diffusion coefficients of the impurities. Layer thicknesses are typically on the order of 10 μm after 10 minutes of contact. The exact distribution of the impurities can be calculated with the principles set forth in Chapter 6, where we discussed the development of concentration distributions resulting from interdiffusion as a function of time.

When the epitaxial layer has a composition significantly different from that of the substrate (e.g., for GaAlAs on GaAs), one speaks of heteroepitaxy. Many useful systems have this type of heteroepitaxy, but the similarity between the substrate and the layer crystal structure actually makes such systems only pseudo-heterooipitaxial. True heteroepitaxial systems, such as GaP on Si, are more difficult to produce.

LPE is a relatively cheap and easy process, but uniform doping in thicker surface

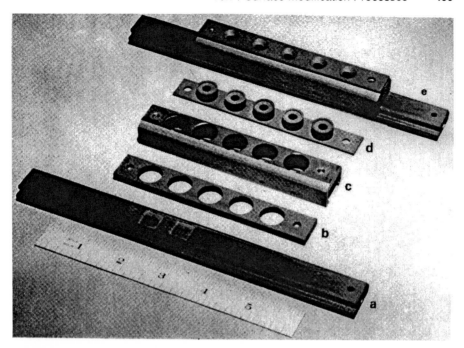

FIGURE 16.16 Graphite slider boat for multilayer liquid-phase epitaxy: (a) substrate holder; (b) wiper; (c) slider proper; (d) cap; (e) assembled boat. (From J. J. Hsieh, "Liquid-Phase Epitaxy," in *Materials, Properties and Preparation*, Vol. 3, S. P. Keller, ed., North-Holland, Amsterdam, 1980, p. 421. Reprinted with permission of Elsevier Science Publishing Co.)

layers is achieved with difficulty, and the surface smoothness is not as good as for films obtained by vapor-phase deposition on polished substrates.

16.7.2 Molecular Beam Epitaxy

Thin, discrete single-crystal films that are epitaxial with a substrate can be produced by molecular beam epitaxy. Figure 16.17 is a sketch of a molecular beam epitaxy (MBE) apparatus. In molecular beam epitaxy, one or more beams of a variety of materials are directed at very low pressure toward a substrate. A fraction of the impinging atoms captured by the surface diffuse across it, producing a nucleus with an epitaxial orientation that grows across the surface. The molecular beams are generated from effusion cells, which are essentially heated crucibles with a lid that has a small hole in it. Typically, the crucibles are high-purity pyrolytic boron nitride or graphite. The material maintains a precisely defined vapor pressure in the effusion cell when the cell temperature is fixed. This is because the small hole in the lid allows only a small rate of vapor escape, so that the equilibrium vapor pressure is maintained inside the cell. Such cells are also called *Knudsen cells*. To prevent any cross-contamination, the source cells are usually separated by liquid nitrogen cooled panels. The very low pressure in the high-vacuum chamber (typically 10^{-10} torr) in which the MBE process is carried out

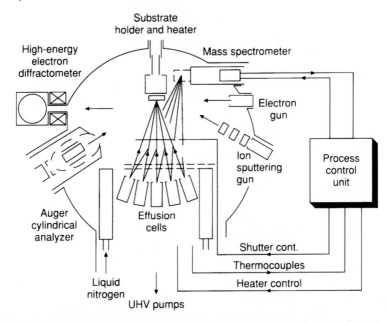

FIGURE 16.17 Schematic diagram of molecular beam epitaxy system. (From L. L. Chang, "Molecular-Beam Epitaxy" in *Materials, Properties and Preparation,* Vol. 3, S. P. Keller, ed., North-Holland, Amsterdam, 1980, p. 567. Reprinted with permission of Elsevier Science Publishing Co.)

allows the escaping molecular gas to travel in a straight line (or beam) toward the substrate, where it forms the deposit. The substrate itself can also be heated. The use of shutters, and exact control of the temperature of the effusion cells and the substrate, makes it possible to deposit films of precise concentration and structure. The films are formed epitaxially on a suitable substrate; for example, precisely doped epitaxial layers of GaAs can be formed on single-crystalline GaAs substrates. In some cases it is desirable to generate alternating layers of different film materials, perhaps a few hundred angstroms thick each, to form a structurally and chemically modulated periodic surface layer. Such films are also called *artificial superlattices* and can have interesting electronic and optical properties. The production of artificial superlattices is now one of the major applications of MBE. One of the major advantages of MBE is relatively low substrate temperature, which minimizes interdiffusion, so that the intermingling at the interfaces between the layers is minimal. This is particularly useful in optical devices that use such modulated surface films.

The MBE technique requires the highest degree of purity, to produce films with well-controlled, low doping levels. As a result, one MBE machine. which costs as much as $1 million (1988) and must be operated by a specialist. needs to be dedicated to the production of one specific type of surface film. The film growth rates tend to be very low, typically below 10 angstroms (10^{-10} m) per second. This would be an impractically low growth rate if thick surface layers were needed, but the MBE apparatus is used only in thin-film applications. The slow way in which the films build is then actually advantageous because it allows for control of the film structure. The ultrahigh vacuum permits the machine to be equipped with a variety of sophisticated surface analytical tools such as described in Chapter 5. This allows for precise process control but adds greatly to the cost of the MBE production method. Films deposited by MBE are among the most expensive materials per unit of weight that have ever been produced.

A variant of MBE is now coming into use; this is *chemical beam epitaxy (CBE)*. The species to be deposited are introduced into the CBE apparatus in the form of a gas or volatile liquid, at very low rates. For example, doped silicon epitaxial films may be grown starting from silane (SiH_4)-containing dopant gases, which thermally decompose on a heated substrate surface. The method again involves ultrahigh vacuum, and distinguishes itself in this way from chemical vapor deposition, described earlier.

16.7.3 Plasma Processes

In many modern film deposition techniques, gas plasmas are utilized. A plasma is a cloud of gas consisting of ionized atoms or ionized molecules and electrons. Overall, the plasma is neutral, but continued energy input can maintain this otherwise unstable mixture. To understand the major factors that affect production processes that involve plasmas, a few concepts must be considered. These are the *mean free path* of the electrons, and the temperature of the electrons and atoms as a function of energy input and gas pressure.

The mean free path of an electron, λ_e, can be shown to be related to the collision cross section of the atomic or molecular species in the plasma and to the number, N, of these per unit volume. It is given by

$$\lambda_e = \frac{1}{N\sigma_a} \qquad (16.6)$$

where σ_a is the collision cross section of the atom (if the atoms are considered as simple hard spheres). Energy exchange during collision of an electron with an atom or ion is, however, not like the elastic collision of two billiard balls, since it may involve atom excitations, ionization, and so on, besides elastic interactions. The collision cross section therefore depends on the energy of the species. At low electron energy the collision is primarily elastic momentum exchange, while at high energies, the important process becomes ionization of the atom (or further ionization of an already ionized atom). Equation (16.6) is only approximate but is sufficient for the present purpose. If the gas atoms are considered to be ideal, equation (16.6) might also be written

$$\lambda_e = \frac{AT}{P\sigma_a} \qquad (16.7)$$

where A is a proportionality constant on the order of the gas constant R divided by Avogadro's number, and P is the gas pressure. A first guess at the collision cross section would be the geometrical cross section of the atomic species in the plasma; the actual cross section depends on the energy of the electron and is shown for an argon plasma in Fig. 16.18. For a cross section of about 15×10^{-16} cm^2, at 0°C and 1 atm, equation (16.7) yields an electron mean free path of about 2.5×10^{-5} cm.

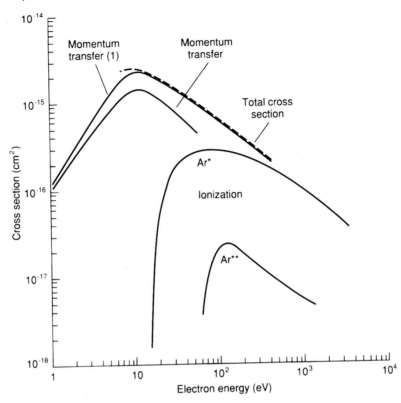

FIGURE 16.18 Collision cross section for electrons in argon gas. (From J. A. Thornton and A. S. Penfold, in *Thin Film Processes*, J. L. Vossen and W. Kern, eds., Academic Press, New York, 1978, p. 75.)

At a pressure of 10^{-10} atm, as employed in MBE, for example, the mean free path would be on the order of meters, much larger than the experimental apparatus, while at an altitude of 1000 km above the earth's surface the mean free path is well over 10^6 m. From equation (16.7) it is then evident that in very low-pressure processes a gas beam can be directed without fanning out due to atom collisions.

The energies of the free electrons and of the ions in a plasma can be very different. To understand this, we consider the plasma in a steady state, in which the energy that is put into the plasma by an externally applied electric field, E, is equal to the energy lost by the electron in colliding with the ions. The energy gained by an electron in the time period between two successive collisions is

$$E_e = eE\lambda_e \tag{16.8}$$

where e *is* the electron charge. This expression follows directly from the definition of work: force *(eE)* times displacement (λ_e). If the energy loss for a typical 90° elastic collision, in which the electron loses its momentum, is now considered, then from the mechanics of such collisions, determined by the overall conservation of momentum, the loss E_{col} of the electron's energy would be given by

$$E_{col} = \frac{2m_e}{m_a} \Delta E \tag{16.9}$$

where ΔE is the difference in the kinetic energy between the electron, E_e, and the atom, E_a, or ion before the collision:

$$\Delta E = E_e - E_a \tag{16.10}$$

m_e is the mass of the electron, 9. 11 x 10^{-31} kg, while m_a is the mass of the atom or ion and is $1.67 \times 10^{-27} \times$ atomic mass number. The steady-state condition is reached when, on average, an electron gains as much energy between collisions as it loses on collision. Thus the right-hand sides of (16.8) and (16.9) become equal and, eliminating λ_e with equation (16.6), it is readily found that

$$\Delta E = \frac{m_a}{2m_e} \frac{eE}{N\sigma_a} \tag{16.11}$$

This indicates that the electron energy can be substantially higher than the energy of the heavy particles, the ions or atoms, in a plasma that is maintained by an externally applied electric field. A leV difference, for example, corresponds to a temperature difference of nearly 12,000°C! The calculation above was, however, based on the assumption that only elastic momentum exchange was involved in the collisions, while actually, with increasing energy input and hence increasing electron energy, additional losses arise. These losses limit the actual temperature that the electrons can reach.

The temperature differences between the electrons and the atomic species in the plasma clearly depend on the gas pressure, as contemplation of equation (16.11) shows, and at higher pressures the collisions of the electrons and the gas atoms are sufficiently frequent that temperature equilibration between the gas atoms and the electrons is approached. This is shown in Fig. 16.19. The low-pressure plasmas are therefore "cold" plasmas, since the gas atoms remain cool and the plasma does not heat up the confinement vessel. At higher pressures, however, the gas temperature is dramatically elevated, and the overall plasma temperature can reach several thousand degrees Celsius with sufficient energy input. The high-pressure plasmas can be used in plasma spraying applications, for example, where a powdered metal or ceramic is fed into the plasma, producing tiny droplets that splatter onto a cold substrate. Such processes are used to produce thermal barrier coatings of zirconia on metal parts to protect them from exposure to heat. At low pressures, where the plasma remains cold, the high electron temperature allows chemical reactions characteristic of high temperatures to occur in a cold gas. These glow discharge plasmas can be used in a wide variety of applications, such as display devices. or more important here, the production of ionized or reactive species in a variety of surface deposition methods.

The plasma may be powered by either dc or ac (typically. at RF frequency) power sources. In addition, magnetic fields may be applied to the plasma to prevent the escape of the electrons. We now discuss briefly the use of plasmas in sputtering and ion plating processes.

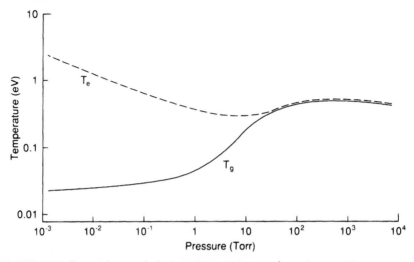

FIGURE 16.19 Dependence of electron temperature and gas temperature on pressure in electric arc. (Arc discharge current = constant.) (Reprinted with permission from J. A. Thornton, "Plasmas in Deposition Processes" in *Deposition Techniques for Films and Coatings*, Noyes Publications, Park Ridge, NJ, 1982, p. 23.)

16.7.4 Sputtering

The principle of this coating method is shown in Fig. 16.20. A target is positioned above the substrate that is to be coated, and a plasma is maintained near the target by a power supply. Typically, the sputtering gas is inert (e.g., argon). The target is also the negative electrode, toward which the positive ions in the plasma get accelerated by the applied voltage. This voltage is somewhere between 500 and 5000 V. The collisions of the argon ions with the target are so vigorous that they knock atoms out of the surface—this is the sputtering action. For dc sputtering, intense plasmas are difficult to produce, and reasonable production rates can only be obtained at gas pressures between 3 to 10 Pa. The sputtered atoms in this case reach the substrate by diffusion rather than by the projection that was characteristic at very low pressures. Deposition rates in the simple two-electrode (diode) setup are on the order of 1000 angstroms per minute or less. The plasma may also be produced by an independent power source and anode-cathode combination, instead of using the target and the substrates as the electrodes. This allows for maintaining more intense plasmas at lower pressures and for increased deposition rates. For metals, the power supply is a dc source, but for nonconductors, such as ceramics, an RF power source must be used.

The basic sputtering process has a large number of variants, including the possibility of using targets sequentially to produce layered coatings or the introduction of one of the coating components as a gas into the plasma. The latter process is known as reactive sputtering. Various bias voltages can also be applied between the substrate and the target, for example, to clean off the substrate surface or to affect the structure and the deposition rate of the deposited film. This process is known as *bias sputtering*. In modern plasma sputtering machines, the plasma is often confined by a magnetic field. This

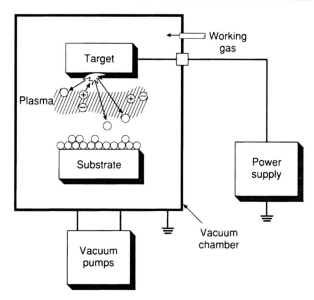

FIGURE 16.20 Schematic diagram of sputter coating process. (Reprinted with permission from J. A. Thornton, "Plasmas in Deposition Processes" in *Deposition Techniques for Films and Coatings*, Noyes Publications, Park Ridge, NJ, 1982)

can prevent the plasma from heating the substrate surface and also allows for depositions rates that are significantly higher than the two-electrode dc or RF sputtering devices. Deposition rates of 10,000 angstroms per minute can be achieved, and very large area (a few square meters) sputtering machines have been constructed.

In the sputtering of targets consisting of compounds rather than single elements, an additional effect appears. The sputtering rate of the different elements is most often not identical. Nevertheless, after awhile a steady-state sputtering is achieved. This will be the result of the enrichment in the target surface of the low-yield elements, such as the metal ions in oxide targets. After the steady state has been achieved, the material removed by sputtering will then have a composition identical to that of the target. It is thus important in sputtering of compounds not to expose the substrate before the steady-state sputtering has been achieved. This is most conveniently done by using shutter arrangements. Multilayer compound surface films can also be produced this way by manipulating the shutters of continuously sputtered targets.

An important variable in sputtering and related processes is the substrate temperature. Increased temperatures will promote the formation of larger grains or make epitaxy easier. The temperature of the substrate is obviously limited by the reevaporation rate or the melting of the deposit, and by the need to avoid chemical reaction or interdiffusion of the deposit with the substrate.

16.7.5 Ion Plating

Ion plating is akin to sputtering, but now the substrate is subjected to bombardment of ions extracted from an inert gas plasma, while the surface film is being deposited from

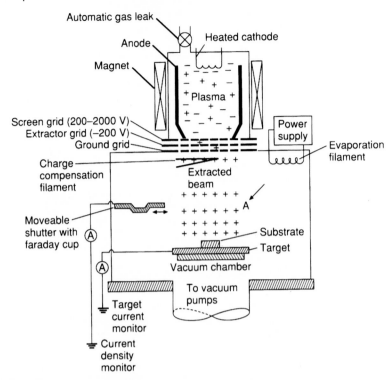

FIGURE 16.21 Ion gun ion-plating system with a resistive heater for the source of depositing material. (Reprinted with permission from D. M. Mattox, "Ion Plating Technology" in *Deposition Techniques for Films and Coatings*, Noyes Publications, Park Ridge, NJ, 1982, p. 246.)

another source, such as a heated filament. The vapor produced by this source mingles with the impinging plasma. Obviously, conditions have to be such that the rate of deposition of the film material exceeds the sputtering rate. Usually the substrate is first cleaned off by the impinging high-energy plasma, leaving it very clean, and then the evaporation source is turned on while keeping the plasma going. The combination of the impinging plasma ions and the vapor deposition allows for simultaneous modification of the structure of the surface and of the deposited film. The surface temperatures in this case can be high, promoting film recrystallization and diffusional bonding with the substrate. A schematic of one type of ion plating system is shown in Fig. 16.21. The effects of the surface temperature on the structure of a polycrystalline film have been studied and are often represented in a zone model developed by Movchan and Demchishin (1969), shown in Fig. 16.22. The tendency to form columnar crystals is a result of preferential growth and leads to surface roughness. As the substrate temperature increases, recrystallization is more prevalent, and larger-grained films with smoother surfaces result.

Often, deposited films will have residual stresses, since their structure can be considerably off-equilibrium. Where evaporated films tend to have residual tensile stresses,

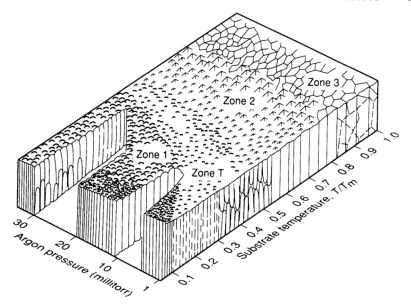

FIGURE 16.22 Schematic diagram of the influence of substrate temperature and argon pressure on the structure of metal coatings deposited by sputtering using cylindrical magnetron sources. T is the substrate temperature and T_m is the melting point of the coating material. (Reprinted with permission from D. M. Mattox, "Adhesion and Surface preparation" in *Deposition Techniques for Films and Coatings*, Noyes Publications, Park Ridge, NJ, 1982, p. 67.)

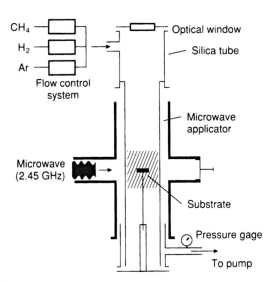

FIGURE 16.23 Diagram of microwave-plasma-assisted chemical vapor deposition (MPACVD) diamond film growth system. [From K. E. Spear, "Diamond-Ceramic Coating of the Future", *J. Am. Ceram. Soc. 72*(2), 172 (1989). Reprinted with permission of American Ceramic Society, Columbus, OH.]

sputtered films tend to have residual compressive stresses. Sometimes these stresses are so large that delamination or film fracture will result. The deposition conditions must then be altered to decrease these stresses (e.g., by raising the substrate temperature or by lowering the sputtering rate).

16.7.6 Diamondlike Surface Films

Many of the properties of diamond are unique, such as its hardness, thermal conductivity, chemical resistance, wear resistance, and so on, and therefore it could serve as an outstanding surface coating in applications where these properties can be exploited. Although not the stable form of graphite at ordinary temperatures and pressures, diamond coatings can be produced by vapor-phase methods on a variety of substrate materials. The principle is relatively simple. A carbon-containing gas, such as methane with a large excess of hydrogen (more than 95%), at a pressure between 1 and 10 kPa, is activated by a plasma or by a very high-temperature filament, and is then passed over a substrate that is kept between 800 and 1000°C. It has been found that diamondlike or actual diamond films deposit on the substrates under these conditions. A schematic of the process is shown in Fig. 16.23 and may be regarded as a *microwave-plasma-assisted chemical vapor deposition* (MPACVD) technique. Depending on the conditions, either isolated microscopic diamonds or continuous diamond or diamondlike films can be formed.

LIST OF SYMBOLS

A	proportionality constant
e	charge on electron
E	applied electric field
E_e, E_a	kinetic energy of electron, of atom
E_{col}	energy loss of electron on collision
E_E	energy gained between two collisions
m_e, m_a	mass of electron, atom
N	number of atoms or molecules per unit volume of plasma
P	pressure
T	temperature
λ_e	mean free path of electron
σ_a	collision cross section of atom or molecule

PROBLEMS

16.1 Perform a mass balance on the solute in the molten pool of a zone refining operation. Assume that the pool (of constant length Z) is well mixed and in equilibrium with the solid at the *solidification* front. Section 10.2.1 may be useful here. Show that the mass balance leads to the equation

$$\frac{Z}{K}\frac{dC_s}{dx} = C_l(x) - C_s$$

where

K = partition ratio
C_s = solute concentration after the pass
x = distance from one end of the solid
C_l = a function of x, is the solution concentration before the pass.

16.2 Write a computer program to calculate the solute distribution given by the equation in Problem 16.1 for any arbitrary starting distribution C_l (x). Assume that at $x = 0$, $C_s = KC_l$ (0). The equation does not describe the final stage of solidification when the pool reaches the end of the crucible. Modify equation (10.18) and incorporate it in your program to treat the final transient. Run the program repeatedly to calculate the type of profiles appearing in Fig. 16.7. Modify the program to simulate zone leveling and use the program to show that macrosegregation can be reduced by this technique.

16.3 In a CBE apparatus a silicon layer is to be grown at 5 µm/h from silane introduced into the vacuum system a distance of 30 cm from the substrate. A line drawn from the point of silane introduction to the center of the substrate surface makes an angle of 60° with that surface. Calculate the rate of introduction of silane (in millimoles per hour) if the silane molecules travel in straight lines from their point of entry into the vacuum system to the substrate surface (and to all other exposed surfaces in the apparatus). Assume that the silane molecules radiate from their point of introduction with a distribution that is independent of direction, and that the number of molecules arriving at the substrate surface by reflection from other surfaces is negligible.

16.4 The apparatus described in Problem 16.3 is to be used for growing a thin film on a large wafer (positioned as described in that question). A question arises as to whether the film will be uniform in thickness. Consider the line on the substrate surface that is the projection onto that surface of the line joining the silane introduction point to the substrate center. Calculate and plot the growth rate as a function of position along the line considered.

16.5 How many atoms are there in an "atomized" droplet of steel of 10 µm diameter?

16.6 In a rapid solidification apparatus, a steel strip 1 m wide and 2 mm thick is cast at 1 m/s. How many casters of this type would be required to accommodate the production of one BOF producing 300 tons per hour of steel?

16.7 The following equation gives the mass transfer coefficient, h, between a gas stream and a substrate across which it is passing in laminar flow.

$$\frac{hL}{D} = 0.664 \, Re^{1/2} \, Sc^{1/3}$$

L is the dimension of the substrate in the direction of gas flow, and D the diffusivity of the species transported to the substrate surface.

$$Re \equiv \frac{LV\rho}{\mu}$$

where V is the gas velocity, ρ is the gas density, and μ is the gas viscosity.

$$Sc \equiv \frac{\mu}{\rho D} \quad \text{with } D \text{ the diffusivity.}$$

Use the equation to calculate the maximum rate of chemical vapor deposition (μm/h) of a material of density 2 g/cm^3 from a gas phase at atmospheric pressure and 500 K. The gas is a carrier gas plus 2 mol/m^3 of a species A, 1 mol of which decomposes to yield 1 mol of deposit (the molecular weight of the deposit being 100). Additional data:

$$L = 150 \text{ mm}$$
$$V = 100 \text{ m/s}$$
$$\mu = 2 \times 10^{-5} \text{ kg/m·s} \quad D = 5 \times 10^{-5} \text{ m}^2/\text{s}$$
$$\rho = 2 \text{ kg/m}^3$$

16.8 Prepare a table comparing CVD, LPCVD, MBE, and CBE.

16.9 You are the new research engineer hired by Vince Overhand Research Enterprises, nestled in the valley near Humpty Dumpty, Kansas. Through the good offices of his local member of Congress, Vince has obtained a multimillion-dollar contract from the federal Humpty Dumpty Ceramic Test Facility to develop a thin (thickness L) composite material to be produced by chemical vapor infiltration of silicon nitride into fibrous preforms. As discussed in this chapter, the objective is to obtain a uniform deposition of silicon nitride throughout the preform. Because VORE's attorney has not yet discovered a way to sidestep certain patents, it will be necessary to rely on diffusion (rather than forced flow) to get gaseous reactants into the preform. Preliminary experiments indicate that the heterogeneous reaction is first order with a chemical rate constant k. Experiments are set up that measure the diffusion coefficient D of the gaseous species within the porous preform. Vince has had the benefit of an undergraduate education at a distinguished university on the west coast, where he came across dimensionless numbers. (Furthermore, he has found dimensionless numbers useful in the rather obscure accounting system that he uses in his business, but that's another story.) He is convinced that the deposit uniformity will be governed by a dimensionless number. What is your suggestion for a dimensionless group? Justify your suggestion on the grounds of what would be expected on physical grounds as the various quantities in the group are changed. Should your group be kept high or low to ensure uniformity?

16.10 A dc plasma deposition process aims to use krypton gas at a pressure of 10^{-2} torr. Will this be a hot or a cold plasma? The energy input is 6 eV. The bell jar in which the plasma is generated is kept at 30°C. Using the diameter of the krypton atoms as the collision cross section, what is the mean free path of the electrons? What is the electron temperature?

REFERENCES AND FURTHER READINGS

AGHAJANIAN, M. K., et al., "A New Infiltration Process for the Fabrication of Metal Matrix Composites," in *Tomorrow's Materials Today,* International SAMPE Symposium and Exhibition Series, Vol. 34, 1989, pp. 917-923.

BUNSHAH, R., ed., *Deposition Technologies for Films and Coatings,* Noyes Publications, Park Ridge, NJ, 1982.

CHANG, L., "Molecular Beam Epitaxy," in *Handbook on Semiconductors,* Vol. 3, S. Keller, ed., North-Holland, New York, 1980, pp. 563ff.

CULLEN, G. W., ed., *Proceedings of the Tenth International Conference on Chemical Vapor Deposition,* Electrochemical Society, Pennington, NJ, 1987.

FROES, F. W., and S. J. SAVAGE, eds., *Processing of Structural Metals by Rapid Solidification,* American Society for Metals, Metals Park, OH, 1987.

HSIEH, J., "Liquid Phase Epitaxy," in *Handbook on Semiconductors,* Vol. 3, S. Keller, ed., North-Holland, New York, 1980, pp. 415-497.

JOHNSON, W. B., T. D. CLAAR, and G. H. SCHIROKY, Preparation and Processing of Platelet Reinforced Ceramics by the Directed Reaction of Zirconium with Boron Carbide, *Ceram. Eng. Sci. Proc., 9 (7-8)* (1989).

MATTOX, D., "Ion Plating Technology," in *Handbook on Semiconductors,* Vol. 3, S. Keller, ed., North-Holland, New York, 1980, p. 244ff.

MOVCHAN, B., and A. DEMCHISHIN, *Phys. Met. Metallogr. 28,* 2773 (1969).

NEWKIRK, M. S., et al., Formation of Lanxide Ceramic Composite Materials, *J. Mater. Res. 1(1),* 81 (1986).

NEWKIRK, M. S., et al., Preparation of Lanxide Ceramic Matrix Composites: Matrix Formation by the Directed Oxidation of Molten Metals, *Ceram. Eng. Sci. Proc. 8* (7-8), 879 (1987).

SPEAR, K., Diamond-Ceramic Coating of the Future, *J. Am. Ceram. Soc. 72,* 171 (1989).

THORNTON, J., "Coating Deposition by Sputtering," in *Handbook on Semiconductors,* Vol. 3, S. Keller, ed., North-Holland, New York, 1980, p. 170ff.

THORNTON, J., "Plasmas in Deposition Process," in *Handbook on Semiconductors,* Vol. 3, S. Keller, ed., North-Holland, New York, 1980, p. 19ff.

Appendix

Block diagrams for the production of several important materials.

Steel

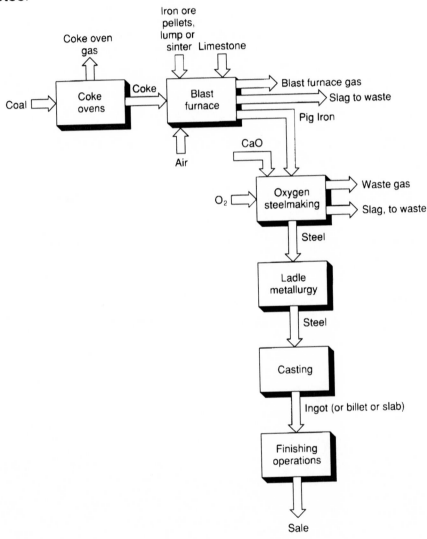

Production of Steel from Ingot to Finished Product

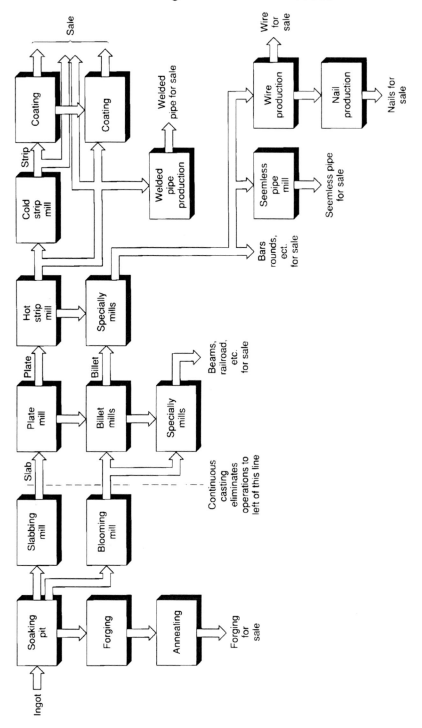

Aluminum

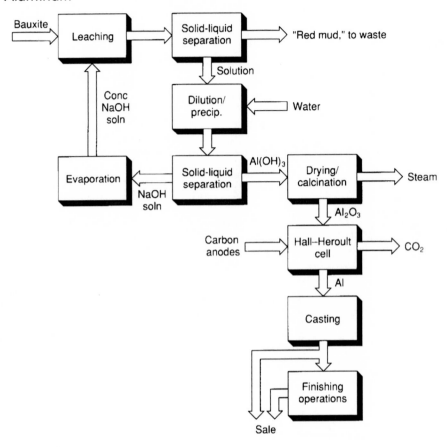

Copper (from Sulfide Ores, Traditional Route)

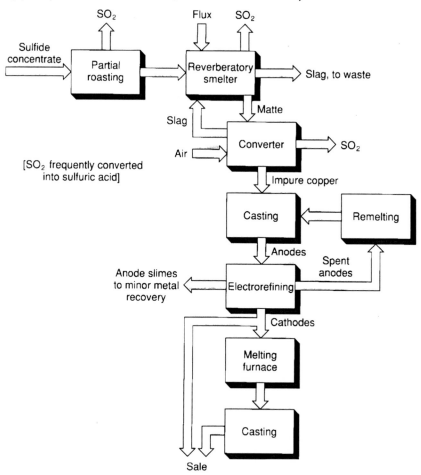

Copper (from Oxide Ores, Representative Hydrometallurgical Route)

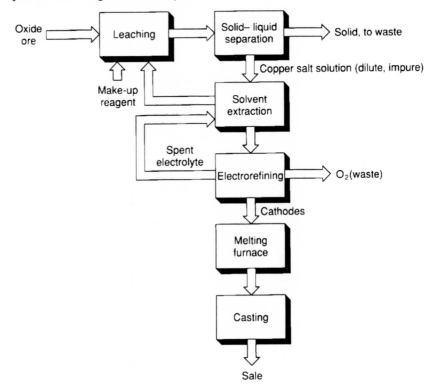

Zinc

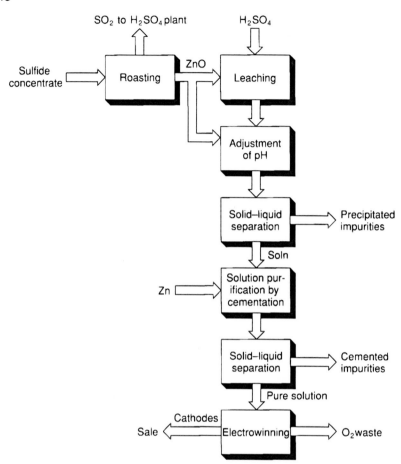

Gold

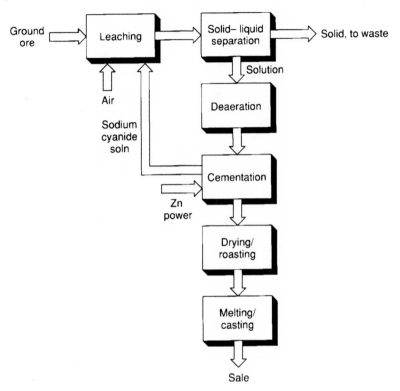

Production of Advanced Ceramic Parts

Parts would be evaluated by nondestructive techniques at various points in this scheme (e.g., following green machining). Granulation (e.g., by spray drying) sometimes follows wet milling.

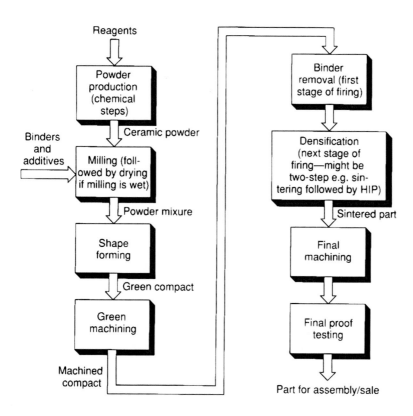

Single Crystal Silicon (via Siemens Process)

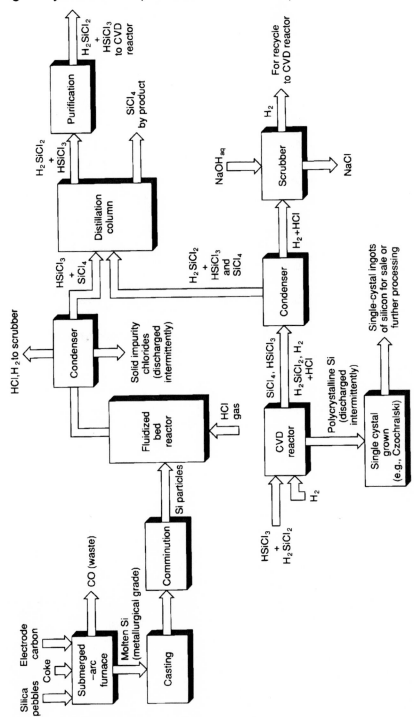

INDEX

("CD" refers to the compact disc of thermodynamic data and worked examples provided by Prof. Arthur Morris and distributed with this text)

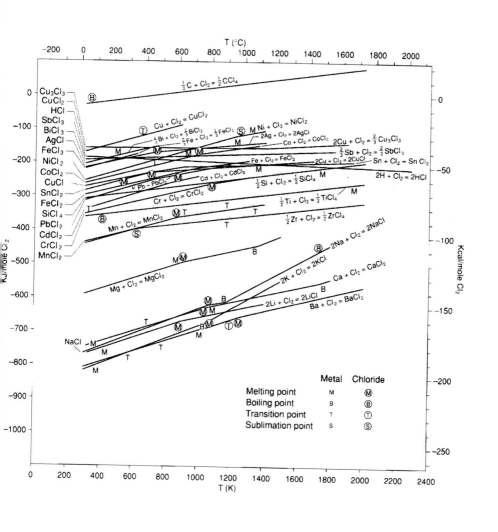

CD Containing
Thermodynamic Data and Examples of the Use of Excel©

prepared by
Dr. Arthur E. Morris, Thermart Software

The CD contains data and descriptive material for making detailed calculations involving materials processes. The contents of the CD are described in the text file *CD_INTRODUCTION.PDF*, which you should read and print before trying to use the material on the CD.

There are two Excel© workbooks on the disk: DATATABLES.XLS and XMPLES.XLS. They contain thermodynamic data and examples of their use by Excel to solve problems and examples similar to those in the text. The CD contains a document describing these examples, XMPLEDESCRIPTION.PDF, which is in Portable Document Format (PDF).

You will need Adobe Acrobat Reader to view and print these files; if you do not have Reader, you can install it using the appropriate installer contained on the CD, or it can be downloaded from the following location:

http://www.tms.org/AdobeAcrobat.html

Lightning Source UK Ltd.
Milton Keynes UK
UKOW04f2352230615

254023UK00003B/169/P